Universitext

Winfried Schirotzek

Nonsmooth Analysis

With 31 Figures

 Springer

Winfried Schirotzek

Institut für Analysis
Fachrichtung Mathematik
Technische Universität Dresden
01062 Dresden
Germany
e-mail: winfried.schirotzek@tu-dresden.de

Mathematics Subject Classification (2000): 49-01, 49-02, 49J50, 49J52, 49J53, 49K27, 49N15, 58C06, 58C20, 58E30, 90C48

Library of Congress Control Number: 2007922937

ISBN-10: 3-540-71332-8 Springer Berlin Heidelberg New York
ISBN-13: 978-3-540-71332-6 Springer Berlin Heidelberg New York

Springer is a part of Springer Science+Business Media
springer.com
© Springer-Verlag Berlin Heidelberg 2007

Cover design: WMXDesign, Heidelberg
Typesetting by the authors and SPi using a Springer LaTeX macro package

Printed on acid-free paper   SPIN: 12029495   41/2141/SPi   5 4 3 2 1 0

TO THE MEMORY OF MY PARENTS

Preface

One of the sources of the classical differential calculus is the search for minimum or maximum points of a real-valued function. Similarly, nonsmooth analysis originates in extremum problems with nondifferentiable data. By now, a broad spectrum of refined concepts and methods modeled on the theory of differentiation has been developed.

The idea underlying the presentation of the material in this book is to start with simple problems treating them with simple methods, gradually passing to more difficult problems which need more sophisticated methods. In this sense, we pass from convex functionals via locally Lipschitz continuous functionals to general lower semicontinuous functionals. The book does not aim at being comprehensive but it presents a rather broad spectrum of important and applicable results of nonsmooth analysis in normed vector spaces. Each chapter ends with references to the literature and with various exercises.

The book grew out of a graduate course that I repeatedly held at the Technische Universität Dresden. Susanne Walther and Konrad Groh, participants of one of the courses, pointed out misprints in an early script preceding the book. I am particularly grateful to Heidrun Pühl and Hans-Peter Scheffler for a time of prolific cooperation and to the latter also for permanent technical support. The Institut für Analysis of the Technische Universität Dresden provided me with the facilities to write the book. I thank Quji J. Zhu for useful discussions and two anonymous referees for valuable suggestions. I gratefully acknowledge the kind cooperation of Springer, in particular the patient support by Stefanie Zoeller, as well as the careful work of Nandini Loganathan, project manager of Spi (India).

My warmest thanks go to my wife for everything not mentioned above.

Dresden, December 2006   Winfried Schirotzek

Contents

Introduction

Minimizing or maximizing a function subject to certain constraints is one of the most important problems of real life and consequently of mathematics. Among others, it was this problem that stimulated the development of differential calculus.

Given a real-valued function f on a real normed vector space E and a nonempty subset A of E, consider the following problem:

(Min) Minimize $f(x)$ subject to $x \in A$.

Let $\bar{x}$ be a local solution of (Min). If $\bar{x}$ is an interior point of A and f is differentiable (in some sense) at $\bar{x}$, then $\bar{x}$ satisfies the famous *Fermat rule*

$$f'(\bar{x}) = o,$$

which is thus a necessary optimality condition. If $\bar{x} \in A$ is not an interior point of A but f goes on to be differentiable at $\bar{x}$, then a necessary optimality condition still holds as a *variational inequality*, which for A convex reads

$$\langle f'(\bar{x}), x - \bar{x} \rangle \geq 0 \quad \text{for any } x \in A. \tag{0.1}$$

The assumption that f be differentiable at $\bar{x}$ is not intrinsic to problem (Min). Consider, for example, the classical problem of Chebyshev approximation, which is (Min) with $E := \mathrm{C}[a,b]$, the normed vector space of all continuous functions $x : [a,b] \to \mathbb{R}$, and

$$f(x) := \|x - z\|_\infty := \max_{a \leq t \leq b} |x(t) - z(t)|,$$

where $z \in \mathrm{C}[a,b] \setminus A$ is given. In this case the functional f fails to be (Gâteaux) differentiable at "most" points $x \in \mathrm{C}[a,b]$ and so the above-mentioned approach no longer works.

However, if f is a *convex* functional, as is the functional in the Chebyshev approximation problem, then a useful substitute for a nonexisting derivative

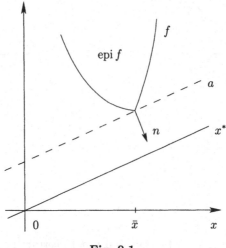

Fig. 0.1

is a subgradient. A *subgradient* of f at $\bar{x}$ is a continuous linear functional x^* on E satisfying

$$\langle x^*, y \rangle \le f(\bar{x} + y) - f(\bar{x}) \quad \text{for all } y \in E.$$

Geometrically this means that the "parallel" affine functional $a(x) := \langle x^*, x - \bar{x} \rangle + f(\bar{x})$ satisfies $a(x) \le f(x)$ for any $x \in E$ and $a(\bar{x}) = f(\bar{x})$ (see Fig. 0.1). Typically a function admits many subgradients (if any) at a point of nondifferentiability. The set of all subgradients of f at $\bar{x}$ is called *subdifferential* of f at $\bar{x}$ and is denoted $\partial f(\bar{x})$.

Initiated by Rockafellar [177] and Moreau [148], a rich theory of convex sets and functions including optimality conditions in terms of subdifferentials has been developed. This theory is known as *convex analysis* after Rockafellar's now classical book [180]. In the convex case the notion of the *conjugate functional* is the basis for associating with a given optimization problem another problem, called the *dual problem*. The study of the relationship between the two problems gives significant insight into convex optimization.

Problems involving functionals that are *neither differentiable nor convex* are more difficult to handle. During the last three decades, however, considerable progress has been made also in this area. One starting point is the observation that in the convex case a subgradient x^* of f at $\bar{x}$ can also be characterized by the inequality

$$\langle x^*, y \rangle \le f_G(\bar{x}, y) \quad \text{for all } y \in E. \tag{0.2}$$

Here, $f_G(\bar{x}, y)$ denotes the directional Gâteaux derivative of f at $\bar{x}$ in the direction y, i.e.,

$$f_G(\bar{x}, y) := \lim_{\tau \downarrow 0} \frac{f(\bar{x} + \tau y) - f(\bar{x})}{\tau},$$

which is a convex functional of y. If f is not convex, then in general the functional $f_G(\bar{x}, \cdot)$ is not convex either and a calculus finally resulting in optimality conditions is not available. The idea now is to replace $f_G(\bar{x}, \cdot)$ by a functional that behaves better. Pshenichnyi [170] and Neustadt [150] considered *upper convex approximations*, i.e., convex functionals that majorize certain directional derivatives such as $f_G(\bar{x}, \cdot)$. A crucial step was Clarke's doctoral thesis [33], which gives an intrinsic (nonaxiomatic) construction of a convex local approximation of f at $\bar{x}$. Another such construction is due, among others, to Michel and Penot [129, 130]. An effective calculus as well as applicable optimality conditions in terms of these constructs can be developed for the class of locally Lipschitz continuous functionals. This is elaborated and applied to problems in the calculus of variations and optimal control in the monograph [36] by Clarke, which also coined the notion *nonsmooth analysis*.

For non-Lipschitz functionals, various derivative-like concepts have been proposed as local approximations. For lower semicontinuous functionals, two concepts are particularly promising: On the one hand smooth local approximations from below which led to the concept of *viscosity subdifferentials* and in particular to *proximal subdifferentials* and on the other hand suitable directional limits that led to *Fréchet subdifferentials*. Crucial progress was reached when it turned out that in *Fréchet smooth Banach spaces* (Banach spaces admitting an equivalent norm that is Fréchet differentiable at each nonzero point) the two concepts coincide; this applies in particular to any reflexive Banach space. Substantial contributions to this theory are due to Rockafellar, Clarke, Borwein, Ioffe, and others.

We return to the problem (Min). Beside local approximations of the functional f one also needs local approximations of the set A near the minimum point $\bar{x}$. For various classes of optimization problems, *tangent cones* are adequate local approximations. If $f_*(\bar{x}, \cdot)$ and $\mathrm{T}_*(A, \bar{x})$ are a directional derivative of f at $\bar{x}$ and a tangent cone to A at $\bar{x}$, respectively, that "fit together," then the variational inequality

$$f_*(\bar{x}, y) \geq 0 \quad \text{for any } y \in \mathrm{T}_*(A, \bar{x}) \tag{0.3}$$

is a necessary optimality condition generalizing (0.1). Another method of locally approximating the set A is via *normal cones*. A conventional way to define a normal cone is as (negative) polar of a tangent cone, which always yields a *convex* cone. Polar cones can also be defined as subdifferentials of suitable functionals, which was done, among others, by Clarke [36] and Ioffe [96]. A third way to define normal cones is stimulated by the observation that $x^* \in E^*$ is a subgradient of the convex functional f at $\bar{x}$ if and only if $(x^*, -1)$ is a normal vector to the epigraph of f (see the vector n in Fig. 0.1). In the nonconvex case Mordukhovich [132] first defined normal cones by set limiting operations and then subdifferentials via normal cones to epigraphs.

In terms of some normal cone $N_*(A, \bar{x})$ and an associated subdifferential $\partial_* f(\bar{x})$, a necessary optimality condition corresponding to (0.3) reads

$$-x^* \in N_*(A, \bar{x}) \quad \text{for some } x^* \in \partial_* f(\bar{x}).$$

This condition is the stronger the smaller $N_*(A, \bar{x})$ and $\partial_* f(\bar{x})$ are. Under this aspect, it turns out that convex normal cones may be too large.

In general, Fréchet subdifferentials only admit an *approximate* (or *fuzzy*) calculus and consequently approximate optimality conditions. It turned out that the limiting normal cones and subdifferentials studied by Mordukhovich, Ioffe, and others admit a rich exact (nonapproximate) calculus at least in *Asplund spaces* (a class of Banach spaces containing all Fréchet smooth Banach spaces), provided the functionals and sets involved satisfy suitable compactness assumptions in the infinite dimensional case.

At this point, we must turn to the basic tools necessary for the respective theory. The principal tool for treating problems with convex and, to a great extent, locally Lipschitz continuous functionals are *separation theorems* and related results such as sandwich theorems. For general lower semicontinuous functionals, *variational principle* take over the part of separation theorems. A variational principle says, roughly speaking, that near a point that "almost" minimizes a functional f, there exists a point z that actually minimizes a slightly perturbed functional. In the first variational principle, discovered by Ekeland [56], the perturbed functional is not differentiable at z. Borwein and Preiss [19] were the first to establish a *smooth variational principle*. By now various smooth variational principles have been derived. A third fundamental tool are *extremal principles* discovered by Mordukhovich [132]. An extremal principle provides a necessary condition for a point to be extremal (in a certain sense) with respect to a system of sets. Mordukhovich largely develops his theory with the aid of extremal principles. They work especially well in Asplund spaces and in fact, Asplund spaces can be characterized by an extremal principle. Therefore Asplund spaces are the appropriate framework for variational analysis.

The book first presents the theory of convex functionals (Chaps. 2–6). All the following chapters are devoted to the analysis of nonconvex nondifferentiable functionals and related objects such as normal cones. A subdifferential of f, however it may be defined, associates with each $\bar{x} \in E$ a (possibly empty) *subset* of the dual space E^* and thus is a set-valued mapping or a *multifunction*. Therefore, nonsmooth analysis includes (parts of) multifunction theory; this is contained in Chap. 10. Each chapter ends with references to the literature and exercises. The latter partly require to carry out proofs of results given in the text, partly contain additional information and examples. Since nonsmooth analysis is still a rapidly growing field of research, many aspects had to be omitted. The Appendix indicates some of them.

1

Preliminaries

1.1 Terminology

General Convention. Throughout these lectures, *vector space* always means *real* vector space, and *topological vector space* always means *Hausdorff* topological vector space. Unless otherwise specified, E will denote a normed vector space.

Let $\mathbb{N}$, $\mathbb{R}$, and $\mathbb{R}_+$ denote the set of all positive integers, of all real numbers and of all nonnegative real numbers, respectively. Further set $\overline{\mathbb{R}} := \mathbb{R} \cup \{-\infty, +\infty\}$. The operations in $\overline{\mathbb{R}}$ are defined as usual; in addition, we set $0 \cdot (\pm\infty) = (\pm\infty) \cdot 0 := 0$, but we do not define $(+\infty) - (+\infty)$. In Remark 1.3.5 below, we shall explain that it is reasonable to allow a functional to attain values in $\overline{\mathbb{R}}$ and not only in $\mathbb{R}$.

The zero element of any vector space except $\mathbb{R}$ will be denoted o. If A and B are nonempty subsets of a vector space E and if $\alpha, \beta \in \mathbb{R}$, we write

$$\alpha A + \beta B := \{\alpha x + \beta y \mid x \in A, \, y \in B\}, \quad \mathbb{R}_+ A := \{\lambda x \mid \lambda \in \mathbb{R}_+, \, x \in A\}.$$

In particular we write $A + y := A + \{y\}$. If one of A, B is empty, we set $A + B := \emptyset$.

A nonempty subset A of the vector space E is said to be *convex* if $x, y \in A$ and $\lambda \in (0, 1)$ imply $\lambda x + (1 - \lambda)y \in A$. The empty set is considered to be convex also.

Let E be a topological vector space. A subset A of E is said to be *circled* if $\alpha A \subseteq A$ whenever $\alpha \in \mathbb{R}$ and $|\alpha| \leq 1$. Recall that each topological vector space admits a neighborhood base of zero, $\mathfrak{U}$, consisting of closed circled sets. A neighborhood base of an arbitrary point $x \in E$ is then given by $x + U$, where $U \in \mathfrak{U}$.

The interior, the closure, and the boundary of a subset A of a topological vector space are denoted $\operatorname{int} A$, $\operatorname{cl} A$ and $\operatorname{bd} A$, respectively. If A is convex, so are $\operatorname{int} A$ and $\operatorname{cl} A$.

A *locally convex (vector) space* is a topological vector space such that each neighborhood of zero includes a convex neighborhood of zero. It follows that a locally convex space possesses a neighborhood base of zero consisting of open convex sets, and also a neighborhood base of zero consisting of closed circled convex sets.

If E is a topological vector space, E^* denotes the *topological dual* of E, i.e., the vector space of all continuous linear functionals $x^* : E \to \mathbb{R}$. If $x^* \in E^*$, we write $\langle x^*, x \rangle := x^*(x)$ for each $x \in E$.

Now let E be a normed vector space with norm $\| \cdot \|$. If $\bar{x} \in E$ and $\epsilon > 0$, we set

$$\mathrm{B}_E(\bar{x}, \epsilon) := \{x \in E \mid \|x - \bar{x}\| \le \epsilon\}, \quad \mathrm{B}_E := \mathrm{B}_E(o, 1),$$
$$\mathring{\mathrm{B}}_E(\bar{x}, \epsilon) := \{x \in E \mid \|x - \bar{x}\| < \epsilon\}, \quad \mathring{\mathrm{B}}_E := \mathring{\mathrm{B}}_E(o, 1).$$

If it is clear from the context, we simply write $\mathrm{B}(\bar{x}, \epsilon)$, B, $\mathring{\mathrm{B}}(\bar{x}, \epsilon)$, and $\mathring{\mathrm{B}}$, respectively. A normed vector space is a locally convex space with respect to the topology induced by the norm. The inner product of a Hilbert space will be denoted as $(x \mid y)$.

1.2 Convex Sets in Normed Vector Spaces

Let E be a normed vector space. A nonempty subset A of E is said to be *absorbing* if for each $y \in E$ there exists $\alpha_0 > 0$ such that $\alpha y \in A$ whenever $|\alpha| \le \alpha_0$. The *core* of A, written $\mathrm{cr}\, A$, is the set of all $x \in A$ such that $A - x$ is absorbing, i.e.,

$$\mathrm{cr}\, A := \{x \in A \mid \forall y \in E \, \exists \alpha_0 > 0 \, \forall \alpha \in [-\alpha_0, \alpha_0] : \; x + \alpha y \in A\}.$$

Each convex neighborhood of zero in E is an absorbing set. In a Banach space, we have the following converse.

Proposition 1.2.1 *If E is a Banach space, then each closed convex absorbing subset A of E is a neighborhood of zero.*

Proof. For each $n \in \mathbb{N}$ let $A_n := nA$. Each A_n is closed, and since A is absorbing, we have $E = \bigcup_{n=1}^{\infty} A_n$. By the Baire category theorem it follows that $\mathrm{int}(A_n)$ is nonempty for at least one n and so the set $\mathrm{int}\, A$ is nonempty. Hence there exist $x \in A$ and $\rho > 0$ such that $\mathrm{B}(x, \rho) \subseteq A$. Since A is absorbing, there further exists $\alpha > 0$ such that $-\alpha x = \alpha(-x) \in A$. Define $\epsilon := \frac{\alpha \rho}{1 + \alpha}$. We show that $\mathrm{B}(o, \epsilon) \subseteq A$. Let $y \in \mathrm{B}(o, \epsilon)$ and set $z := x + \frac{1 + \alpha}{\alpha} y$. Then $z \in \mathrm{B}(x, \rho)$ and so $z \in A$. Since A is convex, we obtain

$$y = \frac{1}{1 + \alpha}(-\alpha x) + \frac{\alpha}{1 + \alpha} z \in A. \quad \square$$

We strengthen the concept of a convex set. A nonempty subset A of E is said to be *cs-closed* (for *convex series closed*) if $\lambda_i \ge 0$ $(i \in \mathbb{N})$, $\sum_{i=1}^{\infty} \lambda_i = 1$, $x_i \in A$ $(i \in \mathbb{N})$ and $x := \sum_{i=1}^{\infty} \lambda_i x_i \in E$ imply $x \in A$. It is clear that each cs-closed set is convex. Lemma 1.2.2 provides examples of cs-closed sets.

Lemma 1.2.2 *Closed convex sets and open convex sets in a normed vector space are cs-closed.*

Proof. See Exercise 1.8.1.   □

It is obvious that int $A \subseteq$ cr A for each subset A of E. The following result describes the relationship between core and interior more precisely.

Proposition 1.2.3 *Let A be a nonempty subset of the normed vector space E.*

(a) *If A is convex and int A is nonempty, then* cr $A =$ int A.
(b) *If E is a Banach space and A is cs-closed, then* cr $A =$ int $A =$ int(cl A).

Proof.

(a) We first show that for any $\alpha \in [0, 1)$,

$$\alpha \operatorname{cl} A + (1 - \alpha) \operatorname{int} A \subseteq \operatorname{int} A. \tag{1.1}$$

Choose some $x_0 \in$ int A. Then $(1-\alpha)(\operatorname{int} A - x_0)$ is an open neighborhood of o and so

$$\alpha \operatorname{cl} A = \operatorname{cl}(\alpha A) \subseteq \alpha A + (1 - \alpha)(\operatorname{int} A - x_0) \subseteq A - (1 - \alpha)x_0.$$

Since this holds for any $x_0 \in$ int A, the left-hand side of (1.1), which is obviously open, is contained in A. This verifies (1.1). To prove cr $A =$ int A, it suffices to show that cr $A \subseteq$ int A. Let $x \in$ cr A. Then there exists $\lambda > 0$ such that $y := x + \lambda(x - x_0) \in A$. By (1.1), it follows that

$$x = \frac{1}{1+\lambda}y + \frac{\lambda}{1+\lambda}x_0 \in \operatorname{int} A.$$

(b) (I) First we verify, following Borwein and Zhu [24], that int $A =$ int(cl A). For this, it suffices to show that int(cl A) $\subseteq$ int A under the additional assumption that int(cl A) $\neq \emptyset$. Let $y_0 \in$ int(cl A). Then there exists $\epsilon > 0$ such that $B(y_0, \epsilon) \subseteq$ cl A and so $B \subseteq \frac{1}{\epsilon}(\operatorname{cl} A - y_0)$. Since together with A, the set $\frac{1}{\epsilon}(A - y_0)$ is also cs-closed, we may assume that $y_0 = o$ and $B \subseteq \operatorname{cl} A \subseteq A + \frac{1}{2}B$. From this inclusion we obtain for $i = 1, 2, \ldots,$

$$\frac{1}{2^i}B \subseteq \frac{1}{2^i}A + \frac{1}{2^{i+1}}B \quad \text{and so} \quad \frac{1}{2}B \subseteq \frac{1}{2}A + \frac{1}{2^2}A + \cdots + \frac{1}{2^i}A + \frac{1}{2^{i+1}}B.$$

Thus, if $y \in \frac{1}{2}B$, there exist $x_1, \ldots, x_i \in A$ such that

$$y \in \frac{1}{2}x_1 + \frac{1}{2^2}x_2 + \cdots + \frac{1}{2^i}x_i + \frac{1}{2^{i+1}}B.$$

It follows that $y = \sum_{i=1}^{\infty} x_i/2^i$ and so $y \in A$ as A is cs-closed. Hence $\frac{1}{2}B \subseteq A$ and therefore $y_0 = o \in$ int A.

(II) In order to verify cr $A =$ int A, it suffices to show cr $A \subseteq$ int A under the additional assumption that cr $A \neq \emptyset$. Let $z_0 \in$ cr $A \subseteq$ cr(cl A). Since cl $A - z_0$ is closed and absorbing, the Baire category theorem implies that int(cl A) $\neq \emptyset$ (cf. the proof of Proposition 1.2.1), and step (I) shows that int A is nonempty. As the set A is cs-closed, it is also convex. Therefore the assertion follows from (a).   □

1.3 Convex Functionals: Definitions and Examples

Convention. In this section, unless otherwise specified, E denotes a vector space and M denotes a nonempty subset of E.

Definition 1.3.1 If $f : M \to \overline{\mathbb{R}}$, then we call

$$\operatorname{dom} f := \{x \in M \mid f(x) < +\infty\} \quad \textit{effective domain of } f,$$

$$\operatorname{epi} f := \{(x,t) \in M \times \mathbb{R} \mid f(x) \leq t\} \; \textit{epigraph of } f.$$

Further, f is said to be *proper* (in German: *eigentlich*) if $\operatorname{dom} f \neq \emptyset$ and $f(x) > -\infty$ for each $x \in M$.

Definition 1.3.2 is crucial for these lectures.

Definition 1.3.2 Let $M \subseteq E$ be nonempty and convex. The functional $f : M \to \overline{\mathbb{R}}$ is called *convex* if

$$f\big(\lambda x + (1 - \lambda)y\big) \leq \lambda f(x) + (1 - \lambda)f(y) \tag{1.2}$$

holds for all $x, y \in M$ and all $\lambda \in (0, 1)$ for which the right-hand side is defined, i.e., is not of the form $(+\infty) + (-\infty)$ or $(-\infty) + (+\infty)$. If (1.2) holds with $<$ instead of $\leq$ whenever $x \neq y$, then f is called *strictly convex*.

Lemma 1.3.3 *Let $M \subseteq E$ be nonempty and convex and let $f : M \to \overline{\mathbb{R}}$:*

(a) *f is convex if and only if $\operatorname{epi} f$ is a convex set.*
(b) *If f is convex, then for each $\lambda \in \mathbb{R}$, the set $\{x \in M \mid f(x) \leq \lambda\}$ is convex.*

Proof. See Exercise 1.8.2.   □

Lemma 1.3.4 *Let E be a topological vector space and let $f : E \to \overline{\mathbb{R}}$ be convex. If $f(x_0) > -\infty$ for some $x_0 \in \operatorname{int} \operatorname{dom} f$, then $f(x) > -\infty$ for each $x \in E$.*

Proof. Assume there exists $x_1 \in E$ satisfying $f(x_1) = -\infty$. Since $x_0 \in \operatorname{int} \operatorname{dom} f$, we have $x_2 := x_0 + \lambda(x_0 - x_1) \in \operatorname{dom} f$ for each sufficiently small $\lambda \in (0, 1)$. Since $x_0 = \frac{\lambda}{1+\lambda}x_1 + \frac{1}{1+\lambda}x_2$ and f is convex, we obtain

$$-\infty < f(x_0) \leq \frac{\lambda}{1 + \lambda}f(x_1) + \frac{1}{1 + \lambda}f(x_2),$$

which is contradictory.   □

Remark 1.3.5 According to Lemma 1.3.4, it is quite pathological for a convex functional on a topological vector space to attain the value $-\infty$. The following construction shows that this is not so with the value $+\infty$. Let $M \subseteq E$

be nonempty and convex, and let $g : M \to \mathbb{R}$ be convex. Then the functional $f : E \to \overline{\mathbb{R}}$ defined by

$$f(x) := \begin{cases} g(x) & \text{if } x \in M, \\ +\infty & \text{if } x \in E \setminus M \end{cases}$$

is proper and convex. Conversely, if $f : E \to \overline{\mathbb{R}}$ is proper and convex, then $M := \text{dom } f$ is nonempty and convex and the restriction of f to M is finite and convex. Therefore, we may assume that a convex functional is of the form $f : E \to \overline{\mathbb{R}}$.

We consider important classes of convex functionals.

Example 1.3.6 Let $M \subseteq E$ be nonempty. The functional $\delta_M : E \to \overline{\mathbb{R}}$ defined by

$$\delta_M(x) := \begin{cases} 0 & \text{if } x \in M, \\ +\infty & \text{if } x \in E \setminus M \end{cases}$$

is called *indicator functional* of M. Obviously, δ_M is proper and convex if and only if M is nonempty and convex.

Example 1.3.7 Let E be a topological vector space and A a nonempty subset of E^*. Then the *support functional* $\sigma_A : E \to \overline{\mathbb{R}}$ of A defined by

$$\sigma_A(x) := \sup_{x^* \in A} \langle x^*, x \rangle, \quad x \in E,$$

is proper and convex. An analogous remark applies to the support functional $\sigma_M : E^* \to \overline{\mathbb{R}}$ of a nonempty subset M of E.

Example 1.3.8 Let $u : E \to \mathbb{R}$ be linear and $c \in \mathbb{R}$. The functional $f : E \to \mathbb{R}$ defined by $f(x) := u(x) + c$ for $x \in E$, which is called *affine*, is convex.

Example 1.3.9 The functional $f : E \to \overline{\mathbb{R}}$ is said to be *sublinear* if f is proper, nonnegatively homogeneous (i.e., $f(\lambda x) = \lambda f(x)$ for any $x \in E$ and any $\lambda \geq 0$), and subadditive (i.e., $f(x + y) \leq f(x) + f(y)$ for all $x, y \in E$). If f is sublinear, then f is also convex, and in particular $f(o) = 0$ (recall that $0 \cdot (+\infty) := 0$). The norm functional of a normed vector space is sublinear and so convex.

Example 1.3.10 Let E be a Hilbert space, let $T : E \to E$ be linear and self adjoint, let $x_0 \in E$, and let $c \in \mathbb{R}$. Consider the *quadratic* functional

$$f(x) := \frac{1}{2}(Tx \,|\, x) - (x_0 \,|\, x) + c, \quad x \in E.$$

Then

$$f \text{ is convex} \iff T \text{ is positively semidefinite (i.e. } (Tx \,|\, x) \geq 0 \,\forall x \in E).$$

We indicate the proof of this statement.

$\Longrightarrow$: This follows immediately from

$$f\left(\frac{1}{2}x\right) \le \frac{1}{2}f(x) + \frac{1}{2}f(o).$$

$\Longleftarrow$: Since $x \mapsto -(x_0 \,|\, x) + c$, $x \in E$, is affine and so convex, it remains to show that $g(x) := (Tx \,|\, x)$, $x \in E$, is convex. Let $x, y \in E$ and $0 < \lambda < 1$. Then

$$g\big(\lambda x + (1 - \lambda)y\big)$$
$$= \lambda^2\big(T(x - y) \,|\, x - y\big) + \lambda\big(T(x - y) \,|\, y\big) + \lambda\big(Ty \,|\, x - y\big) + \big(Ty \,|\, y\big).$$

Since $\lambda^2 < \lambda$ and, by assumption, $\big(T(x - y) \,|\, x - y\big) \ge 0$, we have

$$\lambda^2\big(T(x - y) \,|\, x - y\big) \le \lambda\big(T(x - y) \,|\, x - y\big).$$

Using this and the self adjointness of T, we immediately obtain

$$g\big(\lambda x + (1 - \lambda)y\big) \le \lambda g(x) + (1 - \lambda)g(y).$$

Example 1.3.11 Let E be a normed vector space. If $M \subseteq E$, then the *distance functional* $x \mapsto \mathrm{d}_M(x)$ is defined by

$$\mathrm{d}_M(x) := \mathrm{d}(M, x) := \mathrm{d}(x, M) := \inf_{y \in M} \|x - y\|, \ x \in E.$$

Recall that $\inf \emptyset := +\infty$. If M is nonempty and convex, then d_M is easily seen to be proper and convex.

Example 1.3.12 Let E be a topological vector space, let $M \subseteq E$ be nonempty and set

$$p_M(x) := \inf\{\lambda > 0 \,|\, x \in \lambda M\}, \ x \in E.$$

The functional $p_M : E \to \overline{\mathbb{R}}$ is called *Minkowski functional* or *gauge*.

Lemma 1.3.13 summarizes important properties of p_M.

Lemma 1.3.13 *Let E be a topological vector space. If $M \subseteq E$ is nonempty and convex and $o \in \mathrm{int}\, M$, then:*

(a) $0 \le p_M(x) < +\infty$ *for all $x \in E$.*
(b) p_M *is sublinear and continuous.*
(c) $\mathrm{int}\, M = \{x \in E \,|\, p_M(x) < 1\} \subseteq M \subseteq \{x \in E \,|\, p_M(x) \le 1\} = \mathrm{cl}\, M$.
(d) *If, in addition, M is symmetric (i.e., $x \in M \Longrightarrow -x \in M$) and bounded, then p_M is a norm on E that generates the topology of E.*

Proof. See, for instance, Aliprantis and Border [2]. $\square$

1.4 Continuity of Convex Functionals

In this section, we study continuity properties of convex functionals.

Recall that if E is a normed vector space, then the proper functional $f : E \to \bar{\mathbb{R}}$ is said to be *locally Lipschitz continuous*, or briefly *locally L-continuous*, around $\bar{x} \in \operatorname{dom} f$ if there exist $\epsilon > 0$ and $\lambda > 0$ such that

$$|f(x) - f(y)| \leq \lambda \|x - y\| \quad \forall x, y \in B(\bar{x}, \epsilon).$$

Moreover, f is called locally L-continuous on the open subset D of E if f is locally L-continuous around each $\bar{x} \in D$.

Theorem 1.4.1 *Let E be a topological vector space and let $f : E \to \bar{\mathbb{R}}$ be proper, convex and bounded above on some nonempty open subset G of E. Then f is continuous on $\operatorname{int} \operatorname{dom} f$. If, in particular, E is a normed vector space, then f is locally L-continuous on $\operatorname{int} \operatorname{dom} f$.*

Proof.

(I) By assumption, there exists $a > 0$ such that $f(y) < a$ for each $y \in G$. In particular, $G \subseteq \operatorname{int} \operatorname{dom} f$. We first show that f is continuous on G.

(Ia) To prepare the proof, we show the following: If $\bar{x} \in \operatorname{dom} f$, $z \in E$, $\bar{x} + z \in \operatorname{dom} f$, $\bar{x} - z \in \operatorname{dom} f$, and $\lambda \in [0, 1]$, then

$$|f(\bar{x} + \lambda z) - f(\bar{x})| \leq \lambda \max\{f(\bar{x} + z) - f(\bar{x}), \, f(\bar{x} - z) - f(\bar{x})\}. \quad (1.3)$$

Verification of (1.3): Since $\bar{x} + \lambda z = (1 - \lambda)\bar{x} + \lambda(\bar{x} + z) \in \operatorname{dom} f$, we have $f(\bar{x} + \lambda z) \leq (1 - \lambda)f(\bar{x}) + \lambda f(\bar{x} + z)$, hence

$$f(\bar{x} + \lambda z) - f(\bar{x}) \leq \lambda\big(f(\bar{x} + z) - f(\bar{x})\big). \quad (1.4)$$

Analogously, with z replaced by $-z$, we obtain

$$f(\bar{x} - \lambda z) - f(\bar{x}) \leq \lambda\big(f(\bar{x} - z) - f(\bar{x})\big). \quad (1.5)$$

From $\bar{x} = \frac{1}{2}(\bar{x} + \lambda z) + \frac{1}{2}(\bar{x} - \lambda z)$ we conclude that $f(\bar{x}) \leq \frac{1}{2}f(\bar{x} + \lambda z) + \frac{1}{2}f(\bar{x} - \lambda z)$ and so $f(\bar{x}) - f(\bar{x} + \lambda z) \leq f(\bar{x} - \lambda z) - f(\bar{x})$, which together with (1.5) gives $f(\bar{x}) - f(\bar{x} + \lambda z) \leq \lambda\big(f(\bar{x} - z) - f(\bar{x})\big)$. From this and (1.4) we obtain (1.3).

(Ib) Now let $\bar{x} \in G$ and let $\epsilon > 0$. Then there exists a circled neighborhood U of zero such that $\bar{x} + U \subseteq G$. Let $\lambda \in (0, 1]$ be such that $\lambda(a - f(\bar{x})) < \epsilon$. Then (1.3) implies $|f(y) - f(\bar{x})| < \epsilon$ for all y in the neighborhood $\bar{x} + \lambda U$ of $\bar{x}$. Hence f is continuous at $\bar{x}$.

(II) We show that f is continuous at $\bar{y} \in \operatorname{int} \operatorname{dom} f$. Let $\bar{x}$ and U be as in step (Ib). Then there exists $\delta > 0$ such that $z := \bar{y} + \delta(\bar{y} - \bar{x}) \in \operatorname{dom} f$ (see Fig. 1.1). Set $\lambda := \delta/(1 + \delta)$. If $y \in U$, then $\bar{y} + \lambda y = \lambda \bar{x} + (1 - \lambda)z$ and so

$$f(\bar{y} + \lambda y) \leq \lambda f(\bar{x} + y) + (1 - \lambda)f(z) < \lambda a + (1 - \lambda)f(z).$$

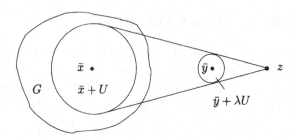

Fig. 1.1

Since $f(z) < +\infty$, we see that f is bounded above on the neighborhood $\bar{y} + \lambda U$ of $\bar{y}$ and so on the open set $\operatorname{int}(\bar{y} + \lambda U)$ that also contains $\bar{y}$. By step (I), f is continuous at $\bar{y}$.

(III) Now let E be a normed vector space and let $\bar{y} \in \operatorname{int} \operatorname{dom} f$. By step (II) there exist $c, \rho > 0$ satisfying

$$|f(x)| \leq c \quad \forall x \in \mathrm{B}(\bar{y}, 2\rho). \tag{1.6}$$

Assume, to the contrary, that f is not locally L-continuous around $\bar{y}$. Then there exist $x, y \in \mathrm{B}(\bar{y}, \rho)$ such that

$$f(x) - f(y) > \tfrac{2c}{\rho} \|x - y\|.$$

Set $\alpha := \frac{\rho}{\|x-y\|}$ and $\hat{x} := x + \alpha(x - y)$. Then we have $x = \frac{1}{1+\alpha}\hat{x} + \frac{\alpha}{1+\alpha}y$ which implies $f(x) \leq \frac{1}{1+\alpha}f(\hat{x}) + \frac{\alpha}{1+\alpha}f(y)$ and so

$$f(\hat{x}) - f(x) \geq \alpha\big(f(x) - f(y)\big) > 2c.$$

But this contradicts (1.6) as $\hat{x} \in \mathrm{B}(\bar{y}, 2\rho)$. $\square$

Convex functions on a *finite*-dimensional normed vector space have a remarkable continuity behavior.

Corollary 1.4.2 *If $f : \mathbb{R}^n \to \overline{\mathbb{R}}$ is proper and convex and $\operatorname{int} \operatorname{dom} f$ is nonempty, then f is locally L-continuous on $\operatorname{int} \operatorname{dom} f$.*

Proof. We shall show that f is bounded above on a nonempty open subset G of $\mathbb{R}^n$; the assertion then follows from Theorem 1.4.1. Let $\bar{x} = (\bar{x}_1, \ldots, \bar{x}_n) \in \operatorname{int} \operatorname{dom} f$. Then there exists $\alpha > 0$ such that $\operatorname{int} \operatorname{dom} f$ contains the closed cube C with center $\bar{x}$ and edge length α. Let

$$G := \{(x_1, \ldots, x_n) \in \mathbb{R}^n \,\big|\, \bar{x}_i < x_i < \bar{x}_i + \tfrac{\alpha}{n}\}.$$

Then G is nonempty, open, and contained in C (Fig. 1.2). Let $\big(e^{(1)}, \ldots, e^{(n)}\big)$ denote the standard basis of $\mathbb{R}^n$. If $x \in G$, then we have

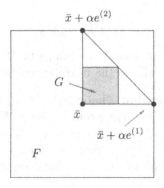

Fig. 1.2

$$x = \sum_{i=1}^{n} x_i e^{(i)} = \sum_{i=1}^{n} \frac{x_i - \bar{x}_i}{\alpha}\left(\bar{x} + \alpha e^{(i)}\right) + \left(1 - \sum_{i=1}^{n} \frac{x_i - \bar{x}_i}{\alpha}\right)\bar{x}$$

and so

$$f(x) \leq \sum_{i=1}^{n} \frac{x_i - \bar{x}_i}{\alpha} f\left(\bar{x} + \alpha e^{(i)}\right) + \left(1 - \sum_{i=1}^{n} \frac{x_i - \bar{x}_i}{\alpha}\right) f(\bar{x})$$
$$\leq \max\{f\left(\bar{x} + \alpha e^{(1)}\right), \ldots, f\left(\bar{x} + \alpha e^{(n)}\right), f(\bar{x})\}. \quad \square$$

By Corollary 1.4.2, each $\mathbb{R}$-valued convex function on $\mathbb{R}^n$ is locally L-continuous. Notice that, in contrast to this, on each *infinite*-dimensional normed vector space there always exist even linear functionals that are discontinuous.

1.5 Sandwich and Separation Theorems

In this and in Sect. 1.6 we repeat, in a form appropriate to our purposes, some facts from Functional Analysis that will be frequently needed in the sequel.

A nonempty subset P of a vector space E is called *cone* if $x \in P$ and $\lambda > 0$ imply $\lambda x \in P$. Moreover, P is called *convex cone* if P is a cone and a convex set. Obviously, P is a convex cone if and only if $x, y \in P$ and $\lambda, \mu > 0$ imply $\lambda x + \mu y \in P$. By definition, the empty set is also a cone.

If P is a nonempty convex cone in E, then

$$x \leq_P y \quad :\Longleftrightarrow \quad y - x \in P$$

defines a relation $\leq_P$ on E that is transitive and compatible with the vector space structure of E, i.e., for all $x, y, z \in E$ and $\lambda \in \mathbb{R}$ we have

$$x \leq_P y \text{ and } y \leq_P z \quad \Longrightarrow \quad x \leq_P z,$$
$$x \leq_P y \quad \Longrightarrow \quad x + z \leq_P y + z,$$
$$x \leq_P y \text{ and } \lambda > 0 \quad \Longrightarrow \quad \lambda x \leq_P \lambda y.$$

Moreover, we have

$$P = \{x \in E \,|\, o \leq_P x\}.$$

Notice that the zero element o need not belong to P and so the relation $\leq_P$ need not be reflexive. We call $\leq_P$ the *preorder* generated by P.

Proposition 1.5.1 (Extension Theorem) *Let M be a linear subspace of the vector space E and P a nonempty convex cone in E satisfying $P \subseteq M - P$. Suppose further that $u : M \to \mathbb{R}$ is linear and satisfies $u(x) \geq 0$ for any $x \in M \cap P$. Then there exists a linear functional $v : E \to \mathbb{R}$ such that $v(x) = u(x)$ for every $x \in M$ and $v(x) \geq 0$ for every $x \in P$.*

Proof. Set $E_1 := \operatorname{span}(M \cup P) = M + P - P$ and

$$p(x) := \inf\{u(y) \,|\, y \in M, y - x \in P\}, \quad x \in E_1.$$

Since $P \subset M - P$, the functional p is easily seen to be finite on E_1. Further p is sublinear and satisfies $u(x) \leq p(x)$ for any $x \in M$. By the Hahn–Banach theorem, there exists a linear functional $v_1 : E_1 \to \mathbb{R}$ such that $v_1(x) = u(x)$ for any $x \in M$ and $v_1(x) \leq p(x)$ for any $x \in E_1$. Setting $v(x) := v_1(x)$ for every $x \in E_1$ and $v(x) := 0$ for every x in the algebraic complement of E_1 in E completes the proof.   □

The following result will be a useful tool in the analysis of convex functionals.

Theorem 1.5.2 (Sandwich Theorem) *Let E be a topological vector space and let $p, q : E \to \overline{\mathbb{R}}$ be proper, convex and such that $-q(x) \leq p(x)$ for all $x \in E$. Suppose further that*
(A1) $(\operatorname{int} \operatorname{dom} p) \cap \operatorname{dom} q \neq \emptyset$ *and p is continuous at some point of $\operatorname{int} \operatorname{dom} p$*
or
(A2) $\operatorname{dom} p \cap (\operatorname{int} \operatorname{dom} q) \neq \emptyset$ *and q is continuous at some point of $\operatorname{int} \operatorname{dom} q$.*
Then there exist $v \in E^*$ and $c \in \mathbb{R}$ such that

$$-q(x) \leq \langle v, x \rangle + c \leq p(x) \quad \forall x \in E \quad (\text{see Fig. 1.3 for } E = \mathbb{R}).$$

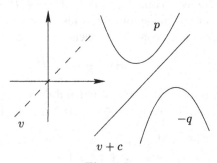

Fig. 1.3

Proof. (I) We set

$$F := E \times \mathbb{R} \times \mathbb{R}, \quad M := \{o\} \times \{0\} \times \mathbb{R},$$
$$u(o, 0, \rho) := -\rho \quad \forall \rho \in \mathbb{R},$$
$$P := \{\beta(y, 1, t) - \alpha(x, 1, s) \mid \alpha, \beta \geq 0; \ s, t \in \mathbb{R};$$
$$x \in \operatorname{dom} p; \ y \in \operatorname{dom} q; \ p(x) \leq s; \ -q(y) \geq t\}.$$

Then we have:

(a) u is a linear functional on the linear subspace M of F satisfying $u(z) \geq 0$ for any $z \in M \cap P$.

(b) P is a convex cone in F.

(c) $P \subset M - P$.

It is easy to verify (a) and (b). We prove (c). Let $z \in P$, thus $z = \beta(y, 1, t) - \alpha(x, 1, s)$. Suppose assumption (A1) holds. Then there exists $z_0 \in (\operatorname{int} \operatorname{dom} p) \cap \operatorname{dom} q$. For $\delta > 0$ define

$$z_\delta := z_0 + \delta\big(\beta y - \alpha x - (\beta - \alpha)z_0\big).$$

It follows that $z_\delta \in \operatorname{dom} p$ if δ is sufficiently small. Hence, with δ and ρ appropriately chosen, we obtain

$$z = \tfrac{1}{\delta}\big(z_\delta, 1, p(z_\delta)\big) - \big(\tfrac{1}{\delta} - \beta + \alpha\big)\big(z_0, 1, -q(z_0)\big) + (o, 0, \rho) \in -P + M.$$

(II) By Proposition 1.5.1 there exists a linear functional $w : F \to \mathbb{R}$ satisfying $w(z) = u(z)$ for all $z \in M$ and $w(z) \geq 0$ for all $z \in P$. In particular, we have $w(o, 0, 1) = -1$. Define

$$v(x) := w(x, o, 0) \quad \forall x \in E, \quad c := w(o, 1, 0).$$

Then $v : E \to \mathbb{R}$ is linear and $w(x, s, t) = v(x) + cs - t$ for all $(x, s, t) \in F$. If $x \in \operatorname{dom} p$, then $z := -\big(x, 1, p(x)\big) \in P$. It follows that

$$0 \leq w(z) = -v(x) - c + p(x) \text{ and so } v(x) + c \leq p(x). \tag{1.7}$$

Analogously we obtain $-q(x) \leq v(x) + c$. Since p is continuous at some point of $\operatorname{int} \operatorname{dom} p$, the second inequality in (1.7) implies that v is bounded above on a nonempty open set and so, by Theorem 1.4.1, is continuous.

(III) A similar argument applies under the assumption (A2).   □

The sandwich theorem describes the separation of a *concave* functional $-q$ and a *convex* functional p by a continuous affine functional. Now we consider the separation of convex sets by a hyperplane.

Recall that a *hyperplane* in the topological vector space E is a set of the form

$$[x^* = \alpha] := \{x \in E \mid \langle x^*, x \rangle = \alpha\}, \tag{1.8}$$

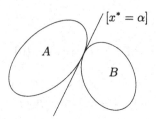

Fig. 1.4

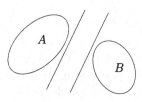

Fig. 1.5

where $x^* \in E^*$, $x^* \neq o$, and $\alpha \in \mathbb{R}$. The hyperplane $[x^* = \alpha]$ separates the space E into the *half-spaces* $[x^* \leq \alpha]$ and $[x^* \geq \alpha]$, where the notation is analogous to (1.8).

The subsets A and B of E are said to be *separated* if there exists a hyperplane $[x^* = \alpha]$ such that (see Fig. 1.4)

$$A \subseteq [x^* \leq \alpha] \quad \text{and} \quad B \subseteq [x^* \geq \alpha]. \tag{1.9}$$

Further we say that the subsets A and B of E are *strongly separated* if there exist a hyperplane $[x^* = \alpha]$ and $\epsilon > 0$ such that (see Fig. 1.5)

$$A \subseteq [x^* \leq \alpha - \epsilon] \quad \text{and} \quad B \subseteq [x^* \geq \alpha + \epsilon]. \tag{1.10}$$

Theorem 1.5.3 (Weak Separation Theorem) *Let A and B be nonempty convex subsets of the topological vector space E and assume that* int $A \neq \emptyset$. *Then the following statements are equivalent:*

(a) *A and B are separated.*
(b) $(\text{int } A) \cap B = \emptyset$.

Proof. (a) $\Longrightarrow$ (b): Assume that there exists a nonzero $x^* \in E^*$ such that (1.9) holds. Suppose, to the contrary, that there exists $\bar{x} \in (\text{int } A) \cap B$. Let $y \in E$ be given. Then for $\rho > 0$ sufficiently small, we have $y_{\pm} := \bar{x} \pm \rho(y - \bar{x}) \in A$ and so

$$\alpha \geq \langle x^*, y_{\pm} \rangle = \langle x^*, \bar{x} \rangle \pm \rho \langle x^*, y - \bar{x} \rangle.$$

Since $\langle x^*, \bar{x} \rangle = \alpha$ it follows that $\pm \langle x^*, y - \bar{x} \rangle \leq 0$ and so $\langle x^*, y \rangle = \langle x^*, \bar{x} \rangle$. Since $y \in E$ is arbitrary and x^* is linear, we conclude that $x^* = o$: a contradiction.

(b) $\implies$ (a): Let $x_0 \in \text{int } A$ and let p denote the Minkowski functional of $A - x_0$. Further define $q : E \to \bar{\mathbb{R}}$ by $q(x) := -1$ for all $x \in B - x_0$ and $q(x) := +\infty$ for all $x \in E \setminus (B - x_0)$. Applying Lemma 1.3.13, it is easy to see that the assumptions of Theorem 1.5.2 are satisfied. Hence there exist $x^* \in E^*$ and $c \in \mathbb{R}$ such that

$$\langle x^*, x \rangle + c \leq p(x) \; \forall x \in E \quad \text{and} \quad \langle x^*, x \rangle + c \geq 1 \; \forall x \in B - x_0.$$

It follows that $x^* \neq o$ (consider $x = o$) and that the hyperplane $[x^* = \alpha]$, where $\alpha := \langle x^*, x_0 \rangle + 1 - c$, separates A and B. $\qquad \square$

Corollary 1.5.4 *Let A be a convex cone in E, B a nonempty convex subset of E, and assume that int $A \neq \emptyset$ but $(\text{int } A) \cap B = \emptyset$. Then there exists $x^* \in E^*$, $x^* \neq o$, such that*

$$\langle x^*, x \rangle \leq 0 \; \forall x \in A \quad \text{and} \quad \langle x^*, y \rangle \geq 0 \; \forall y \in B.$$

Proof. See Exercise 1.8.3. $\qquad \square$

Let $A \subseteq E$ and $\bar{x} \in A$. The hyperplane $[x^* = \alpha]$ is said to be a *supporting hyperplane* of A at $\bar{x}$ if

$$\bar{x} \in [x^* = \alpha] \quad \text{and}$$
$$A \subseteq [x^* \leq \alpha] \quad \text{or} \quad A \subseteq [x^* \geq \alpha] \quad \text{(Fig. 1.6)}.$$

A point $\bar{x} \in A$ admitting a supporting hyperplane of A is said to be a *support point* of A. There is an obvious relationship to the support functional σ_A of A. If $\bar{x}$ is a support point of A and $[x^* = \alpha]$ is a supporting hyperplane such that $A \subseteq [x^* \leq \alpha]$, then $\sigma_A(x^*) = \langle x^*, \bar{x} \rangle$. An immediate consequence of Theorem 1.5.3 is:

Corollary 1.5.5 *Let A be closed convex subset of the topological vector space E and assume that int A is nonempty. Then any boundary point of A is a support point of A.*

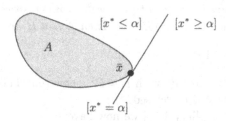

Fig. 1.6

We supplement the preceding results by two statements presented without proof that refer to the case int $A = \emptyset$.

Theorem 1.5.6 (Bishop–Phelps Theorem) *If A is a closed convex subset of the Banach space E, then the set of support points of A is dense in the boundary of A.*

Proof. See, for instance, Phelps [165].   □

Proposition 1.5.7 *Let E be a finite-dimensional normed vector space, and let A and B be nonempty convex subsets of E. Then the following statements are equivalent:*

(a) *A and B are separated.*
(b) *$A \cap B = \emptyset$.*

Proof. See, for instance, Holmes [92].   □

Notice that the condition (b) of Theorem 1.5.3 is equivalent to $o \notin (\mathrm{int}\, A) - B$. Strengthening this condition, we obtain a result on strong separation, provided we restrict ourselves to locally convex spaces.

Theorem 1.5.8 (Strong Separation Theorem 1) *Assume that E is a locally convex space and that A and B are nonempty convex subsets of E. Then the following statements are equivalent:*

(a) *A and B are strongly separated.*
(b) *$o \notin \mathrm{cl}\,(A - B)$.*

Proof. (a) $\Longrightarrow$ (b): Exercise 1.8.4.
(b) $\Longrightarrow$ (a): By virtue of (b) there exists an open convex neighborhood U of zero such that $A \cap (B + U) = \emptyset$. Moreover, $B + U$ is open. Hence, by Theorem 1.5.3, there exists a closed hyperplane $[x^* = \alpha']$ such that $A \subseteq [x^* \leq \alpha']$ and $B + U \subseteq [x^* \geq \alpha']$. Choose some $z_0 \in -U$ such that $\langle x^*, z_0 \rangle > 0$, which exists since $x^* \neq o$. Set $\alpha := \sup_{x \in A}\langle x^*, x \rangle + \frac{1}{2}\langle x^*, z_0 \rangle$ and $\epsilon := \frac{1}{2}\langle x^*, z_0 \rangle$. It follows that $A \subseteq [x^* \leq \alpha - \epsilon]$ and $B \subseteq [x^* \geq \alpha + \epsilon]$.   □

The following is a frequently used special case of Theorem 1.5.8.

Theorem 1.5.9 (Strong Separation Theorem 2) *Assume that A and B are nonempty convex subsets of the locally convex space E such that A is closed, B is compact, and $A \cap B = \emptyset$. Then A and B are strongly separated.*

Proof. See Exercise 1.8.5.   □

Simple examples, already with $E = \mathbb{R}^2$, show that Theorem 1.5.9 may fail if "compact" is replaced by "closed."

In analogy to Corollary 1.5.4 we now have:

Corollary 1.5.10 *Let A be a closed convex cone in E, B a nonempty compact convex subset of E, and assume that $A \cap B = \emptyset$. Then there exists $x^* \in E^*$, $x^* \neq o$, and $\epsilon > 0$ such that*

$$\langle x^*, x \rangle \leq 0 \quad \forall x \in A \quad and \quad \langle x^*, y \rangle \geq \epsilon \quad \forall y \in B.$$

Proof. See Exercise 1.8.6. □

1.6 Dual Pairs of Vector Spaces

For results in this section stated without proof, see the references at the end of the chapter. The following notion will be the basis for the duality theory of convex optimization.

Definition 1.6.1 Let E and F be vector spaces.

(a) A functional $a : E \times F \to \mathbb{R}$ is said to be *bilinear* if $a(\cdot, u)$ is linear on E for each fixed $u \in F$ and $a(x, \cdot)$ is linear on F for each fixed $x \in E$.
(b) Let $a : E \times F \to \mathbb{R}$ be a bilinear functional with the following properties:

$$\begin{aligned} \text{If } a(x, u) = 0 \text{ for each } x \in E, \text{ then } u = o, \\ \text{if } a(x, u) = 0 \text{ for each } u \in F, \text{ then } x = o. \end{aligned} \tag{1.11}$$

Then (E, F) is called *dual pair of vector spaces* with respect to a.

Example 1.6.2 Let E be a locally convex space. Then (E, E^*) is a dual pair with respect to the bilinear functional

$$a(x, x^*) := \langle x^*, x \rangle \quad \forall x \in E \quad \forall x^* \in E^*.$$

We verify (1.11). It is clear by definition that $\langle x^*, x \rangle = 0$ for each $x \in E$ implies $x^* = o$. Now let $\langle x^*, x \rangle = 0$ for each $x^* \in E^*$. Assume that $x \neq o$. Then there exist a closed convex (circled) neighborhood U of zero not containing x. By Theorem 1.5.9, the closed set U and the compact set $\{x\}$ are strongly separated. Hence there exist $x^* \in E^* \setminus \{o\}$, $\alpha \in \mathbb{R}$ and $\epsilon > 0$ such that

$$\langle x^*, y \rangle \leq \alpha - \epsilon \quad \forall y \in U \quad and \quad \langle x^*, x \rangle \geq \alpha + \epsilon.$$

Considering $y = o$, we see that $\alpha - \epsilon \geq 0$, and it follows that $\langle x^*, x \rangle \geq \alpha + \epsilon > \alpha - \epsilon \geq 0$: a contradiction.

We shall now show that this example provides the prototype of a dual pair. Let (E, F) be a dual pair with respect to the bilinear functional a. The *weak topology* $\sigma(E, F)$ on E is by definition the weakest topology on E such that each linear functional $x \mapsto a(x, u)$ is continuous; here u varies over F. In other words, each $u \in F$ defines an element x_u^* of E^* via

$$\langle x_u^*, x \rangle := a(x, u) \quad \forall \, x \in E. \tag{1.12}$$

By (1.11), if $u \neq v$, then $x_u^* \neq x_v^*$. Hence we can identify x_u^* with u, so that we obtain $F \subseteq E^*$. Proposition 1.6.3 states, among others, that each $x^* \in E^*$ is of the form x_u^* for some $u \in F$. If we want to indicate the topology, say τ, of a topological vector space E, we write $E[\tau]$ instead of E. Similarly, we use $E[\|\cdot\|]$.

Proposition 1.6.3 *If (E, F) is a dual pair of vector spaces with respect to the bilinear functional a, then:*

(a) *The weak topology $\sigma(E, F)$ is a locally convex topology on E, a neighborhood base of zero being formed by the sets*

$$U(u_1, \ldots, u_m) := \{ x \in E \, | \, |a(x, u_i)| < 1 \text{ for } i = 1, \ldots, m \},$$

where $m \in \mathbb{N}$ and $u_1, \ldots, u_m \in F$.

(b) *For each $x^* \in E^*$ there exists precisely one $u \in F$ such that $x^* = x_u^*$ (see (1.12)). The dual E^* of the locally convex space $E[\sigma(E, F)]$ can thus be identified with F.*

Of course, the same holds true with the roles of E and F exchanged, i.e., one defines analogously the weak topology $\sigma(F, E)$ on F. Then the dual F^* of $F[\sigma(F, E)]$ can be identified with E.

Let (E, F) be a dual pair with respect to some bilinear functional. A locally convex topology τ_E on E is said to be *compatible* with the dual pair (E, F) if $E[\tau_E]^* = F$ in the sense described above, analogously for a locally convex topology on F. If (E, F) is a dual pair of vector spaces and if τ_E, τ_F are compatible topologies on E and F, respectively, then $(E[\tau_E], F[\tau_F])$ is called *dual pair of locally convex spaces*. A complete characterization of compatible topologies (not needed in this book) is given by the *Mackey–Arens theorem*.

Remark 1.6.4

(a) Let $E[\tau]$ be a locally convex space. Since (E, E^*) is a dual pair of vector spaces (see Example 1.6.2), we can consider the weak topology $\sigma(E, E^*)$ on E, which is weaker than τ. On E^*, we have to distinguish between the topologies $\sigma(E^*, E^{**})$ and $\sigma(E^*, E)$. The latter is called *weak star topology* or *weak* topology*. Among others, we have the following dual pairs of locally convex spaces:

$$\big(E[\tau],\ E^*[\sigma(E^*, E)]\big) \quad \text{and} \quad \big(E[\sigma(E, E^*)],\ E^*[\sigma(E^*, E)]\big).$$

If A is a subset of E^*, we denote by $\mathrm{cl}^* A$ the $\sigma(E^*, E)$-closure of A and by $\overline{\mathrm{co}}^* A$ the $\sigma(E^*, E)$-closed convex hull of A. A sequence (x_k) in E that is $\sigma(E, E^*)$-convergent to $x \in E$ is said to be *weakly convergent to x*, written $x_k \xrightarrow{w} x$ as $k \to \infty$; this means that $\lim_{k \to \infty}\langle x^*, x_k \rangle = \langle x^*, x \rangle$ for any $x^* \in E^*$. Analogously, a sequence (x_k^*) in E^* is *weak* convergent* to $x^* \in E^*$, written

$$x_k^* \xrightarrow{w^*} x^* \quad \text{as } k \to \infty,$$

if and only if $\lim_{k \to \infty}\langle x_k^*, x \rangle = \langle x^*, x \rangle$ for any $x \in E$.

(b) In view of Proposition 1.6.3, we usually denote any dual pair of vector spaces by (E, E^*). Moreover, when there is no need to specify the topology, we shall use (E, E^*) also to denote a dual pair of locally convex spaces, tacitly assuming that E and E^* are equipped with topologies compatible with the dual pair.

(c) Now let $E[\,\|\cdot\|\,]$ be a normed vector space and let $\|\cdot\|_*$ denote the associated norm on E^*. The weak topology $\sigma(E, E^*)$ is weaker than the topology generated by the norm $\|\cdot\|$; the two topologies coincide if and only if E is finite dimensional. In general, the topology generated by the norm $\|\cdot\|_*$ on E^* is not compatible with the dual pair (E, E^*). However, if E is a reflexive Banach space, then

$$\left(E[\|\cdot\|], \; E^*[\|\cdot\|_*] \right)$$

is a dual pair of locally convex spaces.

(d) If E is an infinite-dimensional normed vector space, then the weak topology $\sigma(E, E^*)$ does not admit a countable base of neighborhoods of zero. Hence properties referring to the weak topology can in general, not be characterized by sequences. For instance, each weakly closed subset of E is weakly sequentially closed but not conversely. In this connection, a subset A of E is said to be weakly sequentially closed if each limit point of a weakly convergent sequence in A is an element of A.

It turns out that certain topological properties important in the following depend on the dual pair only.

Proposition 1.6.5 *Let (E, E^*) be a dual pair of vector spaces, and let τ_1 and τ_2 be locally convex topologies on E compatible with the dual pair. Further let A be a convex subset of E. Then: A is τ_1-closed if and only if A is τ_2-closed.*

Corollary 1.6.6 *In a normed vector space, any convex closed subset is weakly sequentially closed.*

We conclude this section with two important results.

Theorem 1.6.7 (Eberlein–Šmulian Theorem) *In a reflexive Banach space, any bounded weakly sequentially closed subset (in particular, any convex bounded closed subset) is weakly sequentially compact.*

Theorem 1.6.8 (Krein–Šmulian Theorem) *If E is a Banach space, then a convex subset M of E^* is weak* closed if and only if $M \cap \rho B_{E^*}$ is weak* closed for any $\rho > 0$.*

1.7 Lower Semicontinuous Functionals

Lower semicontinuous functionals will play an important part in these lectures. We first repeat the definition.

Definition 1.7.1 Let E be a topological vector space, M a nonempty subset of E, and $f : M \to \overline{\mathbb{R}}$:

(a) The functional f is called *lower semicontinuous (l.s.c.)* at $\bar{x} \in M$ if either $f(\bar{x}) = -\infty$ or for every $k < f(\bar{x})$ there exists a neighborhood U of $\bar{x}$ such that
$$k < f(x) \quad \forall x \in M \cap U \quad \text{(cf. Fig. 1.7)}.$$
f is said to be *lower semicontinuous on M* if f is l.s.c. at each $\bar{x} \in M$. Moreover, f is called *upper semicontinuous* at $\bar{x}$ if $-f$ is l.s.c. at $\bar{x}$.

(b) The functional f is said to be *sequentially lower semicontinuous* at $\bar{x} \in M$ if for each sequence (x_n) in M satisfying $x_n \to \bar{x}$ as $n \to \infty$ one has $f(\bar{x}) = \liminf_{n\to\infty} f(x_n)$.

Lemma 1.7.2 characterizes these properties by properties of appropriate sets.

Lemma 1.7.2 *Let E be a topological vector space, M a nonempty subset of E, and $f : M \to \overline{\mathbb{R}}$. Then the following statements are equivalent:*

(a) *f is l.s.c. on M.*
(b) *For each $\lambda \in \mathbb{R}$, the set $M_\lambda := \{x \in M \mid f(x) \le \lambda\}$ is closed relative to M.*
(c) *epi f is closed relative to $M \times \mathbb{R}$.*

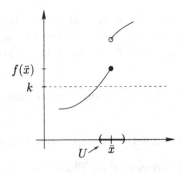

Fig. 1.7

Proof. We only verify (a) $\Longrightarrow$ (c), leaving the proof of the remaining assertions as Exercise 1.8.7. Thus let (a) hold. We show that $(M \times \mathbb{R}) \setminus \mathrm{epi}\, f$ is open. Let $(\bar{x}, \bar{t}) \in (M \times \mathbb{R}) \setminus \mathrm{epi}\, f$ be given. Then $f(\bar{x}) > \bar{t}$. Set $\delta := \frac{1}{2}(f(\bar{x}) - \bar{t})$. Then $f(\bar{x}) > \bar{t} + \delta$. By (a) there exists a neighborhood U of $\bar{x}$ in E such that $f(x) > \bar{t} + \delta$ for each $x \in M \cap U$. Then $V := (M \cap U) \times (\bar{t} - \delta, \bar{t} + \delta)$ is a neighborhood of $(\bar{x}, \bar{t})$ relative to $M \times \mathbb{R}$. For any $(x, t) \in V$ we have $f(x) > \bar{t} + \delta > t$ and so $(x, t) \notin \mathrm{epi}\, f$. $\qquad\square$

Proposition 1.7.3 *Assume that E is a normed vector space, $M \subseteq E$ is nonempty convex and closed, and $f : E \to \overline{\mathbb{R}}$ is proper and convex. Then the following assertions are equivalent:*

(a) *f is l.s.c. on M.*
(b) *f is weakly l.s.c. on M.*
(c) *f is weakly sequentially l.s.c. on M.*

Proof. We only verify (a) $\Longrightarrow$ (c), leaving the verification of the remaining assertions as Exercise 1.8.8. Thus let (a) hold. Suppose, to the contrary, that f is not weakly sequentially l.s.c. at some $\bar{x} \in M$. Then there exists a sequence (x_n) in M satisfying $x_n \xrightarrow{w} \bar{x}$ and $f(\bar{x}) > \lim_{n \to \infty} f(x_n)$. Choose $\lambda \in \mathbb{R}$ and $n_0 \in \mathbb{N}$ such that $f(\bar{x}) > \lambda \geq f(x_n)$ for all $n \geq n_0$. Since M is closed and M_λ is closed relative to M (Lemma 1.7.2), the latter set is closed. In addition, M_λ is convex (Lemma 1.3.3). By Corollary 1.6.6, M_λ is weakly sequentially closed. Hence $x_n \xrightarrow{w} \bar{x}$ implies $\bar{x} \in M_\lambda$, which is a contradiction to $f(\bar{x}) > \lambda$. $\qquad\square$

Proposition 1.7.4 *Let E be a Banach space. If $f : E \to \overline{\mathbb{R}}$ is proper convex l.s.c. and $\mathrm{int}\,\mathrm{dom}\, f$ is nonempty, then f is continuous on $\mathrm{int}\,\mathrm{dom}\, f$.*

Proof. We may assume that $o \in \mathrm{int}\,\mathrm{dom}\, f$. Choose a number $\lambda > f(o)$ and set $A := \{x \in E \mid f(x) \leq \lambda\}$. Then A is convex and closed. We show that A is also absorbing. Let $x \in E$, $x \neq o$, be given. By Corollary 1.4.2 the restriction of f to the one-dimensional subspace $\{\tau x \mid \tau \in \mathbb{R}\}$ is continuous at zero. This and $f(o) < \lambda$ imply that there exists $\alpha_0 > 0$ such that $f(\alpha x) < \lambda$ and so $\alpha x \in A$ whenever $|\alpha| \leq \alpha_0$. By Proposition 1.2.1 the set A is a neighborhood of zero. Now Theorem 1.4.1 completes the proof. $\qquad\square$

Let A be a nonempty subset of E. A set $M \subseteq A$ is called *extremal subset* of A if the following holds:

$$\forall \lambda \in (0, 1) \quad \forall x, y \in A : \quad \lambda x + (1 - \lambda)y \in M \quad \Longrightarrow \quad x, y \in M.$$

Further, $\bar{x} \in A$ is called *extreme point* of A if $\{\bar{x}\}$ is an extremal subset of A.

If A is convex, then $\bar{x} \in A$ is an extreme point of A if and only if the set $A \setminus \{\bar{x}\}$ is also convex. We write $\mathrm{ep}\, A$ for the set of all extreme points of A.

Example 1.7.5 In $\mathbb{R}^n$, the set $\mathrm{ep}(\mathrm{B}(o, 1))$ consists of the boundary of $\mathrm{B}(o, 1)$ for the Euclidean norm and of finitely many points (which?) for the l^1 norm.

Lemma 1.7.6 *If M is an extremal subset of A, then $\mathrm{ep}(M) = M \cap \mathrm{ep}(A)$.*

Proof. See Exercise 1.8.9.   □

We recall an important result of Functional Analysis. In this connection, $\overline{\mathrm{co}}\, B$ denotes the closed convex hull of the set $B \subseteq E$, i.e., the intersection of all closed convex sets containing B.

Theorem 1.7.7 (Krein–Milman Theorem) *Let E be a locally convex space and A a nonempty subset of E:*

(a) *If A is compact, then $\mathrm{ep}\, A \neq \emptyset$.*
(b) *If $A \subseteq E$ is compact and convex, then $A = \overline{\mathrm{co}}(\mathrm{ep}\, A)$.*

Proposition 1.7.8 (Bauer's Maximum Theorem) *Let E be a locally convex space, A a nonempty compact convex subset of E, and $g : A \to \mathbb{R}$ an upper semicontinuous convex functional. Then there exists $\bar{x} \in \mathrm{ep}\, A$ such that $g(\bar{x}) = \max_{x \in A} g(x)$.*

Proof. Let

$$M := \{\bar{x} \in A \mid g(\bar{x}) = \max_{x \in A} g(x)\}.$$

We have to prove that $M \cap \mathrm{ep}\, A \neq \emptyset$. Let $m := \max_{x \in A} g(x)$. We first show that M is an extremal subset of A. In fact, if $\lambda \in (0,1)$; $x, y \in A$ and $\lambda x + (1 - \lambda)y \in M$, then

$$m = g\big(\lambda x + (1 - \lambda)y\big) \leq \lambda g(x) + (1 - \lambda)g(y) \leq \lambda m + (1 - \lambda)m = m$$

and so $\lambda g(x) + (1 - \lambda)g(y) = m$. Consequently, $x, y \in M$. Hence M is an extremal subset of A, and Lemma 1.7.6 implies that $M \cap \mathrm{ep}\, A = \mathrm{ep}\, M$. The hypotheses on A and g entail that M is nonempty and compact. By Theorem 1.7.7(a), we have $\mathrm{ep}\, M \neq \emptyset$.   □

1.8 Bibliographical Notes and Exercises

Concerning separation theorems and dual pairs, we refer to standard textbooks on Functional Analysis, for example Aliprantis and Border [2], Holmes [92], Werner [215], or Yosida [218]. A good introductory presentation of convex functionals give Roberts and Varberg [174]. For cs-closed sets see Jameson [103]. The present form of the sandwich theorem (Theorem 1.5.2) is due to Landsberg and Schirotzek [116]. Ernst et al. [61] establish results on the strict separation of disjoint closed sets in reflexive Banach spaces.

Exercise 1.8.1 Prove Lemma 1.2.2.

Exercise 1.8.2 Verify Lemma 1.3.3.

Exercise 1.8.3 Prove Corollary 1.5.4.

Exercise 1.8.4 Verify the implication (a) $\Longrightarrow$ (b) in Theorem 1.5.8.

Exercise 1.8.5 Prove Theorem 1.5.9.

Exercise 1.8.6 Verify Corollary 1.5.10.

Exercise 1.8.7 Prove the remaining assertions of Lemma 1.7.2

Exercise 1.8.8 Verify the remaining assertions of Proposition 1.7.3.

Exercise 1.8.9 Prove Lemma 1.7.6.

Exercise 1.8.10 Show that the norm functional on the normed vector space E is weakly l.s.c. and the norm functional on E^* is weak* l.s.c.

Exercise 1.8.11 Let E be a normed vector space, let $f : E \to \overline{\mathbb{R}}$ be a proper functional, and define the *lower semicontinuous closure* $\underline{f} : E \to \overline{\mathbb{R}}$ of f by

$$\underline{f}(x) := \liminf_{y \to x} f(y), \quad x \in E.$$

Show that $\underline{f}$ is the largest l.s.c. functional dominated by f.

2

The Conjugate of Convex Functionals

2.1 The Gamma Regularization

Convention. Throughout this chapter unless otherwise specified, (E, E^*) denotes any dual pair of locally convex spaces (cf. Remark 1.6.4 and Proposition 1.6.5).

In this section, we show that a l.s.c. convex functional is the upper envelope of continuous affine functionals.

Definition 2.1.1 For $f : E \to \overline{\mathbb{R}}$ let

$$\mathcal{A}(f) := \{a : E \to \mathbb{R} \mid a \text{ is continuous and affine, } a \leq f\}.$$

The functional $f^\Gamma : E \to \overline{\mathbb{R}}$ defined by

$$f^\Gamma(x) := \sup\{a(x) \mid a \in \mathcal{A}(f)\}, \quad x \in E,$$

is called *Gamma regularization* of f. Recall that $\sup \emptyset := -\infty$.

Figure 2.1 suggests the following result.

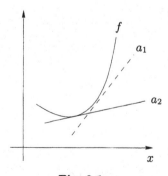

Fig. 2.1

Proposition 2.1.2 *If $f : E \to \overline{\mathbb{R}}$ is proper, then the following statements are equivalent:*

(a) $f = f^{\Gamma}$.
(b) f *is l.s.c. and convex.*

Proof.

(a) $\Longrightarrow$ (b): See Exercise 2.5.2.
(b) $\Longrightarrow$ (a): It is clear that $f^{\Gamma} \le f$. Assume now that, for some $x_0 \in E$ and some $k \in \mathbb{R}$, we had $f^{\Gamma}(x_0) < k < f(x_0)$. We shall show that there exists $a \in \mathcal{A}(f)$ satisfying $k < a(x_0)$ (cf. Fig. 2.2), which would imply the contradiction $f^{\Gamma}(x_0) > k$.

Since f is l.s.c., epi f is closed (Lemma 1.7.2). Furthermore, epi f is convex (Lemma 1.3.3) and $(x_0, k) \notin$ epi f. By the strong separation theorem 2 (Theorem 1.5.9) applied with $A :=$ epi f and $B := \{(x_0, k)\}$, there exist $w \in (E \times \mathbb{R})^{*}$ and $\alpha \in \mathbb{R}$ such that

$$w(x, t) \le \alpha \ \forall\, (x, t) \in \text{epi}\, f \quad \text{and} \quad w(x_0, k) > \alpha. \tag{2.1}$$

We have

$$w(x, t) = \langle x^{*}, x \rangle + ct, \quad \text{where } \langle x^{*}, x \rangle := w(x, 0), \ c := w(o, 1). \tag{2.2}$$

It is obvious that $x^{*} \in E^{*}$. Further, since $(x, t) \in$ epi f entails $(x, t') \in$ epi f for each $t' > t$, we obtain

$$c \le \frac{\alpha - \langle x^{*}, x \rangle}{t'} \quad \forall\, t' > \max\{0, t\}$$

and so, letting $t' \to +\infty$, $c \le 0$. Now we distinguish two cases.

Case 1. Assume that $f(x_0) < +\infty$. Then (2.1) with $t := f(x_0)$ and (2.2) imply

$$\langle x^{*}, x_0 \rangle + cf(x_0) \le \alpha < \langle x^{*}, x_0 \rangle + ck.$$

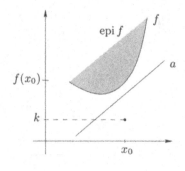

Fig. 2.2

Since $k < f(x_0)$, it follows that $c < 0$. The functional $a : E \to \mathbb{R}$ defined by

$$a(x) := \tfrac{\alpha}{c} - \tfrac{1}{c}\langle x^*, x \rangle, \quad x \in E,$$

is continuous and affine. If $x \in \operatorname{dom} f$, we have by (2.1),

$$a(x) = \tfrac{1}{c}\big(\alpha - w(x, f(x))\big) + f(x) \leq f(x).$$

If $x \notin \operatorname{dom} f$, then $a(x) < +\infty = f(x)$. Hence $a \in \mathcal{A}(f)$. Moreover, we have

$$a(x_0) = \tfrac{1}{c}\big(\alpha - \langle x^*, x_0 \rangle\big) > k.$$

Case 2. Assume that $f(x_0) = +\infty$. If $c < 0$, then define a as in Case 1. Now suppose that $c = 0$. Since f is proper, there exists $y_0 \in \operatorname{dom} f$. According to Case 1, with y_0 instead of x_0, there exists $a_0 \in \mathcal{A}(f)$. Define $a : E \to \mathbb{R}$ by

$$a(x) := a_0(x) + \rho\big(\langle x^*, x \rangle - \alpha\big), \quad \text{where } \rho := \frac{|k - a_0(x_0)|}{\langle x^*, x_0 \rangle - \alpha} + 1.$$

Then a is continuous and affine. Further we have $a(x) \leq a_0(x) \leq f(x)$ for each $x \in \operatorname{dom} f$ and so $a \in \mathcal{A}(f)$. Finally, noting that $a_0(x_0) + |k - a_0(x_0)| \geq k$ and $\langle x^*, x_0 \rangle > \alpha$, we obtain

$$a(x_0) = a_0(x_0) + |k - a_0(x_0)| + \langle x^*, x_0 \rangle - \alpha > k,$$

and the proof is complete. □

2.2 Conjugate Functionals

The concept of the conjugate functional, which has its roots in the *Legendre transform* of the calculus of variations, will be crucial for the duality theory in convex optimization to be developed later.

Definition 2.2.1 Let $f : E \to \overline{\mathbb{R}}$. The functional $f^* : E^* \to \overline{\mathbb{R}}$ defined by

$$f^*(x^*) := \sup_{x \in E}\big(\langle x^*, x \rangle - f(x)\big), \quad x^* \in E^*,$$

is called the *Fenchel conjugate* (or briefly, the *conjugate*) of f.

If f is proper, the definition immediately implies the *Young inequality*

$$\langle x^*, x \rangle \leq f(x) + f^*(x^*) \quad \forall\, x \in E \ \forall x^* \in E^*. \tag{2.3}$$

Geometric Interpretation

Let $x^* \in E^*$ be such that $f^*(x^*) \in \mathbb{R}$. Then the function $a : E \to \mathbb{R}$ defined by

$$a(x) := \langle x^*, x \rangle - f^*(x^*), \quad x \in E,$$

belongs to $\mathcal{A}(f)$. For each $\epsilon > 0$ there exists $x_\epsilon \in E$ satisfying $\langle x^*, x_\epsilon \rangle - f(x_\epsilon) > f^*(x^*) - \epsilon$ and so $a(x_\epsilon) > f(x_\epsilon) - \epsilon$. Hence, for $E = \mathbb{R}$, a may be interpreted as a *tangent* to f, and we have $a(o) = -f^*(x^*)$ (Fig. 2.3).

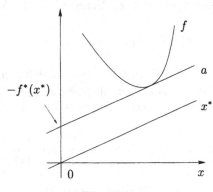

Fig. 2.3

Example 2.2.2 Let $p \in (1, +\infty)$ be given. Define $f : \mathbb{R} \to \mathbb{R}$ by $f(x) := \frac{|x|^p}{p}$. We compute f^*. For $E = \mathbb{R}$ we have $E^* = \mathbb{R}$. With $x^* \in \mathbb{R}$ fixed, set $\varphi(x) := x^*x - f(x)$. The function $\varphi : \mathbb{R} \to \mathbb{R}$ is concave (i.e., $-\varphi$ is convex) and differentiable. Hence φ has a unique maximizer x_0 which satisfies $\varphi'(x_0) = 0$, i.e., $x^* - \operatorname{sgn}(x_0)|x_0|^{p-1} = 0$. It follows that

$$f^*(x^*) = \varphi(x_0) = \frac{|x^*|^q}{q}, \quad \text{where } \frac{1}{p} + \frac{1}{q} = 1.$$

Thus in this case the Young inequality (2.3) is just the classical Young inequality for real numbers:

$$x^*x \le \frac{|x|^p}{p} + \frac{|x^*|^q}{q}.$$

Proposition 2.2.3 *If $f : E \to \overline{\mathbb{R}}$, then the following holds:*

(a) *f^* is convex and l.s.c.*
(b) *If $\operatorname{dom} f \ne \emptyset$, then $f^*(x^*) > -\infty$ for every $x^* \in E^*$.*
(c) *If f is proper, convex, and l.s.c., then f^* is proper, convex, and l.s.c.*

Proof.

(a) It is easy to see that f^* is convex. To prove the second assertion, notice that for each $x \in E$, the functional $\varphi_x(x^*) := \langle x^*, x \rangle - f(x)$, $x^* \in E^*$, is continuous and so $f^* = \sup_{x \in E} \varphi_x$ is l.s.c.

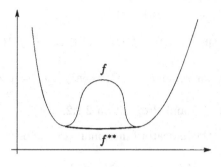

Fig. 2.4

(b) is obvious.

(c) Since f is proper, we have $\mathcal{A}(f) \neq \emptyset$ (see the proof of Proposition 2.1.2). Hence there exist $x^* \in E^*$ and $c \in \mathbb{R}$ such that $\langle x^*, x \rangle + c \leq f(x)$ for each $x \in E$. It follows that

$$f^*(x^*) = \sup_{x \in E}\left(\langle x^*, x \rangle - f(x)\right) \leq -c < +\infty \quad \text{and so } x^* \in \text{dom } f^*. \quad \square$$

For $g : E^* \to \overline{\mathbb{R}}$, the conjugate $g^* : E \to \overline{\mathbb{R}}$ is defined analogously by

$$g^*(x) := \sup_{x^* \in E^*}\left(\langle x^*, x \rangle - g(x^*)\right), \quad x \in E.$$

In this connection, we make use of the fact that, since (E, E^*) is a dual pair, the dual of E^* can be identified with E. Now let $f : E \to \overline{\mathbb{R}}$ be given. Applying the above to $g := f^*$, we obtain the *biconjugate* $f^{**} : E \to \overline{\mathbb{R}}$ of f:

$$f^{**}(x) = \sup_{x^* \in E^*}\left(\langle x^*, x \rangle - f^*(x^*)\right), \quad x \in E.$$

Geometrically, Theorem 2.2.4 says that f^{**} can be interpreted as a *convexification* of f from below (Fig. 2.4).

Theorem 2.2.4 (Biconjugation Theorem) *Let* $f : E \to \overline{\mathbb{R}}$.

(a) *One always has* $f^{**} = f^\Gamma \leq f$.

(b) *If f is proper, then* $f^{**} = f$ *if and only if f is convex and l.s.c.*

Proof.

(a) It is clear that $f^\Gamma \leq f$. We show that $f^{**} = f^\Gamma$. For $x^* \in E^*$ and $c \in \mathbb{R}$, we have

$$\langle x^*, x \rangle + c \leq f(x) \; \forall \, x \in E \quad \Longleftrightarrow \quad f^*(x^*) = \sup_{x \in E}\left(\langle x^*, x \rangle - f(x)\right) \leq -c$$

and so

$$f^\Gamma(x) = \sup\{\langle x^*, x \rangle + c \,|\, x^* \in E^*, \; c \in \mathbb{R}, \; c \leq -f^*(x^*)\}. \tag{2.4}$$

If $f^*(x^*) > -\infty$ for all $x^* \in E^*$, then

$$f^\Gamma(x) \underset{(2.4)}{=} \sup\{\langle x^*, x\rangle - f^*(x^*) \mid x^* \in E^*\} = f^{**}(x) \quad \forall\, x \in E.$$

If $f^*(x^*) = -\infty$ for some $x^* \in E^*$, then $f^\Gamma(x) = +\infty = f^{**}(x)$ for all $x \in E$.

(b) This follows from (a) and Proposition 2.1.2. $\qquad\qquad\qquad\qquad\qquad\square$

Example 2.2.5 For the indicator functional δ_A of a nonempty subset A of E, we have

$$\delta_A^*(x^*) = \sigma_A(x^*) \quad \forall\, x^* \in E^*,$$

where $\sigma_A : E^* \to \overline{\mathbb{R}}$ denotes the support functional of A. If, in particular, E is a normed vector space and $A = \mathrm{B}_E$ (the closed unit ball in E), then

$$\delta_{\mathrm{B}_E}^*(x^*) = \sigma_{\mathrm{B}_E}(x^*) = \sup_{\|x\| \le 1} \langle x^*, x\rangle = \|x^*\| \quad \forall\, x^* \in E^*.$$

Example 2.2.6 Let again E be a normed vector space and $f(x) = \|x\|$, $x \in E$. Consider the dual pair $(E[\|\cdot\|], E^*[\sigma(E^*, E)])$. We want to determine f^*:

(I) Let $x^* \in E^*$, $\|x^*\| \le 1$. Then $\langle x^*, x\rangle \le \|x^*\|\,\|x\| \le \|x\|$ for all $x \in E$ and $\langle x^*, o\rangle = 0 = \|o\|$. Hence

$$f^*(x^*) = \sup_{x \in E}\big(\langle x^*, x\rangle - \|x\|\big) = 0.$$

(II) Let $x^* \in E^*$, $\|x^*\| > 1$. Then there exists $x_0 \in E$ such that $\alpha := \langle x^*, x_0\rangle - \|x_0\| > 0$. For each $\rho > 0$ we have $\langle x^*, \rho x_0\rangle - \|\rho x_0\| = \alpha\rho$. Letting $\rho \to +\infty$, we see that $f^*(x^*) = +\infty$. We conclude that $f^* = \delta_{\mathrm{B}_{E^*}}$. Therefore we obtain

$$\|x\| = f(x) = f^{**}(x) = \delta_{\mathrm{B}_{E^*}}^*(x) = \sup_{\|x^*\| \le 1} \langle x^*, x\rangle \quad \forall\, x \in E;$$

here, the second equation follows from Theorem 2.2.4 and the last follows by applying Example 2.2.5 to E^* instead of E. As a consequence of the Hahn–Banach theorem the supremum is attained, i.e.,

$$\|x\| = \max_{\|x^*\| \le 1} \langle x^*, x\rangle \quad \forall\, x \in E.$$

The following operation is a useful device for calculating the conjugate of a sum.

Definition 2.2.7 Let $f_0, f_1 : E \to \overline{\mathbb{R}}$ be proper. The functional $f_0 \oplus f_1 : E \to \overline{\mathbb{R}}$ defined by

$$f_0 \oplus f_1(x) := \inf_{y \in E}\big(f_0(x - y) + f_1(y)\big) \quad \forall\, x \in E$$

is called *infimal convolution* of f_0 and f_1.

Theorem 2.2.8 (Sum Theorem) *Let* $f_0, f_1 : E \to \overline{\mathbb{R}}$ *be proper.*

(a) *One always has*

$$f_0^* + f_1^* = (f_0 \oplus f_1)^* \quad and \quad (f_0 + f_1)^* \le f_0^* \oplus f_1^*.$$

(b) *Suppose, in addition, that* f_0 *and* f_1 *are convex and that there exists* $\bar{x} \in$ dom $f_0 \cap$ int dom f_1 *such that* f_1 *is continuous at* $\bar{x}$. *Then one has*

$$(f_0 + f_1)^* = f_0^* \oplus f_1^*.$$

Moreover, for each $x^* \in$ dom $(f_0 + f_1)^*$ *there exist* $x_i^* \in$ dom f_i^* *for* $i = 0, 1$ *such that*

$$x^* = x_0^* + x_1^* \quad and \quad (f_0 + f_1)^*(x^*) = f_0^*(x_0^*) + f_1^*(x_1^*), \qquad (2.5)$$

i.e., the infimum in $(f_0^* \oplus f_1^*)(x^*)$ *is attained.*

Proof.

(a) See Exercise 2.5.3.
(b) Let $x^* \in E^*$ and $\alpha := (f_0 + f_1)^*(x^*)$. If $\alpha = +\infty$, then (a) implies $(f_0^* \oplus f_1^*)(x^*) = +\infty$ and so we have equality. Now suppose that $\alpha < +\infty$. We have to show that $(f_0^* \oplus f_1^*)(x^*) \le \alpha$. Since $\bar{x} \in$ dom$(f_0 + f_1)$, we have $\alpha > -\infty$ (Proposition 2.2.3) and so $\alpha \in \mathbb{R}$. Set $p := f_0$ and $q := f_1 - x^* + \alpha$. If $x \in$ dom $f_0 \cap$ dom f_1, then $\alpha \ge \langle x^*, x \rangle - (f_0 + f_1)(x)$, which implies $p(x) \ge -q(x)$ for each $x \in E$. Moreover, q is continuous at $\bar{x}$. By the sandwich theorem (Theorem 1.5.2) there exist $x_0' \in E^*$ and $c \in \mathbb{R}$ satisfying

$$\langle x_0', x \rangle + c \le f_0(x) \quad and \quad c + \langle x_0', x \rangle \ge \langle x^*, x \rangle - f_1(x) - \alpha \quad \forall x \in E. \quad (2.6)$$

From the first inequality in (2.6) we obtain $f_0^*(x_0') \le -c$ and so $x_0' \in$ dom f_0^*. For $x_1' := x^* - x_0'$, the second inequality of (2.6) implies $f_1^*(x_1') - c \le \alpha$. It follows that $x_1' \in$ dom f_1^* and

$$f_0^*(x_0') + f_1^*(x_1') \le \alpha. \qquad (2.7)$$

We have $x^* = x_0' + x_1'$ and so

$$(f_0^* \oplus f_1^*)(x^*) = \inf_{y' \in E^*} \left(f_0^*(x^* - y') + f_1^*(y') \right) \le \alpha.$$

This finally yields

$$(f_0 + f_1)^*(x^*) = \alpha = (f_0^* \oplus f_1^*)(x^*) \underset{(2.7)}{=} f_0^*(x_0') + f_1^*(x_1'). \quad \square$$

As a first simple application of conjugate functionals we derive a duality formula of approximation theory. Recall that if $A \subseteq E$, we write $d_A(x) := \inf_{y \in A} \|x - y\|$.

Proposition 2.2.9 *If $A \subseteq E$ is nonempty and convex, then*

$$d_A(x) = \max_{\|x^*\| \leq 1} \left(\langle x^*, x \rangle - \sup_{y \in A} \langle x^*, y \rangle \right) \quad \forall x \in E.$$

Proof. Setting $f_0(x) := \|x\|$ and $f_1(x) := \delta_A(x)$, we have

$$(f_0 \oplus f_1)(x) = \inf_{y \in E} \left(\|x - y\| + \delta_A(y) \right) = d_A(x).$$

We further set $B := B_{E^*}$. Since the distance function $x \mapsto d_A(x)$ is proper, convex and continuous on E, we obtain

$$\begin{aligned}
d_A(x) = (f_0 \oplus f_1)^{**}(x) &= \left(f_0^* + f_1^* \right)^*(x) \\
&= \sup_{x^* \in E^*} \left(\langle x^*, x \rangle - \left(\delta_B(x^*) + \sup_{y \in A} \langle x^*, y \rangle \right) \right) \\
&= \sup_{x^* \in B} \left(\langle x^*, x \rangle - \sup_{y \in A} \langle x^*, y \rangle \right).
\end{aligned}$$

Since, with respect to $\sigma(E^*, E)$, the set B is compact (Alaoglu Theorem) and the functional $x^* \mapsto \langle x^*, x \rangle - \sup_{y \in A} \langle x^*, y \rangle$ is upper semicontinuous, the supremum over B is attained and so is a maximum. $\qquad \square$

2.3 A Theorem of Hörmander and the Bipolar Theorem

The following result states, roughly speaking, that convex closed subsets of E^* can be described by sublinear l.s.c. functionals on E and vice versa. In this connection, recall the convention at the beginning of this chapter. Also recall that the support functional σ_M of the set $M \subseteq E^*$ is defined by

$$\sigma_M(x) := \sup_{x^* \in M} \langle x^*, x \rangle, \ x \in E.$$

Theorem 2.3.1 (Hörmander Theorem)

(a) *Let M be a nonempty, convex, closed subset of E^*. Then the support functional σ_M is proper, sublinear, and l.s.c., and one has*

$$M = \{x^* \in E^* \mid \langle x^*, x \rangle \leq \sigma_M(x) \ \forall x \in E\}. \tag{2.8}$$

(b) *Let $p : E \to \overline{\mathbb{R}}$ be proper, sublinear, and l.s.c. Then the set*

$$M_p := \{x^* \in E^* \mid \langle x^*, x \rangle \leq p(x) \ \forall x \in E\}$$

is nonempty, convex, and closed, and one has $\sigma_{M_p} = p$.

(c) *If M_1 and M_2 are nonempty, convex, closed subsets of E^*, then*

$$M_1 \subseteq M_2 \quad \Longleftrightarrow \quad \sigma_{M_1}(x) \leq \sigma_{M_2}(x) \quad \forall x \in E.$$

Proof.

(a) It is easy to see that σ_M is proper, sublinear, and l.s.c. We show that (2.8) holds. By Theorem 2.2.4 and Example 2.2.5 (with E^* in place of E) we obtain

$$\delta_M = \left(\delta_M^*\right)^* = \sigma_M^*,$$
$$\sigma_M^*(x^*) = \sup_{x \in E}\left(\langle x^*, x\rangle - \sigma_M(x)\right) = \delta_{M_p}(x^*) \,\forall\, x^* \in E^*, \quad \text{where } p := \sigma_M.$$

Hence $\delta_M = \delta_{M_p}$ and so $M = M_p$.

(b) (I) $M_p \neq \emptyset$: Since p is l.s.c. and $p(o) = 0$, there exists a neighborhood U of zero in E such that $-1 < p(x)$ for each $x \in U$. Define $q : E \to \overline{\mathbb{R}}$ by $q(x) := 1$ for $x \in U$ and $q(x) := +\infty$ for $x \in E \setminus U$. By the sandwich theorem (Theorem 1.5.2) there exist $x^* \in E^*$ and $c \in \mathbb{R}$ satisfying

$$\langle x^*, x\rangle + c \le p(x) \,\forall\, x \in E \quad \text{and} \quad -1 \le \langle x^*, x\rangle + c \,\forall\, x \in U.$$

Since p is sublinear, it follows that $\langle x^*, x\rangle \le p(x)$ for each $x \in E$ and so $x^* \in M_p$.

(II) M_p is closed: For each $x \in E$, let $\varphi_x(x^*) := \langle x^*, x\rangle$, $x^* \in E^*$. Then φ_x is continuous. Hence the set

$$M_x := \{x^* \in E^* \,|\, \langle x^*, x\rangle \le p(x)\} = \varphi_x^{-1}\left(-\infty, p(x)\right]$$

is closed and so is $M_p = \bigcap_{x \in E} M_x$.

(III) $\sigma_{M_p} = p$: We have

$$p^*(x^*) = \sup_{x \in E}\left(\langle x^*, x\rangle - p(x)\right) = \delta_{M_p}$$

and so, using Theorem 2.2.4 and Example 2.2.5,

$$p = p^{**} = \delta_{M_p}^* = \sigma_{M_p}.$$

(c) The implication $\Longrightarrow$ holds by definition, and $\Longleftarrow$ is a consequence of (2.8). $\qquad\square$

We shall now derive a well-known result of Functional Analysis, the bipolar theorem, which in turn will yield statements on inequality systems.

If $A \subseteq E$ is nonempty, then

$$A^\circ := \{x^* \in E^* \,|\, \langle x^*, x\rangle \le 0 \,\forall\, x \in A\}$$

is a convex cone, the (negative) *polar cone* of A. Furthermore, the *bipolar cone* of A is

$$A^{\circ\circ} := \{x \in E \,|\, \langle x^*, x\rangle \le 0 \,\forall\, x^* \in A^\circ\}.$$

We denote by $\overline{\mathrm{cc}}\,A$ the intersection of all closed convex cones containing A, analogously for nonempty subsets of E^*.

Lemma 2.3.2

(a) If $A \subseteq E$ is nonempty, then $\left(\overline{cc}\, A\right)^{\circ} = A^{\circ}$.

(b) If $A \subseteq E$ is nonempty and convex, then $\overline{cc}\, A = \mathrm{cl}\left(\bigcup_{\lambda \geq 0} \lambda A\right)$.

Proof. See Exercise 2.5.4.   □

Proposition 2.3.3 (Bipolar Theorem) *If* $A \subseteq E$ *is nonempty, then* $A^{\circ\circ} = \overline{cc}A$.

Proof. Let $C := \overline{cc}\, A$. By Theorem 2.3.1(a) with the roles of E and E^* exchanged, we have

$$C = \{x \in E \mid \langle x^*, x \rangle \leq \sigma_C(x^*) \, \forall x^* \in E^*\}.$$

Since C is a cone containing o, a direct calculation gives $\sigma_C = \delta_{C^{\circ}}$. Hence $C = C^{\circ\circ}$, and Lemma 2.3.2(a) shows that $C^{\circ\circ} = A^{\circ\circ}$.   □

2.4 The Generalized Farkas Lemma

In this section, we characterize solutions of systems of equations and inequalities. We make the following hypotheses:

(H) (E, E^*) and (F, F^*) are dual pairs of locally convex spaces.
   $P \subseteq E$ and $Q \subseteq F$ are closed convex cones.
   $T : E \to F$ is a continuous linear mapping.

   Recall that the *adjoint* $T^* : F^* \to E^*$ of T is defined by

$$\langle T^* y^*, x \rangle = \langle y^*, Tx \rangle \quad \forall x \in E \quad \forall y^* \in F^*.$$

Lemma 2.4.1 *There always holds* $\mathrm{cl}(P^{\circ} + T^*(Q^{\circ})) = (P \cap T^{-1}(Q))^{\circ}$.

Proof. For any $x \in E$ we have

$$x \in (P^{\circ} + T^*(Q^{\circ}))^{\circ}$$
$$\Longleftrightarrow \quad \langle y', x \rangle + \langle z^*, Tx \rangle \leq 0 \quad \forall (y', z^*) \in P^{\circ} \times Q^{\circ}$$
$$\Longleftrightarrow \quad (x, Tx) \in (P^{\circ} \times Q^{\circ})^{\circ}.$$

Since $(P^{\circ} \times Q^{\circ})^{\circ} = P^{\circ\circ} \times Q^{\circ\circ} = P \times Q$ (the latter by the bipolar theorem), we see that

$$(P^{\circ} + T^*(Q^{\circ}))^{\circ} = P \cap T^{-1}(Q)$$

and another application of the bipolar theorem proves the assertion.   □

   An immediate consequence of this lemma is:

Proposition 2.4.2 (Generalized Farkas Lemma) *If* $P^{\circ} + T^*(Q^{\circ})$ *is closed, then for each* $x^* \in E^*$ *the following statements are equivalent:*

(a) $\exists\, z^* \in Q^\circ :\ x^* - T^*z^* \in P^\circ$.
(b) $\forall\, x \in P :\ Tx \in Q \implies \langle x^*, x \rangle \le 0$.

Frequently the result is formulated in terms of the negation of (b) which is:

($\bar{\text{b}}$) $\exists\, x \in P :\ Tx \in Q$ and $\langle x^*, x \rangle > 0$.

Proposition 2.4.2 then states that *either* (a) *or* ($\bar{\text{b}}$) holds. In this formulation, the result is called *theorem of the alternative*.

If, in particular, $P = E$ and so $P^\circ = \{o\}$, then Proposition 2.4.2 gives a necessary and sufficient condition for the linear operator equation $T^* z^* = x^*$ to have a solution z^* in the cone Q°.

In view of Proposition 2.4.2, it is of interest to have sufficient conditions for $P^\circ + T^*(Q^\circ)$ to be closed.

Proposition 2.4.3 *If E and F are Banach spaces and $T(E) - Q = F$, then $P^\circ + T^*(Q^\circ)$ is $\sigma(E^*, E)$-closed.*

Proof. In the following, topological notions in E^* and F^* refer to the weak* topology. We show that for any $\rho > 0$ the set

$$K := (P^\circ + T^*(Q^\circ)) \cap B_{E^*}(o, \rho)$$

is closed in $B_{E^*}(o, \rho)$; the assertion then follows by the Krein–Šmulian theorem (Theorem 1.6.8). Thus let (z_α^*) be a net (generalized sequence) in K converging to some $z^* \in B_{E^*}(o, \rho)$. Then there exist nets (x_α^*) in P° and (y_α') in Q° such that $z_\alpha^* = x_\alpha^* + T^* y_\alpha'$ for any α. Now let $z \in F$ be given. Since $T(P) - Q = F$, there exist $x \in P$ and $y \in Q$ satisfying $z = Tx - y$. Hence for any α we have

$$\langle y_\alpha', z \rangle = \langle y_\alpha', Tx \rangle - \langle y_\alpha', y \rangle \ge \langle T^* y_\alpha', x \rangle = \langle z_\alpha^*, x \rangle - \langle x_\alpha^*, x \rangle \ge \langle z_\alpha^*, x \rangle \ge -\rho \|x\|.$$

Analogously there exists $\tilde{x} \in P$ such that $\langle y_\alpha', -z \rangle \ge -\rho\|\tilde{x}\|$ for any α. Hence the net (y_α') is pointwise bounded and so, by the Banach–Steinhaus theorem, norm bounded in F^*. The Alaoglu theorem now implies that some subnet $(y_{\alpha'}')$ of (y_α') has a limit y'. Since Q° is closed, we have $y' \in Q^\circ$. Since $x_{\alpha'}^* = z_{\alpha'}^* - T^* y_{\alpha'}'$ and $z_{\alpha'}^* \to z^*$, it follows that $(x_{\alpha'}^*)$ converges to $z^* - T^* y'$, and so $z^* - T^* y' \in P^\circ$. We conclude that $z^* \in K$. $\square$

We supplement Proposition 2.4.3 by a sufficient condition for $T(P) - Q = F$.

Lemma 2.4.4 *If $T(P) \cap \operatorname{int} Q \ne \emptyset$, then $T(P) - Q = F$.*

Proof. By assumption, there exist $x_0 \in P$ such that $Tx_0 \in \operatorname{int} Q$. Let V be a neighborhood of zero in F such that $Tx_0 + V \subseteq Q$. Now let $y \in F$ be given. Then $\rho(-y) \in V$ for some $\rho > 0$ and so $z := Tx_0 - \rho y \in Q$. It follows that $y = T(\frac{1}{\rho} x_0) - \frac{1}{\rho} z \in T(P) - Q$. $\square$

Concerning the finite-dimensional case, let

$$E = P = \mathbb{R}^m, \; F = \mathbb{R}^n, \; Q = -\mathbb{R}^n_+.$$

Further let A be the matrix representation of $T : \mathbb{R}^m \to \mathbb{R}^n$ with respect to the standard bases. Then $T^*Q^\circ = A^\top \mathbb{R}^n_+$ is a polyhedral cone (as the nonnegative hull of the column vectors of A) and so is closed (see, for example, Bazaraa and Shetty [12] or Elster et al. [59]). Hence Proposition 2.4.2 implies

Corollary 2.4.5 (Classical Farkas Lemma) *Let A be an (n, m)-matrix. Then for each $x^* \in \mathbb{R}^m$, the following statements are equivalent:*

(a) $\exists z^* \in \mathbb{R}^n_+ \,:\, A^\top z^* = x^*$.
(b) $\forall x \in \mathbb{R}^m \,:\, Ax \in -\mathbb{R}^n_+ \implies \langle x^*, x \rangle \leq 0$.

2.5 Bibliographical Notes and Exercises

Concerning conjugate functionals we refer to the Bibliographical Notes at the end of Chap. 4. For theorems of the alternative in finite-dimensional spaces, see Borwein and Lewis [18], Elster et al. [59], and the references therein. Concerning related results in infinite-dimensional spaces, for linear and nonlinear mappings, we refer to Craven [43, 44], Craven et al. [45], Ferrero [68], Giannessi [70], Glover et al. [75], Jeyakumar [104], Lasserre [117], and Schirotzek [193, 195]. Proposition 2.4.3 is due to Penot [160]; for further closedness conditions of this kind, see Schirotzek [196, p. 220 ff]. We shall come back to this subject in Sect. 10.2.

Exercise 2.5.1 Calculate the conjugate of the following functions $f : \mathbb{R} \to \overline{\mathbb{R}}$:

(a) $f(x) := e^x, \; x \in \mathbb{R}$.
(b) $f(x) := \ln x$ if $x > 0$, $f(x) := +\infty$ if $x \leq 0$.
(c)

$$f(x) := \begin{cases} +\infty & \text{if } x < -2, \\ -x & \text{if } -2 \leq x < 0, \\ x(2-x) & \text{if } 0 \leq x < 2, \\ 2x & \text{if } x \geq 2. \end{cases}$$

Exercise 2.5.2 Verify the implication *(a)* $\implies$ *(b)* in Proposition 2.1.2.

Exercise 2.5.3 Prove assertion *(a)* of Theorem 2.2.8.

Exercise 2.5.4 Prove Lemma 2.3.2.

3

Classical Derivatives

3.1 Directional Derivatives

Convention. Throughout this chapter, unless otherwise specified, we assume that E and F are normed vector spaces, $D \subseteq E$ is nonempty and open, $\bar{x} \in D$, and $f : D \to F$.

We will recall some classical concepts and facts. To start with, we consider directional derivatives. We write

$$\Delta f(\bar{x}, y) := f(\bar{x} + y) - f(\bar{x}) \quad \forall\, y \in D - \bar{x}.$$

We use the following abbreviations: G-derivative for *Gâteaux derivative*, H-derivative for *Hadamard derivative*, F-derivative for *Fréchet derivative*.

Definition 3.1.1 Let $y \in E$. We call

$$f_G(\bar{x}, y) := \lim_{\tau \downarrow 0} \tfrac{1}{\tau} \Delta f(\bar{x}, \tau y) \quad \textit{directional G-derivative,}$$

$$f_G^s(\bar{x}, y) := \lim_{\substack{\tau \downarrow 0 \\ x \to \bar{x}}} \tfrac{1}{\tau} \Delta f(x, \tau y) \quad \textit{strict directional G-derivative,}$$

$$f_H(\bar{x}, y) := \lim_{\substack{\tau \downarrow 0 \\ z \to y}} \tfrac{1}{\tau} \Delta f(\bar{x}, \tau z) \quad \textit{directional H-derivative,}$$

$$f_H^s(\bar{x}, y) := \lim_{\substack{\tau \downarrow 0 \\ x \to \bar{x} \\ z \to y}} \tfrac{1}{\tau} \Delta f(x, \tau z) \quad \textit{strict directional H-derivative}$$

of f at $\bar{x}$ in the direction y, provided the respective limit exists.

Lemma 3.1.2

(a) *If $f_H(\bar{x}, y)$ exists, then $f_G(\bar{x}, y)$ also exists and both directional derivatives coincide.*
(b) *If f is locally L-continuous around $\bar{x}$, then $f_H(\bar{x}, y)$ exists if and only if $f_G(\bar{x}, y)$ exists.*

Proof. (a) is obvious. We verify (b). Assume that $f_G(\bar{x}, y)$ exists. Let $\epsilon > 0$ be given. Then there exists $\delta_1 > 0$ such that

$$\left\|\tfrac{1}{\tau}\Delta f(\bar{x}, \tau y) - f_G(\bar{x}, y)\right\| < \tfrac{\epsilon}{2} \quad \text{whenever } 0 < \tau < \delta_1.$$

Since f is locally L-continuous around $\bar{x}$, there further exist $\lambda > 0$ and $\delta_2 > 0$ such that

$$\|f(x_1) - f(x_2)\| \le \lambda \|x_1 - x_2\| \quad \forall x_1, x_2 \in B(\bar{x}, \delta_2).$$

Now set

$$\delta_3 := \frac{\epsilon}{2\lambda} \quad \text{and} \quad \delta_4 := \min\left\{\delta_1, \frac{\delta_2}{\delta_3 + \|y\|}\right\}.$$

If $z \in B(y, \delta_3)$ and $0 < \tau < \delta_4$, then we obtain $\|\tau y\| < \delta_2$ and

$$\|\tau z\| \le \tau(\|z - y\| + \|y\|) \le \tau(\delta_3 + \|y\|) \le \delta_2$$

and so

$$\left\|\tfrac{1}{\tau}\Delta f(\bar{x}, \tau z) - f_G(\bar{x}, y)\right\|$$
$$\le \tfrac{1}{\tau}\|f(\bar{x} + \tau z) - f(\bar{x} + \tau y)\| + \left\|\tfrac{1}{\tau}\Delta f(\bar{x}, \tau y) - f_G(\bar{x}, y)\right\|$$
$$\le \lambda\|z - y\| + \tfrac{\epsilon}{2} < \epsilon.$$

We conclude that $f_H(\bar{x}, y)$ exists and equals $f_G(\bar{x}, y)$. $\square$

Lemma 3.1.3 *If the directional H-derivative $f_H(\bar{x}, \cdot)$ exists in a neighborhood of $y_0 \in E$, then it is continuous at y_0.*

Proof. Let $\rho_0 > 0$ be such that $f_H(\bar{x}, y)$ exists for each $y \in B(y_0, \rho_0)$. Let $\epsilon > 0$ be given. Then there exists $\rho \in (0, \rho_0)$ such that

$$\left\|\tfrac{1}{\tau}\Delta f(\bar{x}, \tau y) - f_H(\bar{x}, y_0)\right\| \le \epsilon \quad \text{whenever } 0 < \tau < \rho \text{ and } y \in B(y_0, \rho).$$

Letting $\tau \downarrow 0$, we obtain

$$\|f_G(\bar{x}, y) - f_H(\bar{x}, y_0)\| \le \epsilon \quad \forall y \in B(y_0, \rho).$$

By Lemma 3.1.2 we have $f_G(\bar{x}, y) = f_H(\bar{x}, y)$ for each $y \in B(y_0, \rho_0)$, and the assertion follows. $\square$

Now we consider a proper function $f : E \to \overline{\mathbb{R}}$. If $D := \operatorname{int} \operatorname{dom} f$ is nonempty, then of course the above applies to $f|D$. In addition, we define the following directional derivatives:

$$\overline{f}_G(\bar{x}, y) := \limsup_{\tau \downarrow 0} \tfrac{1}{\tau}\Delta f(\bar{x}, \tau y) \quad \textit{upper directional G-derivative,}$$

$$\underline{f}_G(\bar{x}, y) := \liminf_{\tau \downarrow 0} \tfrac{1}{\tau}\Delta f(\bar{x}, \tau y) \quad \textit{lower directional G-derivative,}$$

$$\overline{f}_H(\bar{x}, y) := \limsup_{\substack{\tau \downarrow 0 \\ z \to y}} \tfrac{1}{\tau}\Delta f(\bar{x}, \tau z) \quad \textit{upper directional H-derivative,}$$

$$\underline{f}_H(\bar{x}, y) := \liminf_{\substack{\tau \downarrow 0 \\ z \to y}} \tfrac{1}{\tau}\Delta f(\bar{x}, \tau z) \quad \textit{lower directional H-derivative.}$$

Notice that these directional derivatives generalize the *Dini derivates* of a function $f : I \to \mathbb{R}$ ($I \subseteq \mathbb{R}$ an interval) which are defined at $\bar{x} \in \operatorname{int} I$ by

$$D^+ f(\bar{x}) := \limsup_{h \downarrow 0} \frac{f(\bar{x} + h) - f(\bar{x})}{h}, \quad D_+ f(\bar{x}) := \liminf_{h \downarrow 0} \frac{f(\bar{x} + h) - f(\bar{x})}{h},$$

$$D^- f(\bar{x}) := \limsup_{h \uparrow 0} \frac{f(\bar{x} + h) - f(\bar{x})}{h}, \quad D_- f(\bar{x}) := \liminf_{h \uparrow 0} \frac{f(\bar{x} + h) - f(\bar{x})}{h}.$$

If $\bar{x} \in I$ is the left boundary point of I, then $D^+ f(\bar{x})$ and $D_+ f(\bar{x})$ still make sense; an analogous remark applies to the right boundary point of I. Notice that, among others, $D^+ f(\bar{x}) = \overline{f}_G(\bar{x}, 1)$. If $D^+ f(\bar{x}) = D_+ f(\bar{x})$, then this common value is called the *right derivative* of f at $\bar{x}$ and is denoted $f'_+(\bar{x})$. The *left derivative* $f'_-(\bar{x})$ is defined analogously.

3.2 First-Order Derivatives

Our aim in this section is to recall various kinds of derivatives. For this, the following notion will be helpful.

Definition 3.2.1 A nonempty collection β of subsets of E is called *bornology* if the following holds:

$$\text{each } S \in \beta \text{ is bounded and } \bigcup_{S \in \beta} S = E,$$
$$S \in \beta \quad \Longrightarrow \quad -S \in \beta,$$
$$S \in \beta \text{ and } \lambda > 0 \quad \Longrightarrow \quad \lambda S \in \beta,$$
$$S_1, S_2 \in \beta \quad \Longrightarrow \quad \exists S \in \beta : S_1 \cup S_2 \subset S.$$

In particular:

- The *G-bornology* β_G is the collection of all finite sets.
- The *H-bornology* β_H is the collection of all compact sets.
- The *F-bornology* β_F is the collection of all bounded sets.

We set

$\mathrm{L}(E, F) :=$ vector space of all continuous linear mappings $T : E \to F$.

Definition 3.2.2 Let β be a bornology on E.

(a) The mapping $f : D \to F$ is said to be *β-differentiable* at $\bar{x}$ if there exists $T \in \mathrm{L}(E, F)$, the *β-derivative* of f at $\bar{x}$, such that

$$\lim_{\tau \to 0} \sup_{y \in S} \left\| \tfrac{1}{\tau} \big(f(\bar{x} + \tau y) - f(\bar{x}) \big) - T(y) \right\| = 0 \quad \forall S \in \beta. \tag{3.1}$$

(b) The mapping $f : D \to F$ is said to be *strictly β-differentiable* at $\bar{x}$ if there exists $T \in L(E, F)$, the *strict β-derivative* of f at $\bar{x}$, such that

$$\lim_{\substack{\tau \to 0 \\ x \to \bar{x}}} \sup_{y \in S} \left\| \tfrac{1}{\tau} \big(f(x + \tau y) - f(x) \big) - T(y) \right\| = 0 \quad \forall S \in \beta. \qquad (3.2)$$

(c) In particular, f is said to be *G-differentiable* or *strictly G-differentiable* if (3.1) or (3.2), respectively, holds with $\beta = \beta_G$. In this case, T is called *(strict) G-derivative* of f at $\bar{x}$. Analogously we use *(strictly) H-differentiable* if $\beta = \beta_H$ and *(strictly) F-differentiable* if $\beta = \beta_F$. In the respective case, T is called *(strict) H-derivative* or *(strict) F-derivative* of f at $\bar{x}$.

Remark 3.2.3

(a) If the β-derivative T of f at $\bar{x}$ exists for some bornology β, then

$$T(y) = \lim_{\tau \to 0} \tfrac{1}{\tau} \big(f(\bar{x} + \tau y) - f(\bar{x}) \big) = f_G(\bar{x}, y) \quad \forall y \in E.$$

Hence if two of the above derivatives exist, then they coincide. This justifies denoting them by the same symbol; we choose

$$f'(\bar{x}) := T.$$

Condition (3.1) means that we have

$$\lim_{\tau \to 0} \left\| \tfrac{1}{\tau} \big(f(\bar{x} + \tau y) - f(\bar{x}) \big) - f'(\bar{x}) y \right\| = 0 \quad \text{uniformly in } y \in S \text{ for each } S \in \beta.$$

An analogous remark applies to (3.2). Here and in the following we write $f'(\bar{x}) y$ instead of $f'(\bar{x})(y)$. If $f : D \to \mathbb{R}$, then $f'(\bar{x}) \in E^*$ and as usual we also write $\langle f'(\bar{x}), y \rangle$ instead of $f'(\bar{x})(y)$.

(b) Now let E be a (real) Hilbert space with inner product $(x \mid y)$ and $f : E \to \mathbb{R}$ a functional. If f is G-differentiable at $\bar{x} \in E$, then the G-derivative $f'(\bar{x})$ is an element of the dual space E^*. By the Riesz representation theorem, there is exactly one $z \in E$ such that $\langle f'(\bar{x}), y \rangle = (z \mid y)$ for all $y \in E$. This element z is called gradient of f at $\bar{x}$ and is denoted $\nabla f(\bar{x})$. In other words, we have

$$\langle f'(\bar{x}), y \rangle = (\nabla f(\bar{x}) \mid y) \quad \forall y \in E.$$

Proposition 3.2.4 says that f is G-differentiable at $\bar{x}$ if and only if the directional G-derivative $y \mapsto f_G(\bar{x}, y)$ exists and is linear and continuous on E. An analogous remark applies to strict G-differentiability as well as (strict) H-differentiability. Recall that if $g : E \to F$, then

$$g(x) = \mathbf{o}(\|x\|), \ x \to o \quad \text{means} \quad \lim_{x \to o} \frac{g(x)}{\|x\|} = o.$$

Proposition 3.2.4

(i) f is G-differentiable at $\bar{x}$ if and only if there exists $f'(\bar{x}) \in L(E, F)$ such that $f'(\bar{x})y = f_G(\bar{x}, y)$ for all $y \in E$.

(ii) f is H-differentiable at $\bar{x}$ if and only if there exists $f'(\bar{x}) \in L(E, F)$ such that $f'(\bar{x})y = f_H(\bar{x}, y)$ for all $y \in E$.

(iii) The following assertions are equivalent:
   (a) f is strictly H-differentiable at $\bar{x}$.
   (b) There exists $f'(\bar{x}) \in L(E, F)$ such that $f'(\bar{x})y = f_H^s(\bar{x}, y)$ for $y \in E$.
   (c) f is locally L-continuous around $\bar{x}$ and strictly G-differentiable at $\bar{x}$.

(iv) f is F-differentiable at $\bar{x}$ if and only if there exists $f'(\bar{x}) \in L(E, F)$ such that
$$\big(f(\bar{x} + z) - f(\bar{x})\big) - f'(\bar{x})z = \mathbf{o}(\|z\|), \ z \to o.$$

(v) f is strictly F-differentiable at $\bar{x}$ if and only if there exists $f'(\bar{x}) \in L(E, F)$ such that
$$\big(f(x + z) - f(x)\big) - f'(\bar{x})z = \mathbf{o}(\|z\|), \ z \to o, \ x \to \bar{x}.$$

Proof. We only verify (iii), leaving the proof of the remaining assertions as Exercise 3.8.4.

(a) $\implies$ (c): Let (a) hold. We only have to show that f is locally L-continuous around $\bar{x}$. Assume this is not the case. Then there exist sequences (x_n) and (x_n') in $\mathrm{B}(\bar{x}, \frac{1}{n})$ such that

$$\|f(x_n) - f(x_n')\| > n\|x_n - x_n'\| \quad \forall n \in \mathbb{N}. \tag{3.3}$$

Setting $\tau_n := \sqrt{n}\|x_n - x_n'\|$ and $y_n := \frac{1}{\tau_n}(x_n' - x_n)$, we obtain as $n \to \infty$,

$$0 \le \tau_n \le \sqrt{n}(\|x_n - \bar{x}\| + \|\bar{x} - x_n'\|) < \frac{2}{\sqrt{n}} \to 0 \quad \text{and} \quad \|y_n\| = \frac{1}{\sqrt{n}} \to 0.$$

By (3.3) we have

$$\left\|\tfrac{1}{\tau_n}\Delta f(x_n, \tau_n y_n)\right\| > \tfrac{1}{\tau_n} \cdot n\|\tau_n y_n\| = \sqrt{n} \quad \forall n \in \mathbb{N},$$

and the continuity of $f'(\bar{x})$ implies that, with some $n_0 \in \mathbb{N}$, we obtain $\|f'(\bar{x})\| \cdot \|y_n\| < \frac{1}{2}$ for each $n > n_0$. It follows that

$$\left\|\tfrac{1}{\tau_n}\Delta f(x_n, \tau_n y_n) - f'(\bar{x})y_n)\right\|$$
$$\ge \left\|\tfrac{1}{\tau_n}\Delta f(x_n, \tau_n y_n)\right\| - \|f'(\bar{x})\| \cdot \|y_n\| > \sqrt{n} - \tfrac{1}{2} \quad \forall n > n_0,$$

which contradicts (3.2) for the compact set $S := \{o\} \cup \{y_n \,|\, n > n_0\}$.

(c) $\implies$ (b): Let $y \in E$ and $\epsilon > 0$ be given. Since f is strictly G-differentiable at $\bar{x}$, there exists $\delta_1 > 0$ such that

$$\left\|\tfrac{1}{\tau}\Delta f(x, \tau y) - f'(\bar{x})y\right\| < \epsilon \quad \text{whenever } 0 < |\tau| < \delta_1, \ \|x - \bar{x}\| < \delta_1. \tag{3.4}$$

Since f is locally L-continuous around $\bar{x}$, there further exist $\lambda > 0$ and $\delta_2 > 0$ such that

$$\|f(x_1) - f(x_2)\| < \lambda\|x_1 - x_2\| \quad \forall\, x_1, x_2 \in \mathrm{B}(\bar{x}, \delta_2). \qquad (3.5)$$

Setting $x_1 := x + \tau z$ and $x_2 := x + \tau y$, we have the estimates

$$\|x_1 - \bar{x}\| \le \|x - \bar{x}\| + |\tau|(\|z - y\| + \|y\|) \quad \text{and} \quad \|x_2 - \bar{x}\| \le \|x - \bar{x}\| + |\tau|\|y\|$$

which show that $x_1, x_2 \in \mathrm{B}(\bar{x}, \delta_2)$ provided $|\tau|$, $\|z - y\|$, and $\|x - \bar{x}\|$ are sufficiently small. Under this condition, (3.4) and (3.5) imply that

$$\left\|\tfrac{1}{\tau}\Delta f(x, \tau z) - f'(\bar{x})y\right\|$$
$$\le \tfrac{1}{|\tau|}\left\|f(x + \tau z) - f(x + \tau y)\right\| + \left\|\tfrac{1}{\tau}\Delta f(x, \tau y) - f'(\bar{x})y\right\| \le \lambda\|z - y\| + \epsilon.$$

This verifies (b).

(b) $\Longrightarrow$ (a): Let (b) hold and assume that (a) does not hold. Let $T \in \mathrm{L}(E, F)$ be given. Then for some compact subset S of E, the relation (3.2) does not hold. Hence there exist $\epsilon_0 > 0$ as well as sequences $\tau_n \downarrow 0$, $y_n \in S$, and $x_n \to \bar{x}$ such that

$$\left\|\tfrac{1}{\tau}\Delta f(x_n, \tau_n y_n) - T(y_n)\right\| > \epsilon_0 \quad \forall n \in \mathbb{N}.$$

Since S is compact, a subsequence of (y_n), again denoted (y_n), converges to some $y \in S$. It follows that for any $n > n_0$ we have

$$\left\|\tfrac{1}{\tau}\Delta f(x_n, \tau_n y_n) - T(y)\right\|$$
$$\ge \underbrace{\left\|\tfrac{1}{\tau}\Delta f(x_n, \tau_n y_n) - T(y_n)\right\|}_{>\,\epsilon_0} - \underbrace{\|T\| \cdot \|y_n - y\|}_{<\,\epsilon_0/2} > \frac{\epsilon_0}{2}; \qquad (3.6)$$

in this connection, we exploited that T is linear and continuous. However, the relation (3.6) contradicts (b). $\qquad\square$

Proposition 3.2.5 If $f : D \to F$ is H-differentiable at $\bar{x}$, then f is continuous at $\bar{x}$.

Proof. See Exercise 3.8.5. $\qquad\square$

3.3 Mean Value Theorems

We recall a variant of the classical mean value theorem (see, for instance, Walter [212]).

Proposition 3.3.1 (Mean Value Theorem in Terms of Dini Derivates)
Let I and J be intervals in $\mathbb{R}$ and let $A \subseteq I$ be a countable set. Further let $f : I \to \mathbb{R}$ be continuous, let $D \in \{D^+, D_+, D^-, D_-\}$, and assume that

$$Df(x) \in J \quad \forall\, x \in I \setminus A.$$

Then

$$\frac{f(b) - f(a)}{b - a} \in J \quad \forall a, b \in I, \, a \neq b.$$

If $f : [a, b] \rightarrow \mathbb{R}$ is continuous on $[a, b]$ and differentiable on (a, b), then by the intermediate value theorem for derivatives the set $J := \{f'(x) \mid x \in (a, b)\}$ is an interval and so the usual mean value theorem follows from Proposition 3.3.1.

Now we return to the setting described by the convention at the beginning of the chapter. If $x, z \in E$, we write

$$[x, z] := \{\lambda x + (1 - \lambda)z \mid 0 \leq \lambda \leq 1\}.$$

If $f : D \rightarrow F$ is G-differentiable on D (i.e., G-differentiable at any $x \in D$), then we may consider the mapping $f' : x \mapsto f'(x)$ of D to $L(E, F)$.

Definition 3.3.2 Let f be G-differentiable on D. The mapping f' is said to be *radially continuous* if for all $x, y \in E$ such that $[x, x+y] \subseteq D$, the function $\tau \mapsto f'(x + \tau y)y$ is continuous on $[0, 1]$.

Proposition 3.3.3 (Mean Value Theorem in Integral Form) *Let* f : $D \rightarrow \mathbb{R}$ *be G-differentiable and let* f' *be radially continuous. Then for all* $x, y \in D$ *such that* $[x, x+y] \subseteq D$ *one has*

$$f(x + y) - f(x) = \int_0^1 \langle f'(x + \tau y), y \rangle \, d\tau. \tag{3.7}$$

Proof. For $\tau \in [0, 1]$ let $\varphi(\tau) := f(x + \tau y)$. By assumption φ is continuously differentiable and $\varphi'(\tau) = \langle f'(x + \tau y), y \rangle$. The main theorem of calculus gives

$$\varphi(1) - \varphi(0) = \int_0^1 \varphi'(\tau) \, d\tau,$$

which is (3.7). □

The above result is formulated for functionals only, in which case it will be used later. In Proposition 4.3.8 below we shall describe an important class of functionals to which the mean value formula (3.7) applies. We mention that, by an appropriate definition of the Riemann integral, the formula extends to a mapping $f : D \rightarrow F$ provided F is a Banach space.

If β is a bornology of E, we denote by $L_\beta(E, F)$ the vector space $L(E, F)$ equipped with the topology of uniform convergence on the sets $S \in \beta$. In particular, $L_{\beta_F}(E, F)$ denotes $L(E, F)$ equipped with the topology generated by the norm $\|T\| := \sup\{\|Tx\| \mid x \in B_E\}$. In particular we write $E_\beta^* := L_\beta(E, \mathbb{R})$.

Proposition 3.3.4 (Mean Value Theorem in Inequality Form) *Let* $y \in E$ *be such that* $[\bar{x}, \bar{x} + y] \subseteq D$ *and* f *is G-differentiable on* $[\bar{x}, \bar{x} + y]$. *Further let* $T \in L_{\beta_F}(E, F)$. *Then one has*

$$\left\| \big(f(\bar{x} + y) - f(\bar{x}) \big) - Ty \right\| \leq \|y\| \sup_{0 \leq \tau \leq 1} \|f'(\bar{x} + \tau y) - T\|.$$

Proof. Set $g(x) := f(x) - T(x - \bar{x})$, $x \in E$. By the Hahn–Banach theorem, there exists $v \in F^*$ satisfying $\|v\| = 1$ and $\langle v, \Delta g(\bar{x}, y) \rangle = \|\Delta g(\bar{x}, y)\|$. Now define $\varphi(\tau) := \langle v, g(\bar{x} + \tau y) \rangle$, $\tau \in [0, 1]$. It is easy to see that φ is differentiable, and one has

$$\varphi'(\tau) = \langle v, g'(\bar{x} + \tau y)y \rangle = \langle v, f'(\bar{x} + \tau y)y - Ty \rangle. \tag{3.8}$$

By the classical mean value theorem, there exists $\tau \in (0, 1)$ such that $\varphi'(\tau) = \varphi(1) - \varphi(0)$. This together with (3.8) and

$$\left| \langle v, f'(\bar{x} + \tau y)y - Ty \rangle \right| \leq \|v\| \|f'(\bar{x} + \tau y)y - Ty\| \leq \|f'(\bar{x} + \tau y) - T\| \|y\|$$

completes the proof. $\qquad\qquad\qquad\qquad\qquad\qquad\qquad\qquad\qquad\qquad\square$

3.4 Relationship between Differentiability Properties

In this section we will study the interrelations between the various differentiability properties. First we introduce some terminology.

Definition 3.4.1

(a) The mapping $f : D \to F$ is said to be *β-smooth* at $\bar{x}$ if f is β-differentiable for any x in an open neighborhood U of $\bar{x}$ and the mapping $f' : x \mapsto f'(x)$ of U to $L_\beta(E, F)$ is continuous on U.
(b) The mapping $f : D \to F$ is said to be *continuously differentiable* at $\bar{x}$ if f is G-differentiable for any x in an open neighborhood U of $\bar{x}$ and the mapping $f' : x \mapsto f'(x)$ of U to $L_{\beta_F}(E, F)$ is continuous at $\bar{x}$.
(c) If $f : D \to F$ is continuously differentiable at every point of D, then f is said to be a *C^1-mapping* on D, written $f \in C^1(D, F)$.

We shall make use of the following abbreviations:

(G): f is G-differentiable at $\bar{x}$,

(SG): f is strictly G-differentiable at $\bar{x}$,

(CD): f is continuously differentiable at $\bar{x}$.

In analogy to (G), (SG) we use (H), (SH), (F), and (SF).

Proposition 3.4.2 *The following implications hold true:*

$$(CD) \implies (SF) \; \overset{\Longrightarrow}{\underset{\longleftarrow}{}} \; (SH) \; \overset{\Longrightarrow}{\underset{\longleftarrow -}{}} \; (SG)$$

$$\Downarrow \qquad\qquad \Downarrow \qquad\qquad \Downarrow$$

$$(F) \; \overset{\Longrightarrow}{\underset{\longleftarrow}{}} \; (H) \; \overset{\Longrightarrow}{\underset{\longleftarrow -}{}} \; (G)$$

In this connection, $\longleftarrow$ *means implication provided* E *is finite dimensional, and* $\longleftarrow -$ *means implication provided* f *is locally L-continuous around* $\bar{x}$.

Proof. In view of the foregoing, it only remains to verify the implication (CD) $\implies$ (SF). Thus let (CD) hold. Then there exists $\rho > 0$ such that f is G-differentiable on $B(\bar{x}, 2\rho)$. If $x \in B(\bar{x}, \rho)$ and $y \in B(o, \rho)$, then $[x, x + y] \subset B(\bar{x}, 2\rho)$. By Proposition 3.3.4 with $T := f'(\bar{x})$, we obtain

$$\|f(x + y) - f(x) - f'(\bar{x})y\| \le \|y\| \sup_{0 \le \tau \le 1} \|f'(x + \tau y) - f'(\bar{x})\|. \tag{3.9}$$

Now let $\epsilon > 0$ be given. Since f' is continuous at $\bar{x}$, there exists $\delta > 0$ such that

$$\sup_{0 \le \tau \le 1} \|f'(x + \tau y) - f'(\bar{x})\| < \epsilon \quad \forall x \in B(\bar{x}, \delta) \quad \forall y \in B(o, \delta).$$

This together with (3.9) implies (SF). □

Remark 3.4.3 By Proposition 3.4.2 it is clear that if f is continuously differentiable on an open neighborhood U of $\bar{x}$, then f is β-smooth at any $\bar{x} \in U$ for any bornology $\beta \subseteq \beta_F$. In particular, f is F-differentiable at any $\bar{x} \in U$ and the F-derivative f' is continuous from U to $L_{\beta_F}(E, F)$.

Beside E and F let G be another normed vector space. Beside $f : D \to F$ let $g : V \to G$ be another mapping, where V is an open neighborhood of $\bar{z} := f(\bar{x})$ in F. Assume that $f(D) \subset V$. Then the composition $g \circ f : D \to G$ is defined.

Proposition 3.4.4 (Chain Rule) *Assume that* f *and* g *are H-differentiable at* $\bar{x}$ *and* $\bar{z}$, *respectively. Then* $g \circ f$ *is H-differentiable at* $\bar{x}$, *and there holds*

$$(g \circ f)'(\bar{x}) = g'(\bar{z}) \circ f'(\bar{x}).$$

An analogous statement holds true if H-differentiable is replaced by F-differentiable.

The *proof* is the same as in multivariate calculus. An analogous chain rule for G-differentiable mappings does not hold (see Exercise 3.8.3).

3.5 Higher-Order Derivatives

We again use the notation introduced at the beginning of the chapter. Assume that $f \in C^1(D, F)$. If the (continuous) mapping $f' : D \to L_{\beta_F}(E, F)$ is continuously differentiable on D, then f is said to be a *twice continuously differentiable mapping* on D, or a C^2-*mapping* on D, with *second-order derivative* $f'' := (f')'$. The set of all twice continuously differentiable mappings $f : D \to F$ is denoted $C^2(D, F)$.

Notice that f'' maps D into $H := L_{\beta_F}\big(E, L_{\beta_F}(E, F)\big)$. Parallel to H we consider the vector space $B(E, F)$ of all continuous bilinear mappings $b : E \times E \to F$, which is normed by

$$\|b\| := \sup\{\|b(y, z)\| \mid \|y\| \leq 1, \|z\| \leq 1\}. \tag{3.10}$$

If $h \in H$, then

$$b_h(y, z) := h(y)z \quad \forall (y, z) \in E \times E$$

defines an element $b_h \in B(E, F)$. Conversely, given $b \in B(E, F)$, set

$$h(y) := b(y, \cdot) \quad \forall y \in E.$$

Then $h \in H$ and $b_h = b$. Evidently the mapping $h \mapsto b_h$ is an isomorphism of H onto $B(E, F)$. Therefore H can be identified with $B(E, F)$. In this sense, we interpret $f''(\bar{x})$ as an element of $B(E, F)$ and write $f''(\bar{x})(y, z)$ instead of $\big(f''(\bar{x})y\big)z$. If, in particular, $f \in C^2(D, \mathbb{R})$, then $f''(\bar{x})$ is a continuous bilinear form on $E \times E$.

Proposition 3.5.1 (Taylor Expansion) *Assume that D is open and $f \in C^2(D, \mathbb{R})$. Then for all $\bar{x} \in D$, $y \in D - \bar{x}$ one has*

$$f(\bar{x} + y) = f(\bar{x}) + \langle f'(\bar{x}), y \rangle + \tfrac{1}{2} f''(\bar{x})(y, y) + r(y), \quad \text{where } \lim_{y \to o} \frac{r(y)}{\|y\|^2} = o.$$

In particular, there exist $\sigma > 0$ and $\epsilon > 0$ such that

$$f(\bar{x} + y) \geq f(\bar{x}) + \langle f'(\bar{x}), y \rangle - \sigma \|y\|^2 \quad \forall y \in B(o, \epsilon). \tag{3.11}$$

Proof. The first assertion follows readily from the classical Taylor expansion of the function $\varphi(\tau) := f(\bar{x} + \tau y)$, $\tau \in [0, 1]$. From the first result we obtain (3.11) since in view of (3.10) we have

$$|\tfrac{1}{2} f''(\bar{x})(y, y)| \leq \tfrac{1}{2} \|f''(\bar{x})\| \|y\|^2 \quad \forall y \in E,$$

and the limit property of r entails the existence of $\kappa > 0$ such that $|r(y)| \leq \kappa \|y\|^2$ if $\|y\|$ is sufficiently small. □

We only mention that in an analogous manner, derivatives of arbitrary order n, where $n \in \mathbb{N}$, can be defined using n-linear mappings, which leads to higher-order Taylor expansions.

3.6 Some Examples

For illustration and later purposes we collect some examples. Further examples are contained in the exercises.

Example 3.6.1 Let E be a normed vector space and $a : E \times E \to \mathbb{R}$ a bilinear functional. Recall that a is said to be symmetric if $a(x, y) = a(y, x)$ for all $x, y \in E$, and a is said to be bounded if there exists $\kappa > 0$ such that

$$|a(x,y)| \le \kappa \|x\| \, \|y\| \quad \forall \, x, y \in E.$$

Consider the quadratic functional $f : E \to \mathbb{R}$ defined by

$$f(x) := \tfrac{1}{2} a(x, x), \quad x \in E,$$

where a is bilinear, symmetric, and bounded. It is left as Exercise 3.8.6 to show that f is continuously differentiable on E and to calculate the derivative. In particular, if E is a Hilbert space with inner product $(x \,|\, y)$, then the functional

$$g(x) := \frac{1}{2} \|x\|^2 = \frac{1}{2}(x \,|\, x), \quad x \in E,$$

is continuously differentiable on E with $\langle g'(x), y \rangle = (x \,|\, y)$ for all $x, y \in E$. Hence $\nabla g(x) = x$ for any $x \in E$. Finally, concerning the norm functional $\omega(x) := \|x\| = \sqrt{2g(x)}$, the chain rule gives $\nabla \omega(x) = \frac{x}{\|x\|}$ for any $x \neq o$.

Example 3.6.2 Let again E denote a Hilbert space with inner product $(x \,|\, y)$ and define $g : E \to \mathbb{R}$ by

$$g(x) := \left(\delta^2 + 2\delta(u \,|\, x - \bar{x}) - \|x - \bar{x}\|^2 \right)^{1/2},$$

where the positive constant δ and the element $u \in E$ are fixed. Choose $\epsilon > 0$ such that the term $(\cdots)$ is positive for each $x \in \mathring{B}(\bar{x}, \epsilon)$. Define $\psi : (0, +\infty) \to \mathbb{R}$ by $\psi(z) := z^{1/2}$ and $\varphi : \mathring{B}(\bar{x}, \epsilon) \to \mathbb{R}$ by

$$\varphi(x) := \delta^2 + 2\delta(u \,|\, x - \bar{x}) - \|x - \bar{x}\|^2.$$

Then we have $g = \psi \circ \varphi$, and the chain rule implies

$$(g'(x) \,|\, y) = \frac{\delta(u \,|\, y)}{\left(\delta^2 + 2\delta(u \,|\, x - \bar{x}) - \|x - \bar{x}\|^2 \right)^{1/2}} \quad \forall \, x \in \mathring{B}(\bar{x}, \epsilon) \quad \forall \, y \in E.$$

In particular, $(g'(\bar{x}) \,|\, y) = (u \,|\, y)$ for all $y \in E$, which means $\nabla g(\bar{x}) = u$. Moreover, it is easy to see that g is a C^2-mapping on $\mathring{B}(\bar{x}, \epsilon)$. This example will be used later in connection with proximal subdifferentials.

In view of Example 3.6.3, recall that an *absolutely continuous* function $x : [a, b] \to \mathbb{R}$ is differentiable almost everywhere, i.e., outside a Lebesgue null set $N \subseteq [a, b]$. Setting $\dot{x}(t) := 0$ for each $t \in N$, which we tacitly assume from now on, the function $\dot{x} : [a, b] \to \mathbb{R}$ belongs to $\mathcal{L}^1[a, b]$ and one has $\int_{[a,b]} \dot{x}(t)\, dt = x(b) - x(a)$. In this connection, also recall that $\mathcal{L}^p[a, b]$, where $p \in [1, +\infty)$, denotes the vector space of all Lebesgue measurable functions $g : [a, b] \to \mathbb{R}$ such that $|g|^p$ is Lebesgue integrable over $[a, b]$. In addition, $\mathcal{L}^\infty[a, b]$ denotes the vector space of all Lebesgue measurable functions $g : [a, b] \to \mathbb{R}$ such that $\operatorname{ess\,sup}_{x \in [a,b]} |g(x)| < +\infty$. We denote by $AC^\infty[a, b]$ the vector space of all absolutely continuous functions $x : [a, b] \to \mathbb{R}$ such that $\dot{x} \in \mathcal{L}^\infty[a, b]$. Notice that $AC^\infty[a, b]$ is a Banach space with respect to the norm

$$\|x\|_{1,\infty} := \max\{\|x\|_\infty, \|\dot{x}\|_\infty\}.$$

Example 3.6.3 Let $E := AC^\infty[a, b]$, where $a < b$, and consider the *variational functional*

$$f(x) := \int_a^b \varphi\big(t, x(t), \dot{x}(t)\big)\, dt \quad \forall\, x \in AC^\infty[a, b].$$

If $\bar{x} \in AC^\infty[a, b]$ is fixed, we write $\overline{\varphi}(t) := \varphi\big(t, \bar{x}(t), \dot{\bar{x}}(t)\big)$ for any $t \in [a, b]$. Assume that the function $(t, x, v) \mapsto \varphi(t, x, v)$ is continuous on $[a, b] \times \mathbb{R} \times \mathbb{R}$ and has continuous first-order partial derivatives with respect to x and v there. We shall show that the functional f is continuously differentiable at any $\bar{x} \in AC^\infty[a, b]$ and that

$$\langle f'(\bar{x}), y \rangle = \int_a^b \big(\overline{\varphi}_x(t) \cdot y(t) + \overline{\varphi}_v(t) \cdot \dot{y}(t)\big)\, dt \quad \forall\, y \in AC^\infty[a, b]. \tag{3.12}$$

Proof.

(I) *The directional G-derivative $f_G(\bar{x}, y)$ exists for all $\bar{x}, y \in AC^\infty[a, b]$.* In fact, we have

$$f_G(\bar{x}, y) = \frac{\partial}{\partial \tau} f(\bar{x} + \tau y)\Big|_{\tau=0} = \int_a^b \frac{\partial}{\partial \tau}\Big[\varphi\big(t, \bar{x}(t) + \tau y(t), \dot{\bar{x}}(t) + \tau \dot{y}(t)\big)\Big]\Big|_{\tau=0} dt$$

$$= \int_a^b [\bar{\varphi}_x(t) y(t) + \bar{\varphi}_v(t) \dot{y}(t)]\, dt.$$

Notice that the assumptions on φ and $\bar{x}$ imply that the integrand in the last term is bounded, which allows differentiating under the integral sign.

(II) It is easy to verify that the functional $y \mapsto f_G(\bar{x}, y)$ is linear and continuous. Hence the G-derivative is given by (3.12).

(III) *f is continuously differentiable at any $\bar{x} \in AC^\infty[a, b]$.* For arbitrary $x, \bar{x}, y \in AC^\infty[a, b]$ we have

$$[f'(x) - f'(\bar{x})]y$$

$$= \int_a^b [\varphi_x(t, x, \dot{x}) - \varphi_x(t, \bar{x}, \dot{\bar{x}})]y\, dt + \int_a^b [\varphi_v(t, x, \dot{x}) - \varphi_v(t, \bar{x}, \dot{\bar{x}})]\dot{y}\, dt$$

and so

$$\|f'(x) - f'(\bar{x})\| = \sup_{\|y\|_{1,\infty} \le 1} |[f'(x) - f'(\bar{x})]y|$$

$$\le \int_a^b |\varphi_x(t, x, \dot{x}) - \varphi_x(t, \bar{x}, \dot{\bar{x}})| \, dt + \int_a^b |\varphi_v(t, x, \dot{x}) - \varphi_v(t, \bar{x}, \dot{\bar{x}})| \, dt,$$

$$< \epsilon \quad \text{if } \|x - \bar{x}\|_{1,\infty} \text{ is sufficiently small.}$$

Justification of the last line: According to hypothesis, φ_x and φ_v are continuous on $[a, b] \times \mathbb{R} \times \mathbb{R}$, hence uniformly continuous on the compact set

$$\{(t, \xi, \zeta) \in \mathbb{R}^3 \,|\, t \in [a, b], \, |\xi - \bar{x}(t)| \le 1, \, |\zeta - \dot{\bar{x}}(t)| \le 1\}.$$

Thus, for each $\epsilon > 0$ there exists $\delta \in (0, 1)$ such that

$$|\varphi_x(t, x(t), \dot{x}(t)) - \varphi_x(t, \bar{x}(t), \dot{\bar{x}}(t))| < \frac{\epsilon}{2(b - a)}$$

whenever $t \in [a, b]$, $|x(t) - \bar{x}(t)| \le \delta$, and $|\dot{x}(t) - \dot{\bar{x}}(t)| \le \delta$. An analogous estimate holds for φ_v. $\qquad\square$

3.7 Implicit Function Theorems and Related Results

Now we make the following assumptions:

(A) E, F, and G are normed vector spaces.
   U and V are open neighborhoods of $\bar{x} \in E$ and $\bar{y} \in F$, respectively.
   $f : U \times V \to G$.

Define $g_1 : U \to G$ by $g_1(x) := f(x, \bar{y})$, $x \in U$. We denote the derivative (in the sense of Gâteaux, Hadamard, or Fréchet) of g_1 at $\bar{x}$, whenever it exists, by $f_{|1}(\bar{x}, \bar{y})$ or by $D_1 f(\bar{x}, \bar{y})$ and call it *partial derivative* of f, with respect to the first variable, at $(\bar{x}, \bar{y})$. Notice that $f_{|1}(\bar{x}, \bar{y})$ is an element of $\mathrm{L}(E, G)$. If $f_{|1}(x, y)$ exists, say, for all $(x, y) \in U \times V$, then

$$f_{|1} : (x, y) \mapsto f_{|1}(x, y), \quad (x, y) \in U \times V,$$

defines the mapping $f_{|1} : U \times V \to \mathrm{L}(E, G)$. An analogous remark applies to $f_{|2}(x, y)$ and $D_2 f(\bar{x}, \bar{y})$.

As in classical multivariate calculus, we have the following relationship.

Proposition 3.7.1 *Let the assumptions* (A) *be satisfied.*

(a) *If f is G-differentiable at $(\bar{x}, \bar{y})$, then the partial G-derivatives $f_{|1}(\bar{x}, \bar{y})$ and $f_{|2}(\bar{x}, \bar{y})$ exist and one has*

$$f'(\bar{x}, \bar{y})(u, v) = f_{|1}(\bar{x}, \bar{y})u + f_{|2}(\bar{x}, \bar{y})v \quad \forall \, (u, v) \in E \times F. \qquad (3.13)$$

An analogous statement holds for H- and F-derivatives.

(b) *Assume that the partial G-derivatives $f_{|1}$ and $f_{|2}$ exist on $U \times V$ and are continuous at $(\bar{x}, \bar{y})$. Then f is F-differentiable at $(\bar{x}, \bar{y})$ and (3.13) holds true.*

Now we establish two *implicit function theorems*: one under standard hypotheses and one under relaxed differentiability hypotheses but with G finite dimensional.

Theorem 3.7.2 (Classical Implicit Function Theorem) *In addition to* (A), *let the following hold:*

(a) *E, F, and G are Banach spaces.*
(b) *f is continuous at $(\bar{x}, \bar{y})$ and $f(\bar{x}, \bar{y}) = 0$.*
(c) *The partial F-derivative $f_{|2}$ exists on $U \times V$ and is continuous at $(\bar{x}, \bar{y})$.*
(d) *The (continuous linear) mapping $f_{|2}(\bar{x}, \bar{y}) : F \to G$ is bijective.*

Then:

(i)   *There exist neighborhoods $U' \subseteq U$ and $V' \subseteq V$ of $\bar{x}$ and $\bar{y}$, respectively, such that for each $x \in U'$ there is precisely one $\varphi(x) \in V'$ satisfying*

$$f\big(x, \varphi(x)\big) = o \quad \forall x \in U'.$$

(ii) *If f is continuous in a neighborhood of $(\bar{x}, \bar{y})$, then the function $\varphi : x \mapsto \varphi(x)$ is continuous in a neighborhood of $\bar{x}$.*
(iii) *If f is continuously differentiable in a neighborhood of $(\bar{x}, \bar{y})$, then φ is continuously differentiable in a neighborhood of $\bar{x}$ and there holds*

$$\varphi'(x) = -f_{|2}\big(x, \varphi(x)\big)^{-1} \circ f_{|1}\big(x, \varphi(x)\big). \tag{3.14}$$

Concerning the *proof* of the theorem, which is based on the Banach fixed point theorem, see for instance Dieudonné [53], Schirotzek [196], or Zeidler [222]. Observe that the assumptions on $f_{|2}$ guarantee that $f_{|2}\big(x, \varphi(x)\big)^{-1}$ exists as an element of $L(G, F)$ provided $\|x - \bar{x}\|$ is sufficiently small.

Now we relax the differentiability assumptions on $f_{|2}$.

Proposition 3.7.3 *In addition to* (A), *let the following hold:*

(a) *G is finite dimensional.*
(b) *f is continuous in a neighborhood of $(\bar{x}, \bar{y})$ and $f(\bar{x}, \bar{y}) = 0$.*
(c) *The partial F-derivative $f_{|2}(\bar{x}, \bar{y})$ exists and is surjective.*

Then, for each neighborhood $V' \subseteq V$ of $\bar{y}$ there exist a neighborhood $U' \subseteq U$ of $\bar{x}$ and a function $\varphi : U' \to V'$ such that the following holds:

(i) *$f\big(x, \varphi(x)\big) = o \quad \forall x \in U', \quad \varphi(\bar{x}) = \bar{y}.$*
(ii) *φ is continuous at $\bar{x}$.*

Proof.

(I) Without loss of generality we may assume that $\bar{y} = o$. Further we set $T := f_{|2}(\bar{x}, o)$. By assumption, T is a continuous linear mapping of F onto the finite-dimensional space G. Hence there exists a finite-dimensional linear subspace $\tilde{F}$ of F such that the linear mapping $T^{-1} : G \to \tilde{F}$ satisfying $TT^{-1}(z) = z$ for any $z \in G$ is a linear isomorphism. In order to verify the assertions (i) and (ii), we may replace $f : E \times F \to G$ by its restriction to $E \times \tilde{F}$. But $\tilde{F}$ can be identified with G and so we may assume that $F = G$. Then T is a bijective linear mapping of G onto G.

(IIa) Let $\epsilon > 0$ be such that $B_F(o, \epsilon) \subseteq V$ and f is continuous on the neighborhood $B_E(\bar{x}, \epsilon) \times B_F(o, \epsilon)$ of $(\bar{x}, o)$. Let $\alpha \in (0, \epsilon)$ be such that

$$|f(\bar{x}, y) - T(y)| \leq \frac{\alpha}{2\|T^{-1}\|} \quad \forall y \in B_F(o, \alpha). \tag{3.15}$$

Since f is continuous and $B_F(o, \alpha)$ is compact, there further exists $\beta \in (0, \epsilon)$ such that

$$|f(\bar{x}, y) - f(x, y)| \leq \frac{\alpha}{2\|T^{-1}\|} \quad \forall x \in B_E(\bar{x}, \beta) \quad \forall y \in B_F(o, \alpha). \tag{3.16}$$

(IIb) For any $x \in B_E(\bar{x}, \beta)$ define $h_x : B_F(o, \alpha) \to F$ by $h_x(y) := y - T^{-1}f(x, y)$. Notice that h_x is continuous.

(IIc) We now show that h_x maps $B_F(o, \alpha)$ into itself. Let any $y \in B_F(o, \alpha)$ be given. We have

$$\|h_x(y)\| \leq \|y - T^{-1}f(\bar{x}, y)\| + \|T^{-1}(f(\bar{x}, y) - f(x, y))\|. \tag{3.17}$$

Furthermore, we obtain

$$\begin{aligned}\|y - T^{-1}f(\bar{x}, y)\| &= \|T^{-1}(T(y) - f(\bar{x}, y))\| \\ &\leq \|T^{-1}\| \cdot \|T(y) - f(\bar{x}, y)\| \underset{(3.15)}{\leq} \frac{\alpha}{2}\end{aligned} \tag{3.18}$$

as well as

$$\|T^{-1}(f(\bar{x}, y) - f(x, y))\| \underset{(3.16)}{\leq} \|T^{-1}\| \cdot \frac{\alpha}{2\|T^{-1}\|} = \frac{\alpha}{2}.$$

Hence (3.17) shows that h_x maps $B_F(o, \alpha)$ into itself.

(IId) In view of (IIb) and (IIc) the Brouwer fixed-point theorem applies, ensuring that h_x has a fixed point $\psi(x)$ in $B_F(o, \alpha)$. This defines a mapping $\psi : x \mapsto \psi(x)$ of $B_E(\bar{x}, \beta)$ into V satisfying

$$\psi(x) - T^{-1}f(x, \psi(x)) = h_x(\psi(x)) = \psi(x)$$

and so $f(x, \psi(x)) = o$ for any $x \in B_E(\bar{x}, \beta)$.

(III) Let a neighborhood $V' \subseteq V$ of o be given. Choose $\nu \in \mathbb{N}$ such that $B_F(o, \frac{1}{\nu}) \subseteq V'$ and set $V_i := B_F(o, \frac{1}{\nu+i})$ for $i = 1, 2, \ldots$ By step (II) we know that for each i there exist a neighborhood U_i of $\bar{x}$ and a function $\psi_i : U_i \to V_i$ satisfying $f(x, \psi_i(x)) = o$ for any $x \in U_i$. Without loss of generality we may assume that U_{i+1} is a proper subset of U_i for $i = 1, 2, \ldots$ and that $\bigcap_{i=1}^{\infty} U_i = \{\bar{x}\}$. Now let $U' := U_1$ and define $\varphi : U' \to V'$ by

$$\varphi(\bar{x}) := o = \bar{y}, \quad \varphi(x) := \psi_i(x) \quad \text{whenever } x \in U_i \setminus U_{i+1}.$$

Then (i) holds by definition of φ. We verify (ii). Thus let $\eta > 0$ be given. Then we have $V_i \subseteq B_F(o, \eta)$ for some i and $\psi_i : U_i \to V_i$. It follows that

$$\|\varphi(x) - o\| = \|\psi_i(x) - o\| \leq \eta \quad \text{whenever } x \in U_i \setminus U_{i+1}.$$

By the construction of U_i and V_i, we conclude that $\varphi(U_i) \subseteq B_F(o, \eta)$.
$\square$

Theorem 3.7.4 (Halkin's Implicit Function Theorem) *In addition to* (A), *let the following hold:*

(a) *G is finite dimensional.*
(b) *f is continuous in a neighborhood of $(\bar{x}, \bar{y})$ and $f(\bar{x}, \bar{y}) = 0$.*
(c) *f is F-differentiable at $(\bar{x}, \bar{y})$ and the partial F-derivative $f_{12}(\bar{x}, \bar{y})$ is surjective.*

Then there exist a neighborhood U' of $\bar{x}$ and a function $\varphi : U' \to V$ satisfying:

(i) *$f(x, \varphi(x)) = o \quad \forall x \in U', \quad \varphi(\bar{x}) = \bar{y}.$*
(ii) *φ is F-differentiable at $\bar{x}$ and there holds*

$$f_{11}(\bar{x}, \bar{y}) + f_{12}(\bar{x}, \bar{y}) \circ \varphi'(\bar{x}) = o. \tag{3.19}$$

Proof.

(I) With the same argument as in step (I) of the proof of Proposition 3.7.3 we may assume without loss of generality that $F = G$. We may also assume that $\bar{x} = o$ and $\bar{y} = o$. We set $S := f_{11}(o, o)$ and $T := f_{12}(o, o)$. Notice that T is a bijective linear mapping of G onto G.

(II) By Proposition 3.7.3, there exist a neighborhood U' of $\bar{x} = o$ and a function $\varphi : U' \to V$ such that (i) holds and φ is continuous at o. We verify (ii). Since f is F-differentiable at o, there exists a function $r : U' \to F$ such that

$$f'(o, o)(x, \varphi(x)) + r(x, \varphi(x)) = o \quad \forall x \in U', \tag{3.20}$$

$$\lim_{\|x\|+\|y\| \to 0} \frac{r(x, y)}{\|x\| + \|y\|} = o. \tag{3.21}$$

By Proposition 3.7.1, (3.20) passes into

$$S(x) + T(\varphi(x)) + r(x, \varphi(x)) = o \quad \forall x \in U',$$

i.e.,

$$\varphi(x) = -T^{-1}S(x) - T^{-1}r(x, \varphi(x)) \quad \forall x \in U'. \tag{3.22}$$

(III) We estimate $\|\varphi(x)\|$. Let $\sigma > 0$ be such that $B_E(o, \sigma) \subseteq U'$ and

$$\|r(x, y)\| \leq \frac{(\|x\| + \|y\|)}{2\|T^{-1}\|} \quad \text{whenever } \|x\| \leq \sigma, \|y\| \leq \sigma. \tag{3.23}$$

Since φ is continuous at o, there further exists $\alpha \in (0, \sigma)$ such that $\|\varphi(x)\| \leq \sigma$ for all $x \in B_E(o, \alpha)$. It follows that

$$\|\varphi(x)\| \underset{(3.22)}{\leq} \|T^{-1}S\| \cdot \|x\| + \|T^{-1}\| \cdot \|r(x, ,\varphi(x))\|$$

$$\underset{(3.23)}{\leq} (\|T^{-1}S\| + \frac{1}{2}) \cdot \|x\| + \frac{1}{2}\|\varphi(x)\| \quad \forall x \in B_E(o, \alpha)$$

and so

$$\|\varphi(x)\| \leq (2\|T^{-1}S\| + 1) \cdot \|x\| \quad \forall x \in B_E(o, \alpha). \tag{3.24}$$

We also have

$$\|T^{-1}r(x, \varphi(x))\| \leq \|T^{-1}\| \cdot \|r(x, \varphi(x))\|.$$

The latter inequality, (3.21) and (3.24) show that $\|T^{-1}r(x, \varphi(x))\|/\|x\|$ is arbitrarily small for all x in a sufficiently small neighborhood of $\bar{x} = o$. In view of (3.22), we conclude that φ is F-differentiable at o, with derivative $\varphi'(o) = -T^{-1}S$. □

To prepare the next result, recall (once more) that if the mapping $f : E \to G$ is F-differentiable at $\bar{x} \in E$, then with some neighborhood U of $\bar{x}$, one has

$$f(x) = f(\bar{x}) + f'(\bar{x})(x - \bar{x}) + r(x) \quad \forall x \in U,$$

$$\text{where } \lim_{x \to \bar{x}} \frac{r(x)}{\|x - \bar{x}\|} = o.$$

Our aim now is to replace the correction term $r(x)$ for the function values on the right-hand side by a correction term $\rho(x)$ for the argument on the left-hand side:

$$f(x + \rho(x)) = f(\bar{x}) + f'(\bar{x})(x - \bar{x}) \quad \forall x \in U,$$

$$\text{where } \lim_{x \to \bar{x}} \frac{\rho(x)}{\|x - \bar{x}\|} = o. \tag{3.25}$$

Theorem 3.7.5 says that this is possible under appropriate hypotheses.

Theorem 3.7.5 (Halkin's Correction Theorem) *Let E and G be normed vector spaces with G finite dimensional. Further let $f : E \to G$ and $\bar{x} \in E$. Assume the following:*

(a) *f is continuous in a neighborhood of $\bar{x}$.*
(b) *The F-derivative $f'(\bar{x})$ exists and is surjective.*

Then there exist a neighborhood U of $\bar{x}$ and a function $\rho : U \to E$ such that (3.25) holds. The function ρ satisfies $\rho(\bar{x}) = o$ and is F-differentiable at $\bar{x}$ with $\rho'(\bar{x}) = o$.

Proof. Let F be the finite-dimensional linear subspace of E which $f'(\bar{x})$ maps onto G. Define $\tilde{f} : E \times F \to G$ by

$$\tilde{f}(x, y) := f(x + y) - f'(\bar{x})(x - \bar{x}) - f(\bar{x}).$$

Notice that $\tilde{f}$ is F-differentiable at $(\bar{x}, o)$ and that

$$\tilde{f}_{|1}(\bar{x}, o) = o, \quad \tilde{f}_{|2}(\bar{x}, o) = f'(\bar{x}). \tag{3.26}$$

Hence Theorem 3.7.4 applies to $\tilde{f}$ at (x, o). Thus there exist a neighborhood U of $\bar{x}$ and a function $\varphi : U \to F$ that is F-differentiable at $\bar{x}$ and is such that

$$\tilde{f}(x, \varphi(x)) = o \quad \forall x \in U, \quad \varphi(\bar{x}) = o,$$
$$\tilde{f}_{|1}(\bar{x}, o) + \tilde{f}_{|2}(\bar{x}, o) \circ \varphi'(\bar{x}) = o.$$

Setting $\rho := \varphi$, the definition of $\tilde{f}$ gives

$$f(x + \rho(x)) = f(\bar{x}) + f'(\bar{x})(x - \bar{x}) \quad \forall x \in U.$$

Moreover, by (3.26) we have $f'(\bar{x}) \circ \rho'(\bar{x}) = o$. Since $f'(\bar{x}) : F \to G$ is bijective, it follows that $\rho'(\bar{x}) = o$. From this and $\rho(\bar{x}) = o$ we finally deduce that $\rho(x)/\|x - \bar{x}\| \to o$ as $x \to \bar{x}$. $\square$

Theorem 3.7.5 will be a key tool for deriving a multiplier rule for a non-smooth optimization problem in Sect. 12.3.

Theorem 3.7.6 (Halkin's Inverse Function Theorem) *Let E be a finite-dimensional normed vector space. Further let $f : E \to E$ and $\bar{x} \in E$. Assume the following:*

(a) *f is continuous in a neighborhood of $\bar{x}$.*
(b) *The F-derivative $f'(\bar{x})$ exists and is surjective.*

Then there exist a neighborhood U of $\bar{x}$ and a function $\varphi : U \to E$ such that the following holds:

(i) *$f(\varphi(x)) = x \quad \forall x \in U, \quad \varphi(f(\bar{x})) = \bar{x}$.*
(ii) *φ is F-differentiable at $f(\bar{x})$, with $\varphi'(f(\bar{x})) = f'(\bar{x})^{-1}$.*

Proof. Define $\tilde{f} : E \times E \to E$ by $\tilde{f}(u,v); = u - f(v)$ and set $\bar{u} := f(\bar{x})$, $\bar{v} := \bar{x}$. Then $\tilde{f}$ is F-differentiable at $(\bar{u}, \bar{v})$, with $\tilde{f}_{|1}(\bar{u}, \bar{v}) = \mathrm{id}_E$ and $\tilde{f}_{|2}(\bar{u}, \bar{v}) = -f'(\bar{x})$. By Theorem 3.7.4 applied to $\tilde{f}$ at $(\bar{u}, \bar{v})$, there exist a neighborhood U of $\bar{u}$ and a function $\varphi : U \to E$ such that $\tilde{f}(u, \varphi(u)) = o$ for any $u \in U$ and $\varphi(\bar{u}) = \bar{v}$. Moreover, φ is F-differentiable at $\bar{u}$ and satisfies

$$\tilde{f}_{|1}(\bar{u}, \bar{v}) + \tilde{f}_{|2}(\bar{u}, \bar{v}) \circ \varphi'(\bar{u}) = o.$$

It is obvious that φ meets the assertions of the theorem.   □

3.8 Bibliographical Notes and Exercises

The subject of this chapter is standard. We refer to Dieudonné [53], Schirotzek [196], Schwartz [197], and Zeidler [222] for differential calculus in Banach spaces and to Zeidler [221, 224] for differentiability properties of integral functionals on Sobolev spaces. The results from Proposition 3.7.3 to the end of Sect. 3.7 are due to Halkin [82]. See also the Bibliographical Notes to Chap. 4.

Exercise 3.8.1 Define $g : \mathbb{R}^2 \to \mathbb{R}$ by

$$g(x_1, x_2) := \begin{cases} \frac{x_1^3}{x_2} & \text{if } x_2 \neq 0, \\ 0 & \text{if } x_2 = 0. \end{cases}$$

Show that g is G-differentiable but not H-differentiable at $\bar{x} = (0, 0)$.

Exercise 3.8.2 Show that the function

$$f(x) := x^2 \sin(1/x) \text{ if } x \in \mathbb{R} \setminus \{0\}, \quad f(x) := 0 \text{ if } x = 0$$

is F-differentiable but not continuously differentiable at $\bar{x} = 0$.

In Sect. 4.6 we shall show that the maximum norm on $C[a, b]$, where $a < b$, is H-differentiable at certain points but nowhere F-differentiable; compare this and the preceding two examples with Proposition 3.4.2.

Exercise 3.8.3 Define $f : \mathbb{R}^2 \to \mathbb{R}^2$ by $f(x_1, x_2) := (x_1, x_2^3)$ and let $g : \mathbb{R}^2 \to \mathbb{R}$ be the function of Exercise 3.8.1. Then f is F-differentiable (and so G-differentiable) on $\mathbb{R}^2$ and g is G-differentiable at $\bar{x} = (0, 0)$. Is the composite function $g \circ f$ G-differentiable at $\bar{x}$?

Exercise 3.8.4 Carry out the omitted proofs for Proposition 3.2.4.

Exercise 3.8.5 Prove Proposition 3.2.5.

Exercise 3.8.6 Show that the functional $f(x) := \frac{1}{2}a(x, x)$, $x \in E$, where $a :$ $E \times E \to \mathbb{R}$ is bilinear, symmetric, and bounded, is continuously differentiable on E and calculate its derivative (cf. Example 3.6.1).

Exercise 3.8.7 Assume that $\varphi : [a, b] \times \mathbb{R} \times \mathbb{R} \to \mathbb{R}$ is continuous and possesses continuous partial derivatives with respect to the second and the third variable. Modeling the proof in Example 3.6.3, show that the functional $f : C^1[a, b] \times [a, b] \times [a, b]$ defined by

$$f(x, \sigma, \tau) := \int_\sigma^\tau \varphi\big(t, x(t), \dot{x}(t)\big)\, dt, \quad x \in C^1[a, b], \quad \sigma, \tau \in (a, b),$$

is continuously differentiable and calculate its derivative. (Functionals of this kind appear in variable-endpoint problems in the classical calculus of variations.)

4

The Subdifferential of Convex Functionals

4.1 Definition and First Properties

For convex functionals, the following notion provides an appropriate substitute for a nonexisting derivative.

Definition 4.1.1 Let $f : E \to \overline{\mathbb{R}}$ be proper and convex, and let $\bar{x} \in \operatorname{dom} f$. The set

$$\partial f(\bar{x}) := \{x^* \in E^* \mid \langle x^*, x - \bar{x} \rangle \leq f(x) - f(\bar{x}) \, \forall x \in E\}$$

is called *subdifferential* of f at $\bar{x}$ (in the sense of convex analysis). Each $x^* \in \partial f(\bar{x})$ is called *subgradient* of f at $\bar{x}$.

A geometric interpretation is given in Fig. 0.1 in the Introduction.

Remark 4.1.2 The main purpose of the subdifferential is to detect minimum points. We first consider free minimization. If $f : E \to \overline{\mathbb{R}}$ is convex and $\bar{x} \in \operatorname{dom} f$, then we obtain

$$f(\bar{x}) = \min_{x \in E} f(x) \quad \Longleftrightarrow \quad 0 \leq f(x) - f(\bar{x}) \, \forall x \in E \quad \Longleftrightarrow \quad o \in \partial f(\bar{x}).$$

Hence the condition $o \in \partial f(\bar{x})$ is a substitute for the optimality condition $f'(\bar{x}) = o$ in the differentiable case. Concerning constrained minimization, for $A \subseteq E$ nonempty and convex, we have

$$f(\bar{x}) = \min_{x \in A} f(x) \quad \Longleftrightarrow \quad (f + \delta_A)(\bar{x}) = \min_{x \in E}(f + \delta_A)(x) \quad \Longleftrightarrow \quad o \in \partial(f + \delta_A)(\bar{x}).$$

For further exploitation, we need at least a *sum rule* of the form

$$\partial(f_1 + f_2)(\bar{x}) \subseteq \partial f_1(\bar{x}) + \partial f_2(\bar{x}).$$

This, among others, will be derived below.

The subdifferential of a convex functional f can also be characterized by the directional G-derivative of f. This relationship will later be the starting point for defining subdifferentials for certain classes of nonconvex functionals.

Theorem 4.1.3 *Let $f : E \to \overline{\mathbb{R}}$ be proper and convex.*

(a) *If $\bar{x} \in \mathrm{dom} f$ and $y \in E$, the function $\tau \mapsto \frac{1}{\tau}(f(\bar{x}+\tau y)-f(\bar{x}))$ is monotone increasing on $\mathbb{R} \setminus \{0\}$; hence the limit $f_G(\bar{x}, y)$ exists in $\overline{\mathbb{R}}$ and one has*

$$f(\bar{x}) - f(\bar{x} - y) \le f_G(\bar{x}, y) = \inf_{\tau > 0} \frac{f(\bar{x} + \tau y) - f(\bar{x})}{\tau} \le f(\bar{x} + y) - f(\bar{x}).$$
(4.1)

(b) *If $\bar{x} \in \mathrm{dom} f$, the functional $f_G(\bar{x}, \cdot)$ of E to $\overline{\mathbb{R}}$ is sublinear.*
(c) *If $\bar{x} \in \mathrm{int}\, \mathrm{dom}\, f$ and $y \in E$, then $f_G(\bar{x}, y) \in \mathbb{R}$.*
(d) *If $\bar{x} \in \mathrm{int}\, \mathrm{dom}\, f$ and f is continuous at $\bar{x}$, then $f_H(\bar{x}, \cdot)$ exists, is continuous on E, and equals $f_G(\bar{x}, \cdot)$.*

Proof.

(a) Since f is proper and convex, so is $\varphi(\tau) := f(\bar{x} + \tau y), \tau \in \mathbb{R}$. Let $\tau_1 < \tau_2 < \tau_3$ and $\tau_{ik} := \tau_i - \tau_k$ for $i, k = 1, 2, 3$. Since $\tau_2 = \frac{\tau_{32}}{\tau_{31}}\tau_1 + \frac{\tau_{21}}{\tau_{31}}\tau_3$, it follows that

$$\varphi(\tau_2) \le \frac{\tau_{32}}{\tau_{31}}\varphi(\tau_1) + \frac{\tau_{21}}{\tau_{31}}\varphi(\tau_3)$$

and so

$$\frac{\varphi(\tau_2) - \varphi(\tau_1)}{\tau_2 - \tau_1} \le \frac{\varphi(\tau_3) - \varphi(\tau_1)}{\tau_3 - \tau_1} \le \frac{\varphi(\tau_3) - \varphi(\tau_2)}{\tau_3 - \tau_2}.$$
(4.2)

From this, by appropriate choices of τ_i, we obtain the monotonicity of the function

$$\tau \mapsto \frac{1}{\tau}(\varphi(\tau) - \varphi(0)) = \frac{1}{\tau}(f(\bar{x} + \tau y) - f(\bar{x}))$$

as well as the relation (4.1).

(b) For $y, z \in E$, the convexity of f implies

$$f(\bar{x}+\tau(y+z)) = f(\tfrac{1}{2}(\bar{x}+2\tau y) + \tfrac{1}{2}(\bar{x}+2\tau z)) \le \tfrac{1}{2}f(\bar{x}+2\tau y) + \tfrac{1}{2}f(\bar{x}+2\tau z)$$

and so

$$f_G(\bar{x}, y + z) \le f_G(\bar{x}, y) + f_G(\bar{x}, z).$$

It is evident that $f_G(\bar{x}, \cdot)$ is positively homogeneous.

(c) Since $\bar{x} \in \mathrm{int}\, \mathrm{dom}\, f$, there exists $\epsilon > 0$ such that $\bar{x} \pm \epsilon y \in \mathrm{dom}\, f$. Applying (4.1) with ϵy instead of y, we see that $\epsilon f_G(\bar{x}, y) (= f_G(\bar{x}, \epsilon y))$ is finite.

(d) There exists an open neighborhood U of zero in E such that

$$f(\bar{x} + y) - f(\bar{x}) \le 1 \quad \forall y \in U.$$

This and (4.1) imply that the convex functional $f_G(\bar{x}, \cdot)$ is bounded above on U and so, by Theorem 1.4.1, is continuous on $\mathrm{int}\, \mathrm{dom}\, f_G(\bar{x}, \cdot)$ which is equal to E. Likewise by Theorem 1.4.1, f is locally L-continuous. Hence by Lemma 3.1.2, $f_H(\bar{x}, \cdot)$ exists and equals $f_G(\bar{x}, \cdot)$. □

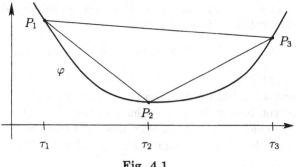

Fig. 4.1

Remark 4.1.4 For a proper convex function $\varphi : \mathbb{R} \to \mathbb{R}$, the inequalities (4.2) have a simple geometric meaning. With the notation of Fig. 4.1, the inequalities say that

$$\mathrm{slope}(P_1 P_2) \leq \mathrm{slope}(P_1 P_3) \leq \mathrm{slope}(P_2 P_3).$$

If φ is a function defined on $\mathbb{R}$, then obviously $\varphi_G(\tau, 1) = \varphi'_+(\tau)$, where $\varphi'_+(\tau)$ denotes the right derivative of φ at τ (cf. Sect. 3.1). Hence Theorem 4.1.3 immediately leads to:

Corollary 4.1.5 *Let $\varphi : \mathbb{R} \to \overline{\mathbb{R}}$ be proper and convex. For any $\tau_0 \in \mathrm{dom}\, \varphi$, the right derivative $\varphi'_+(\tau_0)$ exists in $\overline{\mathbb{R}}$ and satisfies*

$$\varphi'_+(\tau_0) = \inf_{\tau > 0} \frac{\varphi(\tau_0 + \tau) - \varphi(\tau_0)}{\tau}.$$

In particular, if $\tau_0 \in \mathrm{int}\,\mathrm{dom}\, \varphi$, then $\varphi'_+(\tau_0) \in \mathbb{R}$. If $\tau_0 \in \mathrm{dom}\, \varphi$ is the left boundary point of $\mathrm{dom}\, \varphi$, then $\varphi'_+(\tau_0) \in \mathbb{R} \cup \{-\infty\}$.

The subdifferential can be characterized by the directional G-derivative and vice versa.

Proposition 4.1.6 *Let $f : E \to \overline{\mathbb{R}}$ be proper and convex.*

(a) *If $\bar{x} \in \mathrm{dom}\, f$, then*

$$\partial f(\bar{x}) = \{x^* \in E^* \mid \langle x^*, y \rangle \leq f_G(\bar{x}, y) \ \forall y \in E\}. \tag{4.3}$$

(b) *If $\bar{x} \in \mathrm{int}\,\mathrm{dom}\, f$ and f is continuous at $\bar{x}$, then $\partial f(\bar{x})$ is nonempty convex and $\sigma(E^*, E)$-compact, and one has*

$$f_H(\bar{x}, y) = f_G(\bar{x}, y) = \max\{\langle x^*, y \rangle \mid x^* \in \partial f(\bar{x})\} \quad \forall y \in E. \tag{4.4}$$

Proof.

(a) Let $x^* \in \partial f(\bar{x})$. For each $\tau > 0$, we have

$$\tau \langle x^*, y \rangle = \langle x^*, \bar{x} + \tau y \rangle - \langle x^*, \bar{x} \rangle \leq f(\bar{x} + \tau y) - f(\bar{x})$$

and so $\langle x^*, y \rangle \le f_G(\bar{x}, y)$. Conversely, if x^* belongs to the right-hand side of (4.3), then it follows from (4.1) that

$$\langle x^*, \bar{x} + y \rangle - \langle x^*, \bar{x} \rangle = \langle x^*, y \rangle \le f(\bar{x} + y) - f(\bar{x})$$

and so $x^* \in \partial f(\bar{x})$.

(b) By Theorem 4.1.3, the functional $p(y) := f_G(\bar{x}, y)$, $y \in E$, is finite, sublinear, and continuous. By (a), we have $\partial f(\bar{x}) = M_p$; here we use the notation introduced in the Hörmander theorem (Theorem 2.3.1). According to this result, M_p is nonempty, convex, and $\sigma(E^*, E)$-closed, and we have $\sigma_{M_p} = p$, i.e.,

$$\sup\{\langle x^*, y \rangle \mid x^* \in \partial f(\bar{x})\} = f_G(\bar{x}, y) \quad \forall y \in E. \tag{4.5}$$

We show that $\partial f(\bar{x})$ is $\sigma(E^*, E)$-compact. Since f is continuous at $\bar{x}$, the set

$$U := \{y \in E \mid f(\bar{x} + y) - f(\bar{x}) \le 1\}$$

is a neighborhood of zero in E. Setting $U^\circ := \{x^* \in E^* \mid \langle x^*, y \rangle \le 1 \, \forall y \in U\}$, we have $\partial f(\bar{x}) \subseteq U^\circ$. By the Alaoglu theorem, U° is $\sigma(E^*, E)$-compact and so is $\partial f(\bar{x})$ as a $\sigma(E^*, E)$-closed subset. For fixed $y \in E$, the functional $x^* \mapsto \langle x^*, y \rangle$, $x^* \in E^*$, is $\sigma(E^*, E)$-continuous and so the supremum in (4.5) is attained. Finally, by Theorem 4.1.3(d) we may replace f_G by f_H. $\qquad\square$

The representation formula (4.4) can be refined using the concept of extreme point.

Proposition 4.1.7 *If* $f : E \to \overline{\mathbb{R}}$ *is proper, convex, and continuous at* $\bar{x} \in$ int dom f, *then*

$$f_H(\bar{x}, y) = f_G(\bar{x}, y) = \max\{\langle x^*, y \rangle \mid x^* \in \mathrm{ep}(\partial f(\bar{x}))\} \quad \forall y \in E.$$

Proof. By virtue of Proposition 4.1.6(b), we may apply Proposition 1.7.8 to $E^*[\sigma]$ instead of E, $A := \partial f(\bar{x})$, and $g(x^*) := \langle x^*, y \rangle$, $x^* \in E$. $\qquad\square$

Now we characterize G-differentiability of convex functionals.

Proposition 4.1.8 (Differentiability Criterion) *Let* $f : E \to \overline{\mathbb{R}}$ *be proper and convex, and let* $\bar{x} \in$ dom f.

(a) *If* f *is G-differentiable at* $\bar{x}$, *then* $\partial f(\bar{x}) = \{f'(\bar{x})\}$.

(b) *If* f *is continuous at* $\bar{x}$ *and* $\partial f(\bar{x})$ *consists of exactly one element* $x^* \in E^*$, *then* f *is H-differentiable (and so G-differentiable) at* $\bar{x}$ *and* $f'(\bar{x}) = x^*$.

Proof.

(a) On the one hand, we have

$$\langle f'(\bar{x}), \bar{x} + y \rangle - \langle f'(\bar{x}), \bar{x} \rangle = \langle f'(\bar{x}), y \rangle = f_G(\bar{x}, y) \le f(\bar{x} + y) - f(\bar{x}) \quad \forall y \in E$$

and so $f'(\bar{x}) \in \partial f(\bar{x})$. On the other hand, if $x^* \in \partial f(\bar{x})$, then by Proposition 4.1.6 we obtain

$$\langle x^*, y \rangle \leq f_G(\bar{x}, y) = \langle f'(\bar{x}), y \rangle \quad \forall y \in E.$$

This implies, by the linearity of x^* and $f'(\bar{x})$, that $x^* = f'(\bar{x})$.

(b) By Proposition 4.1.6, we conclude that $f_G(\bar{x}, y) = \langle x^*, y \rangle \, \forall y \in E$. Hence the functional $f_G(\bar{x}, \cdot)$ is linear and continuous. Thus f is G-differentiable at $\bar{x}$, and we have $f'(\bar{x}) = x^*$. Moreover, f is locally L-continuous around $\bar{x}$ (Theorem 1.4.1) and so H-differentiable at $\bar{x}$ (Proposition 3.4.2).   □

Proposition 4.1.9 (Semicontinuity Criterion) *Let* $f : E \to \overline{\mathbb{R}}$ *be proper, convex, and G-differentiable at* $\bar{x}$. *Then* f *is lower semicontinuous at* $\bar{x}$.

Proof. Let $k > 0$ be given. Since $f'(\bar{x})$ is continuous at $y = o$ in particular, there exists a neighborhood U of o in E such that

$$k - f(\bar{x}) < \langle f'(\bar{x}), y \rangle \leq f(\bar{x} + y) - f(\bar{x}) \quad \forall y \in U. \quad □$$

4.2 Multifunctions: First Properties

The subdifferential of a convex function $f : E \to \overline{\mathbb{R}}$ associates with each $x \in E$ a (possibly empty) subset $\partial f(x)$ of E^*. The study of this and related objects will be the prominent purpose in the sequel. We now introduce some appropriate concepts.

Definition 4.2.1 Let E and F be vector spaces. A mapping $\Phi : E \to 2^F$, which associates to $x \in E$ a (possibly empty) subset $\Phi(x)$ of F, is called a *multifunction* or *set-valued mapping* and is denoted $\Phi : E \rightrightarrows F$. The *graph* and the *domain* of Φ are defined, respectively, by

$$\text{graph}\, \Phi := \{(x, y) \in E \times F \mid x \in E,\ y \in \Phi(x)\},$$
$$\text{Dom}\, \Phi := \{x \in E \mid \Phi(x) \neq \emptyset\}.$$

Observe that the notation distinguishes the domain of a multifunction from the effective domain $\text{dom}\, f$ of a functional $f : E \to \overline{\mathbb{R}}$. If A is a subset of E, we write $\Phi(A) := \cup_{x \in A} \Phi(x)$.

Remark 4.2.2 A mapping $T : E \to F$ can be identified with the (single-valued) multifunction $\widetilde{T} : E \rightrightarrows F$ defined by $\widetilde{T}(x) := \{Tx\}$, $x \in E$. Concepts defined below for multifunctions will be applied to a mapping $T : E \to F$ according to this identification.

As indicated above, the prototype of a multifunction is the *subdifferential mapping* $\partial f : E \rightrightarrows E^*$ of a convex functional $f : E \to \overline{\mathbb{R}}$, which associates to each $x \in \text{dom}\, f$ the subdifferential $\partial f(x)$ and to each $x \notin \text{dom}\, f$ the empty set.

Definition 4.2.3 Let $\Phi : E \rightrightarrows F$ be a multifunction between locally convex spaces E and F.

(a) Φ is said to be *upper semicontinuous* at $\bar{x} \in \mathrm{Dom}\,\Phi$ if for each open set V in F containing $\Phi(\bar{x})$ there exists an open neighborhood U of $\bar{x}$ such that $\Phi(U) \subseteq V$.

(b) Φ is said to be *lower semicontinuous* at $\bar{x} \in \mathrm{Dom}\,\Phi$ if for each open set V in F such that $V \cap \Phi(\bar{x}) \neq \emptyset$, there exists an open neighborhood U of $\bar{x}$ such that $V \cap \Phi(x) \neq \emptyset$ for any $x \in U$.

(c) Φ is said to be upper [lower] semicontinuous if Φ is upper [lower] semicontinuous at any point $\bar{x} \in \mathrm{Dom}\,\Phi$ (cf. Exercise 4.8.1).

(d) Φ is said to be *locally bounded* at $\bar{x} \in E$ if there exists a neighborhood U of $\bar{x}$ such that $\Phi(U)$ is a bounded subset of F.

(e) A mapping $\varphi : E \to F$ is said to be a *selection* of the multifunction Φ if $\varphi(x) \in \Phi(x)$ for each $x \in \mathrm{Dom}\,\Phi$.

Recall that if $\psi : \mathbb{R} \to \mathbb{R}$ is differentiable, then

$$
\begin{aligned}
\psi \text{ is convex} \quad &\Longleftrightarrow \quad \psi' \text{ is monotone increasing} \\
&\Longleftrightarrow \quad \big(\psi'(y) - \psi'(x)\big) \cdot (y - x) \geq 0 \quad \forall\, x, y \in \mathbb{R}.
\end{aligned}
\tag{4.6}
$$

If we want to generalize this relationship to a G-differentiable functional $f : E \to \mathbb{R}$, we must first define a suitable monotonicity concept for the mapping $f' : E \to E^*$. In view of the nondifferentiable case and the subdifferential mapping, we at once consider multifunctions.

Definition 4.2.4 The multifunction $\Phi : E \rightrightarrows E^*$ is said to be

$$
\begin{aligned}
&\textit{monotone} && \text{if } \langle y^* - x^*, y - x \rangle \geq 0, \\
&\textit{strictly monotone} && \text{if } \langle y^* - x^*, y - x \rangle > 0, \\
&\textit{uniformly monotone} && \text{if } \langle y^* - x^*, y - x \rangle \geq c \cdot \|y - x\|^\gamma;
\end{aligned}
$$

the respective inequality is assumed to hold for all $x, y \in \mathrm{Dom}\,\Phi$, $x \neq y$, $x^* \in \Phi(x)$ and $y^* \in \Phi(y)$. In the last inequality, $c > 0$ and $\gamma > 1$ are constants. If Φ is uniformly monotone with $\gamma = 2$, then Φ is called *strongly monotone*.

According to Remark 4.2.2, a mapping $T : E \to E^*$ is monotone if and only if

$$
\langle T(y) - T(x), y - x \rangle \geq 0 \quad \forall\, x, y \in E.
$$

An analogous remark applies to strict and to uniform monotonicity.

4.3 Subdifferentials, Fréchet Derivatives, and Asplund Spaces

In this section we study the subdifferential mapping $\partial f : E \rightrightarrows E^*$. This will eventually lead to remarkable results on the F-differentiability of continuous convex functionals and, in this connection, to Asplund spaces.

Recall again that, unless otherwise specified, E is a normed vector space and the dual E^* is equipped with the norm topology.

Proposition 4.3.1 *Let $f : E \to \overline{\mathbb{R}}$ be proper and convex. If f is continuous at $\bar{x} \in \operatorname{int} \operatorname{dom} f$, then the subdifferential mapping ∂f is locally bounded at $\bar{x}$.*

Proof. Since f is locally L-continuous at $\bar{x}$ (Theorem 1.4.1), there exist $\epsilon > 0$ and $\lambda > 0$ such that $|f(x) - f(y)| \leq \lambda \|x - y\|$ for all $x, y \in \mathrm{B}(\bar{x}, \epsilon)$. Thus if $x \in \mathrm{B}(\bar{x}, \epsilon)$ and $x^* \in \partial f(x)$, then $\langle x^*, y - x \rangle \leq f(y) - f(x) \leq \lambda \|x - y\|$. It follows that

$$\|x^*\| \leq \lambda \quad \forall x^* \in \partial f(x) \quad \forall x \in \mathrm{B}(\bar{x}, \epsilon), \tag{4.7}$$

which completes the proof. □

Proposition 4.3.2 *Let $f : E \to \overline{\mathbb{R}}$ be proper and convex, and continuous on the nonempty set $\operatorname{int} \operatorname{dom} f$.*

(a) *The subdifferential mapping $\partial f : E \rightrightarrows E^*$ is norm-to-weak* upper semicontinuous on $\operatorname{int} \operatorname{dom} f$.*

(b) *If f is F-differentiable at $\bar{x} \in \operatorname{int} \operatorname{dom} f$, then ∂f is norm-to-norm upper semicontinuous at $\bar{x}$.*

Proof.

(a) Assume that $x \in \operatorname{int} \operatorname{dom} f$ and V is a weak* open subset of E^* containing $\partial f(x)$. It suffices to show that for any sequence (x_n) in $\operatorname{int} \operatorname{dom} f$ with $x_n \to x$ as $n \to \infty$, we have $\partial f(x_n) \subseteq V$ for all sufficiently large n. Suppose this would not hold. Then for some subsequence of (x_n), again denoted (x_n), we could find $x_n^* \in \partial f(x_n) \setminus V$. Since ∂f is locally bounded at x, there exists $c > 0$ such that $\partial f(x_n) \subseteq \mathrm{B}_{E^*}(o, c)$ for all sufficiently large n (compare (4.7)). Since $\mathrm{B}_{E^*}(o, c)$ is weak* compact (Alaoglu theorem), the sequence (x_n^*) admits a weak* cluster point x^*. From $x_n^* \in \partial f(x_n) \setminus V$ we can easily conclude that $x^* \in \partial f(x) \setminus V$ which is a contradiction to $\partial f(x) \subseteq V$.

(b) Let V be an open neighborhood of $f'(\bar{x}) \in E^*$. Assume that for any neighborhood U of $\bar{x}$, we would have $\partial f(U) \setminus V \neq \emptyset$. Then there exist $\epsilon > 0$, a sequence (x_n) in $\operatorname{int} \operatorname{dom} f$, and $x_n^* \in \partial f(x_n)$ such that $x_n \to \bar{x}$ as $n \to \infty$ but $\|x_n^* - f'(\bar{x})\| > 2\epsilon$. The latter implies that there exists a sequence (z_n) in E satisfying $\|z_n\| = 1$ and $\langle x_n^* - f'(\bar{x}), z_n \rangle > 2\epsilon$ for all n. On the other hand, by F-differentiability we have for some $\delta > 0$,

$$f(\bar{x} + y) - f(\bar{x}) - \langle f'(\bar{x}), y \rangle \leq \epsilon \|y\|$$

whenever $y \in E$ and $\|y\| < \delta$. For these y we further obtain

$$\langle x_n^*, (\bar{x} + y) - x_n \rangle \leq f(\bar{x} + y) - f(x_n) \quad \text{and so}$$

$$\langle x_n^*, y \rangle \leq f(\bar{x} + y) - f(\bar{x}) + \langle x_n^*, x_n - \bar{x} \rangle + f(\bar{x}) - f(x_n).$$

Setting $y_n := \delta z_n$, the choice of z_n implies $\|y_n\| = \delta$ and so

$$2\epsilon\delta < \langle x_n^* - f'(\bar{x}), y_n \rangle$$
$$\leq \left(f(\bar{x} + y_n) - f(\bar{x}) - \langle f'(\bar{x}), y_n \rangle \right) + \langle x_n^*, x_n - \bar{x} \rangle + f(\bar{x}) - f(x_n)$$
$$\leq \epsilon\delta + \langle x_n^*, x_n - \bar{x} \rangle + f(\bar{x}) - f(x_n).$$

$$(4.8)$$

Since f is locally bounded at $\bar{x}$, the sequence (x_n^*) is bounded and so $\langle x_n^*, x_n - \bar{x} \rangle \to 0$ as $n \to \infty$. Moreover, since f is continuous at $\bar{x}$, we also have $f(\bar{x}) - f(x_n) \to 0$ as $n \to \infty$. Hence the right-hand side of (4.8) tends to $\epsilon\delta$ as $n \to \infty$ which contradicts the left-hand side.   □

Proposition 4.3.3 *Let $f : E \to \overline{\mathbb{R}}$ be proper and convex, and continuous on the nonempty set $\mathrm{int}\,\mathrm{dom} f$. Then f is F-differentiable [G-differentiable] at $\bar{x} \in \mathrm{int}\,\mathrm{dom} f$ if and only if there exists a selection $\varphi : E \to E^*$ of ∂f which is norm-to-norm continuous [norm-to-weak* continuous] at $\bar{x}$.*

Proof. We verify the statement concerning F-differentiability, leaving the case in brackets as Exercise 4.8.2.

(I) *Necessity.* If f is F-differentiable at $\bar{x}$, then $\partial f(\bar{x})$ is a singleton, and by Proposition 4.3.2 the subdifferential mapping ∂f is upper semicontinuous at $\bar{x}$. Hence any selection of ∂f is continuous at $\bar{x}$.

(II) *Sufficiency.* Let φ be a selection of ∂f that is continuous at $\bar{x}$. Since $\mathrm{int}\,\mathrm{dom} f \subseteq \mathrm{Dom}\,\partial f$, we have $\varphi(\bar{x}) \in \partial f(\bar{x})$ and $\varphi(y) \in \partial f(y)$ for each $y \in \mathrm{int}\,\mathrm{dom} f$. For these y it follows that

$$\langle \varphi(\bar{x}), y - \bar{x} \rangle \leq f(y) - f(x) \quad \text{and} \quad \langle \varphi(y), \bar{x} - y \rangle \leq f(\bar{x}) - f(y).$$

Combining these inequalities, we obtain, again for all $y \in \mathrm{int}\,\mathrm{dom} f$,

$$0 \leq f(y) - f(\bar{x}) - \langle \varphi(\bar{x}), y - \bar{x} \rangle \leq \langle \varphi(y) - \varphi(\bar{x}), y - \bar{x} \rangle \leq \|\varphi(y) - \varphi(\bar{x})\| \cdot \|y - \bar{x}\|.$$

Since φ is continuous at $\bar{x}$, we have $\|\varphi(y) - \varphi(\bar{x})\| \leq 1$ for all y in a neighborhood of $\bar{x}$. Hence the above inequality shows that f is F-differentiable at $\bar{x}$, with derivative $\varphi(\bar{x})$.   □

Corollary 4.3.4 *If $f : E \to \overline{\mathbb{R}}$ is proper and convex, and F-differentiable on the nonempty set $\mathrm{int}\,\mathrm{dom} f$, then f is continuously F-differentiable on $\mathrm{int}\,\mathrm{dom} f$.*

Applying the statement in brackets in Proposition 4.3.3, we obtain an analogous result concerning the norm-to-weak* continuity of the G-derivative. In this connection, the functional f has to be assumed to be continuous on the nonempty set $\mathrm{int}\,\mathrm{dom}\,f$. Below we shall establish a related result under relaxed assumptions (see Proposition 4.3.8).

Proposition 4.3.5 (Convexity Criterion) *If $f : E \to \mathbb{R}$ is G-differentiable, then the following statements are equivalent:*

(a) f *is [strictly] convex.*
(b) f' *is [strictly] monotone.*

Proof. See Exercise 4.8.3. □

Example 4.3.6 Consider the functional

$$f(x) := \tfrac{1}{2}a(x,x), \quad x \in E,$$

where $a : E \times E \to \mathbb{R}$ is bilinear, symmetric, and bounded. By Example 3.6.1, f is continuously differentiable and the derivative satisfies $\langle f'(x), y \rangle = a(x,y)$ for all $x, y \in E$. Assume now that, in addition, a is *strongly positive*, i.e., there exists a constant $c > 0$ such that $a(x,x) \geq c\|x\|^2$ for any $x \in E$. Then we obtain

$$\langle f'(y) - f'(x), y - x \rangle = a(y - x, y - x) \geq c\|y - x\|^2 \quad \forall\, x, y \in E$$

and so f' is strongly monotone. In particular, f' is strictly monotone and so f is strictly convex.

If f is convex but not G-differentiable, we still have the following result.

Proposition 4.3.7 *If $f : E \to \overline{\mathbb{R}}$ is proper and convex, then ∂f is monotone.*

Proof. See Exercise 4.8.4. □

Now we establish the continuity result on the derivative announced after Corollary 4.3.4.

Proposition 4.3.8 *If $f : E \to \overline{\mathbb{R}}$ is proper and convex, and G-differentiable on the nonempty set $\operatorname{int} \operatorname{dom} f$, then f' is radially continuous on $\operatorname{int} \operatorname{dom} f$.*

Proof. For fixed $x, y \in D$, define $\psi_{x,y} : [0,1] \to \mathbb{R}$ by $\psi_{x,y}(\tau) := f(x + \tau(y - x))$ for any $\tau \in [0,1]$. Then we have

$$\psi'_{x,y}(\tau) = \langle f'(x + \tau(y - x)), y - x \rangle$$

and it remains to show that $\psi'_{x,y}$ is continuous on $[0,1]$.

(I) First we show that $\psi'_{x,y}$ is continuous at any $\tau \in (0,1)$. Since $\psi_{x,y}$ is convex, Theorem 4.1.3 applies ensuring that for any ρ, σ satisfying $\rho > \sigma$ we have

$$\psi'_{x,y}(\tau + \sigma) \leq \frac{1}{\rho - \sigma}\left[\psi_{x,y}\big((\tau + \sigma) + (\rho - \sigma)\big) - \psi_{x,y}(\tau + \sigma)\right].$$

Since $\psi_{x,y}$ is continuous on $(0,1)$ (Corollary 1.4.2), we obtain letting $\sigma \downarrow 0$ and then $\rho \downarrow 0$,

$$\lim_{\sigma \downarrow 0} \psi'_{x,y}(\tau+\sigma) \leq \frac{1}{\rho}\left[\psi_{x,y}(\tau+\rho)-\psi_{x,y}(\tau)\right] \quad \text{and} \quad \lim_{\sigma \downarrow 0} \psi'_{x,y}(\tau+\sigma) \leq \psi'_{x,y}(\tau).$$

On the other hand, since $\psi'_{x,y}$ is monotone (increasing), we have $\lim_{\sigma \downarrow 0} \psi'_{x,y}(\tau+\sigma) \geq \psi'_{x,y}(\tau)$. Hence $\psi'_{x,y}$ is right continuous at τ. Analogously it is shown that $\psi'_{x,y}$ is left continuous at τ.

(II) To see that $\psi'_{x,y}$ is continuous on the closed interval $[0,1]$, notice that since D is open, x and y may be replaced in the above argument by $x - \delta(y - x)$ and $y + \delta(y - x)$, respectively, where $\delta > 0$ is sufficiently small.   □

Proposition 4.3.9 *If $f : E \rightarrow \mathbb{R}$ is G-differentiable and f' is uniformly monotone with constants $c > 0$ and $\gamma > 1$, then f is strictly convex and*

$$f(y) - f(x) \geq \langle f'(x), y - x \rangle + \frac{c}{\gamma}\|y - x\|^{\gamma} \quad \forall x, y \in E. \tag{4.9}$$

Proof. By assumption, f' is strictly monotone and so f is strictly convex. Furthermore, by Proposition 4.3.8 the mean value formula (3.7) applies to f. Hence for any $x, y \in E$ we have

$$f(y) - f(x) = \langle f'(x), y - x \rangle + \int_0^1 \langle f'(x + \tau(y - x)) - f'(x), \tau(y - x) \rangle \frac{d\tau}{\tau}$$

$$\geq \langle f'(x), y - x \rangle + \int_0^1 c\tau^{\gamma} \|y - x\|^{\gamma} \frac{d\tau}{\tau},$$

and (4.9) follows.   □

Remark 4.3.10 The above result will later be used to ensure that f has a (unique) global minimum point $\bar{x}$. This point satisfies $f'(\bar{x}) = o$. Hence if (x_n) is a sequence of approximate solutions of $f'(\bar{x}) = o$, then we obtain from (4.9) the error estimate

$$\frac{c}{\gamma}\|x_n - \bar{x}\|^{\gamma} \leq f(x_n) - f(\bar{x}).$$

The next result says that F-differentiability of a continuous convex functional can be characterized without referring to a potential derivative. It will serve us to characterize the set of points where a continuous convex functional is F-differentiable.

Lemma 4.3.11 *Let $f : E \rightarrow \overline{\mathbb{R}}$ be proper, convex, and continuous on the nonempty set $\operatorname{int} \operatorname{dom} f$. Then f is F-differentiable at $\bar{x} \in \operatorname{int} \operatorname{dom} f$ if and only if for each $\epsilon > 0$ there exists $\delta > 0$ such that*

$$f(\bar{x} + \tau y) + f(\bar{x} - \tau y) - 2f(\bar{x}) < \tau\epsilon \tag{4.10}$$

whenever $y \in E$, $\|y\| = 1$ and $0 < \tau < \delta$.

Proof.

(I) *Necessity.* See Exercise 4.8.5.

(II) *Sufficiency.* Assume that the above condition is satisfied and choose some $x^* \in \partial f(\bar{x})$ (which exists by Proposition 4.1.6). Let $y \in E$ satisfying $\|y\| = 1$ be given. For all $\tau > 0$ sufficiently small such that $\bar{x} \pm \tau y \in D$ we have

$$\langle x^*, \tau y \rangle = \langle x^*, (\bar{x} + \tau y) - \bar{x} \rangle \le f(\bar{x} + \tau y) - f(\bar{x}), \tag{4.11}$$

$$- \langle x^*, \tau y \rangle = \langle x^*, (\bar{x} - \tau y) - \bar{x} \rangle \le f(\bar{x} - \tau y) - f(\bar{x}). \tag{4.12}$$

Now let $\epsilon > 0$ be given and choose $\delta > 0$ such that (4.10)–(4.12) hold whenever $\|y\| = 1$ and $0 < \tau < \delta$. Adding the inequalities (4.11) and (4.12), we obtain for these y and τ,

$$0 \le f(\bar{x} + \tau y) - f(\bar{x}) - \langle x^*, \tau y \rangle \le \tau \epsilon.$$

Hence f is F-differentiable at $\bar{x}$. □

Recall that a subset of E is said to be a $\mathbf{G}_\delta$ set if it is the intersection of a countable number of open sets.

Proposition 4.3.12 *Let* $f : E \to \bar{\mathbb{R}}$ *be proper, convex, and continuous on the nonempty set* $D := \operatorname{int} \operatorname{dom} f$. *Then the set* Δ *of all* $x \in D$, *where* f *is F-differentiable, is a (possibly empty)* G_δ *set.*

Proof. For each $n \in \mathbb{N}$ let G_n denote the set of all $x \in D$ for which there exists $\delta > 0$ such that

$$\sup_{\|y\|=1} \frac{f(x + \delta y) + f(x - \delta y) - 2f(x)}{\delta} < \frac{1}{n}.$$

By Theorem 4.1.3, for fixed x and y the functions

$$\tau \mapsto \frac{f(x + \tau(\pm y)) - f(x)}{\tau}$$

are decreasing as $\tau \downarrow 0$. Thus Lemma 4.3.11 shows that $\Delta = \cap_{n=1}^\infty G_n$. It remains to verify that each G_n is open. Let $x \in G_n$ be given. Since f is locally L-continuous (Theorem 1.4.1), there exist $\delta_1 > 0$ and $\lambda > 0$ such that $|f(u) - f(v)| \le \lambda \|u - v\|$ for all $u, v \in \mathrm{B}(x, \delta_1)$, where $\mathrm{B}(x, \delta_1) \subseteq D$. Moreover, since $x \in G_n$, there are $\delta > 0$ and $r > 0$ such that for all $y \in E$ satisfying $\|y\| = 1$ we have $x \pm \delta y \in D$ and

$$\frac{f(x + \delta y) + f(x - \delta y) - 2f(x)}{\delta} \le r < \frac{1}{n}.$$

Now take $\delta_2 \in (0, \delta_1)$ so small that $\mathrm{B}(x, \delta_2) \subseteq D$ and $r + 4\lambda \delta_2/\delta < 1/n$. We are going to show that $\mathrm{B}(x, \delta_2) \subseteq G_n$. Thus let $z \in \mathrm{B}(x, \delta_2)$. Then for any $y \in E$ satisfying $\|y\| = 1$ it follows that

$$\frac{f(z+\delta y)+f(z-\delta y)-2f(z)}{\delta}$$

$$\leq \frac{f(x+\delta y)+f(x-\delta y)-2f(x)}{\delta}+\frac{2|f(z)-f(x)|}{\delta}$$

$$+\frac{|f(z+\delta y)-f(x+\delta y)|}{\delta}+\frac{|f(z-\delta y)-f(x-\delta y)|}{\delta}$$

$$\leq r+\frac{4\lambda\|z-x\|}{\delta}\leq r+\frac{4\lambda\delta_2}{\delta}<\frac{1}{n}$$

and so $z \in G_n$. $\qquad\qquad\qquad\qquad\qquad\qquad\qquad\qquad\qquad\qquad$ $\square$

Definition 4.3.13 A Banach space E is said to be an *Asplund space* if every continuous convex functional defined on a nonempty open convex subset D of E is F-differentiable on a dense subset of D.

Usually a Banach space E is said to be an Asplund space if every continuous convex functional is *generically* F-differentiable, i.e., F-differentiable on a dense G_δ subset of D. Proposition 4.3.12 shows that this is equivalent to the above definition.

Recall that the (infinite dimensional) normed vector space E is said to be *separable* if some countable subset is dense in E. Our aim is to verify that a Banach space with a separable dual is an Asplund space. For this, we need a geometric concept.

Definition 4.3.14

(a) Let $x^* \in E^*$, $x^* \neq o$, and $0 < \alpha < 1$. The set

$$K(x^*,\alpha) := \{x \in E \mid \alpha\|x\|\,\|x^*\| \leq \langle x^*,x\rangle\},$$

which is a closed convex cone, is called *Bishop–Phelps cone* associated with x^* and α.

(b) The set $A \subseteq E$ is said to be *α-cone meager*, where $0 < \alpha < 1$, if for every $x \in A$ and $\epsilon > 0$ there exist $z \in B(x,\epsilon)$ and $x^* \in E^*, x^* \neq o$, such that

$$A \cap \big(z + \operatorname{int} K(x^*,\alpha)\big) = \emptyset.$$

(c) The set $A \subseteq E$ is said to be *angle-small* if for every $\alpha \in (0,1)$, A can be expressed as the union of a countable number of α-cone meager sets.

Example 4.3.15 Consider $\mathbb{R}^n$ with inner product $(x^*\,|\,x)$ and identify $(R^n)^*$ with $\mathbb{R}^n$. Let $x^* \in \mathbb{R}^n$, $x^* \neq o$, and $\alpha \in (0,1)$ be given. For any $x \neq o$ we have

$$x \in K(x^*,\alpha) \quad\Longleftrightarrow\quad \alpha \leq \left(\frac{x^*}{\|x^*\|}\,\bigg|\,\frac{x}{\|x\|}\right),$$

i.e., the projection of the unit vector $x/\|x\|$ in the direction of x^* is at least α. The "ice cream cone" in $\mathbb{R}^3$ is thus a typical example of a Bishop–Phelps cone.

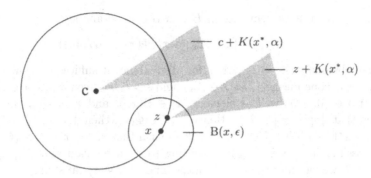

Fig. 4.2

Example 4.3.16 Let $A \subseteq \mathbb{R}^2$ consist of a circle and its center c. Then A is not α-cone meager for any $\alpha \in (0,1)$ but it is the union of two α-cone meager sets and so is angle-small (Fig. 4.2).

Recall that a set $A \subseteq E$ is said to be *nowhere dense* if $\mathrm{cl}\, A$ has empty interior or, equivalently, if $E \setminus \mathrm{cl}\, A$ is dense in E. Further, A is said to be of *first category* (or to be *meager*) if A is the union of a countable number of nowhere dense sets.

Lemma 4.3.17 *If $A \subseteq E$ is α-cone meager for some $\alpha \in (0,1)$, then A is nowhere dense. Hence any angle-small subset of E is of first category.*

Proof. See Exercise 4.8.6. □

The converse of Lemma 4.3.17 does not hold. In fact, an α-cone meager subset of $\mathbb{R}$ contains at most two elements. Hence a subset of $\mathbb{R}$ is angle-small if and only if it is countable. On the other hand, the Cantor set is an example of an uncountable set of first category.

Now we can establish a remarkable result on monotone multifunctions.

Theorem 4.3.18 *Let E be a Banach space with a separable dual. If $\Phi : E \rightrightarrows E^*$ is a monotone multifunction, then there exists an angle-small set $A \subseteq \mathrm{Dom}\, \Phi$ such that Φ is single-valued and upper semicontinuous on $(\mathrm{Dom}\, \Phi) \setminus A$.*

Proof. Set
$$A := \{x \in \mathrm{Dom}\, \Phi \mid \lim_{\delta \downarrow 0} \mathrm{diam}\, \Phi(\mathrm{B}(x,\delta)) > 0\}.$$

(I) It is left as an exercise to show that if $x \in (\mathrm{Dom}\, \Phi) \setminus A$, then $\Phi(x)$ is a singleton and Φ is upper semicontinuous at x.

(II) It remains to show that A is angle-small. We have $A = \cup_{n=1}^{\infty} A_n$, where
$$A_n := \{x \in \mathrm{Dom}\, \Phi \mid \lim_{\delta \downarrow 0} \mathrm{diam}\, \Phi(\mathrm{B}(x,\delta)) > 1/n\}.$$

Let (x_k^*) be a dense sequence in E^*, let $\alpha \in (0,1)$ and set

$$A_{n,k} := \{x \in A_n \mid \mathrm{d}(x_k^*, \Phi(x)) < \alpha/(4n)\}.$$

Then obviously $A_n = \cup_{k=1}^{\infty} A_{n,k}$ for all n. Hence it suffices to show that each $A_{n,k}$ is α-cone meager. Let $x \in A_{n,k}$ and $\epsilon > 0$ be given. Since $x \in A_n$, there exist $\delta \in (0,\epsilon)$ as well as elements $z_i \in \mathrm{B}(x,\delta)$ and $z_i^* \in \Phi(z_i)$ for $i = 1,2$ such that $\|z_1^* - z_2^*\| > 1/n$. Hence if $x^* \in \Phi(x)$, then $\|z_1^* - x^*\| > 1/(2n)$ or $\|z_2^* - x^*\| > 1/(2n)$. Choose $x^* \in \Phi(x)$ such that $\|x_k^* - x^*\| < \alpha/(4n)$ (which is possible since $x \in A_{n,k}$). According to what we said about z_i, z_i^*, where $i = 1,2$, we can find points $z \in \mathrm{B}(x,\epsilon)$ and $z^* \in \Phi(z)$ satisfying

$$\|z^* - x_k^*\| \geq \|z^* - x^*\| - \|x_k^* - x^*\| > 1/(2n) - \alpha/(4n) > 1/(4n).$$

We are going to show that $A_{n,k} \cap (z + \mathrm{int}\, K(z^* - x_k^*), \alpha) = \emptyset$, i.e.,

$$A_{n,k} \cap \{y \in E \mid \langle z^* - x_k^*, y - z \rangle > \alpha\|z^* - x_k^*\| \cdot \|y - z\|\} = \emptyset.$$

Suppose $y \in \mathrm{Dom}\,\Phi$ is such that $\langle z^* - x_k^*, y - z \rangle > \alpha\|z^* - x_k^*\| \cdot \|y - z\|$ and let $y^* \in \Phi(y)$. Then

$$\langle y^* - x_k^*, y - z \rangle = \langle y^* - z^*, y - z \rangle + \langle z^* - x_k^*, y - z \rangle$$
$$\geq \langle z^* - x_k^*, y - z \rangle > \alpha\|z^* - x_k^*\| \cdot \|y - z\| \geq \alpha\|y - z\|/(4n).$$

It follows that $\|y^* - x_k^*\| \geq \alpha/(4n)$ and so $y \notin A_{n,k}$.   □

With the aid of Theorem 4.3.18 we can now easily establish a sufficient condition for a Banach space to be an Asplund space.

Theorem 4.3.19 *If the dual of the Banach space E is separable, then E is an Asplund space.*

Proof. Let $f : E \to \overline{\mathbb{R}}$ be proper and convex, and continuous on the nonempty set $\mathrm{D}:=\mathrm{int}\,\mathrm{dom}\,f$. Then ∂f is monotone (Proposition 4.3.7). By Theorem 4.3.18, there exists an angle-small set A such that ∂f is single-valued and upper semicontinuous on $D \setminus A$ and so any selection of ∂f is continuous on $D \setminus A$. By Proposition 4.3.3, f is F-differentiable on $D \setminus A$. Since A is of first category (Lemma 4.3.17), the set $D \setminus A$ is dense in D (and a G_δ set).   □

Remark 4.3.20

(a) Notice that we actually showed somewhat more than stated, namely that any proper convex continuous functional f is F-differentiable outside an angle-small subset of $\mathrm{int}\,\mathrm{dom}\,f$.

(b) According to Theorem 4.3.19, the following spaces are Asplund spaces: the sequence spaces c_0 and l^p as well as the function spaces $L^p[a,b]$, where $1 < p < \infty$, furthermore any separable reflexive Banach space. It can be shown that *any* reflexive Banach space is an Asplund space (see Deville et al. [50] or Phelps [165]). Notice that c_0 is an example of a nonreflexive Asplund space while l^1 and l^∞ are Banach spaces that are not Asplund spaces.

Recall that for any normed vector space the closed unit ball of the dual space is weak* compact (Alaoglu theorem). It turns out that Asplund spaces have an important additional property.

Theorem 4.3.21 *If E is an Asplund space, then B_{E^*} is weak* sequentially compact.*

Concerning the proof we refer to Stegall [200] and Yost [219]. For a larger class of Banach spaces having the above property see Diestel [52].

4.4 Subdifferentials and Conjugate Functionals

Convention. The dual pair underlying the conjugation will be

$$\left(E[\|\cdot\|],\ E^*[\sigma(E^*,E)]\right)$$

unless we have to refer to the norm on E^* in which case we explicitly assume that E is a reflexive Banach space (cf. Remark 1.6.4).

Recall that the definition of subdifferential and conjugate functional of $f : E \to \overline{\mathbb{R}}$ is given by

$$\partial f(x) := \{x^* \in E^* \mid \langle x^*, y - x \rangle \le f(y) - f(x)\ \forall y \in E\}, \quad x \in \operatorname{dom} f,$$
$$f^*(x^*) := \sup_{x \in E}\left(\langle x^*, x \rangle - f(x)\right), \quad x^* \in E^*.$$

Proposition 4.4.1 *Let $f : E \to \overline{\mathbb{R}}$ be proper and convex, let $x \in \operatorname{dom} f$ and $x^* \in E^*$. Then there holds*

$$\langle x^*, x \rangle \le f(x) + f^*(x^*) \quad \text{(Young inequality)}, \tag{4.13}$$
$$\langle x^*, x \rangle = f(x) + f^*(x^*) \quad \Longleftrightarrow \quad x^* \in \partial f(x). \tag{4.14}$$

Proof. See Exercise 4.8.7. □

Remark 4.4.2 If f and x are as in Proposition 4.4.1, we have

$$f(x) = \min_{y \in E} f(y) \quad \Longleftrightarrow \quad o \in \partial f(x) \quad \underset{(4.14)}{\Longleftrightarrow} \quad f(x) = -f^*(o),$$

i.e., the global minimum of f is $-f^*(o)$.

If the functional f is G-differentiable on E, then we know that $\partial f(\cdot) = \{f'(\cdot)\}$. In this case, there is a close relationship between the Gâteaux derivative and the conjugate functional.

Proposition 4.4.3 *Let E be a reflexive Banach space, further let $f : E \to \mathbb{R}$ be G-differentiable with $f' : E \to E^*$ uniformly monotone. Then $(f')^{-1}$ exists on E^* and is continuous as well as strictly monotone. Moreover, the following holds:*

$$(f^*)' = (f')^{-1},\tag{4.15}$$

$$f(x) = f(o) + \int_0^1 \langle f'(\tau x), x \rangle \, d\tau \quad \forall\, x \in E,\tag{4.16}$$

$$f^*(x^*) = f^*(o) + \int_0^1 \langle x^*, (f')^{-1}(\tau x^*) \rangle \, d\tau \quad \forall\, x^* \in E^*,\tag{4.17}$$

$$f^*(o) = -f((f')^{-1}(o)).\tag{4.18}$$

Proof.

(I) We postpone the verification of the existence of $(f')^{-1} : E^* \to E$ to Sect. 5.4 (see Theorem 5.4.7 and Remark 5.4.8).

(II) Since f' is uniformly monotone, there exist constants $c > 0$ and $\gamma > 1$ such that

$$c\, \|y - x\|^\gamma \leq \langle f'(y) - f'(x), y - x \rangle \leq \|f'(y) - f'(x)\| \, \|y - x\| \quad \forall\, x, y \in E.$$

Setting $x^* = f'(x)$ and $y^* = f'(y)$, we have

$$c\, \|(f')^{-1}(y^*) - (f')^{-1}(x^*)\|^{\gamma - 1} \leq \|y^* - x^*\| \quad \forall\, x^*, y^* \in E^*,$$
$$\langle y^* - x^*, (f')^{-1}(y^*) - (f')^{-1}(x^*) \rangle = \langle f'(y) - f'(x), y - x \rangle > 0 \,\forall\, x \neq y.$$

These two relations show that $(f')^{-1}$ is continuous and strictly monotone.

(III) Now we verify (4.15)–(4.18). First notice that the integrals exist in the Riemann sense since f' is radially continuous by Proposition 4.3.8 and f'^{-1} is continuous by step (II).

Ad (4.15). Let $x, y \in E$, $x^* := f'(x)$, $y^* := f'(y)$. Since f is G-differentiable, we have $\partial f(x) = \{x^*\}$ and so Proposition 4.4.1 yields

$$f^*(x^*) + f(x) = \langle x^*, x \rangle.\tag{4.19}$$

This and an analogous formula with y and y^* leads to

$$f^*(y^*) - f^*(x^*) = f(x) - f(y) - \langle f'(y), x - y \rangle + \langle f'(y) - f'(x), x \rangle \tag{4.20}$$

$$\geq 0 + \langle f'(y) - f'(x), x \rangle = \langle y^* - x^*, (f')^{-1}(x^*) \rangle;\tag{4.21}$$

here, the inequality sign follows from Theorem 4.1.3(a). Interchanging x^* and y^*, we eventually obtain

$$\langle y^* - x^*, (f')^{-1}(x^*) \rangle \leq f^*(y^*) - f^*(x^*) \leq \langle y^* - x^*, (f')^{-1}(y^*) \rangle.$$

This yields for each $z^* \in E^*$,

$$\langle (f^*)'(x^*), z^* \rangle = \lim_{\tau \to 0} \frac{1}{\tau} \left(f^*(x^* + \tau z^*) - f^*(x^*) \right)$$

$$= \langle z^*, (f')^{-1}(x^*) \rangle_E = \langle (f')^{-1}(x^*), z^* \rangle_{E^*};$$

here, we exploited the continuity of the function $\tau \mapsto \langle (f')^{-1}(x^* + \tau z^*), z^* \rangle$ at $\tau = 0$.

Ad (4.18). By (4.19), we have

$$f^*(f'(x)) + f(x) = \langle f'(x), x \rangle \quad \forall x \in E$$

which, with $x := (f')^{-1}(o)$, implies (4.18).

Ad (4.16). This is the mean value formula established in Proposition 3.3.3.

Ad (4.17). The monotonicity of $(f')^{-1}$ and (4.15) entail that $(f^*)'$ is monotone. Therefore (4.17) holds in analogy to (4.16).   □

The formulas (4.17) and (4.18) will turn out to be crucial for calculating the conjugate in connection with boundary value problems.

Equation (4.15) means that for all $x \in E(= E^{**})$ and all $x^* \in E^*$ we have

$$x^* = f'(x) \quad \Longleftrightarrow \quad x = (f')^{-1}(x^*) \quad \Longleftrightarrow \quad x = (f^*)'(x^*). \tag{4.22}$$

The simpler one of these equivalences, namely $x^* = f'(x) \Longleftrightarrow x = (f^*)'(x^*)$, will now be generalized to nondifferentiable convex functionals.

Proposition 4.4.4 *Let $f : E \to \overline{\mathbb{R}}$ be proper and convex.*

(i) *There always holds*

$$x \in \mathrm{dom}\, f, \; x^* \in \partial f(x) \quad \Longrightarrow \quad x^* \in \mathrm{dom}\, f^*, \; x \in \partial f^*(x^*).$$

(ii) *If, in addition, E is reflexive and f is l.s.c., then*

$$x \in \mathrm{dom}\, f, \; x^* \in \partial f(x) \quad \Longleftrightarrow \quad x^* \in \mathrm{dom}\, f^*, \; x \in \partial f^*(x^*).$$

Proof.

(i) If $x \in \mathrm{dom}\, f$ and $x^* \in \partial f(x)$, then Proposition 4.4.1 gives $f^*(x^*) = \langle x^*, x \rangle - f(x)$. Hence $x^* \in \mathrm{dom}\, f^*$. For each $y^* \in E^*$, we obtain using (4.13),

$$\langle y^* - x^*, x \rangle \leq \left(f(x) + f^*(y^*) \right) - \left(f(x) + f^*(x^*) \right) = f^*(y^*) - f^*(x^*)$$

and so $x \in \partial f^*(x^*)$.

(ii) $\Longleftarrow$: By Proposition 2.2.3 and since $x^* \in \mathrm{dom}\, f^*$, we may apply (4.14) with f^* instead of f which gives $\langle x^*, x \rangle = f^*(x^*) + f^{**}(x)$ and so $x \in \mathrm{dom}\, f^{**}$. By Theorem 2.2.4 we have $f^{**} = f$. Hence applying (4.14) again, we obtain $x^* \in \partial f(x)$.   □

4.5 Further Calculus Rules

In this section, we establish computation rules for the subdifferential. The following sum rule will be crucial for deriving optimality conditions (cf. Remark 4.1.2).

Proposition 4.5.1 (Sum Rule) *Let* $f_0, f_1, \ldots, f_n : E \to \overline{\mathbb{R}}$ *be proper and convex. Assume there exists* $\bar{x} \in (\mathrm{dom}\, f_0) \cap (\mathrm{int}\, \mathrm{dom}\, f_1) \cap \cdots \cap (\mathrm{int}\, \mathrm{dom}\, f_n)$ *such that* f_i *is continuous at* $\bar{x}$ *for* $i = 1, \ldots, n$. *Then for each* $x \in (\mathrm{dom}\, f_0) \cap (\mathrm{dom}\, f_1) \cap \cdots \cap (\mathrm{dom}\, f_n)$, *there holds*

$$\partial(f_0 + f_1 + \cdots + f_n)(x) = \partial f_0(x) + \partial f_1(x) + \cdots + \partial f_n(x).$$

Proof. It is easy to see that (even without the continuity assumption) the inclusion $\supseteq$ holds. We now verify the inclusion $\subseteq$ for $n = 1$; for an arbitrary $n \in \mathbb{N}$ the assertion then follows by induction.

Let $x^* \in \partial(f_0 + f_1)(x)$ be given. Set $p := f_0$ and define $q : E \to \overline{\mathbb{R}}$ by

$$q(y) := \begin{cases} \langle x^*, x - y \rangle + f_1(y) - f_1(x) - f_0(x) & \text{if } y \in \mathrm{dom}\, f_1, \\ +\infty & \text{otherwise.} \end{cases}$$

Then all assumptions of the sandwich theorem (Theorem 1.5.2) are fulfilled and this theorem guarantees the existence of $y_0^* \in E^*$ and $c \in \mathbb{R}$ such that

$$-q(y) \le \langle y_0^*, y \rangle + c \le p(y) \quad \forall y \in E.$$

Since $-q(x) = p(x)$, we have $c = p(x) - \langle y_0^*, x \rangle$ and so $y_0^* \in \partial f_0(x)$. For $y_1^* := x^* - y_0^*$ we analogously obtain $y_1^* \in \partial f_1(x)$. Therefore $x^* = y_0^* + y_1^* \in \partial f_0(x) + \partial f_1(x)$. □

Finally we characterize the subdifferential of a functional of the form

$$f(x) := \max_{s \in S} f_s(x), \quad x \in E. \tag{4.23}$$

We denote the directional G-derivative of f_s at $\bar{x}$ by $f_{s,G}(\bar{x}, \cdot)$. We set

$$S(\bar{x}) := \{s \in S \mid f_s(\bar{x}) = f(\bar{x})\}.$$

Notice that if each f_s is convex, then so is f. Recall that $\overline{\mathrm{co}}^* M$ denotes the $\sigma(E^*, E)$-closed convex hull of $M \subseteq E^*$.

Proposition 4.5.2 (Maximum Rule) *Let* S *be a compact Hausdorff space. For any* $s \in S$, *let* $f_s : E \to \mathbb{R}$ *be convex on* E *and continuous at* $\bar{x} \in E$. *Assume further that there exists a neighborhood* U *of* $\bar{x}$ *such that for every* $z \in U$, *the functional* $s \mapsto f_s(z)$ *is upper semicontinuous on* S. *Then the functional* $f : E \to \mathbb{R}$ *defined by* (4.23) *satisfies*

$$f_G(\bar{x}, y) = \sup_{s \in S(\bar{x})} f_{s,G}(\bar{x}, y) \quad \forall y \in E, \tag{4.24}$$

$$\partial f(\bar{x}) = \overline{\mathrm{co}}^* \left(\bigcup_{s \in S(\bar{x})} \partial f_s(\bar{x}) \right). \tag{4.25}$$

Proof.

(I) We verify (4.24). Thus let $y \in E$ be given.

(Ia) If $s \in S(\bar{x})$, then

$$\frac{1}{\tau}\left(f_s(\bar{x} + \tau y) - f_s(\bar{x})\right) \leq \frac{1}{\tau}\left(f(\bar{x} + \tau y) - f(\bar{x})\right)$$

and so $f_{s,G}(\bar{x}, y) \leq f_G(\bar{x}, y)$ for each $s \in S(\bar{x})$.

(Ib) In view of step (Ia), (4.24) is verified as soon as we showed that for each $\epsilon > 0$ there exists $s \in S(\bar{x})$ such that

$$f_G(\bar{x}, y) - \epsilon \leq f_{s,G}(\bar{x}, y). \tag{4.26}$$

Since f is convex, we have $f_G(\bar{x}, y) = \inf_{\tau > 0} \frac{1}{\tau}\left(f(\bar{x} + \tau y) - f(\bar{x})\right)$ and so

$$f_G(\bar{x}, y) - \epsilon < \frac{1}{\tau}\left(f(\bar{x} + \tau y) - f(\bar{x})\right) \quad \forall \tau > 0.$$

Now let $\tau > 0$ be fixed. Then $\epsilon_\tau := \frac{1}{\tau}\left(f(\bar{x} + \tau y) - f(\bar{x})\right) - f_G(\bar{x}, y) + \epsilon$ is positive. By the definition of f, there exists $s \in S$ such that

$$\frac{1}{\tau}f(\bar{x} + \tau y) - \epsilon_\tau \leq \frac{1}{\tau}f_s(\bar{x} + \tau y).$$

Therefore

$$f_G(\bar{x}, y) - \epsilon \leq \frac{1}{\tau}\left(f_s(\bar{x} + \tau y) - f(x)\right). \tag{4.27}$$

In other words, the set S_τ of all $s \in S$ satisfying (4.27) is nonempty.

(Ic) Let $\tau_0 > 0$ be such that $\bar{x} + \tau y \in U$ for each $\tau \in (0, \tau_0)$. Since $s \mapsto f_s(\bar{x} + \tau y)$ is upper semicontinuous, the set S_τ is closed for each $\tau \in (0, \tau_0)$.

(Id) We show that $0 < \sigma < \tau$ implies $S_\sigma \subseteq S_\tau$. Thus let $s \in S_\sigma$ be given. Then

$$f_G(\bar{x}, y) - \epsilon \leq \frac{1}{\sigma}\left(f_s(\bar{x} + \sigma y) - f(\bar{x})\right).$$

Since $\bar{x} + \sigma y = \left(1 - \frac{\sigma}{\tau}\right) + \frac{\sigma}{\tau}(\bar{x} + \tau y)$ and f_s is convex, we obtain

$$f_G(\bar{x}, y) - \epsilon \leq \frac{1}{\sigma}\left[\left(1 - \frac{\sigma}{\tau}\right)f_s(\bar{x}) + \frac{\sigma}{\tau}f_s(\bar{x} + \tau y) - f(\bar{x})\right]$$
$$\leq \frac{1}{\tau}\left[f_s(\bar{x} + \tau y) - f(\bar{x})\right];$$

here, the second inequality follows from $f_s(\bar{x}) \leq f(\bar{x})$. Therefore, $s \in S_\tau$.

(Ie) In view of the above, the Cantor intersection theorem shows that the intersection of all S_τ, where $\tau \in (0, \tau_0)$, is nonempty. If s is any element of this intersection, then

$$\tau\left(f_G(\bar{x}, y) - \epsilon\right) \leq f_s(\bar{x} + \tau y) - f(\bar{x}) \quad \forall \tau \in (0, \tau_0). \tag{4.28}$$

Recalling that f_s is continuous at $\bar{x}$ and letting $\tau \downarrow 0$, we deduce $0 \leq f_s(\bar{x}) - f(\bar{x})$ and so $s \in S(\bar{x})$. Hence (4.28) holds with $f(\bar{x})$ replaced by $f_s(\bar{x})$ and from this we obtain (4.26) on dividing by τ and letting $\tau \downarrow 0$.

(II) We verify (4.25).

(IIa) It is easy to see that $\partial f_s(\bar{x}) \subseteq \partial f(\bar{x})$ for each $s \in S(\bar{x})$. Since $\partial f(\bar{x})$ is convex and $\sigma(E^*, E)$-closed, we conclude that

$$Q := \overline{co}^* \left(\bigcup_{s \in S(\bar{x})} \partial f_s(\bar{x}) \right) \subseteq \partial f(\bar{x}).$$

(IIb) Suppose there exists $x^* \in \partial f(\bar{x}) \setminus Q$. By the strong separation theorem (Theorem 1.5.9) applied to $E^*[\sigma(E^*, E)]$ there exists $z \in E$ such that

$$\langle x^*, z \rangle > \sup_{y^* \in Q} \langle y^*, z \rangle. \tag{4.29}$$

We further have

$$f_G(\bar{x}, z) = \max_{z^* \in \partial f(\bar{x})} \langle z^*, z \rangle \geq \langle x^*, z \rangle,$$

$$\sup_{y^* \in Q} \langle y^*, z \rangle \geq \sup_{s \in S(\bar{x})} \sup_{z^* \in \partial f_s(\bar{x})} \langle z^*, z \rangle = \sup_{s \in S(\bar{x})} f_{s,G}(\bar{x}, z).$$

In view of (4.29) we conclude that $f_G(\bar{x}, z) > \sup_{s \in S(\bar{x})} f_{s,G}(\bar{x}, z)$ which contradicts (4.24). Therefore $\partial f(\bar{x}) = Q$. $\square$

Remark 4.5.3 If, in particular, the set S is finite, then it is a compact Hausdorff space with respect to the discrete topology. For this topology, the function $s \mapsto f_s(z)$ is continuous on S for any f_s and every $z \in E$.

4.6 The Subdifferential of the Norm

With $z \in E$ fixed we consider the functional $\omega_z : E \to \mathbb{R}$ defined by

$$\omega_z(x) := \|x - z\| \quad \forall x \in E. \tag{4.30}$$

We simply write ω for ω_o, i.e., we set

$$\omega(x) := \|x\| \quad \forall x \in E.$$

The results to be derived will reveal interesting properties of the norm functional. In addition, they will later be applied to the *problem of best approximation*. Given $A \subseteq E$ and $z \in E \setminus A$, find $\bar{x} \in A$ satisfying

$$\omega_z(\bar{x}) = \inf_{x \in A} \omega_z(x).$$

For this purpose, we now deduce suitable representations of the subdifferential $\partial \omega_z(\bar{x})$ and the directional H-derivative $\omega_{z,H}(\bar{x}, \cdot)$ of ω_z. Notice that ω_z is continuous and convex. Define

$$S(x) := \{x^* \in E^* \mid \|x^*\| \leq 1, \langle x^*, x \rangle = \|x\|\}, \quad x \in E.$$

Remark 4.6.1 The Hahn–Banach theorem implies that $S(x) \neq \emptyset$ for each $x \in E$. Further, it is easy to see that

$$S(x) = \begin{cases} \{x^* \in E^* \mid \|x^*\| = 1, \langle x^*, x \rangle = \|x\|\} & \text{if } x \neq o, \\ B_{E^*} & \text{if } x = o. \end{cases}$$

Proposition 4.6.2 *The functional ω_z defined by (4.30) satisfies*

$$\partial \omega_z(\bar{x}) = S(\bar{x} - z) \quad \forall \bar{x} \in E, \tag{4.31}$$

$$\omega_{z,H}(\bar{x}, y) = \max\{\langle x^*, y \rangle \mid x^* \in S(\bar{x} - z) \cap \text{ep} \, S(o)\} \quad \forall \bar{x}, y \in E. \tag{4.32}$$

Proof. We only consider the case $\bar{x} \neq z$ which is important for the approximation problem; the verification in the case $\bar{x} = z$ is left as an exercise.
Ad (4.31):

(Ia) Let $x^* \in \partial \omega_z(\bar{x})$. Then we obtain

$$\langle x^*, x - \bar{x} \rangle \leq \|x - z\| - \|\bar{x} - z\| \, \forall x \in E, \text{ in particular } \langle x^*, z - \bar{x} \rangle \leq -\|\bar{x} - z\|.$$

Further we have

$$\langle x^*, \bar{x} - z \rangle = \langle x^*, 2\bar{x} - z \rangle - \langle x^*, \bar{x} \rangle \leq \|(2\bar{x} - z) - z\| - \|\bar{x} - z\| = \|\bar{x} - z\|.$$

It follows that $\langle x^*, \bar{x} - z \rangle = \|\bar{x} - z\|$.

(Ib) Now we show that $\|x^*\| = 1$. Since

$$\langle x^*, x \rangle = \langle x^*, x + \bar{x} \rangle - \langle x^*, \bar{x} \rangle \leq \|(x + \bar{x}) - z\| - \|\bar{x} - z\| \leq \|x\| \quad \forall x \in E,$$

we conclude that $\|x^*\| \leq 1$. Recalling that $\bar{x} \neq z$, we set $x_2 := \frac{\bar{x} - z}{\|\bar{x} - z\|}$. Then (Ia) implies $\langle x^*, x_2 \rangle = 1$. Further we have $\|x_2\| = 1$ and so $\|x^*\| = 1$.

(II) If $x^* \in S(\bar{x} - z)$, then we immediately obtain

$$\langle x^*, x - \bar{x} \rangle = \langle x^*, x - z \rangle - \langle x^*, \bar{x} - z \rangle \leq \|x - z\| - \|\bar{x} - z\|$$

and so $x^* \in \partial \omega_z(\bar{x})$.

Ad (4.32):
By Proposition 4.1.7 we have

$$\omega_{z,H}(\bar{x}, y) = \max\{\langle x^*, y \rangle \mid x^* \in \text{ep} \, S(\bar{x} - z)\} \quad \forall \bar{x}, y \in E.$$

We show that $S(\bar{x} - z)$ is an extremal subset of $S(o)$; then Lemma 1.7.6 implies that $\text{ep} \, S(\bar{x} - z) = S(\bar{x} - z) \cap \text{ep} \, S(o)$ and the assertion follows. Thus, let $x^*, y^* \in S(o)$, $\lambda \in (0, 1)$, and $\lambda x^* + (1 - \lambda) y^* \in S(\bar{x} - z)$. Then

$$\langle \lambda x^*, \bar{x} - z \rangle + \langle (1 - \lambda) y^*, \bar{x} - z \rangle = \|\bar{x} - z\|.$$

Since $\langle x^*, \bar{x} - z \rangle \leq \|x^*\| \cdot \|\bar{x} - z\| \leq \|\bar{x} - z\|$ and analogously for y^*, we deduce that $\langle x^*, \bar{x} - z \rangle = \|\bar{x} - z\|$ and analogously for y^*. Therefore we have $x^*, y^* \in S(\bar{x} - z)$. $\qquad \square$

Remark 4.6.3 We indicate the relationship to the *duality mapping* of E, which is the multifunction $J : E \rightrightarrows E^*$ defined by

$$J(x) := \partial j(x), \quad \text{where } j(x) := \frac{1}{2}\|x\|^2, \quad x \in E.$$

Similarly to the proof of (4.31), it can be shown that $J(x) = \|x\| S(x)$ for any $x \in E$ (see Exercise 4.8.8). This is also an immediate consequence of (4.31) and a chain rule to be established below (Corollary 7.4.6).

Application: The Maximum Norm

Now we want to apply Proposition 4.6.2 to

$$E := \mathrm{C}(T), \quad \text{with norm } \|x\|_\infty := \max_{t \in T}|x(t)|,$$

where T is a compact Hausdorff space.

Recall that $\mathrm{C}(T)$ denotes the vector space of all continuous functions $x : T \to \mathbb{R}$. Also recall that the dual space $\big(\mathrm{C}(T)\big)^*$ is norm isomorphic to, and so can be identified with, the vector space $\mathrm{M}(T)$ of all finite regular signed Borel measures μ on T, with norm $\|\mu\| := |\mu|(T) := \mu^+(T) + \mu^-(T)$ (see, for instance, Elstrodt [60]). Here, μ^+ and μ^- denote the positive and the negative variation of μ, respectively. The isomorphism between $x^* \in \big(\mathrm{C}(T)\big)^*$ and the associated $\mu \in \mathrm{M}(T)$ is given by

$$\langle x^*, x \rangle = \int_T x(t)\,\mathrm{d}\mu(t) \quad \forall x \in \mathrm{C}(T).$$

The signed measure $\mu \in \mathrm{M}(T)$ is said to be *concentrated* on the Borel set $B \subseteq T$ if $|\mu|(T \setminus B) = 0$. Now let $z \in \mathrm{C}(T)$ be fixed. For $\bar{x} \in \mathrm{C}(T)$, we set

$$T^+(\bar{x}) := \{t \in T \mid \bar{x}(t) - z(t) = \|\bar{x} - z\|_\infty\},$$
$$T^-(\bar{x}) := \{t \in T \mid \bar{x}(t) - z(t) = -\|\bar{x} - z\|_\infty\},$$
$$T(\bar{x}) := T^+(\bar{x}) \cup T^-(\bar{x}).$$

As announced, we consider the functional

$$\omega_z(x) := \|x - z\|_\infty, \quad x \in \mathrm{C}(T). \tag{4.33}$$

Proposition 4.6.4 *The functional ω_z defined by (4.33) satisfies*

$$\partial\omega_z(\bar{x}) = \big\{\mu \in \mathrm{M}(T) \mid \|\mu\| = 1, \ \mu^+ \text{ resp. } \mu^- \text{ is concentrated} \atop \text{on } T^+(\bar{x}) \text{ resp. on } T^-(\bar{x})\big\} \quad \forall \bar{x} \in \mathrm{C}(T), \tag{4.34}$$

$$\omega_{z,H}(\bar{x}, y) = \max_{t \in T(\bar{x})}\Big(\mathrm{sgn}\big(\bar{x}(t) - z(t)\big)\,y(t)\Big) \quad \forall \bar{x}, y \in \mathrm{C}(T). \tag{4.35}$$

Proof. Ad (4.34). In view of Proposition 4.6.2 it will do to show that $S(\bar{x} - z)$ equals the right-hand side of (4.34).

(I) Let μ be an element of the latter. Then it follows that

$$\int_T (\bar{x}(t) - z(t)) \, d\mu(t) = \int_T (\cdots) \, d\mu^+(t) - \int_T (\cdots) \, d\mu^-(t)$$

$$= \int_{T^+(\bar{x})} (\cdots) \, d\mu^+(t) - \int_{T^-(\bar{x})} (\cdots) \, d\mu^-(t)$$

$$= \|\bar{x} - z\|_\infty (\mu^+(T) + \mu^-(T)) = \|\bar{x} - z\|_\infty.$$

We also have $\|\mu\| = 1$. Therefore $\mu \in S(\bar{x} - z)$.

(II) Now let $\mu \in S(\bar{x} - z)$. Then $\|\mu\| = 1$ and

$$\|\bar{x} - z\|_\infty = \int_T (\bar{x}(t) - z(t)) \, d\mu^+(t) - \int_T (\bar{x}(t) - z(t)) \, d\mu^-(t)$$

$$\leq \|\bar{x} - z\|_\infty (\mu^+(T) + \mu^-(T)) = \|\bar{x} - z\|_\infty,$$

which implies

$$\int_T (\bar{x}(t) - z(t)) \, d\mu^+(t) = \|\bar{x} - z\|_\infty \mu^+(T), \tag{4.36}$$

$$\int_T (\bar{x}(t) - z(t)) \, d\mu^-(t) = -\|\bar{x} - z\|_\infty \mu^-(T).$$

Assume there exists a Borel set $B \subseteq T$ satisfying $B \cap T^+(\bar{x}) = \emptyset$ and $\mu^+(B) > 0$. Then it follows that

$$\int_T (\bar{x}(t) - z(t)) \, d\mu^+(t) = \int_{T \setminus T^+(\bar{x})} (\cdots) \, d\mu^+(t) + \int_{T^+(\bar{x})} (\cdots) \, d\mu^+(t)$$

$$< \|\bar{x} - z\|_\infty \left[\mu^+(T \setminus T^+(\bar{x})) + \mu^+(T^+(\bar{x})) \right]; \tag{4.37}$$

here the sign $<$ holds since $\mu^+(T \setminus T^+(\bar{x})) \geq \mu^+(B) > 0$. But the relations (4.36) and (4.37) are contradictory. Hence μ^+ is concentrated on $T^+(\bar{x})$. The argument for μ^- is analogous.

Ad (4.35). For $t \in T$ let ϵ_t denote the Dirac measure on T, i.e., for each Borel set $B \subseteq T$ we have

$$\epsilon_t(B) := \begin{cases} 1 & \text{if } t \in B, \\ 0 & \text{if } t \in T \setminus B. \end{cases}$$

It is well known (see, for instance, Köthe [115]) that

$$\text{ep } S(o) = \{\epsilon_t \mid t \in T\} \cup \{-\epsilon_t \mid t \in T\}.$$

This together with (4.31) and (4.34) gives

$$S(\bar{x} - z) \cap \mathrm{ep}\, S(o) = \{\epsilon_t \,|\, t \in T^+(\bar{x})\} \cup \{-\epsilon_t \,|\, t \in T^-(\bar{x})\}.$$

Applying (4.32), we finally obtain

$$\omega_H(\bar{x}, y) = \max_{t \in T(\bar{x})} \left\{ \int_{T^+(\bar{x})} y(s)\, d\epsilon_t(s),\ -\int_{T^-(\bar{x})} y(s)\, d\epsilon_t(s) \right\}$$

$$= \max\Big(\{y(t) \,|\, t \in T^+(\bar{x})\} \cup \{-y(t) \,|\, t \in T^-(\bar{x})\} \Big),$$

and the latter is equal to the right-hand side of (4.35). $\square$

We now consider the special case $E := C[a, b]$, where $a < b$. A function $\bar{x} \in C[a, b]$ is called *peaking function* if there exists $t^* \in [a, b]$ such that $|\bar{x}(t^*)| > |\bar{x}(t)|$ for each $t \neq t^*$. In this case, t^* is called *peak point* of $\bar{x}$.

Proposition 4.6.5 *Let ω be the maximum norm on $C[a, b]$, where $a < b$, and let $\bar{x} \in C[a, b]$. Then:*

(a) *ω is H-differentiable at $\bar{x}$ if and only if $\bar{x}$ is a peaking function. If t^* is the peak point of $\bar{x}$, then one has*

$$\langle \omega'(\bar{x}), y \rangle = \mathrm{sgn}\big(\bar{x}(t^*)\big) y(t^*) \quad \forall y \in C[a, b]. \tag{4.38}$$

(b) *ω is nowhere F-differentiable on $C[a, b]$.*

Proof. (a) We shall utilize the derivative of the function $\xi \mapsto |\xi|$ at $\xi \in \mathbb{R} \setminus \{0\}$:

$$\lim_{h \to 0} \frac{|\xi + h| - |\xi|}{h} = \mathrm{sgn}(\xi) \quad \forall \xi \neq 0.$$

(I) Let ω be H-differentiable at $\bar{x}$. Take $t^* \in [a, b]$ with $\omega(\bar{x}) = |\bar{x}(t^*)|$. Now let $y \in C[a, b]$ and $\tau \neq 0$. Then

$$\omega(\bar{x} + \tau y) - \omega(\bar{x}) \geq |\bar{x}(t^*) + \tau y(t^*)| - |\bar{x}(t^*)|.$$

Dividing by τ, this implies, respectively,

$$\langle \omega'(\bar{x}), y \rangle \geq \mathrm{sgn}(\bar{x}(t^*))\, y(t^*) \quad (\text{letting } \tau \downarrow 0),$$
$$\langle \omega'(\bar{x}), y \rangle \leq \mathrm{sgn}(\bar{x}(t^*))\, y(t^*) \quad (\text{letting } \tau \uparrow 0).$$

Hence (4.38) holds. It remains to show that t^* is unique. Assume, to the contrary, that with some $t_* \neq t^*$ we also had $\omega(\bar{x}) = |\bar{x}(t_*)|$. According to what has already been shown, it follows that

$$\langle \omega'(\bar{x}), y \rangle = \mathrm{sgn}\big(\bar{x}(t_*)\big)\, y(t_*) \quad \forall y \in C[a, b]. \tag{4.39}$$

Choose $y \in C[a,b]$ such that $y(t_*) = 0$ and $y(t^*) = \bar{x}(t^*)$. Then (4.38) and (4.39) are contradictory. Hence $\bar{x}$ is a peaking function.

(II) Now let $\bar{x}$ be a peaking function with peak point t^*. Then $T(\bar{x}) = \{t^*\}$ and so (4.35) passes into

$$\omega_H(\bar{x}, y) = \operatorname{sgn}\big(\bar{x}(t^*)\big)\, y(t^*).$$

Hence the functional $\omega_H(\bar{x}, \cdot)$ is linear and continuous and so is a H-derivative, i.e., (4.38) holds.

(b) Assume, to the contrary, that ω is F-differentiable at some $\bar{x} \in C[a,b]$. Then ω is also H-differentiable at $\bar{x}$ (Proposition 3.4.2). According to (a), $\bar{x}$ is a peaking function with a peak point t^*, and (4.38) holds true. Let (t_n) be a sequence in $[a,b]$ such that $t_n \neq t^*$ for each n and $t_n \to t^*$ as $n \to \infty$. Since t^* is a peak point of $\bar{x}$ and so $\bar{x}(t^*) \neq 0$, we may assume that $\bar{x}(t_n) \neq 0$ for each n. Let $\varphi_n : [a,b] \to [0,1]$ be a continuous function satisfying $\varphi_n(t_n) = 1$ and $\varphi_n(t) = 0$ for each t in a neighborhood of t^* (depending on n). Further let

$$y_n(t) := 2\operatorname{sgn}\big(\bar{x}(t_n)\big)\, |\bar{x}(t_n) - \bar{x}(t^*)|\, \varphi_n(t) \quad \forall t \in [a,b].$$

Then

$$\|y_n\|_\infty = |y_n(t_n)| = 2|\bar{x}(t_n) - \bar{x}(t^*)|. \tag{4.40}$$

It follows that

$$\|\bar{x} + y_n\|_\infty \geq |\bar{x}(t_n) + y_n(t_n)| = \|\bar{x}\|_\infty + |\bar{x}(t_n) - \bar{x}(t^*)| = \|\bar{x}\|_\infty + \frac{1}{2}\|y_n\|_\infty$$

and so

$$\frac{\omega(\bar{x} + y_n) - \omega(\bar{x})}{\|y_n\|_\infty} \geq \frac{1}{2} \quad \forall n. \tag{4.41}$$

On the other hand, from (4.38) and $y_n(t^*) = 0$ we obtain $\langle \omega'(\bar{x}), y_n \rangle = 0$. Since $\omega'(\bar{x})$ is assumed to be an F-derivative and $\|y_n\|_\infty \to 0$ as $n \to \infty$ (see (4.40)), we must have

$$\lim_{n \to \infty} \frac{\omega(\bar{x} + y_n) - \omega(\bar{x}) - 0}{\|y_n\|_\infty} = 0.$$

But this contradicts (4.41).   □

4.7 Differentiable Norms

Let again E be a normed vector space and $z \in E$. Notice that, except for the trivial case $E = \{o\}$, the functional $\omega_z : x \mapsto \|x - z\|$, $x \in E$, cannot be G-differentiable at z since $\tau \mapsto |\tau|$, $\tau \in \mathbb{R}$, is not differentiable at $\tau = 0$. For points different from z, Proposition 4.1.8 and (4.31) immediately yield

Proposition 4.7.1 *For each $\bar{x} \neq z$ the following statements are equivalent:*

(a) ω_z *is G-differentiable at $\bar{x}$.*
(b) ω_z *is H-differentiable at $\bar{x}$.*
(c) $S(\bar{x} - z)$ *consists of exactly one element.*

If one, and so each, of these statements holds true, then $S(\bar{x} - z) = \{\omega_z'(\bar{x})\}$, hence

$$\|\omega_z'(\bar{x})\| = 1, \quad \langle \omega_z'(\bar{x}), \bar{x} - z \rangle = \|\bar{x} - z\|. \tag{4.42}$$

Geometrical Interpretation

Let $\bar{x} \in E$, $\bar{x} \neq o$. By Corollary 1.5.5, the point $\bar{x}$ is a support point of the ball $B(o, \|\bar{x}\|)$, i.e., it admits a supporting hyperplane.

Lemma 4.7.2 *Let $\bar{x} \in E$, $\bar{x} \neq o$.*

(i) *If H is a supporting hyperplane of $B(o, \|\bar{x}\|)$ at $\bar{x}$, then there exists $x^* \in E^*$ such that $H = [x^* = \|\bar{x}\|]$.*
(ii) *If $x^* \in E^*$, $x^* \neq o$, then the following statements are equivalent:*
   (a) $[x^* = \|\bar{x}\|]$ *is a supporting hyperplane of $B(o, \|\bar{x}\|)$ at $\bar{x}$.*
   (b) $[x^* = 1]$ *is a supporting hyperplane of $B(o, 1)$ at $\frac{\bar{x}}{\|\bar{x}\|}$.*
   (c) $x^* \in S(\bar{x})$.

Proof. (i) Let $H = [y^* = \beta]$, where $y^* \in E^*$, $y^* \neq o$, and $\beta \in \mathbb{R}$. We may assume that $\langle y^*, y \rangle \leq \beta$ for each $y \in B(o, \|\bar{x}\|)$ (if $\langle y^*, y \rangle \geq \beta$, we replace y^* and β with $-y^*$ and $-\beta$, respectively). If we had $\beta \leq 0$, then $\langle y^*, y \rangle \leq 0$ for each $y \in E$ and so $y^* = o$, which is not the case. Therefore $\beta > 0$. Set $x^* := \frac{\|\bar{x}\|}{\beta} y^*$. Then we have

$$y \in H \quad \Longleftrightarrow \quad \langle y^*, y \rangle = \beta \quad \Longleftrightarrow \quad y \in [x^* = \|\bar{x}\|].$$

(ii) We only verify (a) $\Longrightarrow$ (c), the remaining implications are immediately clear. So assume that (a) holds. Then $\langle x^*, \bar{x} \rangle = \|\bar{x}\|$ and $\langle x^*, y \rangle \leq \|\bar{x}\|$ for each $y \in B(o, \|\bar{x}\|)$. (Choose $y := o$ to see that we cannot have $\langle x^*, y \rangle \geq \|\bar{x}\|$ for each $y \in B(o, \|\bar{x}\|)$.) It follows that

$$\|\bar{x}\| = \sup\{\langle x^*, y \rangle \mid y \in B(o, \|\bar{x}\|)\} = \|x^*\| \, \|\bar{x}\|.$$

Here the second equation holds according to the definition of $\|x^*\|$. We thus obtain $\|x^*\| = 1$. Hence $x^* \in S(\bar{x})$. $\square$

Roughly speaking, the lemma says that $S(\bar{x})$ contains "as many" elements as there are supporting hyperplanes of $B(o, 1)$ at $\frac{\bar{x}}{\|\bar{x}\|}$. This gives rise to Definition 4.7.3.

Definition 4.7.3 The normed vector space E is said to be *smooth* if $B(o, 1)$ possesses exactly one supporting hyperplane at each boundary point (in other words, if $S(x)$ consists of exactly one element for each $x \neq o$).

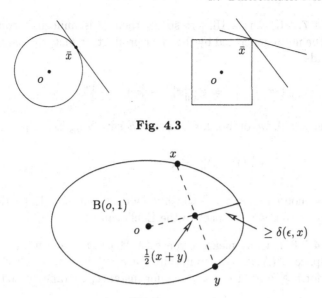

Fig. 4.3

Fig. 4.4

Proposition 4.7.1 (with $z = o$) and Lemma 4.7.2 immediately yield:

Proposition 4.7.4 *The following assertions are equivalent:*

(a) $\| \cdot \|$ *is G-differentiable at each nonzero point.*
(b) $\| \cdot \|$ *is H-differentiable at each nonzero point.*
(c) E *is smooth.*

Example 4.7.5 Figure 4.3 shows $B(o, 1)$ in $\mathbb{R}^2$ for the Euclidean norm and for the maximum norm. The Euclidean norm is G-differentiable at each nonzero point of $\mathbb{R}^2$ while the maximum norm is not G-differentiable at the corner points of $B(o, 1)$. The same holds true in $\mathbb{R}^n$ for $n > 2$.

We now refine the investigation.

Definition 4.7.6 The normed vector space E is said to be *locally uniformly convex* if the following holds (Fig. 4.4):

$$\forall \epsilon \in (0, 2] \quad \forall x \in E, \|x\| = 1 \quad \exists \delta(\epsilon, x) > 0 \quad \forall y \in E :$$

$$\|y\| = 1, \|x - y\| \geq \epsilon \quad \Longrightarrow \quad \|\tfrac{1}{2}(x + y)\| \leq 1 - \delta(\epsilon, x).$$

If $\delta(\epsilon, x)$ can be chosen to be independent of x, then E is said to be *uniformly convex*.

It is clear that with respect to the maximum norm, $\mathbb{R}^n$ for $n \geq 2$ is not locally uniformly convex. However, we have the following positive results.

Example 4.7.7 If E is a Hilbert space, then E is uniformly convex with respect to the norm generated by the inner product. In fact, the parallelogram identity reads

$$\left\|\tfrac{1}{2}(x+y)\right\|^2 = \tfrac{1}{2}\left(\|x\|^2 + \|y\|^2\right) - \tfrac{1}{4}\|x - y\|^2.$$

Hence, given $\epsilon > 0$, we obtain for all $x, y \in E$ satisfying $\|x - y\| \geq \epsilon$,

$$\left\|\tfrac{1}{2}(x+y)\right\|^2 \leq 1 - \tfrac{\epsilon^2}{4} \leq \left(1 - \tfrac{\epsilon^2}{8}\right)^2,$$

i.e., we may choose $\delta(\epsilon, x) := \tfrac{\epsilon^2}{8}$ which is independent of x. In particular, $\mathbb{R}^n$ is uniformly convex with respect to the Euclidean norm.

Example 4.7.8 For any measure space $(X, \mathfrak{A}, \mu)$ and each $p \in (1, +\infty)$, the Lebesgue space $L^p(X, \mathfrak{A}, \mu)$ is uniformly convex (Theorem of Clarkson). We verify this for $p \geq 2$; for $1 < p < 2$ see, for instance, Cioranescu [32].

(I) We first show that for arbitrary $a, b \in \mathbb{R}$ and $p \geq 2$ we have

$$|a + b|^p + |a - b|^p \leq 2^{p-1}(|a|^p + |b|^p). \tag{4.43}$$

Let $\alpha > 0, \beta > 0$, and set $c := \sqrt{\alpha^2 + \beta^2}$. Then $0 < \tfrac{\alpha}{c} < 1$ and $0 < \tfrac{\beta}{c} < 1$, hence

$$\left(\frac{\alpha}{c}\right)^p + \left(\frac{\beta}{c}\right)^p \leq \left(\frac{\alpha}{c}\right)^2 + \left(\frac{\beta}{c}\right)^2 = 1,$$

and so $\alpha^p + \beta^p \leq c^p = (\alpha^2 + \beta^2)^{p/2}$. We deduce that

$$|a + b|^p + |a - b|^p \leq \left(|a + b|^2 + |a - b|^2\right)^{p/2} = 2^{p/2}\left(|a|^2 + |b|^2\right)^{p/2}. \tag{4.44}$$

Since $\tfrac{2}{p} + \tfrac{p-2}{p} = 1$, the Hölder inequality (applied to $\tfrac{p}{2}$ and $\tfrac{p}{p-2}$) yields

$$|a^2 \cdot 1| + |b^2 \cdot 1| \leq \left((a^2)^{p/2} + (b^2)^{p/2}\right)^{2/p} \cdot (1+1)^{\frac{p-2}{p}} = \left(|a|^p + |b|^p\right)^{2/p} \cdot 2^{\frac{p-2}{p}}.$$

This together with (4.44) gives (4.43).

(II) Now we show that $L^p(X, \mathfrak{A}, \mu)$ is uniformly convex. For all $f, g \in L^p(X, \mathfrak{A}, \mu)$ the inequality (4.43) implies

$$\left\|\tfrac{1}{2}(f + g)\right\|_p^p \leq \tfrac{1}{2}\left(\|f\|_p^p + \|g\|_p^p\right) - \tfrac{1}{2^p}\|f - g\|_p^p.$$

Now we can argue as in Example 4.7.7.

Locally uniformly convex spaces have a nice convergence property. To describe it, we introduce the following concept. The norm $\|\cdot\|$ of a Banach space E is said to be a *Kadec norm* if $x_n \xrightarrow{w} x$ and $\|x_n\| \to \|x\|$ as $n \to \infty$ implies $x_n \to x$ as $n \to \infty$.

Lemma 4.7.9 *The norm of a locally uniformly convex Banach space is a Kadec norm.*

Proof. Let $x_n \xrightarrow{w} x$ and $\|x_n\| \to \|x\|$. The conclusion is obvious if $x = o$. So let $x \neq o$ and set $y := \frac{x}{\|x\|}$, $y_n := \frac{x_n}{\|x_n\|}$ which makes sense for all n sufficiently large. It follows that $y_n + y \xrightarrow{w} 2y$. Hence $(y_n + y)$ considered as a sequence in E^{**} is bounded (Banach–Steinhaus theorem). More precisely, we have

$$2 = \|2y\| \leq \liminf_{n\to\infty} \|y_n + y\| \leq \limsup_{n\to\infty} \|y_n + y\| \leq \lim_{n\to\infty} \|y_n\| + \|y\| = 2$$

and so $\lim_{n\to\infty} \|y_n + y\| = 2$. Since E is locally uniformly convex, we conclude that $\lim_{n\to\infty} \|y_n - y\| = 0$ and so (since $\|x_n\| \to \|x\|$) we finally obtain $\|x_n - x\| \to 0$. $\qquad\square$

Proposition 4.7.10 *Let E be reflexive and E^* locally uniformly convex. Then the norm functional ω is continuously differentiable on $E \setminus \{o\}$.*

Proof. (I) Let $x \in E$, $x \neq o$. We show that $S(x)$ contains exactly one element. Assume that $x_1^*, x_2^* \in S(x)$, i.e., $\|x_i^*\| = 1$ and $\langle x_i^*, x \rangle = \|x\|$ for $i = 1, 2$. We then have

$$2 = \|x_1^*\|^2 + \|x_2^*\|^2 = \left\langle x_1^* + x_2^*, \frac{x}{\|x\|} \right\rangle \leq \|x_1^* + x_2^*\| \cdot 1$$

and so $\|\frac{1}{2}(x_1^* + x_2^*)\| \geq 1$. Since E^* is locally uniformly convex, we conclude that $x_1^* = x_2^*$ (otherwise we could choose $\delta(\epsilon, x) > 0$ for $\epsilon := \|x_1^* - x_2^*\|$).

(II) By step (I) and Proposition 4.7.1 we know that ω is G-differentiable at each nonzero point. Now we show that $x_n \to x$ implies $\omega'(x_n) \xrightarrow{w} \omega'(x)$ as $n \to \infty$. Thus let $x_n \to x$. Since $S(x_n) = \{\omega'(x_n)\}$, we have $\|\omega'(x_n)\| = 1$ for each n. Since E is reflexive, some subsequence $(\omega'(x_{n_j}))$ of $(\omega'(x_n))$ is weakly convergent to some $y^* \in E^*$ as $j \to \infty$. For each $y \in E$ we thus obtain

$$\langle y^*, y \rangle = \lim_{j\to\infty} \langle \omega'(x_{n_j}), y \rangle \leq \lim_{j\to\infty} \|\omega'(x_{n_j})\| \cdot \|y\| = \|y\|$$

and so $\|y^*\| \leq 1$. Moreover, we have

$$\langle y^*, x \rangle = \lim_{j\to\infty} \langle \omega'(x_{n_j}), x_{n_j} \rangle \underset{(4.42)}{=} \lim_{j\to\infty} \|x_{n_j}\| = \|x\|.$$

Therefore $y^* \in S(x)$ and step (I) tells us that $y^* = \omega'(x)$. Since each weakly convergent subsequence of $(\omega'(x_n))$ has the same limit $\omega'(x)$, we conclude that

$$\omega'(x_n) \xrightarrow{w} \omega'(x). \tag{4.45}$$

(III) Now let $x_n \to x$. We then have

$$\|\omega'(x_n)\| = \|x_n\| \to \|x\| = \|\omega'(x)\|.$$

This together with (4.45) gives $\omega'(x_n) \to \omega'(x)$ by Lemma 4.7.9.   □

Example 4.7.11 The space $L^p := L^p(X, \mathfrak{A}, \mu)$, where $1 < p < +\infty$, is reflexive and the dual space can be identified with L^q, where $\frac{1}{p} + \frac{1}{q} = 1$. Moreover, by Example 4.7.8, L^q is uniformly convex. Proposition 4.7.10 therefore implies that the norm $\omega := \| \cdot \|_p$ is continuously differentiable away from zero. Explicitly we have

$$\omega'(x) = \frac{\mathrm{sgn}(x)}{\|x\|_p^{p-1}}|x|^{p-1} \quad \forall\, x \in L^p \setminus \{o\}.$$

This is easily verified by showing that the right-hand side is an element of $S(x)$.

Example 4.7.5 showed that the differentiability properties of the norm may change by passing to an equivalent norm. In this connection, the following deep result is of great importance; concerning its proof we refer to Cioranescu [32], Deville et al. [50], or Diestel [51].

Theorem 4.7.12 (Renorming Theorem) *If E is a reflexive Banach space, then there exists an equivalent norm on E such that E and E^* are both locally uniformly convex.*

Definition 4.7.13 The Banach space E is said to be *Fréchet smooth* if it admits an equivalent norm that is F-differentiable on $E \setminus \{o\}$.

Recall that by Corollary 4.3.4 the norm functional is F-differentiable if and only if it is continuously differentiable. Hence Theorem 4.7.12 together with Proposition 4.7.10 leads to:

Proposition 4.7.14 *If E is a reflexive Banach space, then there exists an equivalent norm on E that is continuously differentiable on $E \setminus \{o\}$, and the same holds true for the associated norm on E^*. In particular, every reflexive Banach space is Fréchet smooth.*

For later use we present the following result; for a proof we refer to Deville et al. [50] and Phelps [165].

Proposition 4.7.15 *Every Fréchet smooth Banach space is an Asplund space.*

4.8 Bibliographical Notes and Exercises

The theory of the subdifferential and the conjugate of convex functionals as well as its various applications originated in the work of Moreau and Rockafellar in the early 1960s (see Moreau [148] and Rockafellar [177]). The now classic text on the subject in finite-dimensional spaces is Rockafellar [180]. Concerning finite-dimensional spaces, see also Bazaraa et al. [11], Borwein and Lewis [18], Elster et al. [59], Hiriart-Urruty and Lemaréchal [88,89], and Rockafellar and Wets [189] (comprehensive monograph).

In the infinite-dimensional case, we recommend Aubin [6] (application to mathematical economics), Barbu and Precupanu [9] (application to control problems), Ekeland and Temam [58] (application to variational problems involving partial differential equations), Ioffe and Tikhomirov [101] (application to the calculus of variations and control problems), Pallaschke and Rolewicz [156] (abstract approach with many concrete applications), Pshenichnyi [170] (application to control problems), Schirotzek [196] (application to the calculus of variations), and Zeidler [221] (comprehensive monograph with many applications). For applications of the conjugation to density functionals in quantum physics see Eschrig [62].

The results on Asplund spaces in Sect. 4.3 are mainly taken from Phelps [165]. Theorem 4.3.18 is essentially due to Preiss and Zajíček [167], our presentation follows Phelps [165]. Theorem 4.3.19 was established by Asplund [3]. The famous renorming result of Theorem 4.7.12 is due to Kadec [108] and Troyanski [208]. Concerning differentiability properties of convex functionals, especially of the norm functional, see also Beauzamy [13], Cioranescu [32], Deville et al. [50], Diestel [51], Sundaresan [205], and Yamamuro [216]. For applications to approximation theory see Braess [27], Krabs [112], and Laurent [118].

Exercise 4.8.1 Let $\Phi : E \rightrightarrows F$ be a multifunction between Banach spaces E and F. Verify the following:

(a) Φ is upper semicontinuous if and only if for any open set $V \subseteq F$ the set $\{x \in E \mid \Phi(x) \subseteq V\}$ is open.

(b) Φ is lower semicontinuous if and only if for any open set $V \subseteq F$ the set $\{x \in E \mid \Phi(x) \cap V \neq \emptyset\}$ is open.

Exercise 4.8.2 Let $f : E \to \overline{\mathbb{R}}$ be proper and convex, and continuous on the nonempty set int dom$f$. Show that f is G-differentiable at $\bar{x} \in$ int dom$f$ if and only if there exists a selection $\varphi : E \to E^*$ of ∂f which is norm-to-weak* continuous at $\bar{x}$ (cf. Proposition 4.3.3).

Exercise 4.8.3 Prove Proposition 4.3.5.

Exercise 4.8.4 Verify Proposition 4.3.7.

Exercise 4.8.5 Verify the necessity part of Lemma 4.3.11.

Exercise 4.8.6 Prove Lemma 4.3.17.

Exercise 4.8.7 Prove Proposition 4.4.1.

Exercise 4.8.8 Show that the duality mapping $J : E \rightrightarrows E^*$ satisfies $J(x) = \|x\| \, S(x)$ for any $x \in E$ (cf. Remark 4.6.3).

Exercise 4.8.9 Let E be a Banach space and assume that the duality mapping $J : E \rightrightarrows E^*$ is *linear*, i.e., $x^* \in J(x)$, $y^* \in J(y)$, and $\lambda \in \mathbb{R}$ imply $x^* + y^* \in J(x+y)$ and $\lambda x^* \in J(\lambda x)$. Show the following:

(a) E is a Hilbert space.
(b) J is single-valued and satisfies $\langle J(x), y \rangle = (y \mid x)$ for all $x, y \in E$, i.e., J is the norm isomorphism of E onto E^* according to the Riesz representation theorem.

Exercise 4.8.10 Let L be a linear subspace of the normed vector space E, let $f : E \to \overline{\mathbb{R}}$ be proper and convex, and let $\bar{x} \in L \cap \mathrm{dom}\, f$. Denote by $\partial_L f(\bar{x})$ the subdifferential of $f|L$ (the restriction of f to L) at $\bar{x}$. Verify the following assertions (cf. Singer [199]):

(a) For any $x^* \in \mathrm{ep}(\partial_L f(\bar{x}))$ there exists $y^* \in \mathrm{ep}(\partial f(\bar{x}))$ such that $y^*|L = x^*$.
(b) If $\dim L = n$, then for any $x^* \in \mathrm{ep}(\partial_L f(\bar{x}))$ there exist $y_1^*, \ldots, y_{n+1}^* \in \mathrm{ep}(\partial f(\bar{x}))$ and $\lambda_1, \ldots, \lambda_{n+1} \geq 0$ such that

$$\sum_{i=1}^{n+1} \lambda_i = 1 \quad \text{and} \quad \sum_{i=1}^{n+1} \lambda_i \, y_i^*(x) = x^*(x) \quad \forall x \in L.$$

Hint: By a Theorem of Carathéodory, any point of a compact convex subset C of $\mathbb{R}^n$ is the convex combination of at most $n+1$ extreme points of C.

5

Optimality Conditions for Convex Problems

5.1 Basic Optimality Conditions

We consider the following basic *convex optimization problem*:

(P1) Minimize $f(x)$ subject to $x \in M$,

which we also formally write $f(x) \longrightarrow \min$, $x \in M$. We assume that

$$f : E \to \overline{\mathbb{R}} \text{ is a proper convex functional,}$$
$$M \text{ is a nonempty convex subset of dom } f.$$

Theorem 4.1.3 shows that under this assumption, $f_G(\bar{x}, \cdot)$ exists as an $\overline{\mathbb{R}}$-valued functional.

The point $\bar{x} \in M$ is called a *global minimizer* of f on M, or briefly a *global solution* of (P1), if $f(\bar{x}) \leq f(x)$ for any $x \in M$. Moreover, $\bar{x} \in M$ is called a *local minimizer* of f on M, or briefly a *local solution* of (P1), if there exists a neighborhood U of $\bar{x}$ in E such that $f(\bar{x}) \leq f(x)$ for any $x \in M \cap U$.

Proposition 5.1.1 *The following statements are equivalent:*

(a) $\bar{x}$ *is a global solution of* (P1).
(b) $\bar{x}$ *is a local solution of* (P1).
(c) $f_G(\bar{x}, x - \bar{x}) \geq 0$ *for all* $x \in M$.

Proof. (a) $\Longrightarrow$ (b) is clear.
(b) $\Longrightarrow$ (c): If (b) holds, then for some neighborhood U of $\bar{x}$, we have $f(y) - f(\bar{x}) \geq 0$ for each $y \in M \cap U$. Now let $x \in M$. Since M is convex, we see that

$$y := \bar{x} + \tau(x - \bar{x}) = \tau x + (1 - \tau)\bar{x} \in M \quad \forall \tau \in (0, 1).$$

Further, if $\tau \in (0, 1)$ is sufficiently small, we also have $y \in U$ and so

$$\frac{1}{\tau}\Big(f(\bar{x} + \tau(x - \bar{x})) - f(\bar{x})\Big) \geq 0.$$

Letting $\tau \downarrow 0$, the assertion follows.

(c) $\implies$ (a): Theorem 4.1.3 shows that $f_G(\bar{x}, x - \bar{x}) \leq f(x) - f(\bar{x})$ for any $x \in M$. Hence (c) implies (a). $\qquad\square$

Condition (c) is called *variational inequality* (as x varies over M). Sometimes a variational inequality passes into a *variational equation*.

Corollary 5.1.2 *Assume that $\bar{x} \in$ int M and $f : M \to \mathbb{R}$ is G-differentiable at $\bar{x}$. Then the following conditions are equivalent:*

(a) $f(\bar{x}) = \min_{x \in M} f(x)$ (minimum problem).
(b) $\langle f'(\bar{x}), y \rangle = 0$ $\forall y \in E$ (variational equation).
(c) $f'(\bar{x}) = o$ (operator equation).

Proof. By Proposition 5.1.1, (a) is equivalent to

$$\langle f'(\bar{x}), x - \bar{x} \rangle \geq 0 \quad \forall x \in M. \tag{5.1}$$

Since the implications (b) $\implies$ (5.1) and (b) $\iff$ (c) are obvious, it remains to show that (5.1) implies (b). Thus let $y \in E$ be given. Since $\bar{x}$ is an interior point of M, we have $x := \bar{x} + \tau y \in M$ whenever $\tau \in \mathbb{R}$ is such that $|\tau|$ is sufficiently small. Hence for these τ we deduce from (5.1) that $\tau \langle f'(\bar{x}), y \rangle \geq 0$. Therefore (b) holds. $\qquad\square$

5.2 Optimality Under Functional Constraints

We consider the following convex optimization problem:

(P2) Minimize $f(x)$
    subject to $g_i(x) \leq 0$ $(i = 1, \ldots, m)$, $x \in A$,

which of course means

    minimize $f(x)$ on $M := \{x \in E \mid g_i(x) \leq 0 \ (i = 1, \ldots, m), \ x \in A\}$.

In this connection, the assumptions are:

(A) $f, g_1, \ldots, g_m : E \to \overline{\mathbb{R}}$ are proper convex functionals,
    A is a nonempty convex subset of $D := \text{dom } f \cap \text{dom } g_1 \cap \cdots \cap \text{dom } g_m$.

Our aim is to characterize points $\bar{x} \in M$ that minimize f under the *functional constraints* $g_i(x) \leq 0$ $(i = 1, \ldots, m)$ and the *residual constraint* $x \in A$. We therefore consider the statement

(Min 2) $\bar{x} \in M$ is a global solution of (P2).

We define functionals $\widehat{L} : D \times \mathbb{R}_+^{m+1} \to \mathbb{R}$ and $L : D \times \mathbb{R}_+^m \to \mathbb{R}$ by

$$\widehat{L}(x; \lambda, \mu_1, \ldots, \mu_m) := \lambda f(x) + \sum_{i=1}^{m} \mu_i g_i(x),$$

$$L(x; \mu_1, \ldots, \mu_m) := f(x) + \sum_{i=1}^{m} \mu_i g_i(x).$$

The functionals $\widehat{L}$ and L are called *Lagrange functionals* associated with (P2). Furthermore, the point $(\bar{x}, \bar{\mu}) \in A \times \mathbb{R}_+^m$, where $\bar{\mu} := (\bar{\mu}_1, \ldots, \bar{\mu}_m)$, is called *saddle point* of L with respect to $A \times \mathbb{R}_+^m$ if

$$L(\bar{x}, \mu) \leq L(\bar{x}, \bar{\mu}) \leq L(x, \bar{\mu}) \quad \forall (x, \mu) \in A \times \mathbb{R}_+^m.$$

We consider the following statements:

($\widehat{L}$) $\exists (\bar{\lambda}, \bar{\mu}_1, \ldots, \bar{\mu}_m) \in \mathbb{R}_+^{m+1} \setminus \{o\}$:

$$\widehat{L}(\bar{x}; \bar{\lambda}, \bar{\mu}_1, \ldots, \bar{\mu}_m) = \min_{x \in A} \widehat{L}(x; \bar{\lambda}, \bar{\mu}_1, \ldots, \bar{\mu}_m),$$

$$\bar{\mu}_i \, g_i(\bar{x}) = 0 \quad (i = 1, \ldots, m).$$

(L) $\exists (\bar{\mu}_1, \ldots, \bar{\mu}_m) \in \mathbb{R}_+^m$:

$$L(\bar{x}; \bar{\mu}_1, \ldots, \bar{\mu}_m) = \min_{x \in A} L(x; \bar{\mu}_1, \ldots, \bar{\mu}_m),$$

$$\bar{\mu}_i \, g_i(\bar{x}) = 0 \quad (i = 1, \ldots, m).$$

(SP) $\exists \bar{\mu} \in \mathbb{R}_+^m$: $(\bar{x}, \bar{\mu})$ is a saddle point of L with respect to $A \times \mathbb{R}_+^m$.

Finally we consider the *Slater condition*

$$\exists x_0 \in A : \; g_i(x_0) < 0 \text{ for } i = 1, \ldots, m. \tag{5.2}$$

Theorem 5.2.1 (Global Kuhn–Tucker Theorem) *Let the assumptions* (A) *be satisfied.*

(a) *There always holds*

$$(\text{SP}) \iff (\text{L}) \implies (\text{Min 2}) \implies (\widehat{\text{L}}).$$

(b) *If the Slater condition* (5.2) *is satisfied, then there holds*

$$(\text{SP}) \iff (\text{L}) \iff (\text{Min 2}) \iff (\widehat{\text{L}}).$$

Proof. (a) (SP) $\iff$ (L) $\implies$ (Min 2): Exercise 5.6.1.
  (Min 2) $\implies$ ($\widehat{\text{L}}$): Define

$$K := \{ -(f(x) - f(\bar{x}), g_1(x), \ldots, g_m(x)) \mid x \in A \}.$$

Since the functions $f, g_1, \ldots, g_m$ are convex, K is a convex subset of $\mathbb{R}^{m+1}$. Furthermore, (Min 2) implies $K \cap \text{int } \mathbb{R}_+^{m+1} = \emptyset$. By Corollary 1.5.4, there exists $\bar{\mu} := (\bar{\lambda}, \bar{\mu}_1, \ldots, \bar{\mu}_m) \in \mathbb{R}^{m+1}$ satisfying

$$\bar{\mu} \neq o, \quad (\bar{\mu} \,|\, y) \leq 0 \,\forall\, y \in K, \quad (\bar{\mu} \,|\, z) \geq 0 \,\forall\, z \in \mathbb{R}_+^{m+1}.$$

The assertion now follows immediately.

(b) It suffices to show that $(\widehat{\text{L}})$ implies (L). Assume that $(\widehat{\text{L}})$ holds with $\bar{\lambda} = 0$. Then we have $(\bar{\mu}_1, \ldots, \bar{\mu}_m) \neq o$ and so, by (5.2), $\bar{\mu}_i \, g_i(x_0) < 0$ for at least one $i \in \{1, \ldots, m\}$. On the other hand, the minimum property of $(\widehat{\text{L}})$ entails

$$0 = \sum_{i=1}^{m} \bar{\mu}_i \, g_i(\bar{x}) \leq \sum_{i=1}^{m} \bar{\mu}_i \, g_i(x) \quad \forall\, x \in A,$$

which is contradictory. Hence $\bar{\lambda} \neq 0$ and we may (replacing $\bar{\mu}_i$ by $\bar{\mu}_i \,/\, \bar{\lambda}$ if necessary) assume that $\bar{\lambda} = 1$. Therefore, (L) holds.   □

Remark 5.2.2 (a) Roughly speaking, Theorem 5.2.1 states that the minimization of f under functional and residual side conditions can be replaced by the minimization of the Lagrange functional L or $\widehat{L}$ under the residual side condition $x \in A$ alone, the functional side conditions being integrated into the Lagrange functional.

(b) If $\bar{\lambda} = 0$, then the functional f to be minimized does not appear in the optimality conditions. In this case, the conditions are not well suited for detecting possible minimizers. A condition ensuring that $\bar{\lambda} \neq 0$, and so $(\widehat{\text{L}}) \iff (\text{L})$, is called *regularity condition*. Theorem 5.2.1(b) shows that the Slater condition is a regularity condition. A more thorough study of regularity conditions shows that these are generally conditions on the constraint functionals (as is the Slater condition). Therefore regularity conditions are also called *constraint qualifications*.

We now establish *local* optimality conditions for (Min 2) by using subdifferentials. Consider the following statements.

(J) $\exists\, (\bar{\lambda}, \bar{\mu}_1, \ldots, \bar{\mu}_m) \in \mathbb{R}_+^{m+1} \setminus \{o\}$:

$o \in \bar{\lambda} \partial f(\bar{x}) + \bar{\mu}_1 \partial g_1(\bar{x}) + \cdots + \bar{\mu}_m \partial g_m(\bar{x}) + \partial \delta_A(\bar{x}),$

$\bar{\mu}_i \, g_i(\bar{x}) = 0 \quad (i = 1, \ldots, m).$

(KKT)

$\exists\, (\bar{\mu}_1, \ldots, \bar{\mu}_m) \in \mathbb{R}_+^m$:

$o \in \partial f(\bar{x}) + \bar{\mu}_1 \partial g_1(\bar{x}) + \cdots + \bar{\mu}_m \partial g_m(\bar{x}) + \partial \delta_A(\bar{x}),$

$\bar{\mu}_i \, g_i(\bar{x}) = 0 \quad (i = 1, \ldots, m).$

The conditions (J) and (KKT) are the *(Fritz) John conditions* and the *Karush–Kuhn–Tucker conditions*, respectively.

Theorem 5.2.3 (Local John–Karush–Kuhn–Tucker Theorem) *In addition to the assumptions* (A), *let the functionals* $f, g_1, \ldots, g_m$ *be continuous at some point of* $A \cap \operatorname{int} D$.

(a) *There always holds* (KKT) $\Longrightarrow$ (Min 2) $\Longrightarrow$ (J).

(b) *If the Slater condition* (5.2) *is satisfied, then there holds* (KKT) $\Longleftrightarrow$ (Min 2) $\Longleftrightarrow$ (J).

Proof. See Exercise 5.6.2.

Remark 5.2.4 (a) Since $-\partial \delta_A(\bar{x}) = \{x^* \in E^* \mid \langle x^*, \bar{x} \rangle = \min_{x \in A} \langle x^*, x \rangle\}$, the John condition (J) is equivalent to

$$\exists (\bar{\lambda}, \bar{\mu}_1, \ldots, \bar{\mu}_m) \in \mathbb{R}_+^{m+1} \setminus \{o\} \quad \exists x^* \in \partial f(\bar{x}) \quad \exists y_i^* \in \partial g_i(\bar{x}) \ (i = 1, \ldots, m) :$$

$$\left(\bar{\lambda} x^* + \bar{\mu}_1 y_1^* + \cdots + \bar{\mu}_m y_m^* \right)(\bar{x}) = \min_{x \in A} \left(\bar{\lambda} x^* + \bar{\mu}_1 y_1^* + \cdots + \bar{\mu}_m y_m^* \right)(x),$$

$$\bar{\mu}_i g_i(\bar{x}) = 0 \ (i = 1, \ldots, m).$$

(b) If f is continuous at a point of $A \cap \operatorname{int} \operatorname{dom} f$, then

$$f(\bar{x}) = \min_{x \in A} f(x) \quad \Longleftrightarrow \quad \exists x^* \in \partial f(\bar{x}) : \langle x^*, \bar{x} \rangle = \min_{x \in A} \langle x^*, x \rangle.$$

This follows from Theorem 5.2.3(b) by choosing $m = 1$, $g_1(x) := -1$ if $x \in D$, $g_1(x) := +\infty$ if $x \in E \setminus D$.

(c) If f is continuous at a point of $A \cap \operatorname{int} \operatorname{dom} f$ and G-differentiable at $\bar{x}$, then

$$f(\bar{x}) = \min_{x \in A} f(x) \quad \Longleftrightarrow \quad \langle f'(\bar{x}), x - \bar{x} \rangle \geq 0 \ \forall x \in A.$$

If, in addition, $\bar{x} \in \operatorname{int} A$ (in particular, if $A = E$), then

$$f(\bar{x}) = \min_{x \in A} f(x) \quad \Longleftrightarrow \quad \langle f'(\bar{x}), y \rangle = 0 \ \forall y \in E \quad \Longleftrightarrow \quad f'(\bar{x}) = o.$$

This follows from (b) above noting that now $\partial f(\bar{x}) = \{f'(\bar{x})\}$ by Proposition 4.1.8.

Remark 5.2.4(b) and Proposition 5.1.1 give the following result.

Proposition 5.2.5 *Let* $f : E \to \mathbb{R}$ *be a continuous convex functional, let* A *be a nonempty convex subset of* E, *and let* $\bar{x} \in A$. *Then the following statements are equivalent:*

(a) $f(\bar{x}) = \min_{x \in A} f(x)$.
(b) $\exists x^* \in \partial f(\bar{x}) : \langle x^*, \bar{x} \rangle = \min_{x \in A} \langle x^*, x \rangle$.
(c) $\forall x \in A : f_G(\bar{x}, x - \bar{x}) \geq 0$.

5.3 Application to Approximation Theory

Let A be a nonempty subset of E and $z \in E$. Recall that an element $\bar{x} \in A$ is said to be a *best approximation* of z with respect to A, or a *projection* of z onto A, if

$$\|\bar{x} - z\| = \min_{x \in A} \|x - z\|.$$

We write $\mathrm{proj}_A(z)$ for the set (possibly empty) of all projections of z onto A, i.e., we put

$$\mathrm{proj}_A(z) := \{\bar{x} \in A \mid \|\bar{x} - z\| = \mathrm{d}_A(z)\},$$

where d_A denotes the distance function. This defines the multifunction $\mathrm{proj}_A : E \rightrightarrows E$, called the *projector* associated with the set A. It is clear that $z \in A$ implies $\mathrm{proj}_A(z) = \{z\}$. We now assume that

$$A \subseteq E \text{ is nonempty and convex, and } z \in E \setminus A.$$

Our aim is to characterize $\mathrm{proj}_A(z)$. This can be done by applying the preceding results to the functional

$$f(x) := \|x - z\|, \quad x \in E. \tag{5.3}$$

Best Approximation in a Hilbert Space

Proposition 5.3.1 *Let E be a Hilbert space with inner product $(x \mid y)$. Then one has*

$$\bar{x} \in \mathrm{proj}_A(z) \quad \Longleftrightarrow \quad (z - \bar{x} \mid x - \bar{x}) \leq 0 \quad \forall x \in A.$$

Proof. See Exercise 5.6.3. □

Remark 5.3.2 Recall that, by the Cauchy–Schwarz inequality, the formula

$$\cos \alpha_x := \frac{(z - \bar{x} \mid x - \bar{x})}{\|z - \bar{x}\| \, \|x - \bar{x}\|}$$

defines an angle α_x for any $x \in A$, $x \neq \bar{x}$. Proposition 5.3.1 thus says that $\bar{x}$ is a projection of z onto A if and only if α_x is obtuse (see Fig. 5.1, where $E = \mathbb{R}^2$).

If A is a linear subspace of E, then

$$A^\perp := \{y \in E \mid (y \mid \tilde{x}) = 0 \quad \forall \tilde{x} \in A\}$$

denotes the *orthogonal complement* of A. As an immediate consequence of Proposition 5.3.1 we have:

Corollary 5.3.3 *Let E be a Hilbert space and A a linear subspace of E. Then $\bar{x} \in \mathrm{proj}_A(z)$ if and only if $z - \bar{x} \in A^\perp$ (see Fig. 5.2, where $E = \mathbb{R}^3$).*

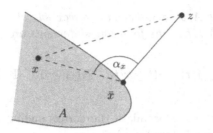

Fig. 5.1

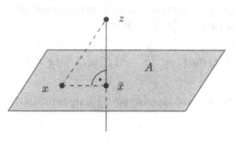

Fig. 5.2

Best Approximation in a Normed Vector Space

We want to characterize best approximations in an arbitrary normed vector space. From Propositions 4.6.2 and 5.2.5 we immediately obtain:

Proposition 5.3.4 (Characterization of Best Approximations) *If E is a normed vector space, the following statements are equivalent:*

(a) $\bar{x} \in \text{proj}_A(z)$.
(b) $\exists\, x^* \in S(\bar{x} - z) : \langle x^*, \bar{x} \rangle = \min_{x \in A} \langle x^*, x \rangle.$
(c) $\forall x \in A : \max\{ \langle y^*, x - \bar{x} \rangle \,|\, y^* \in S(\bar{x} - z) \cap \text{ep}\, S(o) \} \geq 0.$

Best Approximation in $C(T)$

Now we apply the above results to

$$E := C(T), \text{ with norm } \|x\|_\infty := \max_{t \in T} |x(t)|,$$

where T is a compact Hausdorff space. We consider the functional

$$\omega_z(x) := \|x - z\|_\infty, \quad x \in C(T). \tag{5.4}$$

Concerning the terminology, we refer to Sect. 4.6.

Combining Propositions 4.6.4 and 5.3.4(a)$\Leftrightarrow$(c), we obtain the following characterization of best approximations in $C(T)$. Notice that in the Kolmogorov condition the signum operation can now be omitted.

Proposition 5.3.5 *Let A be a nonempty convex subset of $C(T)$, let $z \in C(T) \setminus A$, and let $\bar{x} \in A$. Then the following statements are equivalent:*

(a) $\bar{x} \in \text{proj}_A(z)$.

(b) $\forall x \in A : \max_{t \in T(\bar{x})} (\bar{x}(t) - z(t))(x(t) - \bar{x}(t)) \geq 0$ (Kolmogorov condition).

We leave it to the reader to formulate the corresponding characterization following from Proposition 5.2.5(a)$\Leftrightarrow$(b).

As a special case, we now choose $T = [a, b]$. Recall that each (positive) Borel measure ν on $[a, b]$ is regular and can be represented by a nondecreasing right continuous function $\psi : [a, b] \to \mathbb{R}$ such that

$$\int_{[a,b]} x(t)\, d\nu(t) = \int_{[a,b]} x(t)\, d\psi(t) \quad \forall x \in C[a, b],$$

the integral on the right-hand side being a Riemann–Stieltjes integral. In this connection, we have

$$\psi(x) = \nu((a, x]) \ \forall x \in (a, b], \quad \psi(a) = 0.$$

Moreover, for each Borel set $B \subseteq [a, b]$ we have

$$\nu(B) = \inf \left\{ \sum_{i=1}^{\infty} \big(\psi(b_i) - \psi(a_i)\big) \,\Big|\, B \subset \bigcup_{i=1}^{\infty} (a_i, b_i] \right\}. \tag{5.5}$$

It follows that the Borel measure ν is concentrated on the Borel set $B \subseteq [a, b]$ if and only if the associated function ψ is constant except for jumps at the points $t \in B$.

Let $\mu \in M[a, b]$ be given. Applying what has just been said about ν to μ^+ and μ^-, we obtain nondecreasing right continuous functions φ^+ and φ^- on $[a, b]$. Then the function $\varphi := \varphi^+ - \varphi^-$ is right continuous and of bounded variation on $[a, b]$, and we have

$$\int_{[a,b]} x(t)\, d\mu(t) = \int_{[a,b]} x(t)\, d\varphi(t) \quad \forall x \in C[a, b].$$

Recall that $T(\bar{x}) := \{t \in T \mid |\bar{x}(t) - z(t)| = \|\bar{x} - z\|_\infty\}$. Now we can establish the following:

Proposition 5.3.6 (Classical Chebyshev Approximation) *Let A denote the set of (the restrictions to $[a, b]$ of) all polynomials of degree at most n, where $n \in \mathbb{N}$. Further let $z \in C[a, b] \setminus A$. If $\bar{x} \in A$ is a best approximation of z with respect to A, then the set $T(\bar{x})$ contains at least $n + 2$ points.*

Proof. According to Propositions 4.6.4 and 5.2.5, and by what has been said above, there exists a right continuous function φ of bounded variation on $[a, b]$ that is constant except for jumps on $T(\bar{x})$ (Fig. 5.3) and satisfies

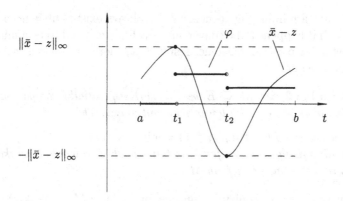

Fig. 5.3

$$\int_{[a,b]} x(t)\, d\varphi(t) = 0 \quad \forall x \in A. \tag{5.6}$$

The latter result follows from the fact that $\langle x^*, \bar{x} \rangle = \min_{x \in A} \langle x^*, x \rangle$ is equivalent to $\langle x^*, \bar{x} \rangle = 0$ since in the present case A is a linear subspace of $C[a,b]$. Now suppose that $T(\bar{x})$ contains only m elements $t_1, \ldots, t_m$, where $m < n + 2$. We may assume that $a < t_1 < t_2 < \cdots < t_m \le b$. For $k \in \{1, \ldots, m\}$ define

$$x(t) := \prod_{\substack{i=1 \\ i \ne k}}^{m} (t - t_i), \ t \in [a,b].$$

Then x is (the restriction to $[a,b]$ of) a polynomial of degree $m - 1$ and so belongs to A. In view of (5.6), we obtain

$$0 = \int_{[a,b]} x(t)\, d\varphi(t) = \sum_{i=1}^{m} x(t_i)\big(\varphi(t_i) - \varphi(t_i - 0)\big) = x(t_k)\big(\varphi(t_k) - \varphi(t_k - 0)\big).$$

As $x(t_k) \ne 0$, it follows that φ is continuous at t_k, which is a contradiction. $\quad\square$

5.4 Existence of Minimum Points and the Ritz Method

In this section and in Sect. 5.5 we digress from the main road of these lectures, briefly discussing the *existence* of minimizers of

$$f : M \to \mathbb{R},$$

where M is a nonempty subset of a normed vector space E. We start with a definition.

Definition 5.4.1 A sequence (x_n) in M is said to be a *minimizing sequence* for f if $\lim_{n \to \infty} f(x_n) = \inf_{x \in M} f(x)$.

It is clear that a minimizing sequence for f always exists but it need not be convergent. If it happens to be convergent, the limit may fail to be a minimizer of f. The starting point for existence results is a generalization of a well-known Weierstrass theorem.

Proposition 5.4.2 *Let $M \subseteq E$ be (weakly) sequentially compact and $f : M \to \mathbb{R}$ (weakly) sequentially lower semicontinuous. Then:*

(a) *There exists $\bar{x} \in M$ satisfying $f(\bar{x}) = \min_{x \in M} f(x)$.*
(b) *Every minimizing sequence for f contains a subsequence that (weakly) converges to a minimizer of f on M.*

Proof. Let (x_n) be a minimizing sequence for f. Since M is (weakly) sequentially compact, some subsequence (x_{n_j}) of (x_n) is (weakly) convergent to some $\bar{x} \in M$. Since f is (weakly) sequentially l.s.c., it follows that

$$f(\bar{x}) \leq \liminf_{j \to \infty} f(x_{n_j}) = \lim_{n \to \infty} f(x_n) = \inf_{x \in M} f(x). \quad \square$$

Concerning the assumptions of Proposition 5.4.2, observe that sequential compactness of M is too restrictive for most applications. In general, even weak sequential compactness is not appropriate. We tackle this problem in the following way:

1. We assume that E is a *reflexive* Banach space. Then by the Eberlein–Šmulian Theorem (Theorem 1.6.7), each bounded, sequentially closed subset M is weakly sequentially compact.
2. In order to get rid of boundedness which is still too restrictive, we replace this hypothesis on M by an hypothesis on f to be defined now.

Definition 5.4.3 The functional $f : M \to \mathbb{R}$ is said to be *coercive* if for any sequence (x_n) in M satisfying $\lim_{n \to \infty} \|x_n\| = +\infty$ one has $\limsup_{n \to \infty} f(x_n) = +\infty$.

We consider the following assumptions:

(A1) E is a reflexive Banach space,
    M is a nonempty, weakly sequentially closed subset of E,
    $f : M \to \mathbb{R}$ is weakly sequentially l.s.c.,
    either M is bounded or f is coercive.

Theorem 5.4.4 *Under the assumptions (A1), the following holds:*

(a) *There exists $\bar{x} \in M$ such that $f(\bar{x}) = \min_{x \in M} f(x)$.*
(b) *Each minimizing sequence for f contains a subsequence that weakly converges to a minimizer of f on M.*

Proof. (I) As discussed above, if M is bounded, the assertion is a consequence of Proposition 5.4.2 and the Eberlein–Šmulian Theorem.

(II) Assume now that M is unbounded and f is coercive. Choose some $x_0 \in M$ and set $M_0 := \{x \in M \mid f(x) \leq f(x_0)\}$. Notice that

$$\inf_{x \in M} f(x) = \inf_{x \in M_0} f(x). \tag{5.7}$$

Since M is weakly sequentially closed and f is weakly sequentially l.s.c. on M, the set M_0 is weakly sequentially closed. Moreover, since f is coercive, M_0 is bounded. By Proposition 5.4.2(a), f has a minimizer on M_0 which by (5.7) is also a minimizer of f on M. This proves (a). Concerning (b), notice that each minimizing sequence for f in M is eventually in M_0 so that Proposition 5.4.2(b) applies with M replaced by M_0.   □

The example $f(x) = e^x$, $x \in \mathbb{R}$, shows that a minimizer may fail to exist if M is unbounded and f is not coercive.

Corollary 5.4.5 *Let E be a finite-dimensional Banach space and M a nonempty closed subset of E. Then $\mathrm{proj}_M(z)$ is nonempty for any $z \in E$.*

Proof. This follows immediately by applying Theorem 5.4.4 to $f(x) := \|x - z\|$, $x \in E$.   □

Now we pass from (A1) to assumptions that are easier to verify:

(A2) E is a reflexive Banach space,
  M is a nonempty, convex, and closed subset of E,
  $f : D \to \mathbb{R}$ (where $D \subseteq E$ is open and contains M) is strictly convex and G-differentiable on M,
  either M is bounded or f is coercive on M.

Theorem 5.4.6 *Under the assumptions (A2), the functional f has precisely one minimizer $\bar{x}$ on M, and each minimizing sequence for f on M is weakly convergent to $\bar{x}$.*

Proof. (I) First we show that the strict convexity of f entails that f has at most one minimizer on M. Suppose, to the contrary, that x_1 and x_2 are minimizers of f on M with $x_1 \neq x_2$ and $f(x_1) = f(x_2) =: a$. Then

$$f\left(\tfrac{1}{2}x_1 + \tfrac{1}{2}x_2\right) < \tfrac{1}{2}f(x_1) + \tfrac{1}{2}f(x_2) = a,$$

which is a contradiction.

(II) By Corollary 1.6.6, M is weakly sequentially closed. By Propositions 1.7.3 and 4.1.9, f is weakly sequentially l.s.c. Hence the assumptions (A1) are satisfied so that Theorem 5.4.4 applies. By statement (b) of that theorem, all weakly convergent subsequences of a minimizing sequence (x_n) for f have the same limit $\bar{x}$; hence the entire sequence (x_n) is weakly convergent to $\bar{x}$.   □

With a given $b^* \in E^*$, we now consider the following statements:

$$f(\bar{x}) - \langle b^*, \bar{x} \rangle = \min_{x \in E}\big(f(x) - \langle b^*, x \rangle\big) \qquad \text{(minimum problem)}, \qquad (5.8)$$

$$\langle f'(\bar{x}) - b^*, y \rangle = 0 \quad \forall y \in E \qquad \text{(variational equation)}, \qquad (5.9)$$

$$f'(\bar{x}) = b^* \qquad \text{(operator equation)}. \qquad (5.10)$$

We specify the assumptions:

(A3) E is a reflexive Banach space, $f : E \to \mathbb{R}$ is G-differentiable,
    f' is uniformly monotone (with constants $c > 0$ and $\gamma > 1$).

Theorem 5.4.7 *If* (A3) *is satisfied, then for each* $b^* \in E^*$ *the following holds:*

(a) *The problems* (5.8)–(5.10) *are mutually equivalent and have precisely one solution* $\bar{x} \in E$.
(b) *Each minimizing sequence* (x_n) *for* $f - b^*$ *is convergent to* $\bar{x}$, *and one has the error estimate*

$$\frac{c}{\gamma}\|x_n - \bar{x}\|^\gamma \le (f - b^*)(x_n) - (f - b^*)(\bar{x}). \qquad (5.11)$$

Proof. (a) Set $g := f - b^*$. Then g is G-differentiable with $g'(x) = f'(x) - b^*$ for any $x \in E$ and so g' is also uniformly monotone with constants c and γ.
(I) The equivalence of (5.8) and (5.9) follows from Corollary 5.1.2. The equivalence of (5.9) and (5.10) is obvious.
(II) *Existence and uniqueness.* By Proposition 4.3.9, g is strictly convex and

$$g(y) - g(x) \ge \langle g'(x), y - x \rangle + \frac{c}{\gamma}\|y - x\|^\gamma \quad \forall x, y \in E. \qquad (5.12)$$

We deduce that

$$g(y) - g(o) \ge \|y\|\Big(-\|g'(x)\| + \frac{c}{\gamma}\|y\|^{\gamma-1}\Big).$$

The term in parentheses is positive if $\|y\|$ is large enough, hence g is coercive. By Theorem 5.4.6, g has precisely one minimizer $\bar{x} \in E$.
(b) We have $g'(\bar{x}) = o$. If (x_n) is any minimizing sequence for g, then (5.12) with $x := \bar{x}$ gives

$$\frac{c}{\gamma}\|x_n - \bar{x}\|^\gamma \le g(x_n) - g(\bar{x}) \to 0 \quad \text{as } n \to \infty. \quad \square$$

Remark 5.4.8 Theorem 5.4.7, in particular, shows that under the assumptions (A3) the G-derivative f' is a bijective mapping of E onto E^*. This observation closes the gap in the proof of Proposition 4.4.3.

The Ritz Method

We now describe a method for *constructing* a minimizing sequence and so for approximately calculating a minimizer. The basic idea is to replace the original problem, which is placed in an infinite-dimensional space, by a sequence of *finite-dimensional* problems.

Suppose that E is an infinite-dimensional normed vector space. A countable subset $\{z_k \mid k \in \mathbb{N}\}$ of E is called *basis* of E if finitely many z_k are always linearly independent and one has

$$E = \text{cl} \bigcup_{n \in \mathbb{N}} E_n, \quad \text{where} \quad E_n := \text{span}\{z_1, \dots, z_n\}.$$

Using transfinite induction, it is easy to prove:

Lemma 5.4.9 *If E is separable, then E possesses a basis.*

The *Ritz method* consists in replacing, for each $n \in \mathbb{N}$, the problems (5.8) and (5.9) by the following *n-dimensional* problems:

$$f(\bar{x}_n) - \langle b^*, \bar{x}_n \rangle = \min_{x_n \in E_n} \left(f(x_n) - \langle b^*, x_n \rangle \right) \quad \text{(\emph{Ritz minimum problem}),}$$

$$\text{(5.8a)}$$

$$\langle f'(\bar{x}_n) - b^*, z_k \rangle = 0 \quad \forall k = 1, \dots, n \quad \text{(\emph{Ritz equations}).}$$

$$\text{(5.9a)}$$

For a given basis $\{z_k \mid k \in \mathbb{N}\}$ of E, each $\bar{x}_n \in E_n$ is representable as $\bar{x}_n = \sum_{j=1}^n \xi_j z_j$, where $\xi_j \in \mathbb{R}$. Hence, (5.9a) is a system of n equations (in general nonlinear) for $(\xi_1, \dots, \xi_n) \in \mathbb{R}^n$ and so can be solved by numerical methods. This yields a numerical approximation of $\bar{x}_n$. We shall not pursue the numerical aspect. Rather we assume that the exact $\bar{x}_n$ is known and ask how it is related to a solution $\bar{x}$ of the original problems (5.8) and (5.9). We shall give a convergence proof for the special case that f is a quadratic functional of the form

$$f(x) := \tfrac{1}{2} a(x, x), \quad x \in E. \tag{5.13}$$

In this connection, we make the following assumptions:

(A4) E is an infinite-dimensional separable reflexive Banach space with basis $\{z_k \mid k \in \mathbb{N}\}$, $a : E \times E \to \mathbb{R}$ is bilinear, symmetric, bounded, and strongly positive, $b^* \in E^*$.

Recall that a is said to be bounded if with some constant $\kappa > 0$,

$$|a(x, y)| \le \kappa \|x\| \|y\| \quad \forall x, y \in E, \tag{5.14}$$

and a is said to be strongly positive if with some constant $c > 0$,

$$a(x, x) \geq c \|x\|^2 \quad \forall x \in E. \tag{5.15}$$

If (A4) holds, then by Example 4.3.6 the functional f defined by (5.13) is continuously differentiable, the derivative is given by $\langle f'(x), y \rangle = a(x, y)$ for all $x, y \in E$, and f' is strongly monotone with constant c. The problems (5.8) and (5.9) now pass, respectively, into

$$\tfrac{1}{2} a(\bar{x}, \bar{x}) - \langle b^*, \bar{x} \rangle = \min_{x \in E} \left(\tfrac{1}{2} a(x, x) - \langle b^*, x \rangle \right), \tag{5.16}$$

$$a(\bar{x}, y) - \langle b^*, y \rangle = 0 \quad \forall y \in E. \tag{5.17}$$

The associated Ritz problems are

$$\tfrac{1}{2} a(\bar{x}_n, \bar{x}_n) - \langle b^*, \bar{x}_n \rangle = \min_{x_n \in E_n} \left(\tfrac{1}{2} a(x_n, x_n) - \langle b^*, x_n \rangle \right), \tag{5.16a}$$

$$a(\bar{x}_n, z_k) - \langle b^*, z_k \rangle = 0 \quad \forall k = 1, \ldots, n. \tag{5.17a}$$

Notice that the minimum problems (5.16) and (5.16a) are *quadratic*, the variational equations (5.17) and (5.17a) are *linear*.

Theorem 5.4.10 *If* (A4) *holds, then:*

(a) *The problems* (5.16) *and* (5.17) *are equivalent and possess precisely one solution* $\bar{x} \in E$.
(b) *For each* $n \in \mathbb{N}$, (5.16a) *and* (5.17a) *are equivalent and possess precisely one solution* $\bar{x}_n \in E_n$.
(c) *The sequence* $(\bar{x}_n)$ *is minimizing for* $f - b^*$, *converges to* $\bar{x}$ *as* $n \to \infty$, *and satisfies*

$$\|\bar{x}_n - \bar{x}\| \leq \tfrac{\kappa}{c} \, d(\bar{x}, E_n) \quad \forall n \in \mathbb{N}.$$

Here, f, κ, and c are as in (5.13), (5.14), *and* (5.15), *respectively.*

Proof. (I) Example 4.3.6 shows that, with f according to (5.13), the assumptions (A3) are satisfied. Therefore assertion (a) is a consequence of Theorem 5.4.7. Analogously, applying Theorem 5.4.7 with E replaced by E_n yields (b).

(II) *Convergence.* From (5.17) with $y = y_n$ and (5.17a) we see that

$$a(\bar{x} - \bar{x}_n, y_n) = 0 \quad \forall y_n \in E_n \tag{5.18}$$

and, in particular, $a(\bar{x} - \bar{x}_n, \bar{x}_n) = 0$. This and (5.18) yield

$$a(\bar{x} - \bar{x}_n, \bar{x} - \bar{x}_n) = a(\bar{x} - \bar{x}_n, \bar{x} - y_n) \quad \forall y_n \in E_n.$$

Since a is strongly positive and bounded, we deduce

$$c \|\bar{x} - \bar{x}_n\|^2 \leq \kappa \|\bar{x} - \bar{x}_n\| \, \|\bar{x} - y_n\| \quad \forall y_n \in E_n$$

and so

$$\|\bar{x} - \bar{x}_n\| \leq \frac{\kappa}{c} \inf_{y_n \in E_n} \|\bar{x} - y_n\| = \frac{\kappa}{c} \, \mathrm{d}(\bar{x}, E_n).$$

Since $E_n \subseteq E_{n+1}$ for all n and $\bigcup_n E_n$ is dense in E, we have $\mathrm{d}(\bar{x}, E_n) \to 0$ as $n \to \infty$. Hence $\bar{x}_n \to \bar{x}$ as $n \to \infty$. This also shows that the sequence $(\bar{x}_n)$ is minimizing for $f - b^*$ as the latter functional is continuous.   □

5.5 Application to Boundary Value Problems

Our aim in this section is to apply Theorems 5.4.7 and 5.4.10 to a boundary value problem. We start by introducing some notation. Let G be a bounded region in $\mathbb{R}^N$, with boundary ∂G and closure $\overline{G} = G \cup \partial G$. Recall that a *region* is a nonempty open connected set. We set

$\mathrm{C}^k(\overline{G}) :=$ set of all continuous functions $u : \overline{G} \to \mathbb{R}$ with continuous partial derivatives up to and including order k on G such that the partial derivatives can be continuously extended to $\overline{G}$.

$\mathrm{C}_c^\infty(G) :=$ set of all continuous functions $u : G \to \mathbb{R}$ having continuous partial derivatives of arbitrary order on G and vanishing outside a compact subset of G (that depends on u).

In particular, $\mathrm{C}(\overline{G}) := \mathrm{C}^0(\overline{G})$ denotes the set of all continuous functions on $\overline{G}$. We write $x = (x_1, \ldots, x_N)$ for the independent variable ranging over G and we denote by D_i differentiation with respect to the ith coordinate, where $i = 1, \ldots, N$. Let $p \in (1, +\infty)$ be fixed. We consider the Lebesgue space $\mathrm{L}^p(G)$ with norm

$$\|u\|_p := \left(\int_G |u|^p \, \mathrm{d}x \right)^{1/p}$$

and the Sobolev spaces

$$\mathrm{W}^{1,p}(G) := \left\{ u \in \mathrm{L}^p(G) \mid \mathrm{D}_i u \text{ exists in } \mathrm{L}^p(G) \text{ for } i = 1, \ldots, N \right\},$$
$$\mathrm{W}_0^{1,p}(G) := \text{closure of } \mathrm{C}_c^\infty(G) \text{ in } \mathrm{W}^{1,p}(G).$$

In this connection, $\mathrm{D}_i u$ now denotes the *generalized derivative* of u with respect to the ith coordinate. By definition, $\mathrm{D}_i u$ is an element of $\mathrm{L}^p(G)$ that satisfies the *integration-by-parts formula*

$$\int_G (\mathrm{D}_i u) \, v \, \mathrm{d}x = - \int_G u \, (\mathrm{D}_i v) \, \mathrm{d}x \quad \forall v \in \mathrm{C}_c^\infty(G). \tag{5.19}$$

The norm in $\mathrm{W}^{1,p}(G)$ is given by

$$\|u\|_{1,p} := \left(\int_G \left(|u|^p + \sum_{i=1}^N |\mathrm{D}_i u|^p \right) \mathrm{d}x \right)^{1/p}.$$

On $W_0^{1,p}(G)$, the following norm $\| \cdot \|_{1,p,0}$ is equivalent to $\| \cdot \|_{1,p}$:

$$\|u\|_{1,p,0} := \left(\int_G \sum_{i=1}^N |D_i u|^p \, \mathrm{d}x \right)^{1/p}.$$

$W^{1,p}(G)$ and $W_0^{1,p}(G)$ are separable reflexive Banach spaces. In particular, $W^{1,2}(G)$ and $W_0^{1,2}(G)$ are Hilbert spaces with respect to the inner product

$$(u \,|\, v)_{1,2} := \int_G \left(uv + \sum_{i=1}^N D_i u \, D_i v \right) \mathrm{d}x.$$

We start with the following *classical boundary value problem*. Given $g \in C(\overline{G})$, find $u \in C^2(\overline{G})$ satisfying

$$-\sum_{i=1}^N D_i \left(|D_i u|^{p-2} D_i u \right) = g \text{ on } G, \quad u = 0 \text{ on } \partial G. \tag{5.20}$$

For $p = 2$ this is the famous *Dirichlet problem* or the first boundary value problem for the *Poisson equation*. Observe that for $p > 2$ the problem is nonlinear. Since the problem may fail to have a solution unless the function g and the boundary ∂G are sufficiently smooth, we pass to a generalized problem. This is obtained heuristically by multiplying the differential equation in (5.20) by $v \in C_c^\infty(G)$ and applying the integration-by-parts formula (5.19):

$$\int_G \sum_{i=1}^N |D_i u|^{p-2} D_i u \, D_i v \, \mathrm{d}x = \int_G gv \, \mathrm{d}x \quad \forall v \in C_c^\infty(G). \tag{5.21}$$

Conversely, if $u \in C^2(\overline{G})$ satisfies (5.21) and $u = 0$ on ∂G, then u is also a solution of (5.20).

The *generalized boundary value problem* associated with (5.20) now reads as follows. Given $g \in L^q(G)$, where $\frac{1}{p} + \frac{1}{q} = 1$, find $u \in W_0^{1,p}(G)$ such that

$$\int_G \sum_{i=1}^N |D_i u|^{p-2} D_i u \, D_i v \, \mathrm{d}x = \int_G gv \, \mathrm{d}x \quad \forall v \in W_0^{1,p}(G). \tag{5.22}$$

In this connection, recall that $C_c^\infty(G)$ is dense in $W_0^{1,p}(G)$ and that the elements of $W_0^{1,p}(G)$ have vanishing generalized boundary values. In view of the above discussion, it is reasonable to say that a solution of (5.22) is a *generalized solution* of (5.20).

Parallel to (5.22) we also consider the *variational problem*

$$\int_G \left(\frac{1}{p} \sum_{i=1}^{N} |D_i u|^p - gu \right) dx \longrightarrow \min, \quad u \in W_0^{1,p}(G). \qquad (5.23)$$

The following result ensures, in particular, that under weak assumptions problem (5.20) has a generalized solution.

Proposition 5.5.1 *Let G be a bounded region in $\mathbb{R}^N$ and let $p \in [2, +\infty)$. Then for any $g \in L^q(G)$, where $\frac{1}{p} + \frac{1}{q} = 1$, the problems (5.22) and (5.23) are equivalent and have precisely one solution $u \in W_0^{1,p}(G)$.*

Before verifying this result, we establish two inequalities.

Lemma 5.5.2 *Let $N \in \mathbb{N}$, $p \geq 2$, and $r > 0$ be given.*

(a) *There exists $c_1 > 0$ (depending on N and r) such that*

$$\left(\sum_{i=1}^{N} |x_i| \right)^r \leq c_1 \sum_{i=1}^{N} |x_i|^r \quad \forall\, (x_1, \ldots, x_N) \in \mathbb{R}^N. \qquad (5.24)$$

(b) *There exists $c_2 > 0$ (depending on p) such that*

$$(|\alpha|^{p-2}\alpha - |\beta|^{p-2}\beta)(\alpha - \beta) \geq c_2 |\alpha - \beta|^p \quad \forall\, \alpha, \beta \in \mathbb{R}. \qquad (5.25)$$

Proof. (a) First let $r \geq 1$. Then $\|x\|_r := \left(\sum_{i=1}^N |x_i|^r \right)^{1/r}$ defines a norm on $\mathbb{R}^N$. Since all norms on $\mathbb{R}^N$ are equivalent, there exists $c_1 > 0$ such that $\|x\|_1 \leq c_1^{1/r} \|x\|_r$ for each $x \in \mathbb{R}^N$. This proves (5.24) if $r \geq 1$. If $0 < r < 1$, then argue similarly with $\| \cdot \|_s$, where $s := 1/r$.

(b) First assume that $0 \leq \beta \leq \alpha$. Then

$$\alpha^{p-1} - \beta^{p-1} = (p-1) \int_0^{\alpha-\beta} (t+\beta)^{p-2}\, dt \geq (p-1) \int_0^{\alpha-\beta} t^{p-2}\, dt = (\alpha-\beta)^{p-1},$$

which implies (5.25) in this case. Now let $\beta \leq 0 \leq \alpha$. Then (5.24) shows that $\alpha^{p-1} + |\beta|^{p-1} \geq c(\alpha + |\beta|)^{p-1}$ and again (5.25) follows. The remaining cases can be reduced to the two considered. $\qquad \square$

Proof of Proposition 5.5.1. Our aim is to show that Theorem 5.4.7 applies to

$$E := W_0^{1,p}(G),$$

$$f(u) := \int_G \frac{1}{p} \sum_{i=1}^{N} |D_i u|^p\, dx, \quad b^*(u) := \int_G gu\, dx, \quad u \in E.$$

(I) First observe that the Sobolev space E is a reflexive Banach space.

(II) It is clear that the integral defining $f(u)$ exists for any $u \in E$.

(III) We show that f is G-differentiable on E. Recall that

$$f_G(u,v) = \frac{\partial}{\partial\tau} f(u+\tau v)\Big|_{\tau=0}.$$

We show that partial differentiation by τ and integration over G can be exchanged. For fixed $u,v \in E$, consider the function

$$\gamma(x,\tau) := \frac{1}{p} \sum_{i=1}^{N} |D_i u + \tau D_i v|^p, \quad x \in G, \quad \tau \in (-1,1).$$

Notice that

$$\frac{\partial}{\partial\tau}\gamma(x,\tau) = \frac{1}{p} \sum_{i=1}^{N} |D_i u + \tau D_i v|^{p-2}(D_i u + \tau D_i v) \cdot D_i v$$

and so

$$\left|\frac{\partial}{\partial\tau}\gamma(x,\tau)\right| \leq \frac{1}{p} \sum_{i=1}^{N} |D_i u + \tau D_i v|^{p-1} \cdot |D_i v| \tag{5.26}$$

$$\underset{(5.24)}{\leq} c \sum_{i=1}^{N} (|D_i u|^{p-1} + |D_i v|^{p-1}) \cdot |D_i v|.$$

Under the last sum sign, the first factor is in $L^q(G)$ (notice that $(p-1)q = p$) and the second factor is in $L^p(G)$. Hence the Hölder inequality shows that the right-hand side of (5.26) is integrable over G and so is the left-hand side. Therefore, measure theory tells us that we may write

$$f_G(u,v) = \int_G \frac{\partial}{\partial\tau}\gamma(x,\tau)\Big|_{\tau=0} dx = \int_G \sum_{i=1}^{N} |D_i u|^{p-2}D_i u \cdot D_i v \, dx. \tag{5.27}$$

The function $v \mapsto f_G(u,v)$ is linear, and the following estimate using the Hölder inequality shows that it is also continuous

$$|f_G(u,v)| \leq \sum_{i=1}^{N} \left(\int_G |D_i u|^{(p-1)q}dx\right)^{1/q} \cdot \left(\int_G |D_i v|^p dx\right)^{1/p} \leq c\,\|v\|_E;$$

here, c is independent of v. Hence f is G-differentiable on E, the G-derivative being given by (5.27).

(IV) We show that f' is uniformly monotone. Let $u, \widehat{u} \in E$ be given. Then

$$\langle f'(u) - f'(\widehat{u}), u - \widehat{u}\rangle =$$

$$\int_G \sum_{i=1}^{N} (|D_i u|^{p-2}D_i u - |D_i\widehat{u}|^{p-2}D_i\widehat{u})(D_i u - D_i\widehat{u}) \, dx$$

$$\geq c \int_G \sum_{i=1}^{N} |D_i u - D_i\widehat{u}|^p \, dx = c\,\|u-\widehat{u}\|_{1,p,0}^p \geq \widetilde{c}\,\|u-\widehat{u}\|_{1,p}^p.$$

In this connection, the first inequality holds by Lemma 5.5.2(b) and the second follows from the equivalence of the two norms on E.

(V) In view of the above, the generalized boundary value problem (5.22) is equivalent to the variational equation $\langle f'(u) - b^*, v \rangle = 0$ for all $v \in E$. Hence the assertion is a consequence of Theorem 5.4.7. $\square$

In the special case $p = 2$ we now apply the Ritz method to the problems (5.22) and (5.23). In other words, setting

$$E := W_0^{1,2}(G), \quad a(u,v) := \int_G \sum_{i=1}^{N} D_i u \, D_i v \, dx, \quad \langle b^*, v \rangle := \int_G g v \, dx, \quad (5.28)$$

we consider the problems (cf. (5.16) and (5.17))

$$\tfrac{1}{2} a(\bar{u}, \bar{u}) - \langle b^*, \bar{u} \rangle = \min_{u \in E} \left(\tfrac{1}{2} a(u, u) - \langle b^*, u \rangle \right), \quad (5.29)$$

$$a(\bar{u}, v) - \langle b^*, v \rangle = 0 \quad \forall v \in E, \quad (5.30)$$

as well as the associated Ritz problems corresponding to a basis $\{ z_k \mid k \in \mathbb{N} \}$ of E (cf. (5.16a) and (5.17a)),

$$\tfrac{1}{2} a(\bar{u}_n, \bar{u}_n) - \langle b^*, \bar{u}_n \rangle = \min_{u_n \in E_n} \left(\tfrac{1}{2} a(u_n, u_n) - \langle b^*, u_n \rangle \right), \quad (5.29a)$$

$$a(\bar{u}_n, z_k) - \langle b^*, z_k \rangle = 0 \quad \forall k = 1, \ldots, n. \quad (5.30a)$$

For all $u, v \in E$ we have the following estimates:

$$|a(u,v)| \leq \sum_{i=1}^{N} \int_G |D_i u \, D_i v| \, dx \leq \sum_{i=1}^{N} \|D_i u\|_2 \cdot \|D_i v\|_2 \leq N \|u\|_{1,2} \cdot \|v\|_{1,2},$$

$$|\langle b^*, v \rangle| \leq \|g\|_2 \cdot \|v\|_2 \leq \|g\|_2 \cdot \|v\|_{1,2},$$

$$a(u, u) = \|u\|_{1,2,0}^2 \geq c \|u\|_{1,2}^2.$$

In the first two lines, we applied the Hölder (or Cauchy–Schwarz) inequality. The inequality in the third line is the Poincaré–Friedrichs inequality (which leads to the equivalence of the norms $\| \cdot \|_{1,2,0}$ and $\| \cdot \|_{1,2}$ on E). The above estimates show that the assumptions (A4) of Sect. 5.4 are satisfied. As a consequence of Theorem 5.4.10 we therefore obtain:

Proposition 5.5.3 *With the notation of (5.28), the following holds:*

(a) *The problems (5.29) and (5.30) are equivalent and possess precisely one solution $\bar{u} \in E$.*

(b) *For each $n \in \mathbb{N}$, (5.29a) and (5.30a) are equivalent and possess precisely one solution $\bar{u}_n \in E_n$.*

(c) *The sequence $(\bar{u}_n)$ satisfies*

$$\|\bar{u}_n - \bar{u}\|_{1,2} \leq \tfrac{N}{c} \, d(\bar{u}, E_n) \quad \forall n \in \mathbb{N}.$$

In particular, $(\bar{u}_n)$ converges to $\bar{u}$ as $n \to \infty$.

5.6 Bibliographical Notes and Exercises

Concerning the subject of this chapter, we refer to the Bibliographical Notes at the end of Chap. 4. The optimality conditions for convex optimization problems are due to John [105], and Kuhn and Tucker [114], with Karush [109] as an early (rather late perceived) predecessor.

The presentation in Sect. 5.5 was strongly influenced by Zeidler [221]. More results on the approximation problem in function spaces offer, among others, Braess [27], Holmes [91], Krabs [112], and Laurent [118].

A result analogous to Theorem 5.4.10 holds under the more general assumptions (A3) (if, in addition, E is infinite dimensional and separable); see, for instance, Zeidler [221] or Schirotzek [196]. The theory of monotone operators provides an important generalization concerning both the solvability of variational equations (or inequalities) and the convergence of an approximation method, the *Galerkin method*. For this and many substantial applications we recommend the comprehensive monograph by Zeidler [223, 224]. Concerning the numerical analysis of the Ritz and Galerkin methods we refer to Atkinson and Han [4], and Großmann and Roos [81].

Standard references on the theory of Sobolev spaces are Adams [1] and Ziemer [228].

Exercise 5.6.1 Prove the implications (SP) $\Longleftrightarrow$ (L) $\Longrightarrow$ (Min 2) of Theorem 5.2.1(a).

Exercise 5.6.2 Verify Theorem 5.2.3.
Hint: To see that (Min 2) implies (J), consider the functional $p : E \to \overline{\mathbb{R}}$ defined by

$$p(x) := \begin{cases} \widehat{L}(x; \bar{\lambda}, \bar{\mu}_1, \ldots, \bar{\mu}_m) + \delta_A(x) & \text{if } x \in D, \\ +\infty & \text{if } x \in E \setminus D \end{cases}$$

Exercise 5.6.3 Prove Proposition 5.3.1.

Exercise 5.6.4 Using Proposition 5.2.5(a)$\Leftrightarrow$(b), formulate and verify a characterization of best approximation in $C(T)$, where T is a compact Hausdorff space.

Exercise 5.6.5 Let E be a Hilbert space and $T : E \to E$ be a continuous linear self-adjoint operator. Define $f(x) := \frac{1}{2}(Tx \mid x)$ for all $x \in E$. Show that any local maximizer or minimizer $\bar{x}$ of f over $\{x \in E \mid \|x\| = 1\}$ is an eigenvector of T with associated eigenvalue $\lambda = 2f(\bar{x})$.

Exercise 5.6.6 Let A be a convex subset of an n-dimensional subspace of E and let $f : E \to \overline{\mathbb{R}}$ be continuous at $\bar{x} \in A$. Show that $\bar{x}$ minimizes f over A if and only if for any $i = 1, \ldots, n+1$ there exist $x_i^* \in \partial f(\bar{x})$ and $\lambda_i \geq 0$ satisfying

$$\sum_{i=1}^{n+1} \lambda_i = 1 \quad \text{and} \quad \sum_{i=1}^{n+1} \lambda_i \langle x_i^*, \bar{x} \rangle = \min_{x \in A} \sum_{i=1}^{n+1} \lambda_i \langle x_i^*, x \rangle.$$

Hint: Recall Exercise 4.8.10.

6

Duality of Convex Problems

The idea of *duality* in convex optimization theory is, roughly speaking, the following. With a given optimization problem, called *primal problem* in this context, one associates another problem, called *dual problem*, in such a way that there is a close relationship between the two problems. The motivation is that the dual problem may be easier to solve than the primal one (which is sometimes really the case) or that at least the dual problem furnishes additional information about the primal one.

6.1 Duality in Terms of a Lagrange Function

Assume that

$A \subseteq E$ is a nonempty convex set,

$f, g_1, \ldots, g_m : A \to \mathbb{R}$ are convex functionals,

$M := \{x \in A \mid g_i(x) \leq 0 \ (i = 1, \ldots, m)\}, \ \bar{x} \in M.$

We continue studying the minimization of f on M (cf. Sect. 5.2). We set

$$\alpha := \inf_{x \in M} f(x). \tag{6.1}$$

Further we define the *Lagrange functional* $L : A \times \mathbb{R}_+^m \to \mathbb{R}$ by

$$L(x, q) := f(x) + \sum_{i=1}^{m} q_i \, g_i(x) \quad \forall x \in A \ \forall q = (q_1, \ldots, q_m) \in \mathbb{R}_+^m. \tag{6.2}$$

We then have

$$\sup_{q \in \mathbb{R}_+^m} L(x, q) = \begin{cases} f(x) & \text{if } x \in M, \\ +\infty & \text{if } x \in A \setminus M \end{cases}$$

and so we obtain

$$\alpha = \inf_{x \in A} \sup_{q \in \mathbb{R}_+^m} L(x, q). \tag{6.3}$$

Parallel to (6.3) we now consider

$$\beta := \sup_{q \in \mathbb{R}_+^m} \inf_{x \in A} L(x, q). \tag{6.4}$$

The original minimization problem, which is described by (6.1) and so by (6.3), is called *primal problem*. The associated problem described by (6.4) is called *dual problem*. We say that $\bar{x} \in A$ is a *solution of the primal problem* (6.3) if $\bar{x} \in M$ and $f(\bar{x}) = \alpha$, thus if

$$\sup_{q \in \mathbb{R}_+^m} L(\bar{x}, q) = \inf_{x \in A} \sup_{q \in \mathbb{R}_+^m} L(x, q).$$

Analogously, $\bar{q} \in \mathbb{R}_+^m$ is said to be a *solution of the dual problem* (6.4) if

$$\inf_{x \in A} L(x, \bar{q}) = \sup_{q \in \mathbb{R}_+^m} \inf_{x \in A} L(x, q).$$

Theorem 6.1.1 establishes relationships between the two problems. In this connection, we again need the Slater condition (cf. (5.2))

$$\exists x_0 \in A : \ g_i(x_0) < 0 \text{ for } i = 1, \ldots, m. \tag{6.5}$$

Theorem 6.1.1 (Lagrange Duality) *Assume that the space E is reflexive, the subset A is nonempty, closed, and convex, and the functionals $f, g_1, \ldots, g_m$ are l.s.c. and convex. Assume further that the Slater condition (6.5) is satisfied. Then:*

(i) *One has $\alpha = \beta$, and the dual problem (6.4) has a solution $\bar{q} \in \mathbb{R}_+^m$.*
(ii) *For $\bar{x} \in M$, the following statements are equivalent:*
   (a) *$\bar{x}$ is a solution of the primal problem (6.3).*
   (b) *The Lagrange function L has a saddle point $(\bar{x}, \bar{q})$ with respect to $A \times \mathbb{R}_+^m$.*
(iii) *If (b) holds, then $\bar{q}$ is a solution of the dual problem (6.4), and one has $\bar{q}_i\, g_i(\bar{x}) = 0$ for $i = 1, \ldots, m$.*

Theorem 6.1.1 will follow from Theorem 6.1.3 below.

Remark 6.1.2

(a) Comparing Theorem 6.1.1 with Theorem 5.2.1, we see that the "dual variable" q is nothing else than the Lagrange multiplier vector, i.e., we have $q_i = \mu_i$ for $i = 1, \ldots, m$.
(b) If the Slater condition is not satisfied, a *duality gap* may occur which means that we have $\alpha > \beta$. A trivial example is $E = \mathbb{R}$, $A = [-1, +\infty)$, $f(x) = -x$, $g(x) = 1$, where we have $\alpha = +\infty$ and $\beta = -\infty$. An example

with finite and different optimal values will be given below in a somewhat different context (Example 6.1.7).

Now we replace the Lagrange functional L of (6.2) by an arbitrary functional $L : A \times B \to \mathbb{R}$ and consider

$$\alpha := \inf_{x \in A} \sup_{q \in B} L(x, q) \quad \text{(primal problem)}, \tag{6.6}$$

$$\beta := \sup_{q \in B} \inf_{x \in A} L(x, q) \quad \text{(dual problem)}. \tag{6.7}$$

Theorem 6.1.3 (General Duality Theorem) *Assume that E, F are reflexive Banach spaces, that $A \subseteq E$ and $B \subseteq F$ are nonempty, closed, and convex, and that $L : A \times B \to \mathbb{R}$ satisfies*

$x \mapsto L(x, q)$ *is l.s.c. and convex on A for each $q \in B$,*

$q \mapsto -L(x, q)$ *is l.s.c. and convex on B for each $x \in A$.*

Then:

(i) *One has $\alpha \geq \beta$ (weak duality).*

(ii) *For $(\bar{x}, \bar{q}) \in A \times B$, the following statements are equivalent:*
   (a) *$\bar{x}$ is a solution of (6.6), $\bar{q}$ is a solution of (6.7), and one has $\alpha = \beta$ (strong duality).*
   (b) *$(\bar{x}, \bar{q})$ is a saddle point of L with respect to $A \times B$.*

(iii) *Assume, in addition, that either A is bounded or $L(x, q_0) \to +\infty$ as $\|x\| \to +\infty$, $x \in A$, for some $q_0 \in B$. Assume further that $\alpha < +\infty$. Then the primal problem (6.6) has a solution $\bar{x} \in A$ and one has $\alpha = \beta$.*

(iv) *Assume, in addition, that either B is bounded or $-L(x_0, q) \to +\infty$ as $\|q\| \to +\infty$, $q \in B$, for some $x_0 \in A$. Assume further that $\beta > -\infty$. Then the dual problem (6.7) has a solution $\bar{q} \in B$ and one has $\alpha = \beta$.*

Remarks on the Proof of Theorem 6.1.3. The proof consists of two main steps. First, the result is verified under the additional assumption that A and B are bounded. In this case, Neumann's minimax theorem can be applied, which in turn is proved with the aid of Brouwer's fixed point theorem. Second, the general case is reduced to the first one. If, say, A is unbounded, then replace A by $A_n := \{x \in A \mid \|x\| \leq n\}$ and L by $L_n(x, q) := L(x, q) + \frac{1}{n}\|x\|^2$. The proof can be found, e.g., in Zeidler [221]. □

Proof of Theorem 6.1.1. Set $F := \mathbb{R}^m$, $B := \mathbb{R}^m_+$, and notice that the Slater condition (6.5) implies

$$L(x_0, q) = f(x_0) + \sum_{i=1}^m q_i\, g_i(x_0) \to -\infty \text{ as } \|q\| \to +\infty, \, q \in \mathbb{R}^m_+. \qquad \square$$

In view of Theorem 6.1.3, we can set up a general dualization principle.

General Dualization Principle

Given the minimum problem $f(x) \to \min$, $x \in M$, find sets A, B and a function $L : A \times B \to \mathbb{R}$ such that

$$\inf_{x \in M} f(x) = \inf_{x \in A} \sup_{q \in B} L(x, q)$$

and consider the following dual problem:

$$\sup_{q \in B} \inf_{x \in A} L(x, q).$$

Special Case

Here, we apply the dualization principle to problems of the form

$$\alpha := \inf_{x \in E} \big(f(x) + h(Tx - a) \big). \tag{6.8}$$

Below we shall see that the problem of linear optimization is of this form. In connection with problem (6.8) we make the following assumptions:

(A) E and F are reflexive Banach spaces,
    $f : E \to \overline{\mathbb{R}}$ and $h : F \to \overline{\mathbb{R}}$ are proper, convex, and l.s.c.,
    $T : E \to F$ is linear and continuous, $a \in F$.

   We set

$$A := \operatorname{dom} f, \ B := \operatorname{dom} h^*,$$
$$L(x, q) := f(x) + \langle q, Tx - a \rangle - h^*(q) \quad \forall x \in A \ \forall q \in B. \tag{6.9}$$

It will turn out that

$$\alpha = \inf_{x \in A} \sup_{q \in B} L(x, q). \tag{6.10}$$

By the general dualization principle, the dual to (6.10) and so to (6.8) is

$$\beta := \sup_{q \in B} \inf_{x \in A} L(x, q). \tag{6.11}$$

This will be shown to coincide with

$$\beta = \sup_{q \in B} \big(-f^*(-T^*q) - h^*(q) - \langle q, a \rangle \big). \tag{6.12}$$

Notice that h^* denotes the conjugate of the functional h while T^* denotes the adjoint of the operator T.

Proposition 6.1.4 *Under the assumptions* (A), *the problem dual to* (6.8) *is* (6.12). *Hence all statements of Theorem 6.1.3 hold for* (6.8) *and* (6.12).

Proof.

(I) By Theorem 2.2.4, we have $h = h^{**}$ and so

$$h(Tx - a) = h^{**}(Tx - a) = \sup_{q \in F^*} \left(\langle q, Tx - a \rangle - h^*(q) \right),$$

which entails

$$\alpha = \inf_{x \in E} \sup_{q \in F^*} \left(f(x) + \langle q, Tx - a \rangle - h^*(q) \right) = \inf_{x \in A} \sup_{q \in B} \left(\dots \right),$$

and the latter term equals (6.10).

(II) We obtain

$$\inf_{x \in A} L(x, q) = - \sup_{x \in A} \left(- \langle T^*q, x \rangle - f(x) + h^*(q) + \langle q, a \rangle \right)$$

$$= -f^*(-T^*q) - h^*(q) - \langle q, a \rangle,$$

which shows that (6.11) coincides with (6.12).   □

Example 6.1.5 We now apply Proposition 6.1.4 to the *problem of linear optimization* which is

$$\alpha := \inf \{ \langle c, x \rangle \mid x \in P_E, \, Tx - a \in P_F \}. \tag{6.13}$$

We make the following assumptions:

$(\tilde{A})$ E and F are reflexive Banach spaces,
P_E and P_F are closed convex cones in E and F, respectively,
$T : E \to F$ is linear and continuous, $c \in E^*$, $a \in F$.

Setting

$$f(x) := \begin{cases} \langle c, x \rangle & \text{if } x \in P_E, \\ +\infty & \text{otherwise,} \end{cases} \quad h(b) := \begin{cases} 0 & \text{if } b \in P_F, \\ +\infty & \text{otherwise,} \end{cases} \tag{6.14}$$

we see that problem (6.13) is of the form (6.8). Hence the dual problem is (6.12), with f and h according to (6.14). In Exercise 6.5.1 it is shown that this dual problem is equivalent to

$$\beta = \sup \{ \langle q, a \rangle \mid q \in -P_F^\circ, \, c - T^*q \in -P_E^\circ \}. \tag{6.15}$$

Recall that P_E° denotes the negative polar cone to P_E. By Proposition 6.1.4 we have the following result.

Corollary 6.1.6 *Under the assumptions $(\tilde{A})$, all assertions of Theorem 6.1.3 apply to the problems (6.13) and (6.15).*

Example 6.1.7 We show that, in the situation of Corollary 6.1.6, a duality gap with finite optimal values may occur (cf. Remark 6.1.2). Let $E = F := \mathbb{R}^3$, $Tx := (0, x_3, x_1)^\top$ for $x = (x_1, x_2, x_3)^\top \in \mathbb{R}^3$, $c := (0, 0, 1)^\top$, and $a := (0, -1, 0)^\top$. Equip $\mathbb{R}^3$ with the Euclidean norm $\|\cdot\|_2$. Further let $P_E = P_F := K$, where

$$K := \{(x_1, x_2, x_3)^\top \in \mathbb{R}^3 \mid x_1 \geq 0,\ x_2 \geq 0,\ 2x_1 x_2 \geq x_3^2\}.$$

It is easy to see that with $z_0 := (1, 1, 0)^\top$, we have

$$K = \{x \in \mathbb{R}^3 \mid z_0^\top x \geq \|z_0\|_2 \|x\|_2 \cos(\pi/4)\}.$$

Hence K consists of all vectors x whose angle with the vector z_0 is not greater than $\pi/4$. Observe that K is a closed convex cone ("ice cream cone") and that $K^\circ = -K$. Further we have $T^*q = (q_3, 0, q_2)^\top$ for $q = (q_1, q_2, q_3)^\top \in \mathbb{R}^3$. Thus the problems (6.13) and (6.15) read, respectively,

$$\alpha = \inf\{x_3 \mid x \in K,\ (0, x_3 - 1, x_1)^\top \in K\},$$

$$\beta = \sup\{-q_2 \mid q \in K,\ (-q_3, 0, 1 - q_2)^\top \in K\}.$$

It follows that $\alpha = 0$ and $\beta = -1$. Notice that the primal problem and the dual problem have infinitely many solutions.

Exercise 6.5.5 presents an example where the optimal values of the primal and the dual problem coincide but the primal problem has no solution.

6.2 Lagrange Duality and Gâteaux Differentiable Functionals

Let $w \in E^*$. With $f(x) := -\langle w, x \rangle$, $x \in E$, and $a := o$, the primal problem (6.8) and the dual problem (6.12), respectively, read as follows:

$$\alpha := \inf_{x \in E} \big(h(Tx) - \langle w, x \rangle\big), \qquad (6.16)$$

$$\beta := \sup_{q \in B} \big(-f^*(-T^*q) - h^*(q)\big).$$

We have

$$f^*(-T^*q) = \sup_{x \in E} \big(\langle -T^*q, x \rangle + \langle w, x \rangle\big) = \begin{cases} 0 & \text{if } w = T^*q, \\ +\infty & \text{otherwise.} \end{cases}$$

The assumptions in Proposition 6.2.1 will entail that $B := \operatorname{dom} h^* = F^*$. Setting

$$K := \{q \in F^* \mid w = T^*q\}, \qquad (6.17)$$

we thus see that the problem dual to (6.16) is

$$\beta = \sup_{q \in K} \big(-h^*(q)\big). \qquad (6.18)$$

Proposition 6.2.1 *Assume that:*

E and F are reflexive Banach spaces, $h : F \to \mathbb{R}$ is G-differentiable,
$h' : F \to F^*$ is uniformly monotone with constants $c > 0$ and $\gamma > 1$,
$T : E \to F$ is linear and isometric (i.e., $\|Tx\| = \|x\| \ \forall x \in E$).

Then:

(a) *Problem (6.16) has precisely one solution $\bar{x} \in E$, problem (6.18) has precisely one solution $\bar{q} \in K$, and one has $\alpha = \beta$.*
(b) *$x = \bar{x}$ is also the unique solution of*

$$\langle h'(Tx), Ty \rangle = \langle w, y \rangle \quad \forall y \in E, \tag{6.19}$$

and $q = \bar{q}$ is also the unique solution of

$$\langle (h')^{-1}(q), p - q \rangle \geq 0 \quad \forall p \in K. \tag{6.20}$$

(c) *One has $\bar{q} = h'(T\bar{x})$.*
(d) *The following error estimates hold for α and $\bar{x}$:*

$$-h^*(q) \leq \alpha \leq h(Tx) - \langle w, x \rangle \quad \forall x \in E \ \forall q \in K, \tag{6.21}$$

$$\frac{c}{\gamma} \|x - \bar{x}\|^\gamma \leq h(Tx) - \langle w, x \rangle + h^*(q) \quad \forall x \in E \ \forall q \in K. \tag{6.22}$$

Proof.

(I) Setting $g(x) := h(Tx)$, $x \in E$, we obtain $g'(x) = T^*h'(Tx)$, $x \in E$, and

$$\begin{aligned}
\langle g'(y) - g'(x), y - x \rangle &= \langle h'(Ty) - h'(Tx), Ty - Tx \rangle \\
&\geq c\|Ty - Tx\|^\gamma = c\|y - x\|^\gamma.
\end{aligned} \tag{6.23}$$

Hence by Theorem 5.4.7, the problem $g(x) - \langle w, x \rangle \to \min$, $x \in E$, and so (6.16) has precisely one solution $\bar{x} \in E$; this is also the unique solution of $\langle g'(\bar{x}) - w, y \rangle = 0$ for any $y \in E$ and so of (6.19).

(II) We show that $\bar{q} := h'(T\bar{x})$ is the unique solution of (6.18). By (6.19), we obtain $\langle \bar{q}, Ty \rangle = \langle w, y \rangle$ for all $y \in E$ and so $T^*\bar{q} = w$ which means that $\bar{q} \in K$. The Young inequality implies

$$h(Tx) + h^*(q) \geq \langle q, Tx \rangle = \langle T^*q, x \rangle = \langle w, x \rangle \quad \forall (x, q) \in E \times K,$$

and it follows that $\alpha \geq \beta$. Since $\{\bar{q}\} = \partial h(T\bar{x})$, we further obtain by Proposition 4.4.1,

$$h(T\bar{x}) + h^*(\bar{q}) = \langle \bar{q}, T\bar{x} \rangle = \langle w, \bar{x} \rangle.$$

We thus conclude that $\alpha = \beta$ and $\bar{q}$ is a solution of (6.18).

(III) The element $\bar{q}$ is the only solution of (6.18) because K is a convex set and h^* is a strictly convex functional (cf. Theorem 5.4.6). Concerning the latter, recall that $(h^*)' = (h')^{-1}$ by Proposition 4.4.3, notice that $(h')^{-1}$ is strictly monotone and apply Proposition 4.3.5 to h^*.

(IV) By Remark 5.2.4(c) (cf. also Proposition 5.2.5), $q \in K$ is a solution of (6.18) if and only if $\langle (h^*)'(q), p - q \rangle \geq 0$ for all $p \in K$ which is equivalent to (6.20).

(V) It is left as Exercise 6.5.2 to verify (d).   □

6.3 Duality of Boundary Value Problems

We adopt the notation introduced at the beginning of Sect. 5.5. Consider the *classical boundary value problem*

$$Au(x) = g(x) \;\forall x \in G, \quad u(x) = 0 \;\forall x \in \partial G, \tag{6.24}$$

where A denotes the linear second-order differential operator defined by

$$Au(x) := -\sum_{i,j=1}^{N} D_i\big(a_{ij}(x)\, D_j u(x)\big).$$

Recall that A is said to be *strongly elliptic* if there exists a constant $c > 0$ such that

$$\sum_{i,j=1}^{N} a_{ij}(x)\, y_i y_j \geq c \sum_{i=1}^{N} y_i^2 \quad \forall x \in G \;\; \forall (y_1, \dots, y_N) \in \mathbb{R}^N.$$

Set

$$E := W_0^{1,2}(G).$$

Then the *generalized problem* associated with (6.24) reads as follows. Find $u \in E$ satisfying

$$\int_G \sum_{i,j=1}^{N} a_{ij}(x) D_j u(x)\, D_i v(x)\, \mathrm{d}x = \int_G g(x)\, v(x)\, \mathrm{d}x \quad \forall v \in E. \tag{6.25}$$

Heuristically, this is obtained from the classical problem by multiplying the differential equation of (6.24) by $v \in C_c^{\infty}(G)$ and partial integration (cf. Sect. 5.5). Parallel to (6.25) we consider the following *minimum problem*:

$$\alpha := \inf_{u \in E} \int_G \left(\frac{1}{2} \sum_{i,j=1}^{N} a_{ij}(x) D_i u(x)\, D_j u(x) - g(x)\, u(x) \right) \mathrm{d}x. \tag{6.26}$$

Our aim is to apply Proposition 6.2.1. In this connection, notice that we now denote the "primal variable" by u instead of x, while we go on denoting the "dual variable" by q. The assumptions in Proposition 6.3.1 will ensure that (6.26) is a convex problem of the form $f(u) \to \min$, $u \in E$, and (6.25) is the equivalent variational equation $\langle f'(u), v \rangle = 0$ for all $v \in E$ (Corollary 5.1.2). Moreover, below we shall show that the problem dual to (6.26) is

$$\beta := \sup_{q \in K} \left(-\int_G \frac{1}{2} \sum_{i,j=1}^{N} a_{ij}^{(-1)}(x)\, q_i(x)\, q_j(x)\, dx \right). \tag{6.27}$$

Here, $\left(a_{ij}^{(-1)}(x)\right)$ denotes the inverse of the matrix $\left(a_{ij}(x)\right)$. Further, we set

$$F := L_N^2(G) := \underbrace{L^2(G) \times \cdots \times L^2(G)}_{N\text{-times}},$$

$$\|q\|_2 := \left(\sum_{i=1}^{N} \int_G |q_i(x)|^2\, dx \right)^{\frac{1}{2}} \quad \forall q = (q_1, \ldots, q_N) \in F,$$

$$K := \left\{ q \in F^* \,\middle|\, \int_G \sum_{i=1}^{N} q_i(x)\, D_i v(x)\, dx = \int_G g(x)\, v(x)\, dx \quad \forall v \in E \right\},$$

$$Tu(x) := \left(D_1 u(x), \ldots, D_N u(x) \right) \quad \forall x \in G \,\forall u \in E,$$

$$\langle w, u \rangle := \int_G g(x)\, u(x)\, dx \quad \forall u \in E,$$

$$h(v) := \frac{1}{2} \int_G \sum_{i,j=1}^{N} a_{ij}(x)\, v_i(x)\, v_j(x)\, dx \quad \forall v \in F.$$

Proposition 6.3.1 *Assume that, for $i, j = 1, \ldots, N$, the functions $a_{ij} : G \to \mathbb{R}$ are continuous, bounded, and symmetric (i.e., $a_{ij} = a_{ji}$). Further assume that the differential operator A is strongly elliptic, with the constant $c > 0$. Let $g \in L^2(G)$ be given. Then:*

(a) *The problem (6.26) has precisely one solution $\bar{u} \in E$, the problem (6.27) has precisely one solution $\bar{q} \in K$, and one has $\alpha = \beta$.*

(b) *The problems (6.25) and (6.26) are equivalent and so $\bar{u}$ is also the unique solution of (6.25).*

(c) *For $i = 1, \ldots, N$, one has $\bar{q}_j = \sum_{i=1}^{N} a_{ij}\, D_i \bar{u}$.*

(d) *The following error estimates hold for α and $\bar{u}$:*

$$-h^*(q) \leq \alpha \leq h(Tu) - \langle w, u \rangle \quad \forall u \in E \,\forall q \in K,$$

$$\frac{c}{2}\|u - \bar{u}\|_{1,2,0}^2 \leq h(Tu) - \langle w, u \rangle + h^*(q) \quad \forall u \in E \,\forall q \in K.$$

Proof. We show that Proposition 6.2.1 applies.

(I) Notice that $T : E \to F$ is linear and isometric. The Hölder inequality shows that $w \in E^*$. Moreover, we obtain

$$\langle h'(v), r \rangle = \frac{\partial}{\partial \tau} h(v + \tau r)\big|_{\tau=0} = \int_G \sum_{i,j=1}^N a_{ij} v_i r_j \, dx \quad \forall v, r \in F, \quad (6.28)$$

which implies that h' is strongly monotone (as A is strongly elliptic).

(II) By Proposition 4.4.3 we have

$$h^*(q) = h^*(o) + \int_0^1 \langle q, (h')^{-1}(\tau q) \rangle \, d\tau \quad \forall q \in F^*, \quad (6.29)$$

$$h^*(o) = -h\big((h')^{-1}(o)\big).$$

Since $h'(o) = o$, it follows that $h^*(o) = -h(o) = 0$. Let $\tau \in [0,1]$ and $q \in F^*$ be given. By the bijectivity of h' there exists $v \in F$ satisfying $v = (h')^{-1}(\tau q)$. In view of (6.28), we conclude that

$$\tau q_j(x) = (\tau q)_j(x) = \sum_{i=1}^N a_{ij}(x) v_i(x) \quad \text{for almost all } x \in G$$

and so

$$v_i(x) = \sum_{j=1}^N a_{ij}^{(-1)}(x) \tau \, q_j(x) \quad \text{for almost all } x \in G \text{ and } i = 1, \dots, N.$$

Inserting this for the ith coordinate of $(h')^{-1}(\tau q)$ into (6.29), we obtain

$$h^*(q) = \int_0^1 \left(\int_G \sum_{i,j=1}^N q_i \, a_{ij}^{(-1)} \, \tau \, q_j \, dx \right) d\tau = \int_G \frac{1}{2} \sum_{i,j=1}^N a_{ij}^{(-1)} q_i \, q_j \, dx.$$

$$(6.30)$$

Finally we have

$$K = \{ q \in F^* \mid \langle q, Tv \rangle_F = \langle w, v \rangle_E \; \forall v \in E \} = \{ q \in F^* \mid T^* q = w \}.$$

The assertions now follow from Proposition 6.2.1.   □

Remark 6.3.2 [Smooth Data] If the boundary ∂G as well as the functions a_{ij} and g are sufficiently smooth, then the solution $\bar{u}$ of (6.25) and (6.26) is an element of $C^2(\overline{G})$ and so is also a classical solution of (6.24). Set

$$R := \{ u \in C^2(\overline{G}) \mid u = o \text{ on } \partial G \},$$

$$S := \{ v \in C^2(\overline{G}) \mid Av = g \text{ on } G \}.$$

Since $R \subseteq E$ and $\bar{u} \in R$, we have $\alpha = \inf_{u \in R} \int_G (\cdots)\, dx$ (cf. (6.26)) and so

$$\alpha = \inf_{u \in R} \big(h(Tu) - \langle w, u \rangle \big). \tag{6.31}$$

Define

$$Q := \Big\{ (q_1, \ldots, q_N) \in F^* \mid q_i = \sum_{j=1}^{N} a_{ij} D_j v \text{ for } i = 1, \ldots, N,\ v \in S \Big\}.$$

If $q \in Q$, then

$$-\sum_{i=1}^{N} D_i q_i = Av = g \tag{6.32}$$

and so $q \in K$; the latter follows by multiplying (6.32) by $\tilde{v} \in C_c^\infty(G)$ and partial integration. Moreover, we have

$$\bar{q} = h'(T\bar{u}) = \sum_{i,j=1}^{N} a_{ij} D_i \bar{u} \in Q;$$

here, the second equality follows by Proposition 6.3.1(c). In view of (6.27), we thus obtain $\beta = \sup_{q \in Q} \big(-h^*(q) \big)$. From (6.30) we deduce that for $q \in Q$ we have $h^*(q) = h(Tv)$ with some $v \in S$ so that we finally obtain

$$\beta = \sup_{v \in S} \big(-h(Tv) \big) = - \inf_{v \in S} h(Tv). \tag{6.33}$$

Hence for smooth data, the statements of Proposition 6.3.1 hold with α and β according to (6.31) and (6.33), respectively. In this connection, we have the error estimates

$$-h(Tv) \le \alpha \le h(Tu) - \langle w, u \rangle \quad \forall u \in R\, \forall v \in S, \tag{6.34}$$

$$\frac{c}{2} \|u - \bar{u}\|_{1,2,0}^2 \le h(Tu) - \langle w, u \rangle + h(Tv) \quad \forall u \in R\, \forall v \in S. \tag{6.35}$$

By applying the Ritz method to (6.33), a sequence $(v^{(n)})$ in S is obtained which can be used in (6.34) to successively improve the lower bound for α according to $\alpha \ge -h\big(Tv^{(n)}\big)$. This is the *Trefftz method*. Notice that the elements $u \in R$ and $v \in S$ only need to satisfy the boundary condition *or* the differential equation.

Example 6.3.3 Consider the differential operator $A = -\Delta = -\sum_{i=1}^{N} D_i D_i$. Then for $u \in C^2(\overline{G})$, we have the boundary value problem

$$-\Delta u = g \text{ on } G, \quad u = o \text{ on } \partial G$$

According to (6.31), the associated minimum problem is

$$\alpha = \inf \left\{ \int_G \left(\frac{1}{2} \sum_{i=1}^{n} (D_i u)^2 - gu \right) dx \,\Big|\, u = o \text{ on } \partial G \right\}$$

and according to (6.34), the dual problem is

$$\beta = \sup \left\{ -\int_G \frac{1}{2} \sum_{i=1}^{N} (D_i v)^2 \, dx \,\Big|\, -\Delta v = g \text{ on } G \right\}.$$

6.4 Duality in Terms of Conjugate Functions

In this section we present another approach to duality, which we first explain in a special case.

Example 6.4.1 As in Example 6.1.5 and with the same notation, we consider the problem of linear optimization

$$\alpha := \inf\{\langle c, x \rangle \,|\, x \in P_E , \, Tx - a \in P_F\}.$$

We now perturb the constraint $Tx - a \in P_F$ by a linear parameter b, i.e., we pass to the perturbed problem

$$S(b) := \inf\{\langle c, x \rangle \,|\, x \in P_E , \, Tx - a - b \in P_F\}.$$

The original problem is $\alpha = S(o)$. Again we formalize the above procedure. Setting

$$\tilde{f}(x) := \begin{cases} \langle c, x \rangle & \text{if } x \in P_E , \\ +\infty & \text{if } x \in E \setminus P_E , \end{cases}$$

$$h(\hat{b}) := \begin{cases} 0 & \text{if } \hat{b} \in P_F , \\ +\infty & \text{if } \hat{b} \in F \setminus P_F , \end{cases}$$

$$f(x) := \tilde{f}(x) + h(Tx - a), \quad x \in E,$$

$$M(x, b) := \tilde{f}(x) + h(Tx - a - b) \quad \forall x \in E \,\forall b \in F,$$

we have

$$S(b) = \inf_{x \in E} M(x, b) \quad \forall b \in F,$$

$$f(x) = M(x, o) \quad \forall x \in E,$$

$$\alpha = S(o) = \inf_{x \in E} f(x).$$

The functional $M : E \times F \to \overline{\mathbb{R}}$ can be interpreted as a perturbation of f.

Now we consider a more general setting. Let the following problem be given

$$\alpha := \inf_{x \in E} f(x), \qquad (6.36)$$

where $f : E \to \overline{\mathbb{R}}$. Choose another normed vector space F and a functional $M : E \times F \to \overline{\mathbb{R}}$ such that $f(x) = M(x,o)$ for each $x \in E$ and consider the following problems:

$$S(b) := \inf_{x \in E} M(x,b), \ b \in F \qquad \text{(perturbed problem)}, \qquad (6.37)$$

$$-\hat{S}(v) := \sup_{q \in F^*} \left(-M^*(v,q)\right), \ v \in E^* \qquad \text{(problem dual to (6.37))}, \qquad (6.38)$$

$$\beta := -\hat{S}(o) = \sup_{q \in F^*} \left(-M^*(o,q)\right) \qquad \text{(problem dual to (6.36))}. \qquad (6.39)$$

Definition 6.4.2 The functional $S : F \to \overline{\mathbb{R}}$ defined by (6.37) is called *value functional* or *marginal functional*. The problems (6.36) and (6.39) are said to be *stable* if $\partial S(o) \neq \emptyset$ and $\partial \hat{S}(o) \neq \emptyset$, respectively.

We make the following assumptions:

(A1) E and F are normed vector spaces, $f : E \to \overline{\mathbb{R}}$ is proper, convex, and l.s.c., $M : E \times F \to \overline{\mathbb{R}}$ is proper, convex, and l.s.c., $M(x,o) = f(x) \ \forall \ x \in E$,
there exist $x_0 \in E$ and $q_0 \in F^*$ satisfying $f(x_0) < +\infty$ and $M^*(o,q_0) < +\infty$.

It turns out that, under the above assumptions, the duality between the problems (6.36) and (6.39) can be completely characterized by means of the value functional S and the perturbation M. Theorem 6.4.3 is one of the central results of duality theory.

Theorem 6.4.3 *Under the assumptions* (A1), *the following holds:*

(i)   *One has* $-\infty < \beta \leq \alpha < +\infty$.
(ii) *The following statements are equivalent:*
      (a) *Problem* (6.36) *has a solution and* $\alpha = \beta$.
      (b) *Problem* (6.39) *is stable.*
(iii) *The following statements are equivalent:*
      (a') *Problem* (6.39) *has a solution and* $\alpha = \beta$.
      (b') *Problem* (6.36) *is stable.*
(iv) *If* $\alpha = \beta$, *then*
         $\partial \hat{S}(o) = $ *solution set of* (6.36), $\partial S(o) = $ *solution set of* (6.39).
(v)   *The following statements are equivalent:*
      (a'') $x \in E$ *is a solution of* (6.36), $q \in F^*$ *is a solution of* (6.39), *and* $\alpha = \beta$.

(b″) $M(x,o) + M^*(o,q) = 0$.
(c″) $(o,q) \in \partial M(x,o)$.

Proof.

(I) First recall that $(E \times F)^*$ can and will be identified with $E^* \times F^*$ according to

$$\langle (v,q), (x,b) \rangle = \langle v,x \rangle + \langle q,b \rangle \quad \forall\, (x,b) \in E \times F \,\forall\, (v,q) \in E^* \times F^*.$$

Using this, we have

$$M^*(v,q) = \sup_{\substack{x \in E \\ b \in F}} \big(\langle v,x \rangle + \langle q,b \rangle - M(x,b) \big). \tag{6.40}$$

Furthermore, we obtain

$$\begin{aligned} S^*(q) &= \sup_{b \in F} \big(\langle q,b \rangle - S(b) \big) = \sup_{b \in F} \sup_{x \in E} \big(\langle q,b \rangle - M(x,b) \big) \\ &= \sup_{\substack{x \in E \\ b \in F}} \big(\langle q,b \rangle - M(x,b) \big) = M^*(o,q) \end{aligned} \tag{6.41}$$

and so

$$S^{**}(o) = \sup_{q \in F^*} \big(o - S^*(q) \big) = \sup_{q \in F^*} \big(-M^*(o,q) \big) = \beta. \tag{6.42}$$

(II) From (6.39) and (6.38) we see that

$$-\beta = \inf_{q \in F^*} M^*(o,q), \tag{6.43}$$

$$\hat{S}(v) = \inf_{q \in F^*} M^*(v,q), \quad v \in E^*. \tag{6.44}$$

Recalling that the dual of $F^*[\sigma(F^*, F)]$ and $E^*[\sigma(E^*, E)]$ can be identified with F and E, respectively, and comparing the relationship between (6.37) and (6.38), we conclude that the problem dual to (6.44) is

$$-\hat{\hat{S}}(b) := \sup_{x \in E} \big(-M^{**}(x,b) \big) = \sup_{x \in E} \big(-M(x,b) \big), \quad b \in F.$$

Therefore, the problem dual to (6.40) and so to (6.39) is

$$-\hat{\hat{S}}(o) = \sup_{x \in E} \big(-M(x,o) \big) = \inf_{x \in E} M(x,o) = \alpha,$$

which is the primal problem.

(III) It is left as an exercise to show that S is convex.

(IV) Ad (i). The hypotheses imply that $\alpha < +\infty$ and $\beta > -\infty$. Applying the Young inequality to M, we obtain

$$M(x,o) + M^*(o,q) \geq \langle o,x \rangle + \langle q,o \rangle = 0$$

which, by passing to the infimum over all $(x,q) \in E \times F^*$, yields $\alpha - \beta \geq 0$.
Ad (v). This follows by applying Proposition 4.4.1 to M.
Ad (iv). Assume that $\alpha = \beta$. Then it follows that

$$q \text{ is a solution of (6.39)} \iff -S^*(q) \underset{(6.41)}{=} -M^*(o,q) \underset{(6.39)}{=} \beta = \alpha = S(o)$$

$$\iff q \in \partial S(o);$$

here, the last equivalence holds by Proposition 4.4.1. According to
step (II), the problem dual to (6.39) is just the original problem (6.36).
Therefore, in analogy to the above, we have

$$x \text{ is a solution of (6.36)} \iff x \in \partial \hat{S}(o).$$

Ad (iii). (a') $\Longrightarrow$ (b'): By (iv), (a') implies that $\partial S(o) \neq \emptyset$ and so (6.36)
is stable.
(b') $\Longrightarrow$ (a'): Let $v \in \partial S(o)$. It follows from Proposition 4.4.1 that

$$S(o) = \langle v, o \rangle - S^*(v) \leq S^{**}(o).$$

On the other hand, we have $S^{**} \leq S$. Hence $\alpha = S(o) = S^{**}(o) = \beta$; here
the last equation is a consequence of (6.42). Moreover, v is a solution of
(6.39) by (iv).
Ad (ii). This follows from (iii) and step (II).   □

Theorem 6.4.3 reveals the importance of the stability concept. Lemma 6.4.4
provides a sufficient condition for stability.

Lemma 6.4.4 *Let the assumptions* (A1) *be fulfilled.*

(a) *If* $b \mapsto M(x_1, b)$ *is continuous at* $b = o$ *for some* $x_1 \in E$, *then the primal
problem* (6.36) *is stable.*
(b) *If* $v \mapsto M^*(v, q_1)$ *is continuous at* $v = o$ *for some* $q_1 \in F^*$, *then the dual
problem* (6.39) *is stable.*

Proof.

(a) By assumption, there exist a number $k > 0$ and a neighborhood U of o in
F such that

$$S(b) = \inf_{x \in E} M(x,b) \leq M(x_1, b) \leq k \quad \forall b \in U.$$

Since the functional S is also convex (cf. the remark in step (III) of the
proof of Theorem 6.4.3), it is continuous at o by Theorem 1.4.1, therefore
$\partial S(o) \neq \emptyset$ by Proposition 4.1.6.
(b) This is proved analogously.   □

Special Case

Now we make the following assumptions:

(A2) $f : E \to \overline{\mathbb{R}}$ and $h : F \to \overline{\mathbb{R}}$ are proper, convex, and l.s.c.,
  $T : E \to F$ is linear and continuous, $a \in F$,
  there exist $x_0 \in E$ and $q_0 \in F^*$ such that $f(x_0)$, $h(Tx_0 - a)$, $f^*(T^*q_0)$,
  and $h^*(-q_0)$ are all $< +\infty$.

As in (6.8) we consider the problem

$$\alpha := \inf_{x \in E} \left(f(x) + h(Tx - a) \right). \tag{6.45}$$

We set (cf. Example 6.4.1 with f instead of $\tilde{f}$)

$$M(x, b) := f(x) + h(Tx - a - b) \tag{6.46}$$

to obtain $\alpha = \inf_{x \in E} M(x, o)$. This gives the following associated problems:

$$S(b) := \inf_{x \in E} M(x, b) \qquad \text{(perturbed problem)}, \tag{6.47}$$

$$-\hat{S}(v) := \sup_{q \in F^*} \left(-M^*(v, q) \right) \qquad \text{(problem dual to (6.47))}, \tag{6.48}$$

$$\beta := -\hat{S}(o) = \sup_{q \in F^*} \left(-M^*(o, q) \right) \qquad \text{(problem dual to (6.45))}. \tag{6.49}$$

A simple calculation shows that

$$-M^*(v, q) = \langle q, a \rangle - f^*(T^*q + v) - h^*(-q)$$

and so

$$\beta = \sup_{q \in F^*} \left(\langle q, a \rangle - f^*(T^*q) - h^*(-q) \right). \tag{6.50}$$

We associate with (6.45) the Lagrange functional

$$L_1(x, q) := f(x) - \langle q, Tx - a \rangle - h^*(-q), \quad x \in A, \ q \in B,$$

where

$$A := \operatorname{dom} f, \ B := \{ q \in F^* \,|\, h^*(-q) < +\infty \}.$$

Notice that $L_1(x, q) = L(x, -q)$, where L denotes the corresponding Lagrange functional in (6.9).

Theorem 6.4.5 *Let the assumptions (A2) be fulfilled:*

(i) *With (6.36) replaced by (6.45), with (6.39) replaced by (6.50) and with M according to (6.46), the statements (i)–(iv) of Theorem 6.4.3 hold.*

(ii) [Modification of Theorem 6.4.3(v)] *The following statements are equivalent:*

(a''') $x \in E$ *is a solution of* (6.45), $q \in F^*$ *is a solution of* (6.50), *and* $\alpha = \beta$.

(b''') (x, q) *is a saddle point of* L_1 *with respect to* $A \times B$.

(c''') $T^*q \in \partial f(x)$ *and* $-q \in \partial h(Tx - a)$.

Proof. See Exercise 6.5.3. $\square$

Lemma 6.4.4 immediately implies the following result.

Lemma 6.4.6 *Let the assumptions* (A2) *be fulfilled:*

(a) *If* h *is continuous at* $Tx_0 - a$, *then the problem* (6.45) *is stable.*

(b) *If* f^* *is continuous at* T^*q_0, *then the problem* (6.50) *is stable.*

Corollary 6.4.7 *Let the assumptions* (A2) *be fulfilled. In addition, assume that* h *is continuous at* $Tx_0 - a$. *Then one has*

$$\inf_{x \in E} \left(f(x) + h(Tx - a) \right) = \max_{q \in F^*} \left(\langle q, a \rangle - f^*(T^*q) - h^*(-q) \right). \tag{6.51}$$

Proof. By Lemma 6.4.6, the problem (6.45) is stable. Hence the assertion follows from Theorem 6.4.5 (cf. Theorem 6.4.3(iii)). $\square$

We further specialize the setting. Let A be a nonempty convex subset of E. Set $f := \delta_A$ and so $f^*(q) = \sup_{x \in A} \langle q, x \rangle$. Moreover, let $F := E$, $T := \mathrm{id}_E$, and $a := o$. Applying Corollary 6.4.7 to these data, we obtain:

Corollary 6.4.8 *Assume that* A *is a nonempty convex subset of* E, h *is proper, convex, l.s.c., and continuous at a point of* A. *Then one has*

$$\inf_{x \in A} h(x) = \max_{q \in E^*} \left(\inf_{x \in A} \langle q, x \rangle - h^*(q) \right). \tag{6.52}$$

Application

Let E be a Hilbert space with scalar product $(v|u)$. We identify E^* with E. Moreover, let $S : E \to \mathbb{R}^n$ be linear and continuous, and let $c \in \mathbb{R}^n$. We consider the problem

$$h(u) := \tfrac{1}{2}\|u\|^2 \to \min, \quad u \in E, \quad Su = c. \tag{6.53}$$

Assume that the set

$$A := \{ u \in E \mid Su = c \}$$

is nonempty. By Corollary 6.4.8 we have

$$\inf_{u \in A} \tfrac{1}{2}\|u\|^2 = \max_{v \in E} \left(\inf_{u \in A} (v|u) - h^*(v) \right). \tag{6.54}$$

We evaluate the dual problem, i.e., the right-hand side of (6.54):
Ad $\inf_{u \in A}(v|u)$. For $Q := \{o\} \subseteq \mathbb{R}^n$ we have $Q^\circ = \mathbb{R}^n$, and $S^*(\mathbb{R}^n)$, as a finite-dimensional linear subspace of $E(= E^*)$, is $\sigma(E^*, E)$-closed. By Lemma 2.4.1 we therefore obtain

$$S^*(\mathbb{R}^n) = \left(S^{-1}\{o\}\right)^\circ = \{v \in E \mid (v|u) = 0 \, \forall u \in \ker S\} =: \left(\ker S\right)^\perp. \quad (6.55)$$

(I) If $v \notin \left(\ker S\right)^\perp$, then there exists $u_0 \in E$ such that $Su_0 = o$ and $(v|u_0) = 1$. Let $u_1 \in A$. Since S is linear, we have $u := \alpha u_0 + u_1 \in A$ for each $\alpha \in \mathbb{R}$. It follows that

$$(v|u) = \alpha + (v|u_1) \to -\infty \quad \text{as } \alpha \to -\infty$$

and so $\inf_{u \in A}(v|u) = -\infty$.

(II) If $v \in \left(\ker S\right)^\perp$, then by (6.55), there exists $a \in \mathbb{R}^n$ satisfying $S^*a = v$, and we obtain

$$\inf_{u \in A}(v|u) = \inf_{u \in A}(S^*a|u) = \inf_{u \in A}(a|Su) = (a|c), \quad (6.56)$$

and this holds for each $a \in \mathbb{R}^n$.

Ad $h^*(v)$. We immediately obtain

$$h^*(v) = \sup_{u \in E}\left((v|u) - \tfrac{1}{2}(u|u)\right) = \tfrac{1}{2}(v|v) = \tfrac{1}{2}\|v\|^2;$$

concerning the second equality, notice that

$$0 \le \tfrac{1}{2}(v - u|v - u) = \tfrac{1}{2}(v|v) - (v|u) + \tfrac{1}{2}(u|u) \quad \forall u, v \in E.$$

With the above, (6.54) passes into

$$\inf_{u \in A} \tfrac{1}{2}\|u\|^2 = \max_{a \in \mathbb{R}^n}\underbrace{\left((a|c) - \tfrac{1}{2}\|S^*a\|^2\right)}_{=:\varphi(a)}. \quad (6.57)$$

Since $\varphi : \mathbb{R}^n \to \mathbb{R}$ is concave and differentiable, we have

$$\varphi(a_0) = \max_{a \in \mathbb{R}^n} \varphi(a) \iff \varphi'(a_0) = o$$
$$\iff (h|c) - (h|SS^*a_0) = 0 \quad \forall h \in \mathbb{R}^n \iff SS^*a_0 = c.$$

Hence we have deduced the following result: *If $a_0 \in \mathbb{R}^n$ is a solution of $SS^*a_0 = c$ (which is a linear equation in $\mathbb{R}^n$), then $v_0 := S^*a_0$ is a solution of the right-hand side of (6.54).*

We leave it as Exercise 6.5.6 to show that $u_0 := v_0$ is a solution of the primal problem $\inf_{u \in A} \tfrac{1}{2}\|u\|^2$.

Example 6.4.9 Consider a dynamical system described by

$$\dot{x}(t) = Fx(t) + bu(t), \quad t \in [0, 1]. \tag{6.58}$$

In this connection, $x : [0, 1] \to \mathbb{R}^n$ is the phase function and $u : [0, 1] \to \mathbb{R}$ is the control function. The (n, n)-matrix F and the vector $b \in \mathbb{R}^n$ are given. We search a control function u that moves the system from $x(0) = o$ to $x(1) = c$ (where $c \in \mathbb{R}^n$ is given) under minimal consumption of energy. We assume that the energy needed for any control function u is

$$h(u) := \tfrac{1}{2} \int_0^1 u^2(t) \mathrm{d}t.$$

We solve the problem in the Hilbert space $E := L^2[0, 1]$ so that we have $h(u) = \tfrac{1}{2}\|u\|^2$. Each solution $x \in L^2[0, 1]$ of (6.58) satisfying $x(0) = o$ is representable as

$$x(t) = \int_0^t \phi(t - \tau)bu(\tau)\,\mathrm{d}\tau, \quad t \in [0, 1];$$

here, ϕ denotes the fundamental matrix associated with (6.58). The operator $S : L^2[0, 1] \to \mathbb{R}^n$ defined by

$$Su := \int_0^1 \phi(1 - \tau)bu(\tau)\,\mathrm{d}\tau,$$

is linear and continuous, and the end condition $x(1) = c$ is equivalent to $Su = c$. Hence, according to what has been said above, the solution of the problem is $u_0 = S^*a_0$, where $a_0 \in \mathbb{R}^n$ is the solution of $SS^*a_0 = c$.

6.5 Bibliographical Notes and Exercises

Principal contributions to duality theory were obtained by Fenchel [65], Moreau [147], Brøndsted [28], and Rockafellar [178, 180]. In particular, Theorem 6.4.3 is due to Fenchel [66] and Rockafellar [179]. Example 6.1.7 is adapted from Ky Fan [64]. The presentation in Sect. 6.3 essentially follows Zeidler [221]. The application at the end of Sect. 6.4 is taken from Luenberger [127]. In addition to the references in the Bibliographical Notes at the end of Chap. 4, we recommend Stoer and Witzgall [201] (finite-dimensional spaces) and Göpfert [76] (applications in locally convex spaces).

Exercise 6.5.1 Show that, with the notation and the assumptions of Example 6.1.5, the problem dual to (6.8) can be written as (6.15).

Exercise 6.5.2 Verify assertion (d) of Proposition 6.2.1.

Exercise 6.5.3 Prove Theorem 6.4.5.

Exercise 6.5.4 Consider Problem (6.13) with $E = P_E := \mathbb{R}^n$ and $F := C[a, b]$ with the maximum norm. Further let P_F be the cone of nonnegative functions in $C[a, b]$. Then (6.13) is a *linear semi-infinite optimization problem*. It is placed in a finite-dimensional space but has infinitely many side conditions of the form $Tx(t) \geq a(t)$ for all $t \in [a, b]$. In this case, E is reflexive but F is not. Check which assertions of the theory developed above still hold (cf. Krabs [112]).

Exercise 6.5.5 Consider the following linear semi-infinite problem (cf. Exercise 6.5.4):

$$\text{Minimize } f(x_1, x_2) := x_2$$
$$\text{subject to } (x_1, x_2) \in \mathbb{R}^2, \quad t^2 x_1 + x_2 \geq t \quad \forall t \in [0, 1].$$

Show that the values of the primal problem and the dual problem coincide, the dual problem has a solution, but the primal problem has no solution.

Exercise 6.5.6 Verify that $u_0 := v_0 = S^* a_0$, where $a_0 \in \mathbb{R}^n$ solves $SS^* a_0 = c$, is a solution of the primal problem (6.53).

Derivatives and Subdifferentials of Lipschitz Functionals

7.1 Preview: Derivatives and Approximating Cones

The aim of this section is to give, in an informal discussion, the motivation for the following sections.

Let $f : E \to \overline{\mathbb{R}}$ be proper, let $A \subseteq E$, and let $\bar{x} \in A$. We consider the following statement:

(Min) $\bar{x}$ is a local minimizer of f on A.

There are two basic approaches to necessary conditions for (Min), which we call *method of tangent directions* and *method of penalty functions*.

Method of Tangent Directions

(I) Suppose that f and A are convex. Then (Min) can be characterized by a variational inequality (see Proposition 5.1.1):

$$(\text{Min}) \quad \Longleftrightarrow \quad f_G(\bar{x}, y) \geq 0 \quad \forall y \in A - \bar{x}. \tag{7.1}$$

(II) Suppose now that f and/or A is not convex. In order to be able to argue similarly as in the convex case, we have to identify "admissible" directions, i.e., directions $y \in E$ that appropriately approximate the set A locally at $\bar{x}$. The set

$$\mathrm{T}_r(A, \bar{x}) := \{y \in E \mid \exists \tau_k \downarrow 0 \, \forall k \in \mathbb{N} : \bar{x} + \tau_k y \in A\},$$

which turns out to be a cone, is called *cone of radial directions* to A at $\bar{x}$ (Fig. 7.1). If (Min) holds and if $y \in \mathrm{T}_r(A, \bar{x})$, then for some sequence $\tau_k \downarrow 0$, we have

$$\tfrac{1}{\tau_k}\big(f(\bar{x} + \tau_k y) - f(\bar{x})\big) \geq 0 \quad \forall k \in \mathbb{N}.$$

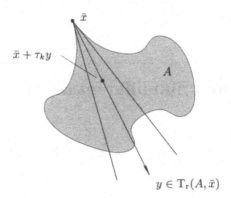

Fig. 7.1

We therefore consider the upper directional G-derivative

$$\overline{f}_G(\bar{x}, y) := \limsup_{\tau \downarrow 0} \tfrac{1}{\tau}\big(f(\bar{x} + \tau y) - f(\bar{x})\big).$$

Using this, we conclude that

$$(\text{Min}) \quad \Longrightarrow \quad \overline{f}_G(\bar{x}, y) \geq 0 \quad \forall y \in \mathrm{T_r}(M, \bar{x}). \tag{7.2}$$

In other words, as a necessary optimality condition we again obtain a variational inequality. This condition is useful as long as $\mathrm{T_r}(A, \bar{x})$ contains "enough" elements. However, if for example M is the kernel of a mapping $h : D \to F$, i.e., if

$$A := \ker h := \{x \in D \,|\, h(x) = o\}, \tag{7.3}$$

then in general we will have $\mathrm{T_r}(A, \bar{x}) = \{o\}$. Since there always holds $\overline{f}_G(\bar{x}, o) = 0$, the optimality condition in (7.2) is trivially satisfied for each $\bar{x} \in A$ and so is not suitable for identifying possible solutions of (Min). Therefore we look for finer local approximations of A at $\bar{x}$.

Let again A be an arbitrary subset of E and $\bar{x} \in A$. The set

$$\mathrm{T}(A, \bar{x}) := \{y \in E \,|\, \exists \tau_k \downarrow 0 \; \exists y_k \to y \; \forall k \in \mathbb{N} : \bar{x} + \tau_k y_k \in A\},$$

which is also a cone, is called *contingent cone* to A at $\bar{x}$ (Fig. 7.2). In analogy to (7.2), we obtain

$$(\text{Min}) \quad \Longrightarrow \quad \overline{f}_H(\bar{x}, y) \geq 0 \quad \forall y \in \mathrm{T}(A, \bar{x}), \tag{7.4}$$

where now the upper directional H-derivative

$$\overline{f}_H(\bar{x}, y) := \limsup_{\tau \downarrow 0,\, z \to y} \tfrac{1}{\tau}\big(f(\bar{x} + \tau z) - f(\bar{x})\big)$$

is the adequate local approximation of f at $\bar{x}$. Since $\mathrm{T_r}(A, \bar{x}) \subseteq \mathrm{T}(A, \bar{x})$, the optimality condition in (7.4) is stronger than that in (7.2).

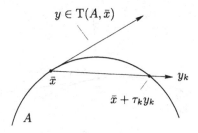

Fig. 7.2

So far, we have not imposed any differentiability hypotheses on the functional f. From (7.2) and (7.4) we immediately deduce the following conditions:

(a) If f is G-differentiable at $\bar{x}$, then

$$\text{(Min)} \quad \Longrightarrow \quad \langle f'(\bar{x}), y \rangle \geq 0 \quad \forall y \in T_r(A, \bar{x}). \tag{7.5}$$

If, in addition, A is convex, then $A - \bar{x} \subset T_r(A, \bar{x})$ and we obtain (cf. (7.1))

$$\text{(Min)} \quad \Longrightarrow \quad \langle f'(\bar{x}), y \rangle \geq 0 \quad \forall y \in A - \bar{x}. \tag{7.6}$$

(b) If f is H-differentiable at $\bar{x}$, then

$$\text{(Min)} \quad \Longrightarrow \quad \langle f'(\bar{x}), y \rangle \geq 0 \quad \forall y \in T(A, \bar{x}). \tag{7.7}$$

The above shows that necessary conditions for (Min) can be obtained by choosing local approximations of the functional f by a (directional) derivative and of the set A by a cone provided these approximations "fit together."

(III) Let A be as in (7.3), with a mapping $h : D \to F$. Our aim now is to derive a necessary condition for (Min) in terms of h. If h is F-differentiable in a neighborhood of $\bar{x}$, h' is continuous at $\bar{x}$, and $h'(\bar{x})$ is surjective, then a well-known theorem of Lyusternik says that

$$T(\ker h, \bar{x}) = \ker h'(\bar{x}), \tag{7.8}$$

i.e., $y \in T(\ker h, \bar{x})$ if and only if $h'(\bar{x})y = o$ (see Fig. 7.3, where $E = \mathbb{R}^n$). This will follow below from a more general result (see Theorem 11.4.2).

If, in addition, the functional f is H-differentiable at $\bar{x}$, then we obtain from (7.7) and (7.8) the following condition:

$$\text{(Min)} \quad \Longrightarrow \quad \left[\forall y \in E : \ h'(\bar{x})y = o \ \Longrightarrow \ \langle -f'(\bar{x}), y \rangle \leq 0. \right]$$

Applying the generalized Farkas lemma of Proposition 2.4.2 to $[\cdots]$, where $T := h'(\bar{x})$, $u := -f'(\bar{x})$, and $Q := \{o\}$, we see that (Min) implies the existence

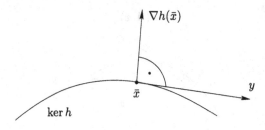

Fig. 7.3

of $v \in F^*$ satisfying $T^* v = u$, i.e., $\langle v, h'(\bar{x})y \rangle = \langle -f'(\bar{x}), y \rangle$ for all $y \in E$ and so

$$f'(\bar{x}) + v \circ h'(\bar{x}) = o. \tag{7.9}$$

Hence as a necessary condition for (Min), we obtain a *Lagrange multiplier rule*, the Lagrange "multiplier" being the functional $v \in F^*$.

(IV) We again consider (Min), where now A is of the form

$$A = \{x \in E \mid g_i(x) \le 0, \ i = 1, \ldots, m\} \tag{7.10}$$

with certain functionals $g_1, \ldots, g_m$. In this case we want to obtain optimality conditions in terms of f and these functionals. Therefore, in view of say (7.4), we want to characterize (a sufficiently large subset of) $T(A, \bar{x})$ in terms of $g_1, \ldots, g_m$. Similarly to (III) this will be achieved under appropriate differentiability assumptions. A multiplier rule is then obtained with the aid of a suitable nonlinear substitute for the generalized Farkas lemma.

Method of Penalty Functions

The idea of this method is to replace the constrained minimum problem

$$f(x) \to \min, \quad x \in A,$$

by a free minimum problem

$$f(x) + p(x) \to \min, \quad x \in E,$$

where the *penalty function* p is such that $p(x) > 0$ if $x \in E \setminus A$, i.e., leaving the set A is "penalized."

(I) Let A be a convex set and f a convex functional that is continuous at some point of A. Then the indicator functional δ_A is an appropriate penalty function. In fact, we have

$$\text{(Min)} \iff (f + \delta_A)(\bar{x}) = \min_{x \in E}(f + \delta_A)(x) \iff o \in \partial(f + \delta_A)(\bar{x}),$$

and the sum rule (Proposition 4.5.1) implies

$$\text{(Min)} \iff o \in \partial f(\bar{x}) + \partial \delta_A(\bar{x}) \iff o \in \partial f(\bar{x}) + (A - \bar{x})^\circ.$$

The cone $N(A, \bar{x}) := T(A, \bar{x})^\circ$ is called *normal cone* to A at $\bar{x}$. Since A is convex, we have $T(A, \bar{x}) = \mathbb{R}_+(A - \bar{x})$ and so $N(A, \bar{x}) = (A - \bar{x})^\circ$. Thus the above equivalence can be written as

$$\text{(Min)} \iff o \in \partial f(\bar{x}) + N(A, \bar{x}). \tag{7.11}$$

Assume now that A is given by (7.10). Then for a further exploitation of (7.11) we need a representation of $N(A, \bar{x})$ in terms of g_i.

(II) In a theory involving locally L-continuous (nonconvex) functionals, the indicator functional is not suitable as a penalty function. In this case, the L-continuous functional $p := \lambda \, d_A$, where $\lambda > 0$ is sufficiently large, will turn out to serve this purpose. Here, as in the convex case, a possible subdifferential mapping $\partial_* f : E \rightrightarrows E^*$ should at least have the following properties:

If $\bar{x}$ is a local minimizer of f on E, then $o \in \partial_* f(\bar{x})$,

$\partial_*(f + g)(\bar{x}) \subseteq \partial_* f(\bar{x}) + \partial_* g(\bar{x})$.

Then it follows that

$$\text{(Min)} \implies o \in \partial_* f(\bar{x}) + \partial_*(\lambda d_A)(\bar{x}) \subseteq \partial_* f(\bar{x}) + N_*(A, \bar{x}),$$

where $N_*(A, \bar{x})$ denotes the $\sigma(E^*, E)$-closure of $\bigcup_{\lambda \geq 0} \partial_*(\lambda d_A)(\bar{x})$.

(III) If A is given by (7.10), then we wish to describe $N_*(A, \bar{x})$ in terms of the functionals $g_1, \ldots, g_m$.

(IV) Finally, analogous investigations are to be done for nonconvex non-Lipschitz functionals.

7.2 Upper Convex Approximations and Locally Convex Functionals

We start carrying out the program indicated in Sect. 7.1.

Definition 7.2.1 Let $f : E \to \overline{\mathbb{R}}$ be proper and let $\bar{x} \in \mathrm{dom}\, f$.

(a) The functional $\varphi : E \to \overline{\mathbb{R}}$ is called *radial upper convex approximation* of f at $\bar{x}$ if φ is proper, sublinear, and satisfies

$$\overline{f}_G(\bar{x}, y) \leq \varphi(y) \quad \forall y \in E.$$

(b) The functional $\varphi : E \to \overline{\mathbb{R}}$ is called *upper convex approximation* of f at $\bar{x}$ if φ is proper, sublinear, and satisfies

$$\overline{f}_H(\bar{x}, y) \leq \varphi(y) \quad \forall y \in E \setminus \{o\}.$$

We write

$$\mathrm{UC_r}(f,\bar{x}) := \text{set of all radial upper convex approximations of } f \text{ at } \bar{x},$$
$$\mathrm{UC}(f,\bar{x}) := \text{set of all upper convex approximations of } f \text{ at } \bar{x}.$$

Since $\overline{f}_G(\bar{x},\cdot) \le \overline{f}_H(\bar{x},\cdot)$, we always have $\mathrm{UC}(f,\bar{x}) \subseteq \mathrm{UC_r}(f,\bar{x})$ (notice that $\overline{f}_G(\bar{x},o) = o = \varphi(o)$).

Proposition 7.2.2 *If $f : E \to \overline{\mathbb{R}}$ is proper and locally L-continuous at $\bar{x} \in$ dom f, then $\overline{f}_H(\bar{x},y) = \overline{f}_G(\bar{x},y)$ for any $y \in E$ and so $\mathrm{UC}(f,\bar{x}) = \mathrm{UC_r}(f,\bar{x})$.*

Proof. See Exercise 7.5.1.   □

Remark 7.2.3 There is a close relationship to quasidifferentiable functionals. The proper functional $f : E \to \overline{\mathbb{R}}$ is said to be *quasidifferentiable* at $\bar{x} \in$ dom f if $f_G(\bar{x},\cdot)$ exists on E and there exist nonempty convex $\sigma(E^*,E)$-compact subsets $\overline{\partial}f(\bar{x})$ and $\underline{\partial}f(\bar{x})$ of E^* such that

$$f_G(\bar{x},y) = \min_{u\in\overline{\partial}f(\bar{x})} \langle u,y \rangle + \max_{v\in\underline{\partial}f(\bar{x})} \langle v,y \rangle \quad \forall y \in E.$$

Now let $f : E \to \overline{\mathbb{R}}$ be quasidifferentiable at $\bar{x} \in$ dom f and set

$$\varphi_u(y) := \max\{\langle u+v,y \rangle \mid v \in \underline{\partial}f(\bar{x})\} \quad \forall y \in E,$$
$$\psi_v(y) := \max\{\langle u-v,y \rangle \mid u \in -\overline{\partial}f(\bar{x})\} \quad \forall y \in E.$$

Then obviously

$$\varphi_u \in \mathrm{UC_r}(f,\bar{x}) \quad \forall u \in \overline{\partial}f(\bar{x}) \quad \text{and} \quad \psi_v \in \mathrm{UC_r}(-f,\bar{x}) \quad \forall v \in \underline{\partial}f(\bar{x}).$$

Definition 7.2.4 describes a special class of functionals admitting (radial) upper convex approximations.

Definition 7.2.4 Let $f : E \to \overline{\mathbb{R}}$ be proper and let $\bar{x} \in$ dom f. The functional f is said to be

- *locally convex* at $\bar{x}$ if $f_G(\bar{x},\cdot)$ exists and is sublinear as a mapping of E to $\overline{\mathbb{R}}$,
- *regularly locally convex* at $\bar{x}$ if $f_H(\bar{x},\cdot)$ exists and is sublinear as a mapping of E to $\mathbb{R}$.

See Exercise 7.5.2 for an example of a functional that is locally convex but not regularly locally convex.

Proposition 7.2.5 *Let $f_1, f_2 : E \to \overline{\mathbb{R}}$ be proper and set $f := f_1 + f_2$. Further let $\bar{x} \in (\mathrm{int}\,\mathrm{dom}f_1) \cap (\mathrm{int}\,\mathrm{dom}f_2)$ and assume that f_1 is continuous at $\bar{x}$ and convex:*

(a) If f_2 is G-differentiable at $\bar{x}$, then f is locally convex at $\bar{x}$ and $f_G(\bar{x},\cdot) = f_{1,H}(\bar{x},\cdot) + f_2'(\bar{x}) \in \mathrm{UC_r}(f,\bar{x})$.

(b) *If f_2 is H-differentiable at $\bar{x}$, then f is regularly locally convex at $\bar{x}$ and*
$$f_H(\bar{x}, \cdot) = f_{1,H}(\bar{x}, \cdot) + f_2'(\bar{x}) \in \mathrm{UC}(f, \bar{x}).$$

Proof. By Theorem 4.1.3 we have $f_{1,H}(\bar{x}, \cdot) \in \mathrm{UC}(f_1, \bar{x})$, and by Remark 3.2.3 it follows that $f_2'(\bar{x}) \in \mathrm{UC}_r(f_2, \bar{x})$ or $f_2'(\bar{x}) \in \mathrm{UC}(f_2, \bar{x})$, respectively. This verifies the assertions.   □

Remark 7.2.6 By Proposition 7.2.5, each functional that is G-differentiable at $\bar{x}$ is locally convex there. Moreover, each continuous convex functional as well as each functional that is H-differentiable at $\bar{x}$ is regularly locally convex. Notice that the proposition also describes regularly locally convex functionals that are neither convex nor G-differentiable; consider, e.g., $f(x) := |x| + x^3$, $x \in \mathbb{R}$, at $\bar{x} = 0$.

We want to derive a maximum rule for the directional G-derivative. For $i = 1, \ldots, m$ let $f_i : E \to \overline{\mathbb{R}}$ and set

$$f(x) := \max_{i=1,\ldots,m} f_i(x), \quad x \in E. \tag{7.12}$$

Let $\bar{x} \in E$, $I := \{1, \ldots, m\}$, and $I(\bar{x}) := \{i \in I \mid f_i(\bar{x}) = f(\bar{x})\}$.

Proposition 7.2.7 (Maximum Rule) *Assume that $\bar{x} \in \bigcap_{i \in I} \mathrm{int}(\mathrm{dom}\, f_i)$, that f_i is continuous at $\bar{x}$ for each $i \in I$, and that $f_{i,G}(\bar{x}, \cdot)$ exists for each $i \in I(\bar{x})$. Then, with f as in (7.12), the directional G-derivative $f_G(\bar{x}, \cdot)$ exists and one has*

$$f_G(\bar{x}, y) = \max_{i \in I(\bar{x})} f_{i,G}(\bar{x}, y), \quad y \in E. \tag{7.13}$$

An analogous result holds with f_G, $f_{i,G}$ replaced by f_H, $f_{i,H}$, respectively.

Proof.

(I) First we show that the directional G-derivative $f_G(\bar{x}, \cdot)$, whenever it exists, depends on the functions f_i for $i \in I(\bar{x})$ only. Let $y \in E$ be given. If $i \in I \setminus I(\bar{x})$, then the continuity of f_i and f imply that there exists $\tau_i > 0$ such that $f_i(\bar{x} + \tau y) < f(\bar{x} + \tau y)$ for each $\tau \in [0, \tau_i]$. Let $\tau_0 := \min\{\tau_i \mid i \in I \setminus I(\bar{x})\}$. For any $\tau \in (0, \tau_0)$ we have

$$\tau^{-1}\big(f(\bar{x} + \tau y) - f(\bar{x})\big) = \max_{i \in I(\bar{x})} \tau^{-1}\big(f_i(\bar{x} + \tau y) - f(\bar{x})\big)$$
$$= \max_{i \in I(\bar{x})} \tau^{-1}\big(f_i(\bar{x} + \tau y) - f_i(\bar{x})\big). \tag{7.14}$$

Hence we may assume that $I(\bar{x}) = I$.

(II) For each $i \in I(= I(\bar{x}))$ we obtain

$$\liminf_{\tau \downarrow 0} \tau^{-1}\big(f(\bar{x} + \tau y) - f(\bar{x})\big) \geq \lim_{\tau \downarrow 0} \tau^{-1}\big(f_i(\bar{x} + \tau y) - f_i(\bar{x})\big) = (f_i)_G(\bar{x}, y). \tag{7.15}$$

We now show that

$$\limsup_{\tau \downarrow 0} \tau^{-1}\big(f(\bar{x} + \tau y) - f(\bar{x})\big) \le \max_{i \in I} f_{i,G}(\bar{x}, y).$$

This and (7.15) verify (7.13). Assume that

$$\limsup_{\tau \downarrow 0} \tau^{-1}\big(f(\bar{x} + \tau y) - f(\bar{x})\big) > \max_{i \in I} f_{i,G}(\bar{x}, y).$$

Then we find $\epsilon > 0$ and a sequence $\tau_k \downarrow 0$ such that

$$\tau_k^{-1}\big(f(\bar{x} + \tau_k y) - f(\bar{x})\big) \ge \max_{i \in I} f_{i,G}(\bar{x}, y) + \epsilon \quad \forall k \in \mathbb{N}.$$

For a subsequence (τ_{k_ν}) of (τ_k) we have $f(\bar{x} + \tau_{k_\nu} y) = f_j(\bar{x} + \tau_{k_\nu} y)$ with a fixed index j. It follows that

$$f_{j,G}(\bar{x}, y) = \lim_{\nu \to \infty} \tau_{k_\nu}^{-1}\big(f_j(\bar{x} + \tau_{k_\nu} y) - f_j(\bar{x})\big) \ge \max_{i \in I} f_{i,G}(\bar{x}, y) + \epsilon,$$

which is a contradiction.

(III) For $f_H(\bar{x}, \cdot)$ see Exercise 7.5.3. $\square$

Proposition 7.2.8 *Assume that $\bar{x} \in \bigcap_{i \in I} \mathrm{int}(\mathrm{dom}\, f_i)$, each f_i is continuous at $\bar{x}$, and f_i is (regularly) locally convex at $\bar{x}$ for each $i \in I(\bar{x})$. Then the functional f defined by (7.12) is (regularly) locally convex at $\bar{x}$.*

Proof. This follows immediately from Proposition 7.2.7. $\square$

Proposition 7.2.8 applies, in particular, to the maximum of a finite number of continuous Gâteaux differentiable (resp. Hadamard differentiable) functionals. Notice that such a maximum functional in general is not Gâteaux differentiable (resp. Hadamard differentiable).

Proposition 4.1.6 stimulates us to define a subdifferential for locally convex functionals.

Definition 7.2.9 *If $f : E \to \overline{\mathbb{R}}$ is locally convex at $\bar{x} \in \mathrm{dom}\, f$, then*

$$\partial_* f(\bar{x}) := \{x^* \in E^* \mid \langle x^*, y \rangle \le f_G(\bar{x}, y) \; \forall y \in E\}$$

is called locally convex subdifferential of f at $\bar{x}$.

If f is convex, then by Proposition 4.1.6 we have $\partial_* f(\bar{x}) = \partial f(\bar{x})$. Notice that if $f_G(\bar{x}, y) = -\infty$ for some $y \in E$, then $\partial_* f(\bar{x})$ is empty.

Proposition 7.2.10 *If $f : E \to \overline{\mathbb{R}}$ is locally convex at $\bar{x} \in \mathrm{dom}\, f$, then the following assertions are equivalent:*

(a) $\partial_* f(\bar{x})$ *is nonempty.*

(b) $f_G(\bar{x}, \cdot)$ *is lower semicontinuous at $y = o$.*

Proof.

(a) $\implies$ (b): Let $x^* \in \partial_* f(\bar{x})$. Further let $k < f_G(\bar{x}, o) = 0$. Since x^* is continuous, there exists a neighborhood W of zero in E such that $k < \langle x^*, y \rangle \leq f_G(\bar{x}, y)$ for each $y \in W$.

(b) $\implies$ (a): By assumption there exists a neighborhood V of zero in E such that $-1 \leq f_G(\bar{x}, y)$ for each $y \in V$. Now let $z \in E$ be given. Then $\eta z \in V$ for some $\eta > 0$. Since $p := f_G(\bar{x}, \cdot)$ is positively homogeneous, it follows that $-\infty < -\frac{1}{\eta} \leq f_G(\bar{x}, z) = p(z)$. Hence p is proper and sublinear. Moreover, let $q(y) := 1$ for $y \in V$ and $q(y) := +\infty$ for $y \in E \setminus V$. Then q is proper, convex, and continuous at zero. We further have $-q(y) \leq p(y)$ for each $y \in E$. The sandwich theorem (Theorem 1.5.2) thus implies that there exists $x^* \in \partial_* f(\bar{x})$. $\square$

The locally convex subdifferential is an appropriate tool for detecting minimizers as is the convex subdifferential (cf. Remark 4.1.2). The following result is an immediate consequence of the definitions.

Proposition 7.2.11 *If $f : E \to \overline{\mathbb{R}}$ is locally convex at $\bar{x} \in \operatorname{dom} f$ and $\bar{x}$ is a local minimizer of f, then $o \in \partial_* f(\bar{x})$.*

7.3 The Subdifferentials of Clarke and Michel–Penot

Convention. Throughout this section, we assume that $D \subseteq E$ is open, $\bar{x} \in D$ and $f : D \to \mathbb{R}$.

Here, we present two intrinsic constructions for upper convex approximations. For comparison, recall that

$$\overline{f}_H(\bar{x}, y) := \limsup_{\substack{\tau \downarrow 0 \\ z \to y}} \frac{1}{\tau} \big(f(\bar{x} + \tau z) - f(\bar{x}) \big).$$

Definition 7.3.1 If $y \in E$, then

$$f^\circ(\bar{x}, y) := \limsup_{\substack{\tau \downarrow 0 \\ x \to \bar{x}}} \frac{1}{\tau} \big(f(x + \tau y) - f(x) \big) \tag{7.16}$$

is called *Clarke directional derivative* of f at $\bar{x}$ in the direction y and

$$f^\diamond(\bar{x}, y) := \sup_{z \in E} \limsup_{\tau \downarrow 0} \frac{1}{\tau} \big(f(\bar{x} + \tau y + \tau z) - f(\bar{x} + \tau z) \big) \tag{7.17}$$

is called *Michel–Penot directional derivative* of f at $\bar{x}$ in the direction y.

Theorem 7.3.2 *Let f be locally L-continuous around $\bar{x}$ with constant $\lambda > 0$. Then:*

(a) *$f^\circ(\bar{x}, \cdot)$ and $f^\diamond(\bar{x}, \cdot)$ are sublinear and (globally) L-continuous with constant λ on E and satisfy*

$$\overline{f}_H(\bar{x}, y) \leq f^\diamond(\bar{x}, y) \leq f^\circ(\bar{x}, y) \leq \lambda \|y\| \quad \forall y \in E. \tag{7.18}$$

In particular, $f^\circ(\bar{x}, \cdot)$ and $f^\diamond(\bar{x}, \cdot)$ are finite upper convex approximations of f at $\bar{x}$.

(b) *For any $y \in E$ one has*

$$f^\circ(\bar{x}, -y) = (-f)^\circ(\bar{x}, y), \quad f^\diamond(\bar{x}, -y) = (-f)^\diamond(\bar{x}, y).$$

Proof.

(a) (Ia) *The third inequality of (7.18) holds.* Let $y \in E$ be fixed. By assumption, we have

$$\frac{1}{\tau}\big(f(x + \tau y) - f(x)\big) \le \frac{1}{\tau}\lambda\|\tau y\| = \lambda\|y\|$$

whenever $\|x - \bar{x}\|$ and $\tau > 0$ are small. Hence $f^\circ(\bar{x}, y) \le \lambda\|y\|$.

(Ib) *The second inequality of (7.18) holds.* Fix $y \in E$. Further let $\epsilon > 0$ be given. For each $z \in E$ there exists $\delta(z) > 0$ such that (consider $x := \bar{x} + \tau z$)

$$\frac{1}{\tau}\big(f(\bar{x} + \tau y + \tau z) - f(\bar{x} + \tau z)\big) < f^\circ(\bar{x}, y) + \epsilon \quad \forall \tau \in (0, \delta(z)).$$

This implies

$$\limsup_{\tau\downarrow 0} \frac{1}{\tau}\big(f(\bar{x} + \tau y + \tau z) - f(\bar{x} + \tau z)\big) \le f^\circ(\bar{x}, y) + \epsilon,$$

which holds for each $z \in E$. We conclude that $f^\diamond(\bar{x}, y) \le f^\circ(\bar{x}, y) + \epsilon$. Letting $\epsilon \downarrow 0$, the assertion follows.

(Ic) *The first inequality of (7.18) holds.* Let $y \in E$ be fixed. Further let $\epsilon > 0$ be given. For all sufficiently small $\tau > 0$ and all $z \in E$ such that $\|y - z\|$ is sufficiently small, we have

$$\frac{1}{\tau}\big(f(\bar{x} + \tau z) - f(\bar{x})\big)$$
$$= \frac{1}{\tau}\big(f(\bar{x} + \tau y + \tau(z - y)) - f(\bar{x} + \tau(z - y))\big)$$
$$+ \frac{1}{\tau}\big(f(\bar{x} + \tau(z - y)) - f(\bar{x})\big)$$
$$\le \frac{1}{\tau}\big(f(\bar{x} + \tau y + \tau(z - y)) - f(\bar{x} + \tau(z - y))\big) + \lambda\|z - y\|$$
$$\le f^\diamond(\bar{x}, y) + \epsilon + \lambda\|z - y\|.$$

Letting $\tau \downarrow 0$, $z \to y$, and finally $\epsilon \downarrow 0$, the first inequality follows.

(IIa) $f^\circ(\bar{x}, \cdot)$ *is sublinear.* It is obvious that $f^\circ(\bar{x}, \cdot)$ is positively homogeneous. We show that it is subadditive. Let $y_1, y_2 \in E$. We have

$$\frac{1}{\tau}\big[f(x + \tau(y_1 + y_2)) - f(x)\big]$$
$$= \frac{1}{\tau}\big[f(\underbrace{(x + \tau y_2)}_{=:\hat{x}\to\bar{x}} + \tau y_1) - f(\underbrace{x + \tau y_2}_{\hat{x}\to\bar{x}})\big] + \frac{1}{\tau}\big[f(x + \tau y_2) - f(x)\big].$$

Passing on both sides to the limit superior for $\tau \downarrow 0$ and $x \to \bar{x}$, we obtain

$$f^\circ(\bar{x}, y_1 + y_2) \le f^\circ(\bar{x}, y_1) + f^\circ(\bar{x}, y_2).$$

(IIb) $f^\diamond(\bar{x}, \cdot)$ *is sublinear.* Again we can restrict ourselves to showing sub-additivity. Let $y_1, y_2 \in E$. Let $\epsilon > 0$ be given. For all $\tau >$ sufficiently small we obtain

$$\frac{1}{\tau}\Big(f\big(\bar{x} + \tau(y_1 + y_2) + \tau z\big) - f\big(\bar{x} + \tau(y_2 + z)\big)\Big)$$
$$\leq f^\diamond(\bar{x}, y_1) + \frac{\epsilon}{2} \quad \forall z \in E,$$
$$\frac{1}{\tau}\Big(f\big(\bar{x} + \tau y_2 + \tau z\big) - f\big(\bar{x} + \tau z\big)\Big) \leq f^\diamond(\bar{x}, y_2) + \frac{\epsilon}{2} \quad \forall z \in E.$$

Adding these inequalities, we get

$$\frac{1}{\tau}\Big(f\big(\bar{x} + \tau(y_1 + y_2) + \tau z\big) - f\big(\bar{x} + \tau z\big)\Big) \leq f^\diamond(\bar{x}, y_1) + f^\diamond(\bar{x}, y_2) + \epsilon \quad \forall z \in E,$$

and finally $f^\diamond(\bar{x}, y_1 + y_2) \leq f^\diamond(\bar{x}, y_1) + f^\diamond(\bar{x}, y_2)$.

(IIIa) $f^\circ(\bar{x}, \cdot)$ *is L-continuous.* Let $y_1, y_2 \in E$. If $\tau > 0$ is small and x is close to $\bar{x}$, we obtain

$$f(x + \tau y_1) - f(x) = \big[f(x + \tau y_2) - f(x)\big] + \big[f(x + \tau y_1) - f(x + \tau y_2)\big]$$
$$\leq \big[f(x + \tau y_2) - f(x)\big] + \tau\lambda\|y_1 - y_2\|$$

and so

$$f^\circ(\bar{x}, y_1) \leq f^\circ(\bar{x}, y_2) + \lambda\|y_1 - y_2\|.$$

By an analogous estimate with y_1 and y_2 interchanged, we see that

$$|f^\circ(\bar{x}, y_1) - f^\circ(\bar{x}, y_2)| \leq \lambda\|y_1 - y_2\|.$$

(IIIb) Analogously, the L-continuity of $f^\diamond(\bar{x}, \cdot)$ is verified.

(b)(IVa) We immediately obtain

$$f^\circ(\bar{x}, -y) = \limsup_{\substack{\tau \downarrow 0 \\ x \to \bar{x}}} \frac{1}{\tau}\big[f(x - \tau y) - f(x)\big]$$
$$= \limsup_{\substack{\tau \downarrow 0 \\ \hat{x} \to \bar{x}}} \frac{1}{\tau}\big[(-f)(\hat{x} + \tau y) - (-f)(\hat{x})\big] = (-f)^\circ(\bar{x}, y);$$

in this connection, $\hat{x}$ stands for $x - \tau y$.

(IVb) For any $z \in E$ we have

$$\limsup_{\tau \downarrow 0} \frac{1}{\tau}\big[f(\bar{x} - \tau y + \tau z) - f(\bar{x} + \tau z)\big]$$
$$= \limsup_{\tau \downarrow 0} \big[(-f)(\bar{x} + \tau y + \tau(z - y)) - (-f)(\bar{x} + \tau(z - y))\big].$$

Taking the supremum over z and $z - y$, respectively, gives the second statement of (b). $\qquad\square$

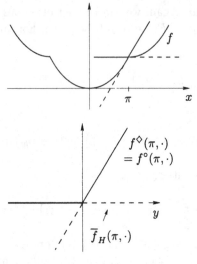

Fig. 7.4

Example 7.3.3 Let $E := \mathbb{R}$, $f(x) := |x| - |\sin x|$, and $\bar{x} := \pi$. Then we have

$$\overline{f}_H(\pi, y) = \begin{cases} 2y & \text{if } y < 0, \\ 0 & \text{if } y \geq 0, \end{cases} \quad f^\Diamond(\pi, y) = f^\circ(\pi, y) = \begin{cases} 0 & \text{if } y < 0, \\ 2y & \text{if } y \geq 0. \end{cases}$$

We see that of the three directional derivatives, the functional $\overline{f}_H(\pi, \cdot)$ is the best local approximation of f at π but it is not convex (Fig. 7.4).

Recall (see Proposition 4.1.6) that if f is convex, then

$$\partial f(\bar{x}) = \{x^* \in E^* \mid \langle x^*, y \rangle \leq f_G(\bar{x}, y) \; \forall y \in E\}.$$

In the nonconvex case, we now give the following:

Definition 7.3.4 If f is locally L-continuous around $\bar{x}$, then

$$\partial_\circ f(\bar{x}) := \{x^* \in E^* \mid \langle x^*, y \rangle \leq f^\circ(\bar{x}, y) \; \forall y \in E\}$$

is called *Clarke subdifferential*, or *Clarke generalized gradient*, of f at $\bar{x}$ and

$$\partial_\Diamond f(\bar{x}) := \{x^* \in E^* \mid \langle x^*, y \rangle \leq f^\Diamond(\bar{x}, y) \; \forall y \in E\}$$

is called *Michel–Penot subdifferential* of f at $\bar{x}$.

Proposition 7.3.5 *If f is locally L-continuous around $\bar{x}$, then for any $\sigma \in \mathbb{R}$ one has*

$$\partial_\circ(\sigma f)(\bar{x}) = \sigma\partial_\circ f(\bar{x}) \quad \text{and} \quad \partial_\Diamond(\sigma f)(\bar{x}) = \sigma\partial_\Diamond f(\bar{x}).$$

Proof. We consider $\partial_o f$; for $\partial_\diamond f$ the proof is analogous. If $\sigma \geq 0$, the formula follows immediately from $(\sigma f)^\circ(\bar{x}, \cdot) = \sigma f^\circ(\bar{x}, \cdot)$. Thus it remains to verify it for $\sigma = -1$. We have

$$
\begin{aligned}
\partial_o(-f)(\bar{x}) &= \{x^* \in E^* \mid \langle x^*, y\rangle \leq (-f)^\circ(\bar{x}, y) \,\forall y \in E\} \\
&= \{x^* \in E^* \mid \langle x^*, -z\rangle \leq f^\circ(\bar{x}, z) \,\forall z \in E\} \\
&= -\partial_o f(\bar{x}).
\end{aligned}
$$

Here the second equation holds by Theorem 7.3.2(b) and with $z := -y$.   □

The importance of the subdifferentials introduced above reveals:

Proposition 7.3.6 *If f is locally L-continuous around $\bar{x}$ and $\bar{x}$ is a local minimizer or a local maximizer of f, then $o \in \partial_o f(\bar{x})$ and $o \in \partial_\diamond f(\bar{x})$.*

Proof. If $\bar{x}$ is a local minimizer of f, then for any $y \in E$ we obtain

$$
\begin{aligned}
0 &\leq \liminf_{\tau \downarrow 0} \frac{1}{\tau}\big(f(\bar{x} + \tau y) - f(\bar{x})\big) \\
&\leq \limsup_{\tau \downarrow 0} \frac{1}{\tau}\big(f(\bar{x} + \tau y) - f(\bar{x})\big) \leq f^\circ(\bar{x}, y).
\end{aligned}
\tag{7.19}
$$

The definition of $\partial_o f(\bar{x})$ now shows that $o \in \partial_o f(\bar{x})$. If $\bar{x}$ is a local maximizer of f, then $\bar{x}$ is a local minimizer of $-f$ and so $o \in \partial_o(-f)(\bar{x}) = -\partial_o f(\bar{x})$ (the latter by Proposition 7.3.5). The proof for $\partial_\diamond f$ is analogous.   □

We establish further properties of the subdifferentials.

Proposition 7.3.7 *Let f be locally L-continuous around $\bar{x}$ with constant $\lambda > 0$. Then:*

(a) *The subdifferentials $\partial_o f(\bar{x})$ and $\partial_\diamond f(\bar{x})$ are nonempty, convex, and $\sigma(E^*, E)$-compact, and satisfy*

$$
\partial_\diamond f(\bar{x}) \subseteq \partial_o f(\bar{x}) \subseteq B_{E^*}(o, \lambda).
$$

(b) *One has*

$$
\begin{aligned}
f^\circ(\bar{x}, y) &= \max\{\langle x^*, y\rangle \mid x^* \in \partial_o f(\bar{x})\} \quad \forall y \in E, \\
f^\diamond(\bar{x}, y) &= \max\{\langle x^*, y\rangle \mid x^* \in \partial_\diamond f(\bar{x})\} \quad \forall y \in E.
\end{aligned}
$$

Proof. Taking Theorem 7.3.2(a) into consideration, the assertions follow as those of Proposition 4.1.6(b).   □

Observe that, beside the lower semicontinuity, it is the sublinearity of $f^\circ(\bar{x}, \cdot)$ and $f^\diamond(x, \cdot)$ that ensures the nonemptyness of the respective subdifferential. If we choose, say, $\overline{f}_H(\bar{x}, \cdot)$ instead, we obtain for the function f of Example 7.3.3,

$$
\{x^* \in E^* \mid \langle x^*, y\rangle \leq \overline{f}_H(\pi, y) \quad \forall y \in E\} = \emptyset.
$$

Proposition 7.3.8 *If $f : E \to \mathbb{R}$ is locally L-continuous on E, then:*

(a) *The functional $(x, y) \mapsto f^\circ(x, y)$ is upper semicontinuous on $E \times E$.*
(b) *Let (x_k) and (x_k^*) be sequences in E and E^*, respectively, such that $x_k^* \in \partial_\circ f(x_k)$ for all $k \in \mathbb{N}$. Assume that (x_k) converges to $\bar{x} \in E$ as $k \to \infty$ and that $x^* \in E^*$ is a $\sigma(E^*, E)$-cluster point of (x_k^*). Then one has $x^* \in \partial_\circ f(\bar{x})$. (That is, graph$(\partial_\circ f)$ is a norm–weak* closed subset of $E \times E^*$.)*
(c) *The subdifferential mapping $\partial_\circ f : E \rightrightarrows E^*$ is norm-to-weak* upper semi-continuous.*

Proof.

(a) Let (x_k) and (y_k) be sequences converging to $\bar{x} \in E$ and $\bar{y} \in E$, respectively. By definition of f°, for each k there exist $z_k \in E$ and $\tau_k > 0$ such that $\|z_k - x_k\| + \tau_k < \frac{1}{k}$ and

$$f^\circ(x_k, y_k) - \frac{1}{k} \leq \frac{f(z_k + \tau_k y_k) - f(z_k)}{\tau_k}$$

$$= \frac{f(z_k + \tau_k \bar{y}) - f(z_k)}{\tau_k} + \frac{f(z_k + \tau_k y_k) - f(z_k + \tau_k \bar{y})}{\tau_k}$$

$$\leq \frac{f(z_k + \tau_k \bar{y}) - f(z_k)}{\tau_k} + \lambda\|y_k - \bar{y}\|;$$

in the last term, $\lambda > 0$ denotes a Lipschitz constant of f near $\bar{x}$. By letting $k \to \infty$, the definition of the upper limit gives $\limsup_{k\to\infty} f^\circ(x_k, y_k) \leq f^\circ(\bar{x}, \bar{y})$. Hence f° is u.s.c. at $(\bar{x}, \bar{y})$.

(b) Let $y \in E$ be given. Some subsequence of $(\langle x_k^*, y \rangle)$, again denoted $(\langle x_k^*, y \rangle)$, satisfies $\langle x_k^*, y \rangle \to \langle x^*, y \rangle$ as $k \to \infty$. By the definition of $\partial_\circ f$ we have $\langle x_k^*, y \rangle \leq f^\circ(x_k, y)$ for all k. Letting $k \to \infty$, we conclude from (a) that $\langle x^*, y \rangle \leq f^\circ(\bar{x}, y)$. Since $y \in E$ was arbitrary, we obtain $x^* \in \partial_\circ f(\bar{x})$.

(c) The proof is analogous to that of Proposition 4.3.2(a). (Notice that by Proposition 7.3.7(a), the multifunction $\partial_\circ f$ is locally bounded at any $\bar{x} \in E$.) □

Statement (b) of Proposition 7.3.8 will be crucial for deriving a chain rule in Sect. 7.4 as well as a multiplier rule in Sect. 12.4. Now we establish relationships between the various notions. Recall the convention at the beginning of this section.

Proposition 7.3.9

(a) *If $f_G(\bar{x}, \cdot)$ exists and is sublinear on E, then $f^\diamond(\bar{x}, y) = f_G(\bar{x}, y)$ for each $y \in E$. In particular, if f is G-differentiable at $\bar{x}$, then $\partial_\diamond f(\bar{x}) = \{f'(\bar{x})\}$.*
(b) *If f is locally L-continuous around $\bar{x}$ and G-differentiable at $\bar{x}$, then $f'(\bar{x}) \in \partial_\circ f(\bar{x})$.*

(c) If f is strictly H-differentiable at $\bar{x}$, then f is locally L-continuous around $\bar{x}$ and $f^\circ(\bar{x}, y) = \langle f'(\bar{x}), y \rangle$ for each $y \in E$ and $\partial_\diamond f(\bar{x}) = \partial_\circ f(\bar{x}) = \{f'(\bar{x})\}$.

(d) If D is convex and f is convex and locally L-continuous around $\bar{x}$, then $f^\diamond(\bar{x}, y) = f^\circ(\bar{x}, y) = f_H(\bar{x}, y)$ for each $y \in E$ and $\partial_\diamond f(\bar{x}) = \partial_\circ f(\bar{x}) = \partial f(\bar{x})$.

Proof.

(a) It is clear that $f_G(\bar{x}, \cdot) \leq f^\diamond(\bar{x}, \cdot)$. We show the reverse inequality. For each $y \in E$ we have

$$f^\diamond(\bar{x}, y)$$

$$\leq \sup_{z \in E}\left(\limsup_{\tau \downarrow 0} \frac{1}{\tau}\big(f(\bar{x} + \tau y + \tau z) - f(\bar{x})\big) + \limsup_{\tau \downarrow 0} \frac{1}{\tau}\big(f(\bar{x}) - f(\bar{x} + \tau z)\big)\right)$$

$$= \sup_{z \in E}\left(f_G(\bar{x}, y + z) - f_G(\bar{x}, z)\right) \leq f_G(\bar{x}, y);$$

here, the last inequality holds because $f_G(\bar{x}, \cdot)$ is subadditive. The second statement follows immediately from the definition of $\partial_\diamond f(\bar{x})$.

(b) If $y \in E$, then

$$\langle f'(\bar{x}), y \rangle = \lim_{\tau \downarrow 0} \frac{1}{\tau}\big[f(\bar{x}+\tau y) - f(\bar{x})\big] \leq \limsup_{\substack{\tau \downarrow 0 \\ x \to \bar{x}}} \frac{1}{\tau}\big[f(x+\tau y) - f(x)\big] = f^\circ(\bar{x}, y)$$

and so $f'(\bar{x}) \in \partial_\circ f(\bar{x})$.

(c) By Proposition 3.2.4, f is locally L-continuous around $\bar{x}$ and strictly G-differentiable at $\bar{x}$. We therefore have

$$\langle f'(\bar{x}), y \rangle = \lim_{\substack{\tau \to 0 \\ x \to \bar{x}}} \frac{1}{\tau}\big[f(x + \tau y) - f(x)\big] = f^\circ(\bar{x}, y) \quad \forall y \in E$$

and so $\partial_\circ f(\bar{x}) = \{f'(\bar{x})\} = \partial_\diamond f(\bar{x})$.

(d) Let $\delta > 0$. Then

$$f^\circ(\bar{x}, y) = \inf_{\epsilon \in (0, \epsilon_0)} \sup_{\substack{\tau \in (0, \epsilon) \\ x \in B(\bar{x}, \delta\epsilon)}} \frac{1}{\tau}\big[f(x + \tau y) - f(x)\big]$$

$$= \inf_{\epsilon \in (0, \epsilon_0)} \sup_{x \in B(\bar{x}, \delta\epsilon)} \frac{1}{\epsilon}\big[f(x + \epsilon y) - f(x)\big];$$

(7.20)

here the first equation is a consequence of the definition of the limit superior, and the second holds by Theorem 4.1.3(a) because f is convex. Denoting the Lipschitz constant of f around $\bar{x}$ by λ, we further obtain

$$\frac{1}{\epsilon}\big|\big[f(x + \epsilon y) - f(x)\big] - \big[f(\bar{x} + \epsilon y) - f(\bar{x})\big]\big|$$

$$\leq \frac{1}{\epsilon}\big(\big|f(x + \epsilon y) - f(\bar{x} + \epsilon y)\big| + \big|f(\bar{x}) - f(x)\big|\big) \leq 2\delta\lambda,$$

provided $\|x - \bar{x}\| < \delta\epsilon$ and ϵ is sufficiently small. With this estimate, (7.20) passes into

$$f^\circ(\bar{x}, y) \leq \inf_{\epsilon \in (0,\epsilon_0)} \frac{1}{\epsilon}\left[f(\bar{x} + \epsilon y) - f(\bar{x})\right] + 2\delta\lambda$$

$$= f_G(\bar{x}, y) + 2\delta\lambda = f_H(\bar{x}, y) + 2\delta\lambda;$$

here the second equation holds by Proposition 4.1.8. Since $\delta > 0$ was arbitrary, it follows that $f^\circ(\bar{x}, y) \leq f_H(\bar{x}, y)$. Since the reverse inequality always holds, we obtain $f^\circ(\bar{x}, y) = f_H(\bar{x}, y)$ and so $\partial_\circ f(\bar{x}) = \partial f(\bar{x})$. The assertion concerning $f^\diamond(\bar{x}, \cdot)$ and $\partial_\diamond f(\bar{x})$ follows from this and (a) since by Theorem 4.1.3(d), $f_H(\bar{x}, \cdot)$ and so $f_G(\bar{x}, \cdot)$ exists and is sublinear. □

Remark 7.3.10

(a) Proposition 7.3.9 shows that the Michel–Penot subdifferential is a generalization of the G-derivative while the Clarke subdifferential generalizes the strict H-derivative.

(b) Since we always have $\partial_\diamond f(\bar{x}) \subseteq \partial_\circ f(\bar{x})$ (Proposition 7.3.7) and the inclusion may be proper (see Example 7.3.11), the necessary optimality condition $o \in \partial_\diamond f(\bar{x})$ (Proposition 7.3.6) is in general stronger than the condition $o \in \partial_\circ f(\bar{x})$.

Example 7.3.11 Let $E := \mathbb{R}$ and $f(x) := x^2 \sin\frac{1}{x}$ for $x \neq 0$, $f(0) := 0$. Then f is locally L-continuous and differentiable at 0, with $f'(0) = 0$. By Proposition 7.3.9 we have $f^\diamond(0, y) = 0$ for each $y \in \mathbb{R}$ and $\partial_\diamond f(0) = \{0\}$. On the other hand, we obtain $f^\circ(0, y) = |y|$ for each $y \in \mathbb{R}$ and so $\partial_\circ f(0) = [-1, 1]$. Notice that f is not strictly differentiable at 0.

Recall that a locally L-continuous function $f : \mathbb{R}^n \to \mathbb{R}$ is differentiable almost everywhere, i.e., outside a set Ω_f of n-dimensional Lebesgue measure zero (Theorem of Rademacher, see, for instance, Evans and Gariepy [63]).

Theorem 7.3.12 *If $f : \mathbb{R}^n \to \mathbb{R}$ is locally L-continuous around $\bar{x}$ and $S \subseteq \mathbb{R}^n$ has n-dimensional Lebesgue measure zero, then one has*

$$\partial_\circ f(\bar{x}) = \mathrm{co}\left\{\lim_{k \to \infty} f'(x_k) \mid x_k \to \bar{x}, \; x_k \notin \Omega_f \cup S\right\}.$$

For a proof of Theorem 7.3.12 we refer to Clarke [36].

7.4 Subdifferential Calculus

The following notion will serve to refine certain computation rules for the subdifferentials.

Definition 7.4.1 The functional $f : D \to \mathbb{R}$ is called *regular* (in the sense of Clarke) at $\bar{x}$ if $f_G(\bar{x}, \cdot)$ and $f^\circ(\bar{x}, \cdot)$ both exist and coincide.

Remark 7.4.2

(a) If f is regular at $\bar{x}$, then together with $f^\circ(\bar{x}, \cdot)$ the functional $f_G(\bar{x}, \cdot)$ is sublinear and so the two functionals also coincide with $f^\Diamond(\bar{x}, \cdot)$ (Proposition 7.3.9).

(b) Let the functional $f : D \to \mathbb{R}$ be locally L-continuous around $\bar{x}$. Then it is regular at $\bar{x}$ if it is strictly H-differentiable or if D and f are convex (Proposition 7.3.9).

Concerning the following computation rules, compare Proposition 4.5.1.

Proposition 7.4.3 (Sum Rule) *Let* $f_0, f_1, \ldots, f_n : D \to \mathbb{R}$ *be locally L-continuous around* $\bar{x}$:

(a) *One has*

$$\partial_\circ \left(\sum_{i=0}^n f_i \right)(\bar{x}) \subseteq \sum_{i=0}^n \partial_\circ f_i(\bar{x}) \quad and \quad \partial_\Diamond \left(\sum_{i=0}^n f_i \right)(\bar{x}) \subseteq \sum_{i=0}^n \partial_\Diamond f_i(\bar{x}).$$
(7.21)

(b) *If* $f_1, \ldots, f_n$ *are strictly H-differentiable at* $\bar{x}$, *then* (7.21) *holds with equality in both cases.*

(c) *If* $f_0, f_1, \ldots, f_n$ *are regular at* $\bar{x}$, *then* $\sum_{i=0}^n f_i$ *is regular at* $\bar{x}$ *and* (7.21) *holds with equality in both cases.*

Proof.

(a) We verify the statement for the Clarke subdifferential, leaving the verification for the Michel–Penot subdifferential as Exercise 7.5.4. Moreover, we confine ourselves to the case $n = 1$; the general case then follows by induction. Since by Proposition 7.3.7, we have $f_i^\circ(\bar{x}, y) = \max\{\langle v, y \rangle \mid v \in \partial_\circ f_i(\bar{x})\}$ for $i = 0, 1$, we conclude that:
   – The support functional of $\partial_\circ(f_0 + f_1)(\bar{x})$ is $(f_0 + f_1)^\circ(\bar{x}, \cdot)$.
   – The support functional of $\partial_\circ f_0(\bar{x}) + \partial_\circ f_1(\bar{x})$ is $f_0^\circ(\bar{x}, \cdot) + f_1^\circ(\bar{x}, \cdot)$.
   From the definition of the Clarke directional derivative we easily obtain

$$(f_0 + f_1)^\circ(\bar{x}, y) \leq f_0^\circ(\bar{x}, y) + f_1^\circ(\bar{x}, y) \quad \forall y \in E. \qquad (7.22)$$

   Hence the assertion follows by the Hörmander theorem (Theorem 2.3.1(c)).

(b) The assumption implies that the functional $f_1 + \cdots + f_n$ is also strictly H-differentiable. Therefore it suffices to consider the case $n = 1$. It is easy to show that

$$(f_0 + f_1)^\circ(\bar{x}, y) = f_0^\circ(\bar{x}, y) + \langle f_1'(\bar{x}), y \rangle = f_0^\circ(\bar{x}, y) + f_1^\circ(\bar{x}, y) \quad \forall y \in E.$$

This together with (a) yields the assertion for the Clarke subdifferential. Again the proof is analogous for the Michel–Penot subdifferential.

(c) By assumption we obtain

$$(f_0 + f_1)_G(\bar{x}, y)$$
$$= f_{0,G}(\bar{x}, y) + f_{1,G}(\bar{x}, y) = f_0^\circ(\bar{x}, y) + f_1^\circ(\bar{x}, y) \underset{(7.22)}{\geq} (f_0 + f_1)^\circ(\bar{x}, y).$$

On the other hand, we always have $(f_0 + f_1)^\circ(\bar{x}, y) \geq (f_0 + f_1)_G(\bar{x}, y)$. Therefore $f_0 + f_1$ is regular at $\bar{x}$. Moreover, by what has just been shown we see that

$$(f_0 + f_1)^\circ(\bar{x}, y) = f_0^\circ(\bar{x}, y) + f_1^\circ(\bar{x}, y) \quad \forall y \in E.$$

From this, the assertion again follows by the Hörmander theorem since the left-hand side is the support functional of $\partial_\circ(f_0 + f_1)(\bar{x})$ and the right-hand side is the support functional of $\partial_\circ f_0(\bar{x}) + \partial_\circ f_1(\bar{x})$. Remark 7.4.2(a) completes the proof. □

Next we establish a mean value theorem. If $x, y \in E$ and $A \subseteq E^*$, we write

$$[x, y] := \{(1 - \tau)x + \tau y \mid \tau \in [0, 1]\},$$
$$(x, y) := \{(1 - \tau)x + \tau y \mid \tau \in (0, 1)\},$$
$$\langle A, x \rangle := \{\langle x^*, x \rangle \mid x^* \in A\}.$$

Theorem 7.4.4 (Mean Value Theorem) *Assume that $f : D \to \mathbb{R}$ is locally L-continuous and $[x, y] \subseteq D$. Then there exists $z \in (x, y)$ satisfying*

$$f(y) - f(x) \in \langle \partial_\circ f(z), y - x \rangle. \tag{7.23}$$

Proof.

(I) For $\lambda \in [0, 1]$ let $x_\lambda := x + \lambda(y - x)$. Define $\varphi, \psi : [0, 1] \to \mathbb{R}$ by

$$\varphi(\lambda) := \psi(\lambda) + \lambda\big(f(x) - f(y)\big), \quad \psi(\lambda) := f(x_\lambda).$$

Since $\varphi(0) = \varphi(1) = f(x)$, the continuous function φ attains a local minimum or a local maximum at some $\lambda_0 \in (0, 1)$. Therefore $0 \in \partial_\circ \varphi(\lambda_0)$ (Proposition 7.3.6). By Propositions 7.3.5 and 7.4.3 we obtain

$$0 \in \partial_\circ \psi(\lambda_0) + \big(f(x) - f(y)\big). \tag{7.24}$$

(II) We show that

$$\partial_\circ \psi(\lambda) \subseteq \langle \partial_\circ f(x_\lambda), y - x \rangle \quad \forall \lambda \in (0, 1). \tag{7.25}$$

Observe that ψ is L-continuous on $(0, 1)$ so that $\partial_\circ \psi$ makes sense. Also notice that the two sets in (7.25) are closed convex subsets of $\mathbb{R}$ and so are intervals. Hence it suffices to prove that

$$\max\big(a\partial_\circ \psi(\lambda)\big) \leq \max\big(a\langle \partial_\circ f(x_\lambda), y - x \rangle\big) \quad \text{for } a = \pm 1.$$

This is verified as follows:

$$\max\big(a\partial_\circ\psi(\lambda)\big) = \psi^\circ(\lambda,a) = \limsup_{\substack{\tau\downarrow 0 \\ \lambda'\to\lambda}}\frac{1}{\tau}\big(\psi(\lambda'+\tau a)-\psi(\lambda')\big)$$

$$= \limsup_{\substack{\tau\downarrow 0 \\ \lambda'\to\lambda}}\frac{1}{\tau}\big[f\big(x+(\lambda'+\tau a)(y-x)\big)-f\big(x+\tau(y-x)\big)\big]$$

$$\le \limsup_{\substack{\tau\downarrow 0 \\ z\to x_\lambda}}\frac{1}{\tau}\big[f\big(z+\tau a(y-x)\big)-f(z)\big]$$

$$= f^\circ(x_\lambda,a(y-x)) = \max\langle\partial_\circ f(x_\lambda),a(y-x)\rangle.$$

In view of (7.24) and (7.25) the proof is complete on setting $z := x_{\lambda_0}$. $\quad\square$

Finally we establish a *chain rule*. We consider the composite function $g \circ h : E \to \mathbb{R}$, where

$$h : E \to \mathbb{R}^n, \quad g : \mathbb{R}^n \to \mathbb{R}, \quad (g\circ h)(x) := g\big(h(x)\big), \quad x \in E. \tag{7.26}$$

We identify $\big(\mathbb{R}^n\big)^*$ with $\mathbb{R}^n$, put $h = (h_1,\dots,h_n)$ with $h_i : E \to \mathbb{R}$, and define for any $a \in \mathbb{R}^n$,

$$h_a(x) := \langle a, h(x)\rangle_{\mathbb{R}^n}, \quad x \in E. \tag{7.27}$$

Recall that $\overline{co}^* A$ denotes the $\sigma(E^*, E)$-closed convex hull of the set $A \subseteq E^*$.

Theorem 7.4.5 (Chain Rule) *Let g and h be as in (7.26). Assume that h is locally L-continuous around $\bar{x} \in E$ and g is locally L-continuous around $h(\bar{x})$. Then:*

(a) *The composite function $g \circ h$ is locally L-continuous around $\bar{x}$, and there holds*

$$\partial_\circ(g\circ h)(\bar{x}) \subseteq \overline{co}^*\{\partial_\circ h_a(\bar{x}) \mid a \in \partial_\circ g(h(\bar{x}))\}. \tag{7.28}$$

(b) *If, in addition, g is regular at $h(\bar{x})$, any h_i is regular at $\bar{x}$, and any $a \in \partial_\circ g\big(h(\bar{x})\big)$ has nonnegative components, then (7.28) holds as equality and $g \circ h$ is regular at $\bar{x}$.*

Proof.

(I) Set $f := g \circ h$. It is easy to see that f is locally L-continuous around $\bar{x}$.

(II) Denote the set on the right-hand side of the inclusion (7.28) by M. The sets $\partial_\circ f(\bar{x})$ and M are nonempty, convex, and $\sigma(E^*.E)$-compact. Therefore (7.28) holds if and only if the associated support functions satisfy $\sigma_{\partial_\circ f(\bar{x})} \le \sigma_M$. By Proposition 7.3.7, we have $\sigma_{\partial_\circ f(\bar{x})} = f^\circ(\bar{x},\cdot)$. Hence the theorem is verified if we can show that

$$f^\circ(\bar{x},y) \le \sigma_M(y) \quad \forall y \in E. \tag{7.29}$$

(III) To verify (7.29), let $y \in E$ be given. We shall construct elements $a \in \partial_o g\big(h(\bar{x})\big)$ and $y^* \in \partial_o h_a(\bar{x})$ satisfying

$$f^\circ(\bar{x}, y) \le \langle y^*, y \rangle. \tag{7.30}$$

Choose sequences $\tau_k \downarrow 0$ and $x_k \to \bar{x}$ such that

$$\lim_{k \to \infty} \frac{1}{\tau_k}\Big(f(x_k + \tau_k y) - f(x_k) \Big) = f^\circ(\bar{x}, y).$$

By the mean value theorem (Theorem 7.4.4) there exist $z_k \in [h(x_k), h(x_k + \tau_k y)]$ and $a_k \in \partial_o g(z_k)$ satisfying

$$
\begin{aligned}
\frac{1}{\tau_k}\Big(f(x_k + \tau_k y) - f(x_k) \Big) &= \frac{1}{\tau_k}\Big[g\big(h(x_k + \tau_k y)\big) - g\big(h(x_k)\big) \Big] \\
&= \left\langle a_k, \frac{1}{\tau_k}\big(h(x_k + \tau_k y) - h(x_k) \big) \right\rangle_{\mathbb{R}^n}.
\end{aligned}
\tag{7.31}
$$

Let λ be a local Lipschitz constant of g at $h(\bar{x})$. Since $z_k \to h(\bar{x})$ as $k \to \infty$, we conclude that for k sufficiently large, λ is also a Lipschitz constant of g at z_k and so $\|a_k\| \le \lambda$ (Proposition 7.3.7(a)). A subsequence of the sequence (a_k), again denoted (a_k), is thus convergent to some $a \in \mathbb{R}^n$. Proposition 7.3.7(c) shows that $a \in \partial_o g\big(h(\bar{x})\big)$.

(IV) Again by the mean value theorem there exist $y_k \in [x_k, x_k + \tau_k y]$ and $y_k^* \in \partial_o h_a(y_k)$ such that

$$\left\langle a, \frac{1}{\tau_k}\big(h(x_k + \tau_k y) - h(x_k) \big) \right\rangle_{\mathbb{R}^n} = \langle y_k^*, y \rangle. \tag{7.32}$$

As above it follows that the sequences (y_k^*) and $(\langle y_k^*, y \rangle)$ are bounded in E^* and $\mathbb{R}$, respectively. Let again denote $(\langle y_k^*, y \rangle)$ a convergent subsequence and let $y^* \in E^*$ be a $\sigma(E^*, E)$-cluster point of (y_k^*). Then $\langle y_k^*, y \rangle \to \langle y^*, y \rangle$ as $k \to \infty$ and $y^* \in \partial_o h_a(\bar{x})$ by Proposition 7.3.7(c).

(V) Combining (7.31) and (7.32) we obtain

$$
\begin{aligned}
&\frac{1}{\tau_k}\Big(f(x_k + \tau_k y) - f(x_k) \Big) \\
&= \left\langle a_k - a, \frac{1}{\tau}\big(h(x_k + \tau_k y) - h(x_k) \big) \right\rangle_{\mathbb{R}^n} + \langle y_k^*, y \rangle.
\end{aligned}
$$

Since h is locally L-continuous around $\bar{x}$ and $x_k \to \bar{x}$, the term $\tau_k^{-1}\big(h(x_k + \tau_k y) - h(x_k) \big)$ is bounded. Moreover, we have $a_k \to a$. Therefore we obtain

$$f^\circ(\bar{x}, y) = \lim_{k \to \infty} \frac{1}{\tau_k}\Big(f(x_k + \tau_k y) - f(x_k) \Big) = \langle y^*, y \rangle,$$

which proves (a). The verification of (b) is left as Exercise 7.5.5.   □

We consider the special case that g is strictly H-differentiable at $h(\bar{x})$. Recall that $D_i g$, where $i = 1, \ldots, n$, denotes the partial derivative of g with respect to the ith variable.

Corollary 7.4.6 *Let g and h be as in (7.26). Assume that h is locally L-continuous around $\bar{x} \in E$ and g is strictly H-differentiable at $h(\bar{x})$. Then the composite function $g \circ h$ is locally L-continuous around $\bar{x}$, and there holds*

$$\partial_o(g \circ h)(\bar{x}) \subseteq \sum_{i=1}^{n} D_i g(h(\bar{x})) \partial_o h_i(\bar{x}). \qquad (7.33)$$

In particular, if $n = 1$, then

$$\partial_o(g \circ h)(\bar{x}) = g'(h(\bar{x})) \partial_o h(\bar{x}). \qquad (7.34)$$

Proof.

(I) By Proposition 7.3.9(c) we have $\partial_o g(h(\bar{x})) = \{g'(h(\bar{x}))\}$ and so, with $a = g'(h(\bar{x}))$,

$$h_a(x) = \sum_{i=1}^{n} D_i g(h(\bar{x})) h_i(x), \quad x \in E.$$

The inclusion (7.33) now follows by applying Theorem 7.4.5 and Propositions 7.3.5 and 7.4.3. The $\overline{\mathrm{co}}^*$ operation is superfluous here since with each $\partial_o h_i(\bar{x})$ the set on the right-hand side of (7.33) is $\sigma(E^*, E)$-compact and convex.

(II) Now let $n = 1$. Define $\gamma : E \to \mathbb{R}$ by

$$\gamma(y) := \max\{g'(h(\bar{x})) \langle x^*, y \rangle \mid x^* \in \partial_o h(\bar{x})\}.$$

For any $y \in E$ we have

$$\gamma(y) = g'(h(\bar{x})) h^\circ(\bar{x}, y) = \limsup_{\substack{\tau \downarrow 0 \\ x \to \bar{x}}} \frac{1}{\tau} g'(h(\bar{x})) (h(x + \tau y) - h(x))$$

$$= \limsup_{\substack{\tau \downarrow 0 \\ x \to \bar{x}}} \frac{1}{\tau} \Big[g(h(x + \tau y)) - g(h(x)) \Big] = (g \circ h)^\circ(\bar{x}, y).$$

Here, the first equation holds by Proposition 7.3.7(b) and the third is a consequence of the strict H-differentiability of g. The assertion (7.34) follows because γ is the support function of $g'(h(\bar{x})) \partial_o h(\bar{x})$ and $(g \circ h)^\circ(\bar{x}, \cdot)$ is the support function of $\partial_o(g \circ h)(\bar{x})$. $\qquad \square$

Given a finite family of functionals $f_i : D \to \mathbb{R}$, $i \in I$, we consider the maximum functional $f : D \to \mathbb{R}$ defined by

$$f(x) := \max\{f_i(x) \mid i \in I\}, \quad x \in D. \qquad (7.35)$$

Let $\bar{x} \in D$. As in Sect. 7.2 we set

$$I(\bar{x}) := \{i \in I \mid f_i(\bar{x}) = f(\bar{x})\}.$$

Proposition 7.4.7 (Maximum Rule) *Let I be a finite set and for all $i \in I$ let $f_i : D \to \mathbb{R}$ be locally L-continuous around $\bar{x}$. Then the functional f defined by (7.35) satisfies*

$$\partial_\circ f(\bar{x}) \subseteq \mathrm{co}\{\partial_\circ f_i(\bar{x}) \mid i \in I(\bar{x})\} \quad and \quad \partial_\diamond f(\bar{x}) \subseteq \mathrm{co}\{\partial_\diamond f_i(\bar{x}) \mid i \in I(\bar{x})\}.$$
$$(7.36)$$

If, in addition, f_i is regular at $\bar{x}$ for any $i \in I(\bar{x})$, then the first inclusion in (7.36) holds with equality and f is regular at $\bar{x}$.

Proof. We verify (7.36) for the Michel–Penot subdifferential, leaving the proof for the Clarke subdifferential as Exercise 7.5.6.

(I)   As in the proof of Proposition 7.2.7 it can be shown that for $i \in I \setminus I(\bar{x})$ the functional f_i does not contribute to $f^\diamond(\bar{x}, \cdot)$.

(II) Let $y \in E$ be given. Choose sequences (z_n) in E and (τ_n) in $(0, +\infty)$ such that $\tau_n \downarrow 0$ and

$$\frac{1}{\tau_n}\left(f(\bar{x} + \tau_n y + \tau_n z_n) - f(\bar{x} + \tau_n z_n)\right) \to f^\diamond(\bar{x}, y) \quad \text{as } n \to \infty.$$

Without loss of generality we may assume that for suitable subsequences, again denoted (z_n) and (τ_n), we have for some $i \in I(\bar{x})$,

$$f(\bar{x} + \tau_n y + \tau_n z_n) = f_i(\bar{x} + \tau_n y + \tau_n z_n).$$

It follows that

$$f_i^\diamond(\bar{x}, y) \geq \limsup_{n \to \infty} \frac{1}{\tau_n}\left(f_i(\bar{x} + \tau_n y + \tau_n z_n) - f_i(\bar{x} + \tau_n z_n)\right)$$

$$\geq \limsup_{n \to \infty} \frac{1}{\tau_n}\left(f(\bar{x} + \tau_n y + \tau_n z_n) - f(\bar{x} + \tau_n z_n)\right) = f^\diamond(\bar{x}, y)$$

and so $f^\diamond(\bar{x}, \cdot) \leq f_i^\diamond(\bar{x}, \cdot) \leq \max_{i \in I(\bar{x})} f_i^\diamond(\bar{x}, \cdot)$. Since $f^\diamond(\bar{x}, \cdot)$ is the support functional of $\partial_\diamond f(\bar{x})$ (Proposition 7.3.7) and so equals $\delta^*_{\partial_\diamond f(\bar{x})}$ (Example 2.2.5), we conclude that, setting $M := \mathrm{co}\{\partial_\diamond f_i(\bar{x}) \mid i \in I(\bar{x})\}$, we have

$$\delta^*_{\partial_\diamond f(\bar{x})} \leq \max_{i \in I(\bar{x})} \delta^*_{\partial_\diamond f_i(\bar{x})} = \delta^*_M;$$

here the equality sign is easily verified. By Proposition 7.3.7 the set M is nonempty, convex, and $\sigma(E^*, E)$-compact and so δ_M is proper, convex, and l.s.c. Hence Theorem 2.2.4 gives $\delta_M = \delta^{**}_M \leq \delta^{**}_{\partial_\diamond f(\bar{x})} = \delta_{\partial_\diamond f(\bar{x})}$, and we conclude that $\partial_\diamond f(\bar{x}) \subseteq M$.   □

7.5 Bibliographical Notes and Exercises

(Radial) upper convex approximations were introduced and studied by Neustadt [150] and Pshenichnyi [168] (see also Scheffler [190] and Scheffler and Schirotzek [192]). (Regularly) locally convex functionals are considered by Ioffe and Tikhomirov [101]. Demyanov and Rubinov [47, 48] studied quasidifferentiable functionals in detail (see also Luderer et al. [126] and the literature cited therein).

Clarke's doctoral thesis [33] marks a breakthrough in that it gives, for Lipschitz functions on $\mathbb{R}^n$, the first intrinsic (nonaxiomatic) approach to generalized directional derivatives. The characterization of $\partial_o f(\bar{x})$ given in Theorem 7.3.12 is Clarke's original definition in the finite-dimensional case. A remarkable generalization of this characterization to Banach spaces with a β-smooth norm is due to Preiss [166].

Clarke [36] systematically elaborated his concept in normed vector spaces. The mean value theorem (Theorem 7.4.4) is due to Lebourg [119]. For further results of this kind see Hiriart-Urruty [86], Penot [163], and Studniarski [202, 203]. A mean value theorem in terms of radial upper convex approximations that encompasses Lebourg's mean value theorem is due to Scheffler [191]. For applications of Clarke's directional derivative and subdifferential to various problems we refer to Clarke [36], Clarke et al. [39], Loewen [123], Mäkelä and Neittaanmäki [149], Panagiotopoulos [157], Papageorgiou and Gasinski [158], and the references in these books. The Michel–Penot directional derivative and subdifferential (see [129, 130]) are considered, among others, by Borwein and Lewis [18] and Ioffe [99].

Exercise 7.5.1 Prove Proposition 7.2.2.
Hint: Compare the proof of Lemma 3.1.2.

Exercise 7.5.2 Define $f : \mathbb{R}^2 \to \mathbb{R}$ by $f(x_1, x_2) := 0$ if $x_1 \geq 0$ and $f(x_1, x_2) := 1$ otherwise. Show that f is locally convex but not regularly locally convex at $\bar{x} := (0, 0)$.

Exercise 7.5.3 Prove Proposition 7.2.7 for the directional H-derivative.

Exercise 7.5.4 Verify Proposition 7.4.3 for the Michel–Penot subdifferential.

Exercise 7.5.5 Verify assertion (b) of Theorem 7.4.5.

Exercise 7.5.6 Prove Proposition 7.4.7 for the Clarke subdifferential.
Hint: Define $g : \mathbb{R}^n \to \mathbb{R}$ and $h : E \to \mathbb{R}^n$ by

$$g(z_1, \ldots, z_n) := \max\{z_1, \ldots, z_n\}, \quad h(x) := \big(f_1(x), \ldots, f_n(x)\big).$$

Observe that $f = g \circ h$ and apply Theorem 7.4.5.

8

Variational Principles

8.1 Introduction

Convention. In this chapter, unless otherwise specified, assume that E is a Banach space and $f : E \to \overline{\mathbb{R}}$ is proper, l.s.c., and bounded below.

The theory of generalized directional derivatives and subdifferentials considered so far for both convex and nonconvex functionals is essentially based on separation and so on convexity arguments; consider, e.g., the proofs of the sum rules (Propositions 4.5.1 and 7.4.3) and the maximum rule (Proposition 7.4.7), where the crucial tool is the sandwich theorem, the Hörmander theorem, and the biconjugation theorem, respectively. These tools were applicable since the derivative-like objects constructed are convex. It turns out that a corresponding theory not enforcing convexity and working beyond the Lipschitz case requires quite different tools. In the following we establish *variational principles* as well as *extremal principles*, which have proved to be adequate for treating lower semicontinuous functionals. (To be precise, Clarke's multiplier rule for Lipschitz functionals to be derived in Sect. 12.4 also hinges on a variational principle.)

First we explain the idea of variational principles. It is clear that a functional f as above may fail to have a global minimizer. However, since f is bounded below, there are points that "almost" minimize f, i.e., for each $\epsilon > 0$ there exists $\bar{x} \in E$ satisfying

$$f(\bar{x}) \leq \inf_{x \in E} f(x) + \epsilon.$$

Ekeland [56] showed that for each such $\bar{x}$ and each $\lambda > 0$ there exists a point $z \in E$ that actually minimizes the slightly perturbed functional

$$\varphi(y) := f(y) + \tfrac{\epsilon}{\lambda}\|z - y\|, \quad y \in E,$$

and is such that $\|z - \bar{x}\| \leq \lambda$. This first variational principle has remarkable applications in quite different areas of nonlinear analysis (see the references at the end of this chapter).

A drawback of Ekeland's variational principle is that the perturbed functional φ is not differentiable at $y = z$ even if the original functional f is differentiable on all of E. The first to overcome this drawback were Borwein and Preiss [19] who established a *smooth variational principle*. Meanwhile several smooth variational principles have been derived. We present below a smooth variational principle due to Loewen and Wang [125] from which the mentioned variational principles will then be deduced in a unified way.

8.2 The Loewen–Wang Variational Principle

We write $\inf_E f$ for $\inf_{x \in E} f(x)$. Recall the notion of a minimizing sequence.

Definition 8.2.1

(a) A point $\bar{x} \in E$ is said to be a *strict minimizer* of f if $f(\bar{x}) < f(x)$ for each $x \in E, x \neq \bar{x}$.
(b) A point $\bar{x} \in E$ is said to be a *strong minimizer* of f if $f(\bar{x}) = \inf_E f$ and each minimizing sequence for f is convergent to $\bar{x}$.

It is clear that each strong minimizer of f is also a strict minimizer. But the converse is false. For example, the point $\bar{x} = 0$ is a strict but not a strong minimizer of the function $f(x) := x^2 e^{-x}$ on $\mathbb{R}$; notice that each sequence (x_n) tending to $+\infty$ as $n \to \infty$ is a minimizing sequence for f.

Recall that $\text{diam}(A) := \sup\{\|x - y\| \,|\, x, y \in A\}$ denotes the *diameter* of the set $A \subseteq E$.

Remark 8.2.2 For $\epsilon > 0$ let

$$\Sigma_\epsilon(f) := \{x \in E \,|\, f(x) \leq \inf_E f + \epsilon\}.$$

It is left as Exercise 8.5.1 to show that the functional f has a strong minimizer on E if and only if

$$\inf\{\text{diam}(\Sigma_\epsilon(f)) \,|\, \epsilon > 0\} = 0.$$

Now we construct a perturbation ρ_∞ of f. The data involved will be specified in Theorem 8.2.3. Let $\rho : E \to [0, +\infty)$ be such that

$$\rho(o) = 0, \quad \eta := \sup\{\|x\| \,|\, x \in E, \rho(x) < 1\} < +\infty \tag{8.1}$$

and set

$$\rho_\infty(x) := \sum_{n=0}^{\infty} \mu_n \, \rho\big((n + 1)(x - z_n)\big). \tag{8.2}$$

Theorem 8.2.3 (Loewen–Wang) *Let $\epsilon > 0$ and let $\bar{x} \in E$ be such that $f(\bar{x}) < \inf_E f + \epsilon$. Assume further that $\rho : E \to [0, +\infty)$ is a continuous function satisfying (8.1) and that (μ_n) is a decreasing sequence in $(0, 1)$ with $\sum_{n=0}^{\infty} \mu_n < +\infty$. Then there exists a sequence (z_n) in E converging to some $z \in E$ such that, with ρ_∞ according to (8.2), the following holds:*

(a) $\rho(\bar{x} - z) < 1$.
(b) $f(z) + \epsilon \rho_\infty(z) \le f(\bar{x})$.
(c) z is a strong minimizer of $f + \epsilon \rho_\infty$ on E. In particular,

$$f(z) + \epsilon \rho_\infty(z) < f(x) + \epsilon \rho_\infty(x) \quad \forall x \in E \setminus \{z\}. \tag{8.3}$$

Proof.

(I) Set $z_0 := \bar{x}$, $f_0 := f$. By induction, z_{n+1} can be chosen (see below) and f_{n+1}, D_n can be defined for $n = 0, 1, \ldots$ in the following way:

$$f_{n+1}(x) := f_n(x) + \epsilon \mu_n \, \rho\big((n+1)(x - z_n)\big), \quad x \in E, \tag{8.4}$$

$$f_{n+1}(z_{n+1}) \le \frac{\mu_{n+1}}{2} f_n(z_n) + \left(1 - \frac{\mu_{n+1}}{2}\right) \inf_E f_{n+1} \le f_n(z_n), \tag{8.5}$$

$$D_n := \left\{ x \in E \,\middle|\, f_{n+1}(x) \le f_{n+1}(z_{n+1}) + \frac{\mu_n \epsilon}{2} \right\}. \tag{8.6}$$

We show that z_{n+1} can be chosen according to (8.5). Note that $\inf_E f_{n+1} \le f_{n+1}(z_n) = f_n(z_n)$. If this inequality is strict, then by the definition of the infimum there exists $z_{n+1} \in E$ such that

$$f_{n+1}(z_{n+1}) < \inf_E f_{n+1} + \frac{\mu_{n+1}}{2} \left(f_n(z_n) - \inf_E f_{n+1} \right).$$

If equality holds, then choose $z_{n+1} := z_n$. In either case, (8.5) is satisfied.

(II) Since f_{n+1} is l.s.c., the set D_n is closed. Moreover D_n is nonempty as $z_{n+1} \in D_n$. Since $\mu_n \in (0, 1)$ and $f_{n+1} \ge f_n$, (8.5) implies

$$f_{n+1}(z_{n+1}) - \inf_E f_{n+1} \le \frac{\mu_{n+1}}{2} \left(f_n(z_n) - \inf_E f_{n+1} \right) \le f_n(z_n) - \inf_E f_n. \tag{8.7}$$

(III) We have $D_n \subseteq D_{n-1}$ for $n = 1, 2 \ldots$. In fact, if $x \in D_n$, then $\mu_{n-1} > \mu_n$ and (8.5) yield

$$f_n(x) \le f_{n+1}(x) \le f_{n+1}(z_{n+1}) + \frac{\mu_n \epsilon}{2} \le f_n(z_n) + \frac{\mu_{n-1} \epsilon}{2}$$

and so $x \in D_{n-1}$.

(IV) We show that $\operatorname{diam}(D_n) \to 0$ as $n \to \infty$. Since $f_{n-1} \le f_n$, (8.5) with n replaced by $n - 1$ implies

$$
\begin{aligned}
f_n(z_n) - \inf_E f_n &\le \frac{\mu_n}{2} \left(f_{n-1}(z_{n-1}) - \inf_E f_n \right) \\
&\le \frac{\mu_n}{2} \left(f_{n-1}(z_{n-1}) - \inf_E f_{n-1} \right) < \frac{\mu_n \epsilon}{2}.
\end{aligned}
\tag{8.8}
$$

The last $<$ follows from (8.7) and $f_0(z_0) - \inf_E f_0 = f(\bar{x}) - \inf_E f < \epsilon$. Now let $x \in D_n$. By the definitions of D_n and f_{n+1} we obtain

$$\mu_n \epsilon \, \rho\big((n+1)(x - z_n)\big)$$
$$\leq f_{n+1}(z_{n+1}) - f_n(x) + \frac{\mu_n \epsilon}{2} \leq f_{n+1}(z_{n+1}) - \inf_E f_n + \frac{\mu_n \epsilon}{2}.$$

This inequality together with $f_{n+1}(z_{n+1}) \leq f_n(z_n)$ (see (8.5)) and (8.8) shows that

$$\rho\big((n+1)(x - z_n)\big) < 1 \quad \forall n = 0, 1, \ldots \tag{8.9}$$

The hypothesis (8.1) therefore implies

$$(n+1)\|x - z_n\| \leq \eta \tag{8.10}$$

and so $\operatorname{diam}(D_n) \leq \frac{2\eta}{n+1} \to 0$ as $n \to \infty$.

(V) In view of (III) and (IV), Cantor's intersection theorem applies to (D_n) ensuring that $\bigcap_{n=0}^{\infty} D_n$ contains exactly one point, say z. For each n we have $z_{n+1} \in D_n$ and $z \in D_n$. Hence $\|z_{n+1} - z\| \to 0$ as $n \to \infty$. Moreover, setting $x = z$ and $n = 0$ in (8.9), we see that $\rho(z - \bar{x}) < 1$. This verifies (a).

(VI) Next we show that $f(z) + \epsilon \rho_\infty(z) \leq f_n(z_n)$ for each n. Let

$$\widetilde{D}_n := \{x \in E \mid f_{n+1}(x) \leq f_{n+1}(z_{n+1})\} \quad \text{for } n = 1, 2, \ldots$$

Since $f_{n+1} \geq f_n$ but $f_{n+1}(z_{n+1}) \leq f_n(z_n)$ (see (8.5)), we have $\widetilde{D}_n \subseteq \widetilde{D}_{n-1}$. Moreover, each $\widetilde{D}_n$ is a nonempty closed subset of D_n. Therefore $\bigcap_{n=1}^{\infty} \widetilde{D}_n = \{z\}$. This together with $f_{n+1}(z_{n+1}) \leq f_n(z_n)$ implies

$$f_k(z) \leq f_k(z_k) \leq f_n(z_n) \leq f_0(z_0) = f(\bar{x}) \quad \forall k > n.$$

From (8.4) we get

$$f_k(x) = f(x) + \epsilon \sum_{j=0}^{k-1} \mu_j \rho\big((j+1)(x - z_j)\big), \quad x \in E. \tag{8.11}$$

Recalling (8.2), we conclude that

$$f(z) + \epsilon \rho_\infty(z) = \lim_{k \to \infty} f_k(z) \leq f_n(z_n) \leq f(\bar{x}). \tag{8.12}$$

This was claimed above and this also verifies (b).

(VII) Let $\tilde{f} := f + \epsilon \rho_\infty$. We show that $\Sigma_{\mu_n \epsilon/2}(\tilde{f}) \subseteq D_n$ for each n. Thus let $x \in \Sigma_{\mu_n \epsilon/2}(\tilde{f})$. Then

$$f_{n+1}(x) \underset{(8.11)}{\leq} \tilde{f}(x) \leq \tilde{f}(z) + \frac{\mu_n \epsilon}{2} \underset{(8.12)}{\leq} f_{n+1}(z_{n+1}) + \frac{\mu_n \epsilon}{2}$$

and so indeed $x \in D_n$.

(VIII) Since the sequence (μ_n) is decreasing, the sequence of the closed sets $\Sigma_{\mu_n \epsilon/2}(\tilde{f})$ is decreasing with respect to inclusion and so (V) and (VII) give $\bigcap_{n=0}^{\infty} \Sigma_{\mu_n \epsilon/2}(\tilde{f}) = \{z\}$. Hence z is a minimizer of $\tilde{f}$. By (IV) and (VII) we have

$$\lim_{n \to \infty} \operatorname{diam}\left(\Sigma_{\mu_n \epsilon/2}(\tilde{f})\right) = 0.$$

Hence Remark 8.2.2 finally tells us that z is even a strong minimizer of $\tilde{f}$. $\square$

As a first corollary to Theorem 8.2.3 we derive a Banach space variant of Ekeland's variational principle, with the additional property that the minimizer of the perturbed functional is strong.

Theorem 8.2.4 (Ekeland) *Let $\epsilon > 0$ and let $\bar{x} \in E$ be such that $f(\bar{x}) < \inf_E f + \epsilon$. Then for any $\lambda > 0$ there exists $z \in E$ such that:*

(a) $\|\bar{x} - z\| < \lambda$.

(b) z *is a strong minimizer of the functional* $x \mapsto f(x) + \frac{\epsilon}{\lambda}\|x - z\|$ *on E. In particular,*

$$f(z) < f(x) + \frac{\epsilon}{\lambda}\|x - z\| \quad \forall x \in E \setminus \{z\}. \tag{8.13}$$

Proof. Set

$$\rho(x) := \frac{\|x\|}{\lambda}, \quad \mu_n := \frac{1}{2^{n+1}(n+1)}.$$

By Theorem 8.2.3 there exist a sequence (z_n) and a point z in E such that, in particular, (8.3) holds true, i.e., for each $x \neq z$ we have

$$\begin{aligned}
f(z) &< f(x) + \epsilon\big(\rho_\infty(x) - \rho_\infty(z)\big) \\
&= f(x) + \frac{\epsilon}{\lambda} \sum_{n=0}^{\infty} \frac{1}{2^{n+1}}(\|x - z_n\| - \|z - z_n\|) \\
&\leq f(x) + \frac{\epsilon}{\lambda} \sum_{n=0}^{\infty} \frac{1}{2^{n+1}}\|x - z\| = f(x) + \frac{\epsilon}{\lambda}\|x - z\|.
\end{aligned} \tag{8.14}$$

This verifies (8.13). Now let (x_n) be a minimizing sequence for $\varphi(x) := f(x) + \frac{\epsilon}{\lambda}\|x - z\|$, i.e., $\varphi(x_n) \to \inf_E \varphi = f(z)$ as $n \to \infty$. Then (8.14) shows that (x_n) is a minimizing sequence for $f + \epsilon\rho_\infty$. By Theorem 8.2.3(c) we conclude that $x_n \to z$. Hence z is a strong minimizer of φ. $\square$

Corollary 8.2.5 *Let $\epsilon > 0$ and let $\bar{x} \in E$ be such that $f(\bar{x}) < \inf_E f + \epsilon$. Assume that f is G-differentiable. Then there exists $z \in \mathrm{B}(\bar{x}, \sqrt{\epsilon})$ satisfying*

$$f(z) < \inf_E f + \epsilon \quad \text{and} \quad \|f'(z)\| < \sqrt{\epsilon}.$$

Proof. See Exercise 8.5.2. $\square$

The corollary states that near an "almost minimum" point of f we can find an "almost critical" point. In particular, there exists a minimizing sequence (x_k) of f such that the sequence $(f'(x_k))$ converges to zero.

Corollary 8.2.6 *Let A be a closed subset of E, let $f : A \to \overline{\mathbb{R}}$ be l.s.c. and bounded below. Let $\epsilon > 0$ and let $\bar{x} \in A$ be such that $f(\bar{x}) < \inf_A f + \epsilon$. Then for any $\lambda > 0$ there exists $z \in A$ such that:*

(a) $\|\bar{x} - z\| < \lambda$.

(b) *z is a strong minimizer of the functional $x \mapsto f(x) + \frac{\epsilon}{\lambda}\|x - z\|$ on A. In particular,*

$$f(z) < f(x) + \frac{\epsilon}{\lambda}\|x - z\| \quad \forall x \in A \setminus \{z\}. \tag{8.15}$$

Proof. See Exercise 8.5.3.   □

Here, we give a geometric application of Ekeland's variational principle. In view of this, recall that by Corollary 1.5.5, every boundary point of a closed convex set $M \subseteq E$ is a support point provided M has interior points. Without the latter condition, the existence of support points cannot be guaranteed. However, the following result due to Bishop and Phelps [15] ensures that M contains support points with respect to certain Bishop–Phelps cones. In this connection, M is not assumed to be convex or to have interior points. In Fig. 8.1, the point y is a support point of $M \subseteq \mathbb{R}^2$ with respect to the Bishop–Phelps cone $K(x^*, \alpha)$ while z is not.

Proposition 8.2.7 *Let E be a Banach space and M be a closed subset of E. Suppose that $x^* \in E^* \setminus \{o\}$ is bounded on M. Then for every $\alpha > 0$ there exists $y \in M$ such that*

$$M \cap \left(y + K(x^*, \alpha)\right) = \{y\}.$$

Proof. See Exercise 8.5.4.   □

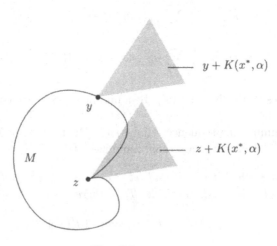

Fig. 8.1

8.3 The Borwein–Preiss Variational Principle

Now we deduce the smooth variational principle of Borwein and Preiss, again with a strong minimizer.

Theorem 8.3.1 (Borwein–Preiss) *Let $\epsilon > 0$ and let $\bar{x} \in E$ be such that $f(\bar{x}) < \inf_E f + \epsilon$. Further let $\lambda > 0$ and $p \geq 1$. Then there exist a sequence (ν_n) in $(0,1)$ with $\sum_{n=0}^{\infty} \nu_n = 1$ and a sequence (z_n) in E converging to some $z \in E$ such that the following holds:*

(a) $\|z - \bar{x}\| < \lambda$.
(b) $f(z) \leq \inf_E f + \epsilon$.
(c) z *is a strong minimizer of* $f + \epsilon \rho_\infty$ *on* E, *where*

$$\rho_\infty(x) := \frac{1}{\lambda^p} \sum_{n=0}^{\infty} \nu_n \|x - z_n\|^p. \tag{8.16}$$

Proof. We set

$$\rho(x) := \frac{\|x\|^p}{\lambda^p}, \quad \mu_n := \frac{1}{2^{n+1}(n+1)\sigma}, \quad \text{where } \sigma := \sum_{n=0}^{\infty} \frac{(n+1)^{p-1}}{2^{n+1}}. \tag{8.17}$$

Then there exists a sequence (z_n) converging to some z as in Theorem 8.2.3. The perturbation functional according to (8.2) here is given by (8.16), where

$$\nu_n := \frac{(n+1)^{p-1}}{2^{n+1}\sigma}.$$

It is left as Exercise 8.5.5 to show that these data meet the assertions.   □

Remark 8.3.2 If, under the assumptions of Theorem 8.3.1 and with $p > 1$, the norm functional $\omega(x) := \|x\|$, $x \in E$, is β-differentiable on $E \setminus \{o\}$ for some bornology β, then the perturbation functional ρ_∞ defined by (8.16) is β-differentiable on all of E and satisfies

$$\rho_\infty'(x) = \frac{p}{\lambda\sigma} \sum_{n=0}^{\infty} \frac{(n+1)^{p-1}}{2^{n+1}} \left\|\frac{x - z_n}{\lambda}\right\|^{p-1} \omega'(x - z_n) \quad \forall x \neq z_n, \ n \in \mathbb{N},$$

$$\rho_\infty'(z_n) = 0 \quad \forall n \in \mathbb{N},$$

$$\|\rho_\infty'(x)\| \leq \frac{p}{\lambda\sigma} \sum_{n=0}^{\infty} \frac{(n+1)^{p-1}}{2^{n+1}} \frac{1}{(n+1)^{p-1}} = \frac{p}{\lambda\sigma} \quad \forall x \in E.$$

The estimation follows with the aid of (8.10), where by (8.17) we have $\eta = \lambda$ and $\|\omega'(z - z_n)\| = 1$ (Proposition 4.7.1).

If E is a Hilbert space, the perturbation functional ρ_∞ in Theorem 8.3.1 can be simplified.

Theorem 8.3.3 *Assume that E is a Hilbert space. Let $\epsilon > 0$ and let $\bar{x} \in E$ be such that $f(\bar{x}) < \inf_E f + \epsilon$. Then for any $\lambda > 0$ there exist $y, z \in E$ such that the following holds:*

(a) $\|z - \bar{x}\| < \lambda, \quad \|z - y\| < \lambda$.

(b) $f(z) \leq \inf_E f + \epsilon$.

(c) z *is a strong minimizer of the functional* $x \mapsto f(x) + \frac{\epsilon}{\lambda^2}\|x - y\|^2$ *on* E.

Proof. We set

$$\rho(x) := \frac{\|x\|^2}{\lambda^2}, \quad \mu_n := \frac{1}{2^{n+1}(n+1)^2}.$$

Then there exists a sequence (z_n) converging to some z as in Theorem 8.2.3. The perturbation functional is

$$\rho_\infty(x) = \frac{1}{\lambda^2} \sum_{n=0}^{\infty} \frac{\|x - z_n\|^2}{2^{n+1}}.$$

Now define

$$y := \sum_{n=0}^{\infty} \frac{z_n}{2^{n+1}}, \quad c := \sum_{n=0}^{\infty} \frac{\|z_n\|^2}{2^{n+1}} - \|y\|^2.$$

A direct calculation using the inner product shows that for each $x \in E$,

$$\rho_\infty(x) = \frac{1}{\lambda^2} \sum_{n=0}^{\infty} \frac{\|x\|^2 - 2(x \mid z_n) + \|z_n\|^2}{2^{n+1}}$$

$$= \frac{1}{\lambda^2}\left(\|x\|^2 - 2(x \mid y) + \sum_{n=0}^{\infty} \frac{\|z_n\|^2}{2^{n+1}}\right) = \frac{1}{\lambda^2}\|x - y\|^2 + \frac{c}{\lambda^2}.$$

Noting that c/λ^2 is constant, the assertions follow from Theorem 8.2.3, except for the estimate $\|z - y\| < \lambda$. The latter is obtained as follows (observe that c is positive). Since

$$f(z) + \frac{\epsilon}{\lambda^2}\|z - y\|^2 \leq f(z) + \epsilon\rho_\infty(z) \leq f(\bar{x}) < \inf_E f + \epsilon,$$

we have

$$\frac{\epsilon}{\lambda^2}\|z - y\|^2 < \inf_E f - f(z) + \epsilon \leq \epsilon,$$

which completes the proof. $\qquad\qquad\qquad\qquad\qquad\qquad\qquad\qquad\square$

8.4 The Deville–Godefroy–Zizler Variational Principle

We prepare the next result. If $g : E \to \mathbb{R}$ is bounded, we write

$$\|g\|_\infty := \sup\{|g(x)| \mid x \in E\}.$$

A functional $b : E \to \mathbb{R}$ is said to be a *bump functional* if b is bounded and the set $\operatorname{supp}(b) := \{x \in E \mid b(x) \neq 0\}$ is nonempty and bounded.

Lemma 8.4.1 *Let E be a Fréchet smooth Banach space and $\|\cdot\|$ be an equivalent norm on E that is F-differentiable on $E \setminus \{o\}$. Then E admits a bump functional $b : E \to \mathbb{R}$ that is L-continuous, continuously differentiable, and such that*

$$b(E) \subseteq [0,1], \quad b(x) = 0 \text{ if } \|x\| \geq 1, \quad b(o) = 1. \tag{8.18}$$

Proof. Let $\varphi : \mathbb{R} \to \mathbb{R}$ be a continuously differentiable function satisfying

$$\varphi(\mathbb{R}) \subseteq [0,1], \quad \varphi(t) = 0 \text{ if } t \leq 1 \text{ and if } t \geq 3, \quad \varphi(2) = 1.$$

Choose $x_0 \in E$ such that $\|x_0\| = 2$ and set

$$b(x) := \varphi(\|5x + x_0\|), \quad x \in E.$$

Together with φ and $\|\cdot\|$, the functional b is L-continuous. Moreover, by Corollary 4.3.4 the norm functional is continuously differentiable on $E \setminus \{o\}$ and so b is continuously differentiable on E; in this connection notice that in a neighborhood of the critical point $x = -(1/5)x_0$ the function φ is zero. Obviously (8.18) holds. $\qquad\square$

We make the following assumptions:

(A1) E is a Banach space, Y is a Banach space (with norm $\|\cdot\|_Y$) of bounded continuous real-valued functions on E.

(A2) $\|g\|_\infty \leq \|g\|_Y$ for any $g \in Y$.

(A3) For any $g \in Y$ and $z \in E$, the function $g_z : E \to \mathbb{R}$ defined by $g_z(x) := g(x + z)$ satisfies $g_z \in Y$ and $\|g_z\|_Y = \|g\|_Y$.

(A4) For any $g \in Y$ and $\alpha \in \mathbb{R}$, the function $x \mapsto g(\alpha x)$ is an element of Y.

(A5) E admits a bump functional $b \in Y$.

Theorem 8.4.2 (Deville–Godefroy–Zizler [49]) *If the assumptions (A1)–(A5) are satisfied, then the set G of all $g \in Y$ such that $f + g$ attains a strong minimum on E is a dense G_δ subset of Y.*

Proof. Given $g \in Y$, define

$$S(g, \alpha) := \{x \in E; \, | \, g(x) \leq \inf_E g + \alpha\},$$

$$U_k := \{g \in Y \mid \exists\, \alpha > 0 : \, \text{diam}\, S(f + g, \alpha) < \frac{1}{k}\}.$$

We will show that each U_k is open and dense in Y and that $\cap_{k=1}^\infty U_k = G$:

(I) We show that each U_k is open. Let $g \in U_k$ be given and let α be an associated positive number. Then for any $h \in Y$ satisfying $\|g - h\|_Y < \alpha/3$ we have $\|g - h\|_\infty < \alpha/3$. If $x \in S(f + h, \alpha/3)$, then

$$(f + h)(x) \leq \inf_E (f + h) + \frac{\alpha}{3}.$$

It follows that

$$(f+g)(x) \le (f+h)(x) + \|g - h\|_\infty \le \inf_E(f+h) + \frac{\alpha}{3} + \|g - h\|_\infty$$

$$\le \inf_E(f+g) + \frac{\alpha}{3} + 2\|g - h\|_\infty \le \inf_E(f+g) + \alpha.$$

Hence $S(f + h, \alpha/3) \subseteq S(f + g, \alpha)$ and so $h \in U_k$.

(II) Next we show that each U_k is dense in Y.

(IIa) Let $g \in Y$ and $\epsilon > 0$ be given. It suffices to find $h \in Y$ such that $\|h\|_Y < \epsilon$ and diam $S(f + g + h, \alpha) < 1/k$ for some $\alpha > 0$. We may assume that the functional b of (A5) satisfies $\|b\|_Y < \epsilon$. By (A3) we may further assume that $b(o) > 0$ and by (A4) that $\mathrm{supp}(b) \subseteq B(o, 1/(2k))$. Set $\alpha := b(o)/2$, choose $\bar{x} \in E$ such that

$$(f+g)(\bar{x}) < \inf_E(f+g) + \frac{b(o)}{2}$$

and define $h : E \to \mathbb{R}$ by $h(x) := -b(x - \bar{x})$. Then (A3) implies that $h \in Y$ and $\|h\|_Y = \|b\|_Y < \epsilon$.

(IIb) We show that
$$S(f + g + h, \alpha) \subseteq B(\bar{x}, 1/(2k)). \tag{8.19}$$

Let $\|x - \bar{x}\| > 1/(2k)$. Since $\mathrm{supp}(h) \subseteq B(\bar{x}, 1/(2k))$, we have $h(x) = 0$. It follows that

$$(f + g + h)(x) = (f + g)(x) \ge \inf_E(f+g) > (f+g)(\bar{x}) - \alpha$$

$$= (f + g + h)(\bar{x}) + b(o) - b(o)/2 \ge \inf_E(f + g + h) + \alpha$$

and so $x \notin S(f + g + h, \alpha)$. This verifies (8.19). By what has been said in step (IIa) we can now conclude that U_k is dense in Y.

(IIc) The Baire category theorem now implies that $\cap_{k=1}^\infty U_k$ is dense in Y.

(III) Finally we show that $\cap_{k=1}^\infty U_k = G$. It is left as Exercise 8.5.6 to verify that $G \subseteq \cap_{k=1}^\infty U_k$. Now let $g \in \cap_{k=1}^\infty U_k$ be given. We will show that $f + g$ attains a strong minimum on E. Given $k \in \mathbb{N}$ choose $\alpha_k > 0$ such that diam $S(f+g, \alpha_k) < 1/k$. Since each set $S(f+g, \alpha_k)$ is closed, the Cantor intersection theorem shows that $\cap_{k=1}^\infty S(f + g, \alpha_k)$ consists of exactly one point $\bar{x}$, which obviously is a minimizer of $f + g$. Now let (x_n) be a sequence in E satisfying $\lim_{n\to\infty}(f+g)(x_n) = \inf_E(f+g)$. For any $k \in \mathbb{N}$ there exists k_0 such that $k \ge k_0$ implies $(f + g)(x_n) \le \inf_E(f+g) + \alpha_k$ and so $x_n \in S(f + g, \alpha_k)$. We conclude that

$$\|x_n - \bar{x}\| \le \mathrm{diam}\, S(f + g, \alpha_k) < \frac{1}{k} \quad \forall n \ge k_0.$$

Hence $x_n \to \bar{x}$ as $n \to \infty$. Therefore, $\bar{x}$ is a strong minimizer of $f + g$ and so $g \in G$. $\qquad \square$

As a consequence of Theorem 8.4.2 we derive:

Theorem 8.4.3 *Let E be a Banach space admitting a continuously differentiable bump functional b such that $\|b\|_\infty$ and $\|b'\|_\infty$ are finite. Further let $f : E \to \mathbb{R}$ be proper, l.s.c., and bounded below. Then there exits a constant $\alpha > 0$ (depending only on E) such that for all $\epsilon \in (0,1)$ and for any $x_0 \in E$ satisfying*

$$f(x_0) < \inf_E f + \alpha\epsilon^2, \tag{8.20}$$

there exist a continuously differentiable function $g : E \to \mathbb{R}$ and $y_0 \in E$ such that:

(a) *y_0 is a strong minimizer of $f + g$.*
(b) *$\max\{\|g\|_\infty, \|g'\|_\infty\} < \epsilon$.*
(c) *$\|y_0 - x_0\| < \epsilon$.*

Proof. Let Y be the vector space of all continuously differentiable functions $g : E \to \mathbb{R}$ such that $\|g\|_\infty$ and $\|g'\|_\infty$ are finite. Equipped with the norm $\|g\|_Y := \max\{\|g\|_\infty, \|g\|_\infty\}$, Y is a Banach space. A construction analogous to that in the proof of Lemma 8.4.1 allows us to assume that b satisfies (8.18), in particular $\|b\|_\infty = 1$. Define

$$\alpha := \frac{1}{4\max\{\|b'\|_\infty, 1\}} \quad \text{and} \quad h(x) := f(x) - 2\alpha\epsilon^2 b\left(\frac{x - x_0}{\epsilon}\right), \ x \in E.$$

By Theorem 8.4.2 there exist $k \in Y$ and $y_0 \in E$ such that $h + k$ attains a strong minimum at y_0 and

$$\|k\|_\infty \le \alpha\epsilon^2/2, \quad \|k'\|_\infty \le \alpha\epsilon^2/2 \le \epsilon/2. \tag{8.21}$$

We have

$$h(x_0) = f(x_0) - 2\alpha\epsilon^2 < \inf_E f - \alpha\epsilon^2,$$

$$h(y) \ge \inf_E f \quad \text{whenever } \|y - x_0\| \ge \epsilon.$$

If (c) would not hold, the above estimate would give

$$\inf_E f + k(y_0) \le (h + k)(y_0) \le (h + k)(x_0) < \inf_E f - \alpha\epsilon^2 + k(x_0)$$

and so $k(y_0) < k(x_0) - \alpha\epsilon^2$, which is a contradiction to (8.21). Hence (c) is verified. Finally set

$$g(x) := k(x) - 2\alpha\epsilon^2 b\left(\frac{x - x_0}{\epsilon}\right), \quad x \in E.$$

It is easy to see that (a) and (b) also hold. $\qquad \square$

8.5 Bibliographical Notes and Exercises

Some references have already been given in the text. For various applications of Ekeland's variational principle we refer to Ekeland [57], Figueiredo [69], Pallaschke [155], and Penot [162]. Vector-valued variants of Ekeland's variational principle have been obtained by Göpfert et al. [78]. Concerning the smooth variational principle of Deville, Godefroy, and Zizler see also Deville et al. [50]. The proof given here follows Borwein and Zhu [24]. This book also contains further variational principles, many applications, and additional references.

Borwein and Preiss [19] show that a result analogous to Theorem 8.3.3 holds in any reflexive Banach space with a Kadec norm and with $\rho_\infty(x) = \frac{1}{\lambda^p}\|x - y\|^p$ for any given $p > 1$. Recall that on each reflexive Banach space there exists an equivalent norm that is locally uniformly convex (Theorem 4.7.12), and each locally uniformly convex norm is a Kadec norm (Lemma 4.7.9). In particular, on each Hilbert space the norm generated by the inner product is (locally) uniformly convex (Example 4.7.7) and so is the initial norm on L^p for $1 < p < +\infty$ (Example 4.7.11).

Exercise 8.5.1 Verify Remark 8.2.2.

Exercise 8.5.2 Prove Corollary 8.2.5.

Exercise 8.5.3 Verify Corollary 8.2.6.

Exercise 8.5.4 Verify Proposition 8.2.7.
Hint: Apply Ekeland's variational principle to the functional $f := -\frac{x^*}{\|x^*\|} + \delta_M$ and with appropriate choices of ϵ and λ.

Exercise 8.5.5 Elaborate the details of the proof of Theorem 8.3.1.

Exercise 8.5.6 Show that, with the assumptions and the notation of Theorem 8.4.2, one has $G \subseteq \cap_{k=1}^\infty U_k$.

9

Subdifferentials of Lower Semicontinuous Functionals

9.1 Fréchet Subdifferentials: First Properties

In this section we study another kind of derivative-like concepts.

Definition 9.1.1 Assume that E is a Banach space, $f : E \to \overline{\mathbb{R}}$ is proper and l.s.c., and $\bar{x} \in \operatorname{dom} f$.

(a) The functional f is said to be *Fréchet subdifferentiable (F-subdifferentiable)* at $\bar{x}$ if there exists $x^* \in E^*$, the *F-subderivative* of f at $\bar{x}$, such that

$$\liminf_{y \to o} \frac{f(\bar{x} + y) - f(\bar{x}) - \langle x^*, y \rangle}{\|y\|} \geq 0. \tag{9.1}$$

(b) The functional f is said to be *viscosity subdifferentiable* at $\bar{x}$ if there exist $x^* \in E^*$, the *viscosity subderivative* of f at $\bar{x}$, and a C^1-function $g : E \to \mathbb{R}$ such that $g'(\bar{x}) = x^*$ and $f - g$ attains a local minimum at $\bar{x}$. If, in particular,
$$g(x) = \langle x^*, x - \bar{x} \rangle - \sigma \|x - \bar{x}\|^2$$
with some positive constant σ, then x^* is called *proximal subgradient* of f at $\bar{x}$. The sets

$$\partial_F f(\bar{x}) := \text{set of all F-subderivatives of } f \text{ at } \bar{x},$$
$$\partial_V f(\bar{x}) := \text{set of all viscosity subderivatives of } f \text{ at } \bar{x},$$
$$\partial_P f(\bar{x}) := \text{set of all proximal subgradients of } f \text{ at } \bar{x}$$

are called *Fréchet subdifferential (F-subdifferential)*, *viscosity subdifferential*, and *proximal subdifferential* of f at $\bar{x}$, respectively.

Remark 9.1.2 Observe that the function g in Definition 9.1.1(b) can always be chosen such that $(f - g)(\bar{x}) = 0$ (cf. Fig. 9.1).

We study the relationship between the different notions.

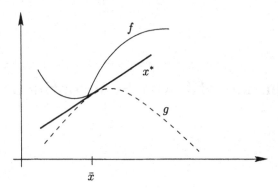

Fig. 9.1

Proposition 9.1.3 *Assume that E is a Banach space, $f : E \to \overline{\mathbb{R}}$ is proper and l.s.c., and $\bar{x} \in \text{dom} f$. Then $\partial_V f(\bar{x}) \subseteq \partial_F f(\bar{x})$.*

Proof. See Exercise 9.8.1. □

Remark 9.1.4 Notice that $\partial_F f(\bar{x})$ and $\partial_V f(\bar{x})$ can be defined as above for any proper, not necessarily l.s.c. functional f. However, if $\partial_F f(\bar{x})$ (in particular, $\partial_V f(\bar{x})$) is nonempty, then in fact f is l.s.c. at $\bar{x}$ (see Exercise 9.8.2).

The next result is an immediate consequence of the definition of the viscosity F-subdifferential and Proposition 9.1.3.

Proposition 9.1.5 (Generalized Fermat Rule) *If the proper l.s.c. functional $f : E \to \overline{\mathbb{R}}$ attains a local minimum at $\bar{x}$, then $o \in \partial_V f(\bar{x})$ and in particular $o \in \partial_F f(\bar{x})$.*

We shall now show that we even have $\partial_V f(\bar{x}) = \partial_F f(\bar{x})$ provided E is a Fréchet smooth Banach space. We start with an auxiliary result.

Lemma 9.1.6 *Let E be a Fréchet smooth Banach space and $\|\cdot\|$ be an equivalent norm on E that is F-differentiable on $E\backslash\{o\}$. Then there exist a functional $d : E \to \mathbb{R}_+$ and a number $\alpha > 1$ such that:*

(a) d is bounded, L-continuous on E and continuously differentiable on $E\backslash\{o\}$.
(b) $\|x\| \leq d(x) \leq \alpha\|x\|$ if $\|x\| \leq 1$ and $d(x) = 2$ if $\|x\| \geq 1$.

Proof. Let $b : E \to \mathbb{R}$ be the bump functional of Lemma 8.4.1. Define $d : E \to \mathbb{R}_+$ by $d(o) := 0$ and

$$d(x) := \frac{2}{s(x)}, \quad \text{where} \quad s(x) := \sum_{n=0}^{\infty} b(nx) \quad \text{for } x \neq o.$$

We show that d has the stated properties:

Ad (b). First notice that the series defining s is locally a finite sum. In fact, if $\bar{x} \neq o$, then we have

$$b(nx) = 0 \quad \forall x \in \mathrm{B}(\bar{x}, \|\bar{x}\|/2) \quad \forall n \geq 2\|\bar{x}\|. \tag{9.2}$$

Moreover, $s(x) \geq b(o) = 1$ for any $x \neq o$. Hence d is well defined. We have

$$d(E) \subseteq [0, 2] \quad \text{and} \quad d(x) = 2 \text{ whenever } \|x\| \geq 1.$$

Further it is clear that

$$[x \neq o \text{ and } b(nx) \neq 0] \implies n < 1/\|x\| \tag{9.3}$$

and so, since $0 \leq b \leq 1$, we conclude that $s(x) \leq 1 + 1/\|x\|$. Hence $d(x) \geq 2\|x\|/(1+\|x\|)$, which shows that $d(x) \geq \|x\|$ whenever $\|x\| \leq 1$. Since $b(o) = 1$ and b is continuous at o, there exists $\eta > 0$ such that $b(x) \geq 1/2$ whenever $\|x\| \leq \eta$. Let $x \in E$ and $m \geq 1$ be such that $\eta/(m+1) < \|x\| \leq \eta/m$. It follows that

$$s(x) \geq \sum_{n=1}^{m} b(nx) \geq \frac{m+1}{2} > \frac{\eta}{2\|x\|}$$

and so $d(x) < (4/\eta)\|x\|$ whenever $\|x\| \leq \eta$. This and the boundedness of d imply that $d(x)/\|x\|$ is bounded on $E \setminus \{o\}$. This verifies (b).

Ad (a). Since by (9.2) the sum defining s is locally finite, the functional d is continuously differentiable on $E \setminus \{o\}$. For any $x \neq o$ we have

$$d'(x) = -2 \left(\sum_{n=0}^{\infty} nb'(nx)\right) \left(\sum_{n=0}^{\infty} b(nx)\right)^{-2} = -\frac{(d(x))^2}{2} \sum_{n=0}^{\infty} nb'(nx).$$

Since b is L-continuous, $\lambda := \sup\{\|b'(x)\| \mid x \in E\}$ is finite and we obtain for any $x \neq o$,

$$\left\|\sum_{n=0}^{\infty} nb'(nx)\right\| \leq \lambda \sum_{n=0}^{[\|x\|^{-1}]} n \leq \lambda \left(1 + \frac{1}{\|x\|}\right)^2;$$

here the first inequality holds by (9.3). This estimate together with (b) yields

$$\|d'(x)\| \leq \lambda \max\{\alpha, 2\}^2 (\|x\| + 1)^2,$$

showing that d' is bounded on $B(o, 1) \setminus \{o\}$. Since d' is zero outside $B(o, 1)$, it follows that d' is bounded on $E \setminus \{o\}$. Hence d is L-continuous on E. This verifies (a). $\qquad \square$

Now we can supplement Proposition 9.1.3.

Theorem 9.1.7 *Let E be a Fréchet smooth Banach space, $f : E \to \overline{\mathbb{R}}$ be a proper l.s.c. functional, and $\bar{x} \in \mathrm{dom}\, f$. Then $\partial_V f(\bar{x}) = \partial_F f(\bar{x})$.*

Proof. In view of Proposition 9.1.3 it remains to show that $\partial_F f(\bar{x}) \subseteq \partial_V f(\bar{x})$. Thus let $x^* \in \partial_F f(\bar{x})$. Replacing f with the functional $\tilde{f} : E \to \mathbb{R}$ defined by

$$\tilde{f}(y) := \sup\{f(\bar{x} + y) - f(\bar{x}) - \langle x^*, y \rangle, -1\}, \quad y \in E,$$

we have $o \in \partial_F \tilde{f}(o)$. We show that $o \in \partial_V \tilde{f}(o)$. Notice that $\tilde{f}(\bar{x}) = 0$ and $\tilde{f}$ is bounded below. By (9.1) we obtain

$$\liminf_{y \to o} \frac{\tilde{f}(y)}{\|y\|} \geq 0. \tag{9.4}$$

Define $\rho : \mathbb{R}_+ \to \mathbb{R}$ by $\rho(t) := \inf\{\tilde{f}(y) \mid \|y\| \leq t\}$. Then ρ is nonincreasing, $\rho(0) = 0$ and $\rho \leq 0$. This and (9.4) give

$$\lim_{t \to 0} \frac{\rho(t)}{t} = 0. \tag{9.5}$$

Define ρ_1 and ρ_2 on $(0, +\infty)$ by

$$\rho_1(t) := \int_t^{et} \frac{\rho(s)}{s}\,ds, \quad \rho_2(t) := \int_t^{et} \frac{\rho_1(s)}{s}\,ds.$$

Since ρ is nonincreasing, we have

$$\rho_1(et) = \int_{et}^{e^2 t} \frac{\rho(s)}{s}\,ds \geq \rho(e^2 t) \int_{et}^{e^2 t} \frac{1}{s}\,ds = \rho(e^2 t). \tag{9.6}$$

Since ρ_1 is also nonincreasing, we obtain analogously $\rho_1(et) \leq \rho_2(t) \leq 0$. This and (9.5) yield

$$\lim_{t \downarrow 0} \frac{\rho_2(t)}{t} = \lim_{t \downarrow 0} \frac{\rho_1(t)}{t} = \lim_{t \downarrow 0} \frac{\rho(t)}{t} = 0. \tag{9.7}$$

Now define $\tilde{g} : E \to \mathbb{R}$ by $\tilde{g}(x) := \rho_2(d(x))$ for $x \neq o$ and $\tilde{g}(o) := 0$, where d denotes the functional in Lemma 9.1.6. Recall that $d(x) \neq 0$ whenever $x \neq o$. Since ρ_1 is continuous on $(0, +\infty)$ and so ρ_2 is continuously differentiable on $(0, +\infty)$, the chain rule implies that $\tilde{g}$ is continuously differentiable on $E \setminus \{o\}$ with derivative

$$\tilde{g}'(x) = \frac{\rho_1(ed(x)) - \rho_1(d(x))}{d(x)} \cdot d'(x), \quad x \neq o.$$

The properties of d and (9.7) further imply that $\lim_{x \to o} \|\tilde{g}'(x)\| = 0$. Therefore it follows as a consequence of the mean value theorem that $\tilde{g}$ is also F-differentiable at o with $\tilde{g}'(o) = o$, and $\tilde{g}'$ is continuous at o. Since ρ is nonincreasing, we have $\rho_2(t) \leq \rho_1(t) \leq \rho(t)$; here, the second inequality follows analogously as (9.6) and the first is a consequence of the second. Let $\|x\| \leq 1$. Then $\|x\| \leq d(x)$, and since ρ_2 is nonincreasing (as ρ_1 is nonincreasing), we obtain

$$(\tilde{f} - \tilde{g})(x) = \tilde{f}(x) - \rho_2(d(x)) \geq \tilde{f}(x) - \rho_2(\|x\|) \geq \tilde{f}(x) - \rho(\|x\|) \geq 0.$$

Since $0 = (\tilde{f} - \tilde{g})(o)$, we see that $\tilde{f} - \tilde{g}$ attains a local minimum at o. Hence $o \in \partial_V \tilde{f}(o)$ and so $x^* \in \partial_V f(\bar{x})$. □

Remark 9.1.8 Let E, f, and $\bar{x}$ be as in Theorem 9.1.7. Further let $x^* \in \partial_V f(\bar{x})$, which by Theorem 9.1.7 is equivalent to $x^* \in \partial_F f(\bar{x})$. Then there exists a *concave* C^1 function $g : E \to \mathbb{R}$ such that $g'(\bar{x}) = x^*$ and $f - g$ attains a local minimum at $\bar{x}$ (cf. Fig. 9.1); see Exercise 9.8.4.

In order to have both the limit definition and the viscosity definition of F-subderivatives at our disposal, we shall in view of Theorem 9.1.7 assume that E is a Fréchet smooth Banach space and we denote the common F-subdifferential of f at $\bar{x}$ by $\partial_F f(\bar{x})$.

The relationship to classical concepts is established in Proposition 9.1.9. In this connection recall that

$$\partial_P f(\bar{x}) \subseteq \partial_F f(\bar{x}). \tag{9.8}$$

Proposition 9.1.9 *Assume that E is a Fréchet smooth Banach space and $f : E \to \overline{\mathbb{R}}$ is proper and l.s.c.*

(a) *If the directional G-derivative $f_G(\bar{x}, \cdot)$ of f at $\bar{x} \in \mathrm{dom}\, f$ exists on E, then for any $x^* \in \partial_F f(\bar{x})$ (provided there exists one),*

$$\langle x^*, y \rangle \leq f_G(\bar{x}, y) \quad \forall y \in E.$$

If, in particular, f is G-differentiable at $\bar{x} \in \mathrm{dom}\, f$, then $\partial_F f(\bar{x}) \subseteq \{f'(\bar{x})\}$.

(b) *If $f \in C^1(U)$, where $U \subseteq E$ is nonempty and open, then $\partial_F f(x) = \{f'(x)\}$ for any $x \in U$.*

(c) *If $f \in C^2(U)$, where $U \subseteq E$ is nonempty and open, then $\partial_P f(x) = \partial_F f(x) = \{f'(x)\}$ for any $x \in U$.*

(d) *If f is convex, then $\partial_P f(x) = \partial_F f(x) = \partial f(x)$ for any $x \in \mathrm{dom}\, f$.*

(e) *If f is locally L-continuous on E, then $\partial_F f(x) \subseteq \partial_o f(x)$ for any $x \in E$.*

Proof.

(a) Let $x^* \in \partial_F f(\bar{x})$ be given. Then there exist a C^1 function g and a number $\epsilon > 0$ such that $g'(\bar{x}) = x^*$ and for each $x \in B(\bar{x}, \epsilon)$ we have

$$(f - g)(x) \geq (f - g)(\bar{x}) \quad \forall x \in B(\bar{x}, \epsilon). \tag{9.9}$$

Now let $y \in E$. Then for each $\tau > 0$ sufficiently small we have $\bar{x} + \tau y \in B(\bar{x}, \epsilon)$ and so

$$\frac{1}{\tau}\big(f(\bar{x} + \tau y) - f(\bar{x})\big) \geq \frac{1}{\tau}\big(g(\bar{x} + \tau y) - g(\bar{x})\big).$$

Letting $\tau \downarrow 0$ it follows that $f_G(\bar{x}, y) \geq \langle g'(\bar{x}), y \rangle = \langle x^*, y \rangle$. If f is G-differentiable at $\bar{x}$, then by linearity the latter inequality passes into $f'(\bar{x}) = x^*$.

(b) It is obvious that $f'(x) \in \partial_F f(x)$ for each $x \in U$. This and (a) imply $\partial_F f(x) = \{f'(x)\}$ for each $x \in U$.

(c) By Proposition 3.5.1 we have $f'(x) \in \partial_P f(x)$, which together with (a) and (9.8) verifies the assertion.

(d) It is evident that $\partial f(\bar{x}) \subseteq \partial_P f(\bar{x}) \subseteq \partial_F f(\bar{x})$ for each $\bar{x} \in \operatorname{dom} f$. Now let $x^* \in \partial_F f(\bar{x})$ be given. As in the proof of (a) let g and ϵ be such that (9.9) holds. Further let $x \in E$. If $\tau \in (0,1)$ is sufficiently small, then $(1-\tau)\bar{x} + \tau x \in B(\bar{x}, \epsilon)$ and we obtain using the convexity of f,

$$(1-\tau)f(\bar{x}) + \tau f(x) \geq f\big((1-\tau)\bar{x} + \tau x\big) \underset{(9.9)}{\geq} f(\bar{x}) + g\big((1-\tau)\bar{x} + \tau x\big) - g(\bar{x}).$$

It follows that

$$f(x) - f(\bar{x}) \geq \frac{g\big(\bar{x} + \tau(x - \bar{x})\big) - g(\bar{x})}{\tau}.$$

Letting $\tau \downarrow 0$, we see that $f(x) - f(\bar{x}) \geq \langle g'(\bar{x}), x - \bar{x} \rangle = \langle x^*, x - \bar{x} \rangle$. Since $x \in E$ was arbitrary, we conclude that $x^* \in \partial f(\bar{x})$.

(e) See Exercise 9.8.5.

$\square$

In Sect. 9.5 we shall establish the relationship between the Fréchet subdifferential and the Clarke subdifferential.

9.2 Approximate Sum and Chain Rules

Convention. Throughout this section, we assume that E is a Fréchet smooth Banach space, and $\|\cdot\|$ is a norm on E that is F-differentiable on $E \setminus \{o\}$.

Recall that we write $\omega_{\bar{x}}(x) := \|x - \bar{x}\|$, and in particular $\omega(x) := \|x\|$, $x \in E$.

One way to develop subdifferential analysis for l.s.c. functionals is to start with sum rules. It is an easy consequence of the definition of the F-subdifferential that we have

$$\partial_F f_1(\bar{x}) + \partial_F f_2(\bar{x}) \subseteq \partial_F (f_1 + f_2)(\bar{x}).$$

But the reverse inclusion

$$\partial_F (f_1 + f_2)(\bar{x}) \subseteq \partial_F f_1(\bar{x}) + \partial_F f_2(\bar{x}) \tag{9.10}$$

does not hold in general.

Example 9.2.1 Let $f_1(x) := |x|$ and $f_2(x) := -|x|$ for $x \in \mathbb{R}$. Then $\partial_F(f_1 + f_2)(0) = \{0\}$ but since $\partial_F f_2(0) = \emptyset$ (see Exercise 9.8.3), we have $\partial_F f_1(0) + \partial_F f_2(0) = \emptyset$.

Yet what we usually need is just (9.10). For instance, if $\bar{x}$ is a local minimizer of the proper l.s.c. function f on the closed subset M of E, then $\bar{x}$ is a local minimizer of $f + \delta_M$ on all of E, which implies

$$o \in \partial_F(f + \delta_M)(\bar{x}). \tag{9.11}$$

Now we would like to conclude that $o \in \partial_F f(\bar{x}) + \partial_F \delta_M(\bar{x})$ which (9.10) would ensure. In a special case we do obtain an exact sum rule.

Proposition 9.2.2 Let $f_1, f_2 : E \to \overline{\mathbb{R}}$ be proper and l.s.c. If f_1 is F-differentiable at $\bar{x}$, then

$$\partial_F(f_1 + f_2)(\bar{x}) = f_1'(\bar{x}) + \partial_F f_2(\bar{x}).$$

Proof. See Exercise 9.8.6. □

In the general case we at least obtain, among others, an *approximate*, or *fuzzy*, sum rule of the following form. If $x^* \in \partial_F(f_1 + f_2)(\bar{x})$, then for each $\sigma(E^*, E)$-neighborhood V of zero in E^* there exist x_1 and x_2 close to $\bar{x}$ such that $f_i(x_i)$ is close to $f_i(\bar{x})$ for $i = 1, 2$ and

$$x^* \in \partial_F f_1(x_1) + \partial_F f_2(x_2) + V.$$

We establish several approximate sum rules, which are then applied to derive a general mean value theorem as well as multiplier rules for constrained optimization problems involving lower semicontinuous data. The first result is nonlocal, meaning that there is no reference point $\bar{x}$.

Theorem 9.2.3 (Nonlocal Approximate Sum Rule) Let $f_1, \ldots, f_n : E \to \overline{\mathbb{R}}$ be proper, l.s.c., bounded below, and such that

$$\liminf_{\rho \downarrow 0} \left\{ \sum_{i=1}^n f_i(y_i) \mid \mathrm{diam}\{y_1, \ldots, y_n\} \leq \rho \right\} < +\infty.$$

Then for any $\epsilon > 0$ there exist $x_i \in E$ and $x_i^* \in \partial_F f_i(x_i)$, $i = 1, \ldots, n$, satisfying

$$\mathrm{diam}\{x_1, \ldots, x_n\} \cdot \max\{1, \|x_1^*\|, \ldots, \|x_n^*\|\} < \epsilon, \tag{9.12}$$

$$\sum_{i=1}^n f_i(x_i) < \liminf_{\rho \downarrow 0} \left\{ \sum_{i=1}^n f_i(y_i) \mid \mathrm{diam}\{y_1, \ldots, y_n\} \leq \rho \right\} + \epsilon, \tag{9.13}$$

$$\left\| \sum_{i=1}^n x_i^* \right\| \leq \epsilon. \tag{9.14}$$

Proof.

(I) For each positive real number κ let

$$\varphi_\kappa(y_1, \ldots, y_n) := \sum_{i=1}^n f_i(y_i) + \kappa \sum_{i,j=1}^n \|y_i - y_j\|^2 \quad \text{and} \quad \alpha_\kappa := \inf_{E^n} \varphi_\kappa.$$

We show that

$$\alpha_\kappa \leq \beta := \liminf_{\rho \downarrow 0} \left\{ \sum_{i=1}^n f_i(y_i) \mid \operatorname{diam}\{y_1, \ldots, y_n\} \leq \rho \right\} \quad \forall \kappa > 0.$$

Assume, to the contrary, that for some κ we had

$$\beta < \alpha_\kappa \leq \sum_{i=1}^n f_i(y_i) + \kappa \sum_{i,j=1}^n \|y_i - y_j\|^2 \quad \forall (y_1, \ldots, y_n) \in E^n.$$

By the definition of β, for each sufficiently large $\nu \in \mathbb{N}$ we find $(z_1, \ldots, z_n) \in E^n$ satisfying

$$\sum_{i=1}^n f_i(z_i) < \alpha_\kappa \quad \text{and} \quad \operatorname{diam}\{z_1, \ldots, z_n\} \leq \frac{1}{\nu}$$

and so

$$\sum_{i=1}^n f_i(z_i) + \kappa \sum_{i,j=1}^n \|z_i - z_j\|^2 \leq \sum_{i=1}^n f_i(z_i) + \frac{\kappa n(n-1)}{2\nu^2}.$$

If ν is large enough, the right-hand side, and so the left-hand side, of the last inequality is smaller than α_κ, but this contradicts the definition of α_κ. Thus we have shown that the generalized sequence $(\alpha_\kappa)_{\kappa>0}$ is bounded (above). Since it is also increasing, the limit $\alpha := \lim_{\kappa \to \infty} \alpha_\kappa$ exists.

(II) Observe that E^n with the Euclidean product norm is also a Fréchet smooth Banach space. By the Borwein–Preiss variational principle (Theorem 8.3.1 and Remark 8.3.2) applied to φ_κ for $\kappa = 1, \ldots, n$ (with $p = 2$ and $\lambda > 0$ sufficiently large), there exist a C^1-function ψ_κ and a point $(z_{1,\kappa}, \ldots, z_{n,\kappa}) \in E^n$ such that $\varphi_\kappa + \psi_\kappa$ attains a local minimum at $(z_{1,\kappa}, \ldots, z_{n,\kappa})$ and that

$$\|\psi'(z_{1,\kappa}, \ldots, z_{n,\kappa})\| < \frac{\epsilon}{n},$$

$$\varphi_k(z_{1,\kappa}, \ldots, z_{n,\kappa}) < \inf_{E^n} \varphi_\kappa + \frac{1}{\kappa} \leq \alpha + \frac{1}{\kappa}.$$

(9.15)

For each $\kappa > 0$ define $\gamma_\kappa : E \to \overline{\mathbb{R}}$ by

$$\gamma_\kappa(y_1, \ldots, y_n) := -\psi_\kappa(y_1, \ldots, y_n) - \kappa \sum_{i,j=1}^{n} \|y_i - y_j\|^2.$$

Then γ_κ is a C^1-function satisfying

$$\sum_{i=1}^{n} f_i(y_i) - \gamma_\kappa(y_1, \ldots, y_n) = \varphi_\kappa(y_1, \ldots, y_n) + \psi_\kappa(y_1, \ldots, y_n). \qquad (9.16)$$

Since for each $i = 1, \ldots, n$ the function

$$y \mapsto \varphi_\kappa(z_{1,\kappa}, \ldots, z_{i-1,\kappa}, y, z_{i+1,\kappa}, \ldots, z_{n,\kappa})$$
$$+ \psi_\kappa(z_{1,\kappa}, \ldots, z_{i-1,\kappa}, y, z_{i+1,\kappa}, \ldots, z_{n,\kappa})$$

attains a local minimum at $y = z_{i,\kappa}$, we conclude from (9.16) that

$$x_{i,\kappa}^* := D_i \gamma_k(z_{1,\kappa}, \ldots, z_{n,\kappa}) \in \partial_F f_i(z_{i,\kappa}) \quad \text{for } i = 1, \ldots, n.$$

Summing over i and recalling the definition of γ_κ gives

$$\sum_{i=1}^{n} x_{i,\kappa}^* = -\sum_{i=1}^{n} D_i \psi_\kappa(z_{1,\kappa}, \ldots, z_{n,\kappa}) - 2\kappa \sum_{i,j=1}^{n} (\omega^2)'(z_{i,\kappa} - z_{j,\kappa}).$$

For symmetry reasons the double sum over i, j vanishes. Moreover, by (9.15) we have $\| -\sum_{i=1}^{n} D_i \psi_\kappa(z_{1,\kappa}, \ldots, z_{n,\kappa})\| \leq \epsilon$. It follows that

$$\left\|\sum_{i=1}^{n} x_{i,\kappa}^*\right\| \leq \epsilon. \qquad (9.17)$$

(III) By the definition of α_κ and φ_κ we conclude that

$$\alpha_{\kappa/2} \leq \varphi_{\kappa/2}(z_{1,\kappa}, \ldots, z_{n,\kappa})$$
$$= \varphi_\kappa(z_{1,\kappa}, \ldots, z_{n,\kappa}) - \frac{\kappa}{2} \sum_{i=1}^{n} \|z_{i,\kappa} - z_{j,\kappa}\|^2 \qquad (9.18)$$
$$\underset{(9.15)}{\leq} \alpha_\kappa + \frac{1}{\kappa} - \frac{\kappa}{2} \sum_{i,j=1}^{n} \|z_{i,\kappa} - z_{j,\kappa}\|^2.$$

Rearranging we obtain

$$\kappa \sum_{i,j=1}^{n} \|z_{i,\kappa} - z_{j,\kappa}\|^2 \leq 2(\alpha_\kappa - \alpha_{\kappa/2} + \tfrac{1}{\kappa})$$

and so $\lim_{\kappa \to \infty} \kappa \sum_{i,j=1}^{n} \|z_{i,\kappa} - z_{j,\kappa}\|^2 = 0$. Hence $\lim_{\kappa \to \infty} \text{diam} \{z_{1,\kappa}, \ldots, z_{n,\kappa}\} = 0$ and, recalling (9.17), we conclude that

$$\lim_{\kappa \to \infty} \text{diam}\{z_{1,\kappa}, \ldots, z_{n,\kappa}\} \cdot \max\{\|x_{1,\kappa}^*\|, \ldots, \|x_{n,\kappa}^*\|\} = 0.$$

(IV) We now obtain

$$\alpha \leq \beta \leq \liminf_{\kappa \to \infty} \sum_{i=1}^{n} f_i(z_{i,\kappa}) = \liminf_{\kappa \to \infty} \varphi_\kappa(z_{1,\kappa}, \ldots, z_{n,\kappa}) \leq \alpha. \tag{9.15}$$

Therefore we have $\alpha = \beta$. In view of (9.15), the assertion follows by setting $x_i := z_{i,\kappa}$ and $x_i^* := x_{i,\kappa}^*$ for $i = 1, \ldots, n$, with κ sufficiently large. $\qquad\square$

As an immediate application of Theorem 9.2.3 we obtain a remarkable density result.

Proposition 9.2.4 *Let* $f : E \to \bar{\mathbb{R}}$ *be proper and l.s.c. Then* $\mathrm{Dom}(\partial_F f)$ *is dense in* $\mathrm{dom}\, f$.

Proof. Let $\bar{x} \in \mathrm{dom}\, f$ and $\eta > 0$ be given. Since f is l.s.c. at $\bar{x}$, there exists $\epsilon > 0$ such that $f(x) > f(\bar{x}) - 1$ for all $x \in \mathrm{B}(\bar{x}, \epsilon)$. We may assume that $\epsilon < \eta$. The functionals f_1, f_2 defined by

$$f_1(x) := \begin{cases} f(x) & \text{if } x \in \mathrm{B}(\bar{x}, \epsilon), \\ +\infty & \text{otherwise,} \end{cases} \qquad f_2(x) := \begin{cases} 0 & \text{if } x = \bar{x}, \\ +\infty & \text{otherwise} \end{cases}$$

satisfy the hypotheses of Theorem 9.2.3. Hence for $i = 1, 2$ there exist $x_i \in E$ and $x_i^* \in \partial_F f_i(x_i)$ such that

$$\|x_1 - x_2\| < \epsilon, \quad o \in x_1^* + x_2^* + \mathrm{B}_{E^*}(o, \epsilon),$$
$$f_1(x_1) + f_2(x_2) < f_1(\bar{x}) + f_2(\bar{x}) + \epsilon = f(\bar{x}) + \epsilon.$$

The last inequality shows that $x_2 = \bar{x}$. Hence $x_1 \in \mathring{\mathrm{B}}(\bar{x}, \epsilon)$, and since f_1 coincides with f on $\mathrm{B}(\bar{x}, \epsilon)$, we obtain $\partial_F f_1(x_1) = \partial_F f(x_1)$. Moreover, $\partial_F f_2(\bar{x}) = \{o\}$. Therefore $x_1^* \in \partial_F f(x_1)$ and we have $\|x_1 - \bar{x}\| < \eta$. $\qquad\square$

Now we turn to local approximate sum rules. In this connection, we shall assume that, for some $\eta > 0$, the reference point $\bar{x}$ has the following property:

$$\sum_{i=1}^{n} f_i(\bar{x}) \leq \lim_{\rho \downarrow 0} \inf \left\{ \sum_{i=1}^{n} f_i(y_i) \,\Big|\, \|y_0 - \bar{x}\| \leq \eta, \ \mathrm{diam}\{y_0, y_1, \ldots, y_n\} \leq \rho \right\}. \tag{9.19}$$

We first give a sufficient condition for (9.19).

Lemma 9.2.5 *Let* $f_i : E \to \bar{\mathbb{R}}$ *be proper and l.s.c. Assume that* $\bar{x} \in \cap_{i=1}^{n} \mathrm{dom}\, f_i$ *is a local minimizer of* $\sum_{i=1}^{n} f_i$ *and that one of the following conditions* (a) *and* (b) *is satisfied:*

(a) *All but one of* f_i *are uniformly continuous in a neighborhood of* $\bar{x}$.
(b) *The restriction of at least one* f_i *to a neighborhood of* $\bar{x}$ *has compact level sets.*

Then (9.19) *holds.*

Proof. See Exercise 9.8.7 $\qquad\square$

Theorem 9.2.6 (Strong Local Approximate Sum Rule) *Assume that* $f_1, \ldots, f_n : E \to \overline{\mathbb{R}}$ *are proper and l.s.c., let* $\bar{x} \in \cap_{i=1}^n \mathrm{dom}\, f_i$, *and assume that there exists* $\eta > 0$ *such that (9.19) holds. Then for any* $\epsilon > 0$ *there exist* $x_i \in \mathrm{B}(\bar{x}, \epsilon)$ *and* $x_i^* \in \partial_F f_i(x_i)$, $i = 1, \ldots, n$, *satisfying*

$$|f_i(x_i) - f_i(\bar{x})| < \epsilon, \quad i = 1, \ldots, n, \tag{9.20}$$

$$\mathrm{diam}\{x_1, \ldots, x_n\} \cdot \max\{\|x_1^*\|, \ldots, \|x_n^*\|\} < \epsilon, \tag{9.21}$$

$$\left\| \sum_{i=1}^n x_i^* \right\| < \epsilon. \tag{9.22}$$

Proof. Obviously we may assume that $\eta < \min\{\epsilon, 1\}$. Since each f_i is l.s.c., we may further assume that

$$f_i(x) > f_i(\bar{x}) - \epsilon/n \quad \forall x \in \mathrm{B}(\bar{x}, \eta), \; i = 1, \ldots, n. \tag{9.23}$$

In view of (9.19), for $\epsilon_1 := \eta^2/(32n^2)$ there exists $\rho \in (0, \eta)$ such that

$$\sum_{i=1}^n f_i(\bar{x}) \le \inf\Big\{ \sum_{i=1}^n f_i(y_i) + \delta_{\mathrm{B}(\bar{x}, \eta)}(y_0) \;\Big|\; \|y_i - y_j\| \le \rho \,\forall i, j = 0, 1, \ldots, n \Big\} + \epsilon_1. \tag{9.24}$$

For $i = 1, \ldots, n$ let

$$\tilde{f}_i(x) := f_i(x) + \|x - \bar{x}\|^2 + \delta_{\mathrm{B}(\bar{x}, \eta)}(x), \quad x \in E.$$

Then $\tilde{f}_i$ is l.s.c., bounded below, and satisfies

$$r := \liminf_{\rho \downarrow 0} \Big\{ \sum_{i=1}^n \tilde{f}_i(y_i) \;\Big|\; \mathrm{diam}\{y_1, \ldots, y_n\} \le \rho \Big\} \le \sum_{i=1}^n f_i(\bar{x}) < +\infty.$$

Applying Theorem 9.2.3 to $\tilde{f}_i$ and $\epsilon_2 \in (0, \min\{\rho, \epsilon_1\})$, we obtain x_i and $y_i^* \in \partial_F \tilde{f}_i(x_i)$, $i = 1, \ldots, n$, such that

$$\sum_{i=1}^n \tilde{f}_i(x_i) < r + \epsilon_2, \tag{9.25}$$

$$\mathrm{diam}\{x_1, \ldots, x_n\} \cdot \{\|y_1^*\|, \ldots, \|y_n^*\|\} < \epsilon_2, \tag{9.26}$$

$$\left\| \sum_{i=1}^n y_i^* \right\| < \epsilon_2. \tag{9.27}$$

Observe that (9.25) implies $x_i \in \mathrm{B}(\bar{x}, \eta)$. Moreover, from (9.24) and (9.26) we deduce that

$$\sum_{i=1}^{n}\left(f_i(\bar{x}) + \|x_i - \bar{x}\|^2\right) - \epsilon_1 \le \sum_{i=1}^{n} \tilde{f}_i(x_i) < r + \epsilon_2$$

$$\le \sum_{i=1}^{n} \tilde{f}_i(\bar{x}) + \epsilon_2 = \sum_{i=1}^{n} f_i(\bar{x}) + \epsilon_2$$

and so

$$\sum_{i=1}^{n} \|x_i - \bar{x}\|^2 \le \epsilon_1 + \epsilon_2 < 2\epsilon_1. \tag{9.28}$$

Define $x_i^* := y_i^* - (\omega_{\bar{x}}^2)'(x_i)$. It follows that

$$x_i^* \in \partial_F \tilde{f}(x_i) - \partial_F \omega_{\bar{x}}^2(x_i) \subseteq \partial_F\left(\tilde{f} - \omega_{\bar{x}}^2\right)(x_i)$$
$$= \partial_F\left(f_i + \delta_{B(\bar{x},\eta)}\right)(x_i) = \partial_F f_i(x_i);$$

here the latter equation holds because by (9.28) the point x_i is in the interior of $B(\bar{x},\eta)$. From (9.25), (9.27), and $\left\|(\omega_{\bar{x}}^2)'(x_i)\right\| \le \epsilon/(2n)$ we obtain (9.21) and (9.22). Finally, the estimate

$$f_i(x_i) \underset{(9.25)}{\le} f_i(\bar{x}) + \sum_{j\ne i}\left(f_j(\bar{x}) - f_j(x_j)\right) + \epsilon_2$$

$$\underset{(9.23)}{<} f_i(\bar{x}) + \frac{(n-1)\epsilon}{n} + \epsilon_2 < f_i(\bar{x}) + \epsilon$$

together with (9.23) verifies (9.20).   □

Theorem 9.2.6 in conjunction with Lemma 9.2.5 is a necessary optimality condition. It can also be interpreted as a *strong* approximate sum rule in that it relates a local minimizer of $\sum_{i=1}^{n} f_i$ to F-subderivatives x_i^* of f_i whose sum is close to zero in the norm topology of E^*. The following *weak* rule relates an F-subderivative x^* of $\sum_{i=1}^{n} f_i$ to F-subderivatives x_i^* of f_i whose sum is close to x^* in the weak* topology of E^*. Notice that in this result, condition (9.19) can be omitted.

Theorem 9.2.7 (Weak Local Approximate Sum Rule) *Assume   that* $f_1,\ldots,f_n : E \to \bar{\mathbb{R}}$ *are proper and l.s.c., let* $\bar{x} \in \cap_{i=1}^{n}\mathrm{dom}\, f_i$, *and let* $x^* \in \partial_F\left(\sum_{i=1}^{n} f_i\right)(\bar{x})$. *Then for any* $\epsilon > 0$ *and any* $\sigma(E^*, E)$-*neighborhood* V *of zero in* E^* *there exist* $x_i \in B(\bar{x},\epsilon)$ *and* $x_i^* \in \partial_F f_i(x_i)$, $i = 1,\ldots,n$, *satisfying*

$$|f_i(x_i) - f_i(\bar{x})| < \epsilon, \quad i = 1,\ldots,n,$$
$$\mathrm{diam}\{x_1,\ldots,x_n\} \cdot \max\{\|x_1^*\|,\ldots,\|x_n^*\|\} < \epsilon,$$
$$x^* \in \sum_{i=1}^{n} x_i^* + V.$$

Proof. Let ϵ and V be given. Then there exist $y_1, \ldots, y_m \in E$ such that

$$\{x^* \in E^* \mid |\langle x^*, y_k \rangle| \leq 1, \ k = 1, \ldots, m\} \subseteq V.$$

Let $L := \operatorname{span}\{\bar{x}, y_1, \ldots, y_m\}$ and $\rho := \left(2 \max\{\|y_1\|, \ldots \|y_m\|\}\right)^{-1}$. Then we have

$$L^\perp + \mathrm{B}(o, 2\rho) \subseteq V.$$

For x^* there exists a C^1-function g such that $g'(\bar{x}) = x^*$ and $\left(\sum_{i=1}^n f_i\right) - g$ attains a local minimum at $\bar{x}$. Choose $\epsilon' > 0$ such that

$$\|x - \bar{x}\| < \epsilon' \quad \Longrightarrow \quad \|g'(x) - g'(\bar{x})\| < \rho. \tag{9.29}$$

We may assume that $\epsilon' < \min\{\epsilon, \rho\}$. The functional $\left(\sum_{i=1}^n f_i\right) - g + \delta_L$ attains a local minimum at $\bar{x}$ and δ_L has locally compact lower level sets. Hence by Lemma 9.2.5 the $(n+2)$-tuple $(f_1, \ldots, f_n, -g, \delta_L)$ satisfies condition (9.19). By Theorem 9.2.6 there exist $x_i \in \mathrm{B}(\bar{x}, \epsilon')$ for $i = 1, \ldots, n+2$ and $x_i^* \in \partial_F f_i(\bar{x})$ for $i = 1, \ldots, n$ as well as $x_{n+1}^* := -g'(x_{n+1})$ and $x_{n+2}^* \in \partial_F \delta_L(x_{n+2})$ satisfying

$$|f_i(x_i) - f_i(\bar{x})| < \epsilon', \quad i = 1, \ldots, n,$$

$$|\delta_L(x_{n+2}) - \delta_L(\bar{x})| < \epsilon',$$

$$\operatorname{diam}\{x_1, \ldots, x_{n+2}\} \cdot \max\{\|x_1^*\|, \ldots, \|x_{n+2}^*\|\} < \epsilon',$$

$$\left\| \sum_{i=1}^n x_i^* - g'(x_{n+1}) + x_{n+2}^* \right\| < \epsilon'.$$

The inequality involving δ_L shows that $x_{n+2}^* \in L$. Moreover, it is easy to see that $\partial_F \delta_L(x_{n+2}) = L^\perp$. Since $\|x_{n+1} - \bar{x}\| < \epsilon'$, (9.29) gives $\|g'(x_{n+1}) - x^*\| < \rho$. We conclude that

$$\left\| \left(\sum_{i=1}^n x_i^* + x_{n+2}^* \right) - x^* \right\|$$

$$\leq \|g'(x_{n+1}) - x^*\| + \left\| \sum_{i=1}^n x_i^* - g'(x_{n+1}) + x_{n+2}^* \right\| < \rho + \epsilon' \leq 2\rho,$$

which implies

$$x^* \in \sum_{i=1}^n x_i^* + L^\perp + \mathrm{B}(o, 2\rho) \subset \sum_{i=1}^n x_i^* + V. \qquad \square$$

Finally we establish an approximate chain rule. For this, we need the following concepts.

Definition 9.2.8 Let $T : E \to F$ be a mapping between Banach spaces E and F.

(a) T is said to be *locally compact* at $\bar{x} \in E$ if there is a neighborhood U of $\bar{x}$ such that $T(C)$ is compact for any closed set $C \subseteq U$.

(b) For any $y^* \in F^*$, the *scalarization* $\langle y^*, T \rangle : E \to \mathbb{R}$ of T is defined by

$$\langle y^*, T \rangle (x) := \langle y^*, T(x) \rangle, \quad x \in E.$$

Recall that if $g : F \to \overline{\mathbb{R}}$ and $T : E \to F$, then $g \circ T : E \to \overline{\mathbb{R}}$ denotes the composition of g and T.

Theorem 9.2.9 (Approximate Chain Rule) *Assume that E and F are Fréchet smooth Banach spaces, $g : F \to \overline{\mathbb{R}}$ is proper and l.s.c., and $T : E \to F$ is locally L-continuous and locally compact at $\bar{x} \in \operatorname{dom} f$. Suppose that $x^* \in \partial_F (g \circ T)(\bar{x})$. Then for any $\epsilon > 0$ there exist $x \in B_E(\bar{x}, \epsilon)$, $y \in B_F(T(\bar{x}), \epsilon)$, $y^* \in \partial_F g(y)$, $z^* \in B_{F^*}(y^*, \epsilon)$, and $\tilde{x}^* \in \partial_F \langle z^*, T \rangle (x)$ such that*

$$|g(y) - g(T(\bar{x}))| < \epsilon,$$
$$\|y - T(x)\| \max\{\|\tilde{x}^*\|, \|y^*\|, \|z^*\|\} < \epsilon, \qquad (9.30)$$
$$\|x^* - \tilde{x}^*\| < \epsilon.$$

Proof. Let $h : E \to \mathbb{R}$ be a C^1 function such that $h'(\bar{x}) = x^*$ and $g \circ T - h$ attains a local minimum at $\bar{x}$. Define

$$f_1(u, y) := g(y) - h(u), \quad f_2(u, y) := \delta_{\operatorname{graph} T}(u, y) \quad \forall (u, y) \in E \times F \quad (9.31)$$

and put $f := f_1 + f_2$. Then f attains a local minimum at $(\bar{x}, T(\bar{x}))$. Moreover, since T is locally compact, condition (b) of Lemma 9.2.5 also holds. By that lemma the hypothesis (9.19) of the approximate sum rule of Theorem 9.2.6 is satisfied. Applying the latter theorem, we find $x, u \in B_E(\bar{x}, \epsilon)$ such that $\|h'(u) - x^*\| < \epsilon/2$ as well as $y \in B_F(T(\bar{x}), \epsilon)$ with $|f(y) - f(T(\bar{x}))| < \epsilon$, $y^* \in \partial_F g(y)$ and

$$(\tilde{x}^*, -z^*) \in \partial_F \delta_{\operatorname{graph} T}(x, T(x)) \qquad (9.32)$$

satisfying (9.30) and

$$\|(-h'(u), y^*) + (\tilde{x}^*, -z^*)\| < \epsilon/2. \qquad (9.33)$$

From (9.32) we conclude that there is a C^1 function $\tilde{h} : E \times F \to \mathbb{R}$ such that $\tilde{h}'(x, T(x)) = (\tilde{x}^*, -z^*)$ and $\tilde{h}$ attains a local minimum 0 at $(x, T(x))$. The latter means that for any u in a neighborhood of x we have

$$0 \geq \tilde{h}(u, T(u)) - \tilde{h}(x, T(x)) = \langle \tilde{x}^*, u - x \rangle - \langle z^*, T(u) - T(x) \rangle + \mathbf{o}(\|u - x\|).$$

This shows that $\tilde{x}^* \in \partial_F \langle z^*, T \rangle (x)$. Applying (9.33) we obtain

$$\|x^* - \tilde{x}^*\| \leq \|x^* - h'(u)\| + \|h'(u) - \tilde{x}^*\| < \epsilon,$$

which completes the proof. $\qquad \square$

9.3 Application to Hamilton–Jacobi Equations

Convention. Throughout this section, assume that E is a Hilbert space.

An equation of the form

$$f(x) + H\big(x, f'(x)\big) = 0 \quad \forall x \in E, \tag{9.34}$$

where $H : E \times E^* \to \mathbb{R}$ is given, is called *Hamilton–Jacobi equation* for f. Equations of this (and a more general) type play a fundamental role in the calculus of variations and in optimal control theory. In this context, (9.34) may fail to have a classical, i.e., continuously differentiable, solution $f : E \to \mathbb{R}$. It turns out that F-subdifferentials can be used to introduce an adequate concept of generalized solutions. First, parallel to the F-subdifferential $\partial_F(\bar{x})$, we now consider the *F-superdifferential* of f at $\bar{x}$ defined by

$$\partial^F f(\bar{x}) := -\partial_F(-f)(\bar{x}).$$

Definition 9.3.1

(a) A function $f : E \to \mathbb{R}$ is said to be a *viscosity supersolution* of (9.34) if f is lower semicontinuous and for any $x \in E$ and any $x^* \in \partial_F f(x)$ one has

$$f(x) + H(x, x^*) \geq 0.$$

(b) A function $f : E \to \mathbb{R}$ is said to be a *viscosity subsolution* of (9.34) if f is upper semicontinuous and for any $x \in E$ and any $x^* \in \partial^F f(x)$ one has

$$f(x) + H(x, x^*) \leq 0.$$

(c) If $f : E \to \mathbb{R}$ is both a viscosity supersolution and a viscosity subsolution of (9.34), then f is said to be a *viscosity solution* of (9.34).

If $f : E \to \mathbb{R}$ is continuously differentiable, then $\partial_F f(x) = \partial^F f(x) = \{f'(x)\}$ for every $x \in E$ (Proposition 9.1.9). Hence any classical solution of (9.34) is also a viscosity solution, and any viscosity solution of (9.34) that is continuously differentiable is also a classical solution.

We consider the following condition on the function H:

$$|H(y, y^*) - H(x, x^*)| \leq \varphi(y - x, y^* - x^*) + c \, \max\{\|x^*\|, \|y^*\|\} \, \|y - x\|. \tag{9.35}$$

Theorem 9.3.2 (Comparison Theorem) *Let $H : E \times E^* \to \mathbb{R}$ be such that, with some constant $c > 0$ and some continuous function $\varphi : E \times E^* \to \mathbb{R}$ satisfying $\varphi(o, o) = 0$, the condition (9.35) holds for any $x, y \in E$ and any $x^*, y^* \in E^*$. If f is a viscosity subsolution of (9.34) and bounded above and g is a viscosity supersolution of (9.34) and bounded below, then $f \leq g$.*

Proof. Let $\epsilon > 0$ be given. By Theorem 9.2.3 applied to $f_1 := g$ and $f_2 := -f$, there exist $x, y \in E$ and $x^* \in \partial_F g(x)$, $y^* \in \partial^F f(y)$ satisfying

$$\|x - y\| < \epsilon, \quad \|x^*\| \, \|x - y\| < \epsilon, \quad \|y^*\| \, \|x - y\| < \epsilon,$$
$$g(x) - f(y) < \inf_E (g - f) + \epsilon, \quad \|x^* - y^*\| \leq \epsilon.$$

From the properties of f and g we deduce

$$g(x) + H(x, x^*) \geq 0, \quad f(y) + H(y, y^*) \leq 0.$$

Combining the above inequalities as well as (9.35), we obtain

$$\inf_E (g - f) > g(x) - f(y) - \epsilon \geq \big(H(y, y^*) - H(x, x^*) \big) - \epsilon$$
$$\geq - \big(\varphi(y - x, y^* - x^*) + c \max\{\|x^*\|, \|y^*\|\} \, \|y - x\| \big) - \epsilon.$$

By the assumptions on φ, the right-hand side of the last inequality converges to 0 as $\epsilon \to 0$. Therefore $\inf_E (g - f) \geq 0$. $\qquad\square$

As an immediate consequence of Theorem 9.3.2 we have:

Corollary 9.3.3 *Under the assumptions of Theorem 9.3.2, (9.34) has at most one continuous bounded viscosity solution.*

9.4 An Approximate Mean Value Theorem

Convention. Throughout this section, assume that E is a Fréchet smooth Banach space.

Theorem 9.4.1 (Approximate Mean Value Theorem) *Let $f : E \to \overline{\mathbb{R}}$ be proper and l.s.c. Further let $a, b \in E$ and $\rho \in \mathbb{R}$ be such that $a \neq b$, $f(a) \in \mathbb{R}$, and $\rho \leq f(b) - f(a)$. Then there exist a point $c \in [a, b]$ as well as sequences (x_n) in E and (x_n^*) in E^* satisfying*

$$(x_n, f(x_n)) \to (c, f(c)) \text{ as } n \to \infty, \quad x_n^* \in \partial_F f(x_n) \, \forall n \in \mathbb{N},$$
$$\liminf_{n \to \infty} \langle x_n^*, c - x_n \rangle \geq 0, \tag{9.36}$$
$$\liminf_{n \to \infty} \langle x_n^*, b - a \rangle \geq \rho, \tag{9.37}$$
$$f(c) \leq f(a) + |\rho|. \tag{9.38}$$

Proof.

(I) Let $z^* \in E^*$ be such that $\langle z^*, a - b \rangle = \rho$ and set

$$g(x) := f(x) + \langle z^*, x \rangle + \delta_{[a,b]}(x), \quad x \in E.$$

Then g attains its minimum on the compact set [a,b] at some c, and we may assume that $c \in [a, b)$ because $g(b) \geq g(a)$. By the strong local approximate sum rule (Theorem 9.2.6) there exist sequences (x_n), (y_n), (x_n^*), and (y_n^*) satisfying

$$(x_n, f(x_n)) \to (c, f(c)) \quad \text{as } n \to \infty,$$
$$y_n \in [a, b] \quad \forall n, \quad y_n \to c \quad \text{as } n \to \infty,$$
$$x_n^* \in \partial_F f(x_n), \quad y_n^* \in \partial_F \delta_{[a,b]}(y_n) \quad \forall n,$$
$$\|x_n^*\| \cdot \|x_n - y_n\| < 1/n, \quad \|y_n^*\| \cdot \|x_n - y_n\| < 1/n \quad \forall n,$$
$$\|x_n^* + z^* + y_n^*\| < 1/n \quad \forall n.$$

Since $y_n \to c$, we have $y_n \in [a, b)$ if n is sufficiently large which we now assume. Since $\delta_{[a,b]}$ is convex, Proposition 9.1.9 gives

$$\partial_F \delta_{[a,b]}(y_n) = \partial \delta_{[a,b]}(y_n) = \{y^* \in E^* \mid \langle y^*, x - y_n \rangle \le 0 \quad \forall x \in [a, b]\}. \tag{9.39}$$

Since $y_n^* \in \partial_F \delta_{[a,b]}(y_n)$ and $c \in [a, b)$, we see that $\limsup_{n \to \infty} (y_n^* \mid c - y_n) \le 0$. Using this, we obtain

$$\liminf_{n \to \infty} (x_n^* \mid c - x_n) = \liminf_{n \to \infty} (x_n^* + u \mid c - x_n)$$
$$= \liminf_{n \to \infty} (-y_n^* \mid c - y_n) = -\limsup_{n \to \infty} (y_n^* \mid c - y_n) \ge 0,$$

which verifies (9.36).

(II) We turn to (9.37). As $y_n \in [a, b)$ for all sufficiently large n, we have for these n,

$$b - a = \lambda_n(b - y_n), \quad \text{where } \lambda_n := \frac{\|b - a\|}{\|b - y_n\|}.$$

It follows that

$$\liminf_{n \to \infty} \langle x_n^* + z^*, b - a \rangle = \liminf_{n \to \infty} \left[\lambda_n \langle x_n^* + z^*, b - y_n \rangle \right]$$
$$= \frac{\|b - a\|}{\|b - c\|} \liminf_{n \to \infty} \langle -y_n^*, b - y_n \rangle \underset{(9.39)}{\ge} 0,$$

which immediately implies (9.37).

(III) To verify (9.38), notice that $g(c) \le g(a)$ and so $f(c) \le f(a) + (u \mid a - c)$. With some $\lambda \in (0, 1]$ we have $c = \lambda a + (1 - \lambda)b$ and it follows that

$$f(c) \le f(a) + (1 - \lambda)(u \mid a - b) \le f(a) + |\rho|. \qquad \square$$

As an application we show:

Proposition 9.4.2 *Let $f : E \to \bar{\mathbb{R}}$ be proper and l.s.c., let U be an open convex subset of E such that $U \cap \operatorname{dom} f \ne \emptyset$. Then for any $\lambda > 0$ the following assertions are equivalent:*

(a) f is L-continuous on U with Lipschitz constant λ.
(b) $\sup\{\|x^*\| \mid x^* \in \partial_F f(x)\} \leq \lambda$ for any $x \in U$.

Proof. (a) $\Longrightarrow$ (b): This is straightforward.
(b) $\Longrightarrow$ (a): Let $a, b \in U$ and $\rho \in \mathbb{R}$ be such that $a \in \text{dom } f$, $a \neq b$, and $\rho \leq f(b) - f(a)$. Further let $\epsilon > 0$ be given. By Theorem 9.4.1 there exist $x \in U$ and $x^* \in \partial_F f(x)$ such that (see (9.37))

$$\rho \leq \langle x^*, b - a \rangle + \epsilon \leq \lambda\|b - a\| + \epsilon.$$

Since $\rho \leq f(b) - f(a)$ and $\epsilon > 0$ are arbitrary, it follows that $f(b) - f(a) \leq \lambda\|b - a\|$ and so, in particular, $f(b) < +\infty$. Exchanging the roles of a and b, we thus obtain $|f(b) - f(a)| \leq \lambda\|b - a\|$. $\qquad\square$

9.5 Fréchet Subdifferential vs. Clarke Subdifferential

For a locally L-continuous functional we now deduce a representation of the Clarke subdifferential in terms of Fréchet subdifferentials.

Proposition 9.5.1 *Let E be a Fréchet smooth Banach space and let $f : E \to \mathbb{R}$ be locally L-continuous on E. Then for any $\bar{x} \in E$ one has*

$$\partial_o f(\bar{x}) = \overline{\text{co}}^* \{x^* \in E^* \mid \exists x_k \in E \; \exists x_k^* \in \partial_F f(x_k) : \; x_k \to \bar{x}, \; x_k^* \xrightarrow{w^*} x^*\}. \tag{9.40}$$

Proof. For convenience we write $x^* = {}^*\lim_{k\to\infty} x_k^*$ if $x_k^* \xrightarrow{w^*} x^*$. In view of Theorem 2.3.1, it suffices to show that the support functionals of the two sets in (9.40) coincide, i.e.,

$$f^o(\bar{x}, y) = \sup\{\langle x^*, y \rangle \mid x^* = {}^*\lim_{k\to\infty} x_k^*, \; x_k^* \in \partial_F f(x_k), \; x_k \to \bar{x}\} \quad \forall y \in E.$$

Since $\partial_F f(x_k) \subseteq \partial_o f(x_k)$ by Proposition 9.1.9(e) and $\partial_o f$ is norm-to-weak* u.s.c. by Proposition 7.3.8(c), we have

$$f^o(\bar{x}, y) \geq \sup\{\langle x^*, y \rangle \mid x^* = {}^*\lim_{k\to\infty} x_k^*, \; x_k^* \in \partial_F f(x_k), \; x_k \to \bar{x}\} \quad \forall y \in E.$$

It remains to verify that

$$f^o(\bar{x}, y) \leq \sup\{\langle x^*, y \rangle \mid x^* = {}^*\lim_{k\to\infty} x_k^*, \; x_k^* \in \partial_F f(x_k), \; x_k \to \bar{x}\} \quad \forall y \in E. \tag{9.41}$$

Let $y \in E$ be fixed. Choose sequences $z_k \to \bar{x}$ and $\tau_k \downarrow 0$ satisfying

$$f^o(\bar{x}, y) = \lim_{k\to\infty} \frac{f(z_k + \tau_k y) - f(z_k)}{\tau_k}.$$

Now let $\epsilon > 0$. By Theorem 9.4.1, for each k there exist $\tilde{z}_k \in [z_k, z_k + \tau_k y)$, $x_k \in B(\tilde{z}_k, \epsilon\tau_k)$, and $x_k^* \in \partial_F f(x_k)$ such that

$$\langle x_k^*, y \rangle \geq \frac{f(z_k + \tau_k y) - f(z_k)}{\tau_k} - \epsilon. \tag{9.42}$$

If $\lambda > 0$ denotes a Lipschitz constant of f around $\bar{x}$, then $x_k^* \in B_{E^*}(o, \lambda)$ for each k (Proposition 7.3.7). Since $B_{E^*}(o, \lambda)$ is weak* compact, we may assume that (x_k^*) is weak* convergent to some $x^* \in B_{E^*}(o, \lambda)$. By letting $k \to \infty$, we conclude from (9.42) that

$$\sup\{\langle x^*, y \rangle \mid x^* =^* \lim_{k \to \infty} x_k^*, \ x_k^* \in \partial_F f(x_k), \ x_k \to \bar{x}\} \geq f^\circ(\bar{x}, y) - \epsilon.$$

Since $\epsilon > 0$ and $y \in E$ are arbitrary, (9.41) follows. $\qquad\square$

Remark 9.5.2 Proposition 9.5.1 may be considered as an infinite-dimensional analogue of Clarke's Theorem 7.3.12. It shows that the Clarke subdifferential is the convexification of weak* limits of Fréchet subdifferentials. In many instances, convexity is a very convenient property as it allows to use techniques of convex analysis (cf. the remarks at the beginning of this chapter). However, as already observed above, the Clarke subdifferential may be too coarse to detect minimizers and so may be the smaller Michel–Penot subdifferential (cf. Remark 7.3.10, Example 7.3.11, and Exercise 9.8.3). The Fréchet subdifferential is qualified as an appropriate derivative-like object not only by its "smallness," but also, in particular, by the rich calculus it admits. This will also allow to derive multiplier rules in terms of Fréchet subdifferentials in Chap. 12. But we already know that the results are of an approximate nature. In Chap. 13 we shall study derivative-like objects that admit exact results.

9.6 Multidirectional Mean Value Theorems

Convention. Throughout this section, E denotes a Fréchet smooth Banach space.

Let $f : E \to \mathbb{R}$ be a G-differentiable l.s.c. functional. As a special case of the mean value inequality of Proposition 3.3.4 we obtain that for any $x, y \in E$ there exists $z \in (x, y)$ such that

$$f(y) - f(x) \leq \langle f'(z), y - x \rangle.$$

Given a compact convex set $S \subseteq E$, we pass to the inequality

$$\min_{\tilde{y} \in S} f(\tilde{y}) - f(x) \leq \langle f'(z), y - x \rangle \quad \forall y \in S. \tag{9.43}$$

We first establish a *multidirectional* version of this inequality by showing that for fixed x the same element z can be chosen in (9.43) while the direction y varies over S, provided E is finite dimensional.

If $x \in E$ and $S \subseteq E$, we write $[x, S] := \text{co}(\{x\} \cup S)$.

Proposition 9.6.1 *Assume that S is a nonempty, bounded, closed, convex subset of $\mathbb{R}^N$, $x \in \mathbb{R}^N$, and $f : \mathbb{R}^N \to \mathbb{R}$ is l.s.c. on $\mathbb{R}^N$ and G-differentiable on a neighborhood of $[x, S]$. Then there exists $z \in [x, S]$ satisfying (9.43).*

Proof. Let U be an open neighborhood of $[x, S]$ on which f is G-differentiable. Set

$$r := \min_{\tilde{y} \in S} f(\tilde{y}) - f(x)$$

and define $g : U \times [0, 1] \to \mathbb{R}$ by

$$g(y, \tau) := f(x + \tau(y - x)) - r\tau.$$

Since g is l.s.c. and $[x, S]$ is compact, g attains its infimum on $S \times [0, 1]$ at some $(\hat{y}, \hat{\tau}) \in S \times [0, 1]$. We show that (9.43) holds with

$$z := \begin{cases} x + \hat{\tau}(\hat{y} - x) & \text{if } \hat{\tau} \in [0, 1), \\ x & \text{if } \hat{\tau} = 1. \end{cases}$$

Now we distinguish three cases:

(I) Assume first that $\hat{\tau} \in (0, 1)$. Then the function $\tau \mapsto g(\hat{y}, \tau)$ attains its infimum on the interval $[0, 1]$ at the interior point $\hat{\tau}$. It follows that

$$0 = \frac{\partial}{\partial t} g(\hat{y}, \tau)\Big|_{\tau = \hat{\tau}} = \langle f'(z), \hat{y} - x \rangle - r. \tag{9.44}$$

Analogously, the function $y \mapsto g(y, \hat{\tau})$ attains its infimum over the set S at $\hat{y}$. Since the function is G-differentiable, the necessary optimality condition (7.6) in Sect. 7.1 implies that

$$0 \leq \left\langle \frac{\partial}{\partial y} g(y, \hat{\tau})\Big|_{y = \hat{y}}, y - \hat{y} \right\rangle = \langle \hat{\tau} f'(z), y - \hat{y} \rangle \quad \forall y \in S$$

and so

$$\langle f'(z), y - \hat{y} \rangle \geq 0 \quad \forall y \in S. \tag{9.45}$$

Combining (9.44) and (9.45), we obtain

$$r \leq \langle f'(z), \hat{y} - x \rangle + \langle f'(z), y - \hat{y} \rangle = \langle f'(z), y - x \rangle;$$

here $y \in S$ is arbitrary. Hence (9.43) is verified in case $\hat{\tau} \in (0, 1)$.

(II) Now assume that $\hat{\tau} = 0$ so that $(\hat{y}, 0)$ is a minimizer of g on $S \times [0, 1]$. It follows directly that

$$f(x) = g(\hat{y}, 0) \leq g(y, \tau) = f(x + \tau(y - x)) - r\tau \quad \forall (y, \tau) \in S \times (0, 1]$$

and so

$$r \leq \lim_{\tau \downarrow 0} \frac{f(x + \tau(y - x)) - f(x)}{\tau} = \langle f'(z), y - x \rangle \quad \forall y \in S,$$

which is equivalent to (9.43).

(III) Finally assume that $\hat{\tau} = 1$. Then, in particular, we have $g(\hat{y}, 1) \le g(\hat{y}, 0) = f(x)$ and on the other hand,

$$g(\hat{y}, 1) = f(\hat{y}) - r = f(\hat{y}) - \min_{\tilde{y} \in S} f(\tilde{y}) + f(x) \ge f(x).$$

Therefore, $f(x) = g(\hat{y}, 1)$ which means that $f(x)$ is the minimum of g on $S \times [0, 1]$. Since $f(x) = g(y, 0)$ for any $y \in S$, we see that any $(y, 0)$ is also a minimizer of g. Hence we can replace $\hat{\tau} = 1$ with $\hat{\tau} = 0$ and refer to case (II).   □

Now we establish a multidirectional mean value theorem in terms of F-subdifferentials in arbitrary Fréchet smooth Banach spaces.

Theorem 9.6.2 (Multidirectional Mean Value Theorem) *Assume that E is a Fréchet smooth Banach space, S is a nonempty closed convex subset of E, $f : E \to \overline{\mathbb{R}}$ is a l.s.c. functional, and $x \in \operatorname{dom} f$. Suppose that for some $\rho > 0$, f is bounded below on $[x, S] + B(o, \rho)$. Let r be such that*

$$r < \liminf_{\eta \downarrow 0} \{ f(y) \mid y \in S + B(o, \eta) \} - f(x). \tag{9.46}$$

Then for any $\epsilon > 0$ there exist $z \in [x, S] + B(o, \epsilon)$ and $z^* \in \partial_F f(z)$ such that

$$f(z) < \liminf_{\eta \downarrow 0} \{ f(y) \mid y \in [x, S] + B(o, \eta) \} + |r| + \epsilon, \tag{9.47}$$

$$r < \langle z^*, y - x \rangle + \epsilon \| y - x \| \quad \forall y \in S. \tag{9.48}$$

Proof.

(I) First we assume that (9.46) holds for $r = 0$ and we consider the case $r = 0$. The functional $\tilde{f} := f + \delta_{[x,S] + B(o,\rho)}$ is bounded below on all of E. By (9.46) there exists $\eta \in (0, \rho/2)$ such that

$$f(x) < \inf \{ f(y) \mid y \in S + B(o, 2\eta) \}.$$

Without loss of generality we may assume that ϵ satisfies

$$0 < \epsilon < \inf \{ f(y) \mid y \in S + B(o, 2\eta) \} - f(x) \quad \text{and} \quad \epsilon < \eta. \tag{9.49}$$

By the nonlocal approximate sum rule (Theorem 9.2.3), applied to $f_1 := \tilde{f}$ and $f_2 := \delta_{[x,S]}$, there exist $z \in \operatorname{dom} f \cap ([x, S] + B(o, \rho))$ and $u \in [x, S]$ such that

$$\| z - u \| < \epsilon, \quad z^* \in \partial_F f_1(z) = \partial_F f(z)$$

as well as $u^* \in \partial_F \delta_{[x,S]}(u)$ satisfying

$$\max \{ \| z^* \|, \| u^* \| \} \cdot \| z - u \| < \epsilon, \tag{9.50}$$

$$f(z) < \liminf_{\eta \downarrow 0} \{ f(y) \mid y \in [x, S] + B(o, \eta) \} + \epsilon \le f(x) + \epsilon, \tag{9.51}$$

$$\| z^* + u^* \| < \epsilon. \tag{9.52}$$

Since $[x, S]$ is a convex set, we have

$$u^* \in \partial_F \delta_{[x,S]}(u) = N_F([x, S], u) = N([x, S], u)$$

and so

$$\langle u^*, w - u \rangle \leq 0 \quad \forall w \in [x, S]. \tag{9.53}$$

From (9.52) and (9.53) we obtain

$$\langle z^*, w - u \rangle = \langle z^* + u^*, w - u \rangle - \langle u^*, w - u \rangle$$
$$\geq \langle z^* + u^*, w - u \rangle > -\epsilon \|w - u\| \quad \forall w \in [x, S], \, w \neq u. \tag{9.54}$$

We show that $d(S, u) \geq \eta$. If we had $d(S, u) < \eta$, it would follow that $\|\tilde{y} - u\| < \eta$ for some $\tilde{y} \in S$, thus

$$\|\tilde{y} - z\| \leq \|\tilde{y} - u\| + \|u - z\| \underset{(9.50)}{<} \eta + \epsilon < 2\eta$$

and so $d(S, z) < 2\eta$. But then

$$f(z) \geq \inf\{f(y) \mid y \in S + B(o, 2\eta)\} \underset{(9.49)}{>} f(x) + \epsilon,$$

which contradicts (9.51). Hence $d(S, u) \geq \eta$. Let $u := x + \hat{\tau}(\hat{y} - x)$ with some $\hat{\tau} \in [0, 1]$ and $\hat{y} \in S$. Then

$$0 < \eta \leq \|\hat{y} - u\| = (1 - \hat{\tau})\|\hat{y} - x\|$$

and so $\hat{\tau} < 1$. Consider any $y \in S$ and set $w := y + \hat{\tau}(\hat{y} - y)$. Then $w \neq u$, otherwise it would follow that $y = x$ which is contradictory because $x \notin S$ (see (9.46)). Inserting this w into (9.54), we finally obtain (9.48) with $r = 0$.

(II) Now we consider the general case (9.46). Equip $E \times \mathbb{R}$ with the Euclidean product norm. Choose $\epsilon' \in (0, \epsilon/2)$ such that

$$r + \epsilon' < \liminf_{\eta \downarrow 0}\{f(y) \mid y \in S + B(o, \eta)\} - f(x).$$

Define $F : E \times \mathbb{R} \to \overline{\mathbb{R}}$ by $F(y, \tau) := f(y) - (r + \epsilon')\tau$. Then F is l.s.c. on $E \times \mathbb{R}$ and bounded below on $[(x, 0), S \times \{1\}] + B_{E \times \mathbb{R}}(o, \rho)$. Furthermore we have

$$0 < \liminf_{\eta \downarrow 0}\{f(y) \mid y \in S + B(o, \eta)\} - (r + \epsilon') - f(x)$$
$$= \liminf_{\eta \downarrow 0}\{F(y, 1) \mid (y, 1) \in (S \times \{1\}) + B_{E \times \mathbb{R}}(o, \eta)\} - F(x, 0).$$

Hence the special case (I) applies with f, x, and S replaced by F, $(x, 0)$, and $S \times \{1\}$, respectively. Consequently there exist $(z, \tau) \in [(x, 0),$

$S \times \{1\}] + B_{E \times \mathbb{R}}(o, \epsilon)$ and $(z^*, \tau^*) \in \partial_F F(z, \tau) \subseteq \partial_F f(z) \times \{-(r + \epsilon')\}$ such that

$$f(z) = F(z, \tau) + (r + \epsilon')\tau$$
$$< \liminf_{\eta \downarrow 0} \{F(y, \tau) \mid (y, \tau) \in [(x, 0), S \times \{1\}] + B_{E \times \mathbb{R}}(o, \eta)\} + \epsilon' + (r + \epsilon')\tau$$
$$\leq \liminf_{\eta \downarrow 0} \{f(y) \mid y \in [x, S] + B(o, \eta)\} + |r| + \epsilon,$$

and for any $(y, 1) \in S \times \{1\}$ we have

$$0 < \langle (z^*, \tau^*), (y, 1) - (x, 0) \rangle + \epsilon' \|(y - x, 1)\|$$
$$\leq \langle z^*, y - x \rangle - (r + \epsilon') + \epsilon'(\|y - x\| + 1)$$
$$= \langle z^*, y - x \rangle - r + \epsilon' \|y - x\| \leq \langle z^*, y - x \rangle - r + \epsilon \|y - x\|.$$

The proof is thus complete. $\qquad \square$

Notice that the set S in Theorem 9.6.2 is not assumed to be bounded (in this context, see Exercise 9.8.8).

To prepare the next result, consider a G-differentiable functional $f : E \to \mathbb{R}$ such that $f'(\bar{x}) \neq o$ for some $\bar{x} \in E$. Then $\bar{x}$ is not a local minimizer of f. Hence for some $r > 0$ we have

$$\inf_{x \in B(\bar{x}, r)} f(x) < f(\bar{x}).$$

Theorem 9.6.3 generalizes this fact to the nondifferentiable case and at the same time quantifies the inequality.

Theorem 9.6.3 (Decrease Principle) *Assume that* $f : E \to \bar{\mathbb{R}}$ *is l.s.c. and bounded below, and* $\bar{x} \in E$. *Assume further that for some* $r > 0$ *and* $\sigma > 0$, *one has*

$$[x \in B(\bar{x}, r) \text{ and } x^* \in \partial_F f(x)] \implies \|x^*\| > \sigma. \tag{9.55}$$

Then

$$\inf_{x \in B(\bar{x}, r)} f(x) \leq f(\bar{x}) - r\sigma. \tag{9.56}$$

Proof. Obviously we may suppose that $\bar{x} \in \text{dom} f$. Let $r' \in (0, r)$. Observe that

$$\inf_{x \in B(\bar{x}, r)} f(x) \leq \lim_{\eta \downarrow 0} \inf_{y \in B(\bar{x}, \eta)} f(y). \tag{9.57}$$

Let $\epsilon \in (0, r - r')$ and set

$$\tilde{r} := \lim_{\eta \downarrow 0} \inf_{y \in B(\bar{x}, \eta)} f(y) - f(\bar{x}) - \epsilon.$$

Applying Theorem 9.6.2 with $S := B(\bar{x}, r')$ and x, r replaced by $\bar{x}, \tilde{r}$, respectively, we conclude that there exist $z \in B(\bar{x}, r') + B(o, \epsilon) \subseteq B(\bar{x}, r)$ and $z^* \in \partial_F f(z)$ such that

$$\tilde{r} < \langle z^*, y - \bar{x} \rangle + \epsilon \|y - \bar{x}\| \quad \forall y \in B(\bar{x}, r'). \tag{9.58}$$

By hypothesis (9.55) we have $\|z^*\| > \sigma$. Hence there exists $z_0 \in E$ satisfying $\|z_0\| = 1$ and $\langle z^*, z_0 \rangle > \sigma$. Inserting $y := -r' z_0 + \bar{x}$ into (9.58) yields

$$\tilde{r} < -r' \langle z^*, z_0 \rangle + r' \epsilon < -r' \sigma + r' \epsilon.$$

Recalling (9.57) and the definition of $\tilde{r}$, we further obtain

$$\inf_{B(\bar{x}, r)} f(x) < f(\bar{x}) - r' \sigma + (r' + 1)\epsilon.$$

Letting $r' \uparrow r$ and so $\epsilon \downarrow 0$ gives (9.56). □

Corollary 9.6.4 is a counterpart of Corollary 8.2.5 for a nondifferentiable functional f.

Corollary 9.6.4 *Let $f : E \to \overline{\mathbb{R}}$ be l.s.c. and bounded below. Assume that $\bar{x} \in E$ and $\epsilon > 0$ are such that $f(\bar{x}) < \inf_E f + \epsilon$. Then for any $\lambda > 0$ there exist $z \in B(\bar{x}, \lambda)$ and $z^* \in \partial_F f(z)$ satisfying*

$$f(z) < \inf_E f + \epsilon \quad \text{and} \quad \|z^*\| < \frac{\epsilon}{\lambda}.$$

Proof. See Exercise 9.8.9. □

9.7 The Fréchet Subdifferential of Marginal Functions

We now establish representations of the F-subdifferential of a marginal functional of the form

$$f(x) := \inf_{y \in F} \varphi(x, y), \quad x \in E. \tag{9.59}$$

The first result will be crucial for deriving an implicit multifunction theorem in Sect. 13.10. Recall that $\underline{f}(x) := \liminf_{y \to x} f(y)$ denotes the lower semicontinuous closure of f (see Exercise 1.8.11). Notice that $\partial_F f(x) \subseteq \partial_F \underline{f}(x)$.

Proposition 9.7.1 *Assume that E and F are Fréchet smooth Banach spaces, $\varphi : E \times F \to \overline{\mathbb{R}}$ is l.s.c., and f is defined by (9.59). Let $x \in E$ and $x^* \in \partial_F \underline{f}(x)$. Then for any sufficiently small $\epsilon > 0$ there exist $(x_\epsilon, y_\epsilon) \in E \times F$ and $(x_\epsilon^*, y_\epsilon^*) \in \partial_F \varphi(x_\epsilon, y_\epsilon)$ such that*

$$\|x - x_\epsilon\| < \epsilon, \quad |\underline{f}(x) - \underline{f}(x_\epsilon)| < \epsilon, \tag{9.60}$$

$$\varphi(x_\epsilon, y_\epsilon) < \underline{f}(x_\epsilon) + \epsilon < f(x_\epsilon) + \epsilon, \tag{9.61}$$

$$\|x_\epsilon^* - x^*\| < \epsilon, \quad \|y_\epsilon^*\| < \epsilon. \tag{9.62}$$

Proof. Let $g : E \to \mathbb{R}$ be a C^1 function such that $g'(x) = x^*$ and for some $\rho \in (0, 1)$ one has

$$(\underline{f} - g)(u) \geq (\underline{f} - g)(x) = 0 \quad \forall u \in B(x, \rho).$$

Let $\epsilon \in (0, \rho)$ and let $\alpha > 0$ be the constant associated with E in the smooth variational principle of Theorem 8.4.3. Choose a positive number $\eta < \min\{\epsilon/5, \sqrt{\epsilon/(5\alpha)}\}$ such that the following holds:

$$\underline{f}(u) \geq \underline{f}(x) - \epsilon/5 \quad \forall u \in B(x, \eta),$$
$$|g(u) - g(\hat{u})| < \epsilon/5 \quad \text{whenever } \|u - \hat{u}\| < \eta,$$
$$\|g'(u) - g'(x)\| < \epsilon/2 \quad \forall u \in B(x, \eta).$$

Now choose $\bar{u} \in B(x, \eta/2)$ close enough to x so that $f(\bar{u}) - g(\bar{u}) < \underline{f}(x) - g(x) + \alpha\eta^2/8$. Then there exists $\bar{v} \in F$ satisfying

$$\varphi(\bar{u}, \bar{v}) - g(\bar{u}) < f(\bar{u}) - g(\bar{u}) + \alpha\eta^2/8 < \underline{f}(x) - g(x) + \alpha\eta^2/4$$
$$\leq \inf_{\substack{u \in B(x, \rho) \\ v \in F}} \big(\varphi(u, v) - g(u)\big) + \alpha\eta^2/4.$$

Applying Theorem 8.4.3 to the functional $(u, v) \mapsto \varphi(u, v) - g(u)$, we find $(x_\epsilon, y_\epsilon) \in B((\bar{u}, \bar{v}), \eta/2) \subseteq B((x, \bar{v}), \eta)$ and a C^1 function $h : E \times F \to \mathbb{R}$ such that $\max\{\|h\|_\infty, \|h'\|_\infty\} < \eta$ and the functional

$$(u, v) \mapsto \varphi(u, v) - g(u) + h(u, v), \quad (u, v) \in E \times F,$$

attains its minimum at (x_ϵ, y_ϵ). It follows that

$$\big(g'(x_\epsilon) - h_{|1}(x_\epsilon, y_\epsilon), -h_{|2}(x_\epsilon, y_\epsilon)\big) \in \partial_F\varphi(x_\epsilon, y_\epsilon).$$

Setting $x_\epsilon^* := g'(x_\epsilon) - h_{|1}(x_\epsilon, y_\epsilon)$ and $y_\epsilon^* := -h_{|2}(x_\epsilon, y_\epsilon)$, we see that (9.62) holds. We further have

$$\varphi(x_\epsilon, y_\epsilon) \leq \varphi(\bar{u}, \bar{v}) + \big(g(x_\epsilon) - g(\bar{u})\big) + h(\bar{u}, \bar{v}) - h(x_\epsilon, y_\epsilon)$$
$$\leq \underline{f}(x) + \alpha\eta^2 + |g(x_\epsilon) - g(\bar{u})| + 2\|h\|_\infty$$
$$< \underline{f}(x_\epsilon) + \epsilon/5 + \alpha\eta^2 + |g(x_\epsilon) - g(\bar{u})| + 2\|h\|_\infty$$
$$\leq \underline{f}(x_\epsilon) + \epsilon \leq f(x_\epsilon) + \epsilon,$$

which verifies (9.61). This inequality together with $\underline{f}(x_\epsilon) \leq f(x_\epsilon) \leq \varphi(x_\epsilon, y_\epsilon)$ shows that (9.60) also holds. $\qquad\square$

The following result strengthens Proposition 9.7.1; it states that each $x^* \in \partial_F\underline{f}(x)$ (in particular, each $x^* \in \partial_F f(x)$) can be approximated by some $u^* \in \partial_{F,1}\varphi(u, y)$, where u is close to x, for *all* y such that $\varphi(x, y)$ is close to $f(x)$. Here, $\partial_{F,1}\varphi(u, y)$ denotes the F-subdifferential of $u \mapsto \varphi(u, y)$.

Proposition 9.7.2 *Assume that E is a Fréchet smooth Banach space and F is an arbitrary nonempty set. For each $y \in F$ let $u \mapsto \varphi(u, y)$, $u \in E$, be proper and l.s.c. Let $f : E \to \bar{\mathbb{R}}$ be defined by (9.59), let $x \in E$, and $x^* \in \partial_F f(x)$. Then there exist a nonnegative continuous function $\gamma : [0, +\infty) \to \mathbb{R}$ with $\gamma(0) = 0$ and a constant $c > 0$ such that for any sufficiently small $\epsilon > 0$, any $y \in F$, and any $\hat{x} \in B(x, \epsilon)$ satisfying $\varphi(\hat{x}, y) < \underline{f}(x) + \epsilon$, there exist $u \in B(x, c\sqrt{\epsilon})$ and $u^* \in \partial_{F,1} \varphi(u, y)$ such that*

$$\varphi(u, y) < \underline{f}(x) + c\sqrt{\epsilon} \quad \text{and} \quad \|u^* - x^*\| \leq \gamma(\epsilon).$$

Proof. Since $x^* \in \partial_F \underline{f}(x)$, there exist $\delta \in (0, 1)$ and a C^1 functional $g : E \to \mathbb{R}$ such that $g'(x) = x^*$ and one has

$$0 = (\underline{f} - g)(x) \leq (\underline{f} - g)(u) \quad \forall u \in B(x, 2\delta). \tag{9.63}$$

Moreover we may assume that δ is so small that g is L-continuous on $B(x, 2\delta)$ with Lipschitz constant $\lambda > 0$ (cf. Propositions 3.2.4 and 3.4.2). Now let $\epsilon \in (0, \delta^2)$ be given. Further let $\hat{x}$ be as in the theorem. It follows that

$$\varphi(\hat{x}, y) - g(\hat{x}) = \big(\varphi(\hat{x}, y) - \underline{f}(x)\big) + \big(\underline{f}(x) - g(\hat{x})\big) < \epsilon + \big(g(x) - g(\hat{x})\big)$$
$$\leq \epsilon + \lambda\|x - \hat{x}\| \leq (1 + \lambda)\epsilon.$$

For any $u \in B(x, 2\sqrt{\epsilon})$ we have $\varphi(u, y) - g(u) \geq 0$ (which follows from (9.63)). This and the foregoing estimate imply

$$\big(\varphi(u, y) - g(u)\big) - \big(\varphi(\hat{x}, y) - g(\hat{x})\big) \geq -(1 + \lambda)\epsilon \quad \forall u \in B(x, 2\sqrt{\epsilon}).$$

Applying Theorem 9.6.2 with S, x, and f replaced by $B(x, \sqrt{\epsilon})$, $\hat{x}$, and $u \mapsto \varphi(u, y) - g(u)$, respectively, we conclude that there exist $u \in B(x, \sqrt{\epsilon}) + B(o, \sqrt{\epsilon}) \subseteq B(x, 2\sqrt{\epsilon})$ and $u^* \in \partial_{F,1} \varphi(u, y)$ such that

$$\varphi(u, y)) < \varphi(\hat{x}, y) + \big(g(u) - g(\hat{x})\big) + \epsilon < \underline{f}(x) + \lambda\|u - \hat{x}\| + 2\epsilon$$
$$< \underline{f}(x) + \lambda(\|u - x\| + \|x - \hat{x}\|) + 2\epsilon \leq \underline{f}(x) + (3\lambda + 2)\sqrt{\epsilon}$$

and

$$\langle u^* - g'(u), v - \hat{x}\rangle \geq -(1 + \lambda)\epsilon \quad \forall v \in B(\hat{x}, \sqrt{\epsilon}); \tag{9.64}$$

in this connection notice that $\partial_{F,1}\big(\varphi(u, y) - g(u)\big) = \partial_{F,1}\varphi(u, y) - \{g'(u)\}$ and employ Exercise 9.8.8. From (9.64) we conclude that $\|u^* - g'(u)\| \leq (1 + \lambda)\sqrt{\epsilon}$ and so

$$\|u^* - x^*\| \leq \|u^* - g'(u)\| + \|g'(u) - g'(x)\| \leq (1 + \lambda)\sqrt{\epsilon} + \|g'(u) - g'(x)\| \leq \gamma(\epsilon),$$

where

$$\gamma(\epsilon) := (1 + \lambda)\sqrt{\epsilon} + \sup\{\|g'(u) - g'(x)\| \mid u \in B(x, 2\sqrt{\epsilon})\} \quad \forall \epsilon \geq 0.$$

Obviously, γ has the required properties. It remains to set $c := 3\lambda + 2$. $\quad\square$

9.8 Bibliographical Notes and Exercises

The concept of the Fréchet subdifferential can be traced back to Bazaraa and Goode [10]. It was developed by Kruger and Mordukhovich [113], Borwein and Strójwas [20], Schirotzek [194], and others. Crucial progress in the study of the Fréchet subdifferential was possible after Deville et al. [50] had shown that it coincides with the viscosity subdifferential in any Fréchet smooth Banach space and so in particular with the proximal subdifferential (see Lemma 9.1.6 and Theorem 9.1.7). The concept of proximal subgradients is due to Rockafellar [185].

The presentation of Chap. 9 owes much to Borwein and Zhu [23, 24]. As these authors, we took the nonlocal approximate sum rule (Theorem 9.2.3), which is due to Zhu [225], as starting point for the theory of the F-subdifferential. Local approximate sum rules go back to Ioffe [95]. Generalizations were obtained among others by Borwein and Ioffe [17] and Borwein and Zhu [22].

The approximate mean value inequality is originally due to Zagrodny [220], its version in terms of F-subdifferentials was established by Loewen [124]. The simple proof of Theorem 9.4.1 via the nonlocal approximate sum rule is adapted from Borwein and Zhu [23]. Zagrodny [220] also shows that Lebourg's mean value theorem (Theorem 7.4.4) can be derived from Theorem 9.4.1.

The multidirectional mean value theorem is due to Clarke and Ledyaev [38]. It is a cornerstone in the subdifferential theory of Clarke et al. [39]. The F-subdifferential version of Theorem 9.6.2 appeared in Zhu's paper [225], the finite-dimensional version of Proposition 9.6.1 was taken from Clarke et al. [39]. For further substantial results and applications in this direction we also recommend [39] (infinite-dimensional spaces) and Rockafellar and Wets [189] (finite-dimensional spaces).

Proposition 9.7.1 is due to Ledyaev and Zhu [120, 121], while Proposition 9.7.2 is taken from Borwein and Zhu [24] who attribute it to Ledyaev and Treiman.

Crandall and Lions [42] introduced the concept of a viscosity solution of the Hamilton–Jacobi equation. Viscosity subsolutions and viscosity supersolutions were first studied by Crandall et al. [41]. Theorem 9.3.2 and its proof are due to Borwein and Zhu [22]. For related comparison results as well as existence theorems in terms of viscosity solutions see also Clarke et al. [39], Deville et al. [50], and Subbotin [204].

Exercise 9.8.1 Verify Proposition 9.1.3.

Exercise 9.8.2 Verify Remark 9.1.4.

Exercise 9.8.3

(a) Let $f(x) := -|x|$, $x \in \mathbb{R}$. Show that $\partial_F f(0) = \emptyset$ but $\partial_\Diamond f(0) = \partial_\circ f(0) = [-1, 1]$.
(b) Let $f(x) := -|x|^{3/2}$, $x \in \mathbb{R}$. Show that $\partial_P f(0) = \emptyset$ but $\partial_F f(0) = \{f'(0)\}$. (Hence the inclusion in (9.8) can be proper.)

Exercise 9.8.4 Verify Remark 9.1.8.
Hint (cf. Borwein and Zhu [23]): Let $\eta(t) := -\rho_2(\alpha t)$ for $t > 0$ and $\eta(0) := 0$; here α is as in Lemma 9.1.6 and ρ_2 is as in the proof of Theorem 9.1.7. Notice that η is differentiable and put

$$\gamma(t) := \int_0^t \beta(s)\, ds, \ t \geq 0, \quad \text{where } \beta(s) := \sup_{0 \leq \sigma \leq s} \eta'(\sigma), \ s \geq 0.$$

Show that the function $g : E \to \mathbb{R}$ defined by

$$g(x) := f(\bar{x}) + \langle x^*, x - \bar{x} \rangle - \gamma(\|x - \bar{x}\|), \quad x \in E,$$

meets the requirement.

Exercise 9.8.5 Prove statement (e) of Proposition 9.1.9.

Exercise 9.8.6 Prove Proposition 9.2.2.

Exercise 9.8.7 Verify Lemma 9.2.5.

Exercise 9.8.8 We refer to Theorem 9.6.2:

(a) Show that in (9.48) the term $\epsilon\|y - x\|$ can be omitted if the set S is bounded.
(b) Consider the example $E = S := \mathbb{R}$ and $f(y) := e^y$ to see that in general the conclusion of Theorem 9.6.2 is false if the term $\epsilon\|y - x\|$ is omitted from (9.48).

Exercise 9.8.9 Verify Corollary 9.6.4.

Multifunctions

10.1 The Generalized Open Mapping Theorem

In this section, let E and F be Banach spaces.

Given a multifunction $\Phi : E \rightrightarrows F$ (cf. Sect. 4.3), we write

$$
\begin{aligned}
&\operatorname{Dom}\Phi := \{x \in E \mid \Phi(x) \neq \emptyset\}, && \textit{domain of } \Phi, \\
&\ker\Phi \; := \{x \in E \mid o \in \Phi(x)\}, && \textit{kernel of } \Phi, \\
&\operatorname{range}\Phi := \bigcup_{x \in E} \Phi(x), && \textit{range of } \Phi, \\
&\operatorname{graph}\Phi := \{(x,y) \in E \times F \mid x \in E,\ y \in \Phi(x)\}, && \textit{graph of } \Phi, \\
&\Phi^{-1}(y) := \{x \in E \mid y \in \Phi(x)\}, \quad y \in F, && \textit{inverse of } \Phi.
\end{aligned}
$$

The multifunction Φ is said to be *closed* or *convex* if $\operatorname{graph}\Phi$ is closed or convex, respectively. We call Φ *closed-valued* or *bounded-valued* if $\Phi(x)$ is, respectively, a closed or a bounded subset of F for any $x \in E$. Notice that a closed multifunction is closed-valued but the converse is not true.

Our aim in this section is to generalize the open mapping theorem from continuous linear mappings to multifunctions. We prepare this with an auxiliary result. Let $p_E : E \times F \to E$ denote the projection onto E defined by $p_E(x,y) := x$ for each $(x,y) \in E \times F$. Analogously define $p_F : E \times F \to F$.

Lemma 10.1.1 *If C is a closed convex subset of $E \times F$ such that $p_E(C)$ is bounded, then*

$$
\operatorname{int}\operatorname{cl}\big(p_F(C)\big) = \operatorname{int}\big(p_F(C)\big).
$$

Proof. We may assume that C is nonempty. The assertion is verified when we have shown that $\operatorname{int}\operatorname{cl}\big(p_F(C)\big) \subseteq p_F(C)$. Thus let $y \in \operatorname{int}\operatorname{cl}\big(p_F(C)\big)$ be given. We shall show that there exists $x \in C$ satisfying $(x,y) \in C$. Let $\epsilon > 0$ be such that $B(y, 2\epsilon) \subseteq \operatorname{cl}\big(p_F(C)\big)$. Choose $(x_0, y_0) \in C$ and define a sequence $\big((x_k, y_k)\big)$ in C recursively in the following way. Assume that (x_k, y_k) is already defined. If $y_k = y$, then set $x_{k+1} := x_k$ and $y_{k+1} := y_k$. In this case, the element $x := x_k$ meets the requirement. If $y_k \neq y$, then set

$$\alpha_k := \frac{\epsilon}{\|y_k - y\|}, \quad z_k := y + \alpha_k(y - y_k).$$

Then $z_k \in B(y, \epsilon) \subseteq \mathrm{cl}(p_F(C))$. Hence there exists $(\tilde{x}, \tilde{y}) \in C$ satisfying $\|\tilde{y} - z_k\| \leq \frac{1}{2}\|y_k - y\|$. Now let

$$(x_{k+1}, y_{k+1}) := \frac{\alpha_k}{1 + \alpha_k}(x_k, y_k) + \frac{1}{1 + \alpha_k}(\tilde{x}, \tilde{y}).$$

Since C is convex, we have $(x_{k+1}, y_{k+1}) \in C$. It follows that

$$\|x_{k+1} - x_k\| = \frac{\|\tilde{x} - x_k\|}{1 + \alpha_k} \leq \frac{\mathrm{diam}\, p_E(C)}{\epsilon}\|y_k - y\|, \tag{10.1}$$

$$\|y_{k+1} - y\| = \frac{\|\tilde{y} - z_k\|}{1 + \alpha_k} \leq \frac{1}{2}\|y_k - y\|. \tag{10.2}$$

From (10.2) we conclude that $\|y_k - y\| \leq 2^{-k}\|y_0 - y\|$. Hence $y_k \to y$ as $k \to \infty$. This together with (10.1) shows that (x_k) is a Cauchy sequence and so is convergent to some $x \in E$. Since C is closed, we have $(x, y) \in C$.   □

Now we can establish the announced result.

Theorem 10.1.2 (Generalized Open Mapping Theorem) *Let E and F be Banach spaces and $\Phi : E \rightrightarrows F$ be a closed convex multifunction. If $y \in \mathrm{int}\,\mathrm{range}(\Phi)$, then $y \in \mathrm{int}\,\Phi(\mathring{B}(x, \rho))$ for each $x \in \Phi^{-1}(y)$ and each $\rho > 0$.*

Remark 10.1.3 Recall that a mapping $T : E \to F$ is said to be *open* if it maps open subsets of E onto open subsets of F. The classical open mapping theorem of Banach states that if T is continuous, linear, and surjective, then T is open. We show that this follows from Theorem 10.1.2. Since T is continuous and linear, the multifunction $\widetilde{T}$ defined by $\widetilde{T}(x) := \{T(x)\}$, $x \in E$, has a closed and convex graph. Moreover, the surjectivity of T is equivalent to $o \in \mathrm{int}\,T(E)$ and so to $o \in \mathrm{int}\,\mathrm{range}\,\widetilde{T}$. Hence applying Theorem 10.1.2 with $x = o$ and $y = o$, we conclude that $o \in \mathrm{int}\,T(\mathring{B}_E)$ which is equivalent to T being open. (Recall that $\mathring{B}_E$ denotes the open unit ball of E.)

Proof of Theorem 10.1.2. We may and do assume that $x = o$, $y = o$ (replace Φ with $\tilde{x} \mapsto \Phi(x - \tilde{x}) - y$ if necessary) and that $\rho = 1$. Set $M := \mathrm{cl}\,\Phi(\frac{1}{2}\mathring{B}_E)$. Then M is nonempty closed and convex. Let $z \in F$. Since by assumption range Φ is a neighborhood of zero, we have $\lambda z \in \mathrm{range}\,\Phi$ for some $\lambda > 0$ and so $\lambda z \in \Phi(x')$ for some $x' \in E$. Furthermore, for each $\alpha \in (0, 1)$ we obtain

$$\alpha\lambda z = \alpha\lambda z + (1 - \alpha)o$$
$$\in \alpha\Phi(x') + (1 - \alpha)\Phi(o) \subseteq \Phi(\alpha x' + (1 - \alpha)o) = \Phi(\alpha x');$$

here the inclusion $\subseteq$ holds since graph Φ is convex. We conclude that $\alpha\lambda z \in \Phi(\frac{1}{2}\mathring{B}_E)$ for sufficiently small $\alpha > 0$. Since this holds for each $z \in F$, it follows that $\Phi(\frac{1}{2}\mathring{B}_E)$, and so M, is absorbing. Thus by Proposition 1.2.1, M is a neighborhood of zero. Therefore, $\rho\mathring{B}_F \subseteq \operatorname{int} M$ for some $\rho > 0$. Now let $C := \operatorname{graph}(\Phi) \cap (\frac{1}{2}B_E \times F)$. Then C is closed and convex and $p_E(C) \subseteq \frac{1}{2}B_E$. Hence Lemma 10.1.1 implies that $\operatorname{int} \operatorname{cl} p_F(C) = \operatorname{int} p_F(C)$. Noting this and $p_F(C) = \Phi(\frac{1}{2}B_E)$, we finally obtain

$$\rho\mathring{B}_F \subseteq \operatorname{int} M \subseteq \operatorname{int} \operatorname{cl} \Phi(\tfrac{1}{2}B_E) = \operatorname{int} p_F(C) = \operatorname{int} \Phi(\tfrac{1}{2}B_E) \subseteq \operatorname{int} \Phi(\mathring{B}_E). \qquad \square$$

10.2 Systems of Convex Inequalities

Let G be a normed vector space and P be a convex cone in G. Recall that we denote by $\leq_P$ the preorder generated by P, i.e., for all $u, v \in G$ we set (see Sect. 1.5)

$$u \leq_P v \quad :\Longleftrightarrow \quad v \in u + P.$$

Now let E be another normed vector space, let K be a convex subset of E, and let $S : K \to G$. Generalizing the notion of a convex functional, we say that the mapping S is P-convex if for all $x, y \in K$ and all $\lambda \in (0, 1)$ we have

$$S\big(\lambda x + (1 - \lambda)y\big) \leq_P \lambda S(x) + (1 - \lambda)S(y),$$

in other words, if we have

$$\lambda S(x) + (1 - \lambda)S(y) \in S\big(\lambda x + (1 - \lambda)y\big) + P.$$

Consider the following assumptions:

(A) E, G, and H are normed vector spaces, $K \subseteq E$ is nonempty and convex. $P \subseteq G$ is a convex cone with $\operatorname{int}(P) \neq \emptyset$, $Q \subseteq H$ is a closed convex cone. $S : K \to G$ is P-convex, $T : K \to H$ is Q-convex.

Furthermore let

$$C := \{(y, z) \in G \times H \mid \exists x \in K : y - Sx \in \operatorname{int} P, \; z - Tx \in Q\}.$$

We shall utilize the condition

$$\exists \bar{y} \in G : (\bar{y}, o) \in \operatorname{int} C. \tag{10.3}$$

We establish a theorem of the alternative for convex mappings.

Proposition 10.2.1 *If* (A) *and* (10.3) *are satisfied, then precisely one of the following statements is true:*
(a) $\exists x \in K : \quad Sx \in -\operatorname{int} P, \; Tx \in -Q$.
(b) $\exists u \in P^\circ \setminus \{o\} \quad \exists v \in Q^\circ \quad \forall x \in K : \quad \langle u, Sx \rangle + \langle v, Tx \rangle \leq 0$.

Proof. It is obvious that (a) and (b) cannot hold simultaneously. Assuming now that (a) is not satisfied, we have to show that (b) holds. It is easy to see that the set C is convex. Moreover, (10.3) implies $\text{int } C \neq \emptyset$, and since (a) does not hold, we have $(o, o) \notin C$. Hence (o, o) and C can be separated by a closed hyperplane, i.e., there exists $(u, v) \in G^* \times H^*$ such that $\langle u, v \rangle \neq o$ and

$$\langle u, y \rangle + \langle v, z \rangle \leq 0 \quad \forall (y, z) \in C. \tag{10.4}$$

Now let $x \in K$, $p \in \text{int } P$, $q \in Q$, $\alpha > 0$, and $\beta > 0$. It follows that $(Sx + \alpha p, Tx + \beta q) \in C$ and so by (10.4),

$$\langle u, Sx \rangle + \alpha \langle u, p \rangle + \langle v, Tx \rangle + \beta \langle v, q \rangle \leq 0. \tag{10.5}$$

Letting $\alpha \downarrow 0$ and $\beta \downarrow 0$, we obtain $\langle u, Sx \rangle + \langle v, Tx \rangle \leq 0$. Moreover, letting $\alpha \to +\infty$ and $\beta \to +\infty$ in (10.5), we get $\langle u, p \rangle \leq 0$ and $\langle v, q \rangle \leq 0$, respectively. Since this holds for all $p \in \text{int } P$ and all $q \in Q$, it follows that $u \in (\text{int } P)^\circ = P^\circ$ and $v \in Q^\circ$. Assume that $u = o$. Then (10.4) implies $\langle v, z \rangle \leq 0$ for each $(y, z) \in C$. By virtue of (10.3), there exists a neighborhood W of zero in F such that $\{\bar{y}\} \times W \subseteq C$. Hence we obtain $\langle v, z \rangle \leq 0$ for each $z \in W$ and so $v = o$. This is a contradiction to $(u, v) \neq o$. $\qquad\square$

To make Proposition 10.2.1 applicable, we need conditions sufficient for (10.3). A simple condition is available if $\text{int } Q$ is nonempty.

Lemma 10.2.2 *Let* (A) *be satisfied. If* $Tx_0 \in -\text{int } Q$ *for some* $x_0 \in K$, *then* (10.3) *holds.*

Proof. Choose $p_0 \in \text{int } P$ and set $\bar{y} := Sx_0 + p_0$. By assumption there exist zero neighborhoods V in G and W in H such that $p_0 + V \subseteq \text{int } P$ and $-Tx_0 + W \subseteq Q$. For each $y \in V$ and each $z \in W$ we thus obtain

$$(\bar{y} + y) - Sx_0 = p_0 + y \in \text{int } P \quad \text{and} \quad z - Tx_0 \in Q.$$

We conclude that $(\bar{y} + V) \times W \subseteq C$ and so $(\bar{y}, o) \in \text{int } C$. $\qquad\square$

Now we drop the assumption that Q has interior points. With a subset A of K we formulate the following conditions:

$$\exists x_0 \in A : \ S \text{ is continuous at } x_0 \text{ and } Tx_0 \in -Q \cap \text{int}\big(T(A) + Q\big). \tag{10.3a}$$

$$\exists x_0 \in \text{int } A : \ S \text{ is continuous at } x_0 \text{ and } Tx_0 \in -Q \cap \text{int}\big(T(E) + Q\big). \tag{10.3b}$$

Proposition 10.2.3 *Let* (A) *be satisfied. Further let* E *and* H *be Banach spaces and* A *be a closed convex subset of* E *such that* $A \subseteq K$.

(a) *If* $T : K \to H$ *is continuous on* A *and* Q-*convex, then* (10.3a) *implies* (10.3).

(b) *If T is defined, continuous, and Q-convex on all of E, then* (10.3b) *implies* (10.3a) *and so* (10.3).

Proof.

(a) Choose $p_0 \in \operatorname{int} P$ and a neighborhood V of zero in G such that $p_0 + V + V \subseteq \operatorname{int} P$. Further set $\bar{y} := Sx_0 + p_0$. Since S is continuous at x_0, there exists a neighborhood U of zero in E such that $Sx_0 - Sx \in V$ for each $x \in K \cap (x_0 + U)$. Define $\Phi : E \rightrightarrows H$ by

$$\Phi(x) := \begin{cases} Tx + Q & \text{if } x \in A, \\ \emptyset & \text{if } x \in E \setminus A. \end{cases} \tag{10.6}$$

It is easy to see that graph Φ is closed and convex. Hence Theorem 10.1.2 applies to Φ. Therefore, since $Tx_0 \in \operatorname{int} \Phi(A) = \operatorname{int} \operatorname{range} \Phi$, we obtain $Tx_0 \in \operatorname{int}\big(T(\mathring{B}(x,\rho)) + Q\big)$ for each $x \in E$ satisfying $Tx - Tx_0 \in -Q$ and any $\rho > 0$. Hence there exists a neighborhood W of zero in H satisfying

$$Tx_0 + W \subseteq T\big(\mathring{B}(x_0,1)\big) + Q. \tag{10.7}$$

We show that
$$(\bar{y} + V) \times W \subseteq C. \tag{10.8}$$

Thus let $y \in V$ and $z \in W$ be given. By (10.7) there exists $x \in \mathring{B}(x_0, 1)$ such that $Tx_0 + z \in Tx + Q$ and so $z - Tx \in -Tx_0 + Q \subseteq Q + Q = Q$. Moreover, we also have

$$(\bar{y} + y) - Sx = p_0 + (Sx_0 - Sx) + y \in p_0 + V + V \subseteq \operatorname{int} P.$$

This shows that $(\bar{y} + y, z) \in C$. Hence (10.8) is verified and it follows that $(\bar{y}, o) \in \operatorname{int} C$.

(b) Let U be a neighborhood of zero in E with $x_0 + U \subseteq A$. The multifunction Φ defined by (10.6) again satisfies the assumptions of Theorem 10.1.2. Hence there exists a neighborhood W of zero in H such that $Tx_0 + W \subseteq T(x_0 + U) + Q$. Since $x_0 + U \subseteq A$, we obtain $Tx_0 \in \operatorname{int}\big(T(A) + Q\big)$. $\square$

For later use we want to reformulate the special case $Q = \{o\}$ of Proposition 10.2.1, incorporating Proposition 10.2.3(a) with $A := K$. We therefore consider the following assumptions:

($\hat{\mathrm{A}}$) E, G, and H are normed vector spaces, with E and H complete. $K \subseteq E$ is nonempty, convex, and closed, $P \subseteq G$ is a convex cone with $\operatorname{int}(P) \neq \emptyset$. $S : K \to G$ is P-convex and continuous, $T : E \to H$ is linear and continuous.

Proposition 10.2.4 *Let* $(\hat{A})$ *be satisfied and assume that* $T(\mathbb{R}_+ K) = H$. *Then precisely one of the following statements is true:*

(a) $\exists x \in K : \quad Sx \in -\operatorname{int} P,\ Tx = o.$
(b) $\exists u \in P^\circ \setminus \{o\} \quad \exists v \in H^* \quad \forall x \in K : \quad \langle u, Sx \rangle + \langle v, Tx \rangle \leq 0.$

Notice that in the case considered in Proposition 10.2.4, the hypothesis $T(\mathbb{R}_+ K) = H$ implies (and in fact is equivalent to) the second condition of (10.3a).

10.3 Metric Regularity and Linear Openness

Convention. Throughout this section, let E and F denote Banach spaces.

In order to motivate the following, we consider a *generalized equation* of the form

$$h(x) \in Q, \quad x \in E, \tag{10.9}$$

where the mapping $h : E \to F$ and the nonempty subset Q of F are given. We seek $x \in E$ that satisfies (10.9). It turns out that much information on the solvability and the solutions of (10.9) can be obtained by studying the *perturbed generalized equation*

$$h(x) \in Q + y, \quad x \in E \tag{10.10}$$

that depends on the parameter $y \in F$. This leads us to considering the multifunction $\Phi_h : E \rightrightarrows F$ defined by

$$\Phi_h(x) := h(x) - Q, \quad x \in E. \tag{10.11}$$

Notice that (10.10) is equivalent to $y \in \Phi_h(x)$. In other words, for a given $y \in F$ the solution set of (10.10) is $\Phi_h^{-1}(y)$.

The idea of stability is the following. Given $(x, y) \in E \times F$, the distance $d(x, \Phi_h^{-1}(y))$ should be small whenever the distance $d(y, \Phi_h(x))$ is small. For an arbitrary multifunction Φ, not necessarily of the form (10.11), the following definition quantifies this idea.

Definition 10.3.1 *The multifunction* $\Phi : E \rightrightarrows F$ *is said to be* metrically regular *around* $(\bar{x}, \bar{y}) \in \operatorname{graph}(\Phi)$ *if there exist a neighborhood* W *of* $(\bar{x}, \bar{y})$ *and a constant* $\kappa > 0$ *such that*

$$d(x, \Phi^{-1}(y)) \leq \kappa\, d(y, \Phi(x)) \quad \forall (x, y) \in W. \tag{10.12}$$

The constant κ *is called* constant of metric regularity.

Parallel to metric regularity we consider the following concept.

Definition 10.3.2 The multifunction $\Phi : E \rightrightarrows F$ is said to be *open at a linear rate* around $(\bar{x}, \bar{y}) \in \text{graph}(\Phi)$ if there exist a neighborhood W of $(\bar{x}, \bar{y})$ and constants $\rho > 0$ and $\tau_0 > 0$ such that

$$y + \rho\tau\,\mathrm{B}_F \subseteq \Phi(x + \tau\,\mathrm{B}_E) \quad \forall\,(x,y) \in \text{graph}(\Phi) \cap W \quad \forall\,\tau \in [0, \tau_0]. \quad (10.13)$$

The constant ρ is called *linear rate of openness.*

To elucidate this concept we consider a mapping $T : E \rightarrow F$. Observe that T is an open mapping if and only if for any $x \in E$ and any $\tau > 0$ there exists $\sigma(x, \tau) > 0$ such that

$$T(x) + \sigma(x,\tau)\mathrm{B}_F \subseteq T(x + \tau\mathrm{B}_E).$$

In contrast to this, T is open at a linear rate around $(\bar{x}, T(\bar{x}))$ if for any $(x, T(x)) \in W$ and any $\tau \in [0, \tau_0]$, we have

$$T(x) + \rho\tau\mathrm{B}_F \subseteq T(x + \tau\mathrm{B}_E).$$

The latter means that in a neighborhood of $(\bar{x}, T(\bar{x}))$ and for $\tau > 0$ sufficiently small, $\sigma(x, \tau)$ can be chosen to be of the form $\rho\tau$, i.e., independent of x and linear in τ. Instead of "openness at a linear rate" we shall sometimes briefly speak of "linear openness."

Our strategy is as follows. We show that metric regularity and linear openness are equivalent (Theorem 10.3.3). If Φ is convex, linear openness can be characterized, using the generalized open mapping theorem, by the condition

$$\bar{y} \in \text{int range}\,\Phi, \quad (10.14)$$

see Theorem 10.3.5. We then proceed to show that metric regularity is stable under Lipschitz continuous perturbations (Theorem 10.3.6). The latter result will be applied below to characterize tangential approximations of sets of the form $h^{-1}(Q)$, which in turn will be crucial for deriving multiplier rules.

Theorem 10.3.3 *If* $\Phi : E \rightrightarrows F$ *is a multifunction and* $(\bar{x}, \bar{y}) \in \text{graph}(\Phi)$, *then the following assertions are equivalent:*

(a) Φ *is metrically regular around* $(\bar{x}, \bar{y})$ *with constant* $\kappa > 0$.
(b) Φ *is open around* $(\bar{x}, \bar{y})$ *at the linear rate* $\rho = \kappa^{-1}$.

Proof. First observe that Φ is metrically regular around $(\bar{x}, \bar{y})$ with constant κ if and only if the multifunction $\Psi : E \rightrightarrows F$ defined by $\Psi(x) := \Phi(x + \bar{x}) - \bar{y}$ is metrically regular around (o, o) with the same constant κ. An analogous remark applies to linear openness. Therefore we may and do assume that $(\bar{x}, \bar{y}) = (o, o)$.

(a) $\Longrightarrow$ (b): Let $(x, y) \in \text{graph}\, \Phi$ be sufficiently close to (o, o) and let $\tau > 0$ be sufficiently small. Then for each $y' \in F$ satisfying $\|y' - y\| < \kappa^{-1}\tau$ we obtain by (a) that

$$d(x, \Phi^{-1}(y')) \le \kappa\, d(y', \Phi(x)) \le \kappa\|y' - y\| < \tau.$$

Hence there exists $x' \in \Phi^{-1}(y')$ such that $\|x - x'\| < \tau$. This implies $y' \in \Phi(x') \subseteq \Phi(x + \tau\, B_E)$. Since $y' \in y + \kappa^{-1}\tau\, B_F$ was arbitrary, we see that (b) holds.

(b) $\Longrightarrow$ (a): We may assume that W, ρ, and τ_0 in Definition 10.3.2 are such that

$$W = \delta_E\, B_E \times \delta_F\, B_F, \quad \tau_0\rho \le \tfrac{1}{2}\delta_F, \tag{10.15}$$

where δ_E and δ_F are positive constants. Further let ϵ_E and ϵ_F be positive constants satisfying

$$\epsilon_E \le \delta_E, \quad \rho\epsilon_E + \epsilon_F \le \tau_0\rho. \tag{10.16}$$

We show now that (a) holds with $\kappa = \rho^{-1}$. Let $x \in \epsilon_E B_E$ and $y \in \epsilon_F B_F$ be given. Applying (10.13) with (o, o) instead of (x, y) and with $\tau := \rho^{-1}\|y\|$, we conclude that there exists $x' \in \Phi^{-1}(y)$ such that $\|x'\| \le \rho^{-1}\|y\|$. It follows that

$$d(x, \Phi^{-1}(y)) \le \|x - x'\| \le \|x\| + \rho^{-1}\|y\| \le \epsilon_E + \rho^{-1}\epsilon_F.$$

Hence, if $d(y, \Phi(x)) \ge \rho(\epsilon_E + \rho^{-1}\epsilon_F) = \rho\epsilon_E + \epsilon_F$, then the assertion is verified. Assume now that $d(y, \Phi(x)) < \rho(\epsilon_E + \rho^{-1}\epsilon_F)$. Then for each sufficiently small $\alpha > 0$ there exists $y_\alpha \in \Phi(x)$ satisfying

$$\|y - y_\alpha\| \le d(y, \Phi(x)) + \alpha < \rho\epsilon_E + \epsilon_F \underset{(10.16)}{\le} \tau_0\rho. \tag{10.17}$$

It follows that

$$\|y_\alpha\| \le \|y_\alpha - y\| + \|y\| < \tau_0\rho + \epsilon_F \underset{(10.16)}{\le} \tau_0\rho + \tau_0\rho \underset{(10.15)}{\le} \delta_F \tag{10.18}$$

and so $(x, y_\alpha) \in \text{graph}(\Phi) \cap W$. In view of (b) there exists $x' \in \Phi^{-1}(y)$ such that $\|x - x'\| \le \rho^{-1}\|y - y_\alpha\|$. We thus obtain

$$d(x, \Phi^{-1}(y)) \le \|x - x'\| \le \rho^{-1}\|y - y_\alpha\| \underset{(10.17)}{\le} \rho^{-1} d(y, \Phi(x)) + \rho^{-1}\alpha.$$

Letting $\alpha \downarrow 0$ we see that (a) holds with $\kappa = \rho^{-1}$. $\qquad\square$

Under additional assumptions on the multifunction we obtain refined results.

Convex Multifunctions

Proposition 10.3.4 *If the multifunction* $\Phi : E \rightrightarrows F$ *is convex, then for each* $(\bar{x}, \bar{y}) \in \text{graph}\, \Phi$ *the following assertions are equivalent:*

(a) Φ *is open at a linear rate around* $(\bar{x}, \bar{y})$.
(b) *There exist constants* $\mu > 0$ *and* $\nu > 0$ *such that*

$$\bar{y} + \nu \, B_F \subseteq \Phi(\bar{x} + \mu \, B_E). \tag{10.19}$$

More precisely, if (b) *holds, then* (10.13) *holds with the following data:*

$$W = \mu \, B_E \times \tfrac{1}{2} \nu \, B_F, \quad \rho = \frac{\nu}{4\mu}, \quad \tau_0 = 2\mu. \tag{10.20}$$

Proof.
(a) $\Longrightarrow$ (b): Set $\mu = \tau_0$ and $\nu = \rho \tau_0$.
(b) $\Longrightarrow$ (a): We may assume that $\bar{x} = o$ and $\bar{y} = o$. Define W as in (10.20).
Now let $(x, y) \in \mathrm{graph}(\Phi) \cap W$. For each $\tilde{\tau} \in [0,1]$ we obtain

$$\begin{aligned}
y + \tfrac{1}{2}\nu\tilde{\tau} \, B_F &= (1 - \tilde{\tau})y + \tilde{\tau}(y + \tfrac{1}{2}\nu \, B_F) \\
&\subseteq (1 - \tilde{\tau})y + \tilde{\tau}\nu \, B_F \\
&\subseteq (1 - \tilde{\tau})\Phi(x) + \tilde{\tau}\Phi(\mu \, B_E) \quad \text{(by (10.19))} \\
&\subseteq \Phi\big((1 - \tilde{\tau})x + \mu\tilde{\tau} \, B_E\big) \quad \text{(since Φ is convex)} \\
&\subseteq \Phi(x + 2\mu\tilde{\tau} \, B_E).
\end{aligned}$$

Hence setting $\tau := 2\mu\tilde{\tau}$ we see that with ρ and τ_0 as in (10.20), condition (10.13) is satisfied. $\qquad\square$

Now we can establish an important result in the theory of generalized equations.

Theorem 10.3.5 (Stability Theorem) *If* $\Phi : E \rightrightarrows F$ *is a closed convex multifunction, then for each* $(\bar{x}, \bar{y}) \in \mathrm{graph}\,\Phi$ *the following assertions are equivalent:*

(a) Φ *is metrically regular around* $(\bar{x}, \bar{y})$.
(b) Φ *is open at a linear rate around* $(\bar{x}, \bar{y})$.
(c) *There exist constants* $\mu > 0$ *and* $\nu > 0$ *such that* $\bar{y} + \nu \, B_F \subseteq \Phi(\bar{x} + \mu \, B_E)$.
(d) $\bar{y} \in \mathrm{int\,range}\,\Phi$.

In particular, assume that (c) *holds and that* $(x, y) \in E \times F$ *satisfies*

$$\|x - \bar{x}\| < \frac{\mu}{2}, \quad \|y - \bar{y}\| < \frac{\nu}{8}. \tag{10.21}$$

Then one has

$$\mathrm{d}(x, \Phi^{-1}(y)) \leq 4\mu\nu^{-1} \, \mathrm{d}(y, \Phi(x)). \tag{10.22}$$

Proof. Concerning (a) $\Longleftrightarrow$ (b) and (b) $\Longleftrightarrow$ (c), see Theorem 10.3.3 and Proposition 10.3.4, respectively.
(c) $\Longrightarrow$ (d): This is obvious.
(d) $\Longrightarrow$ (c): This follows from Theorem 10.1.2.

Now assume that (c) and (10.21) hold. By Proposition 10.3.4, the multifunction Φ is open at a linear rate, the data according to (10.13) being

$$W = \delta_E\, \mathrm{B}_E \times \delta_F\, \mathrm{B}_F, \quad \delta_E := \mu, \quad \delta_F := \frac{1}{2}\nu, \quad \rho = \frac{\nu}{4\mu}, \quad \tau_0 = 2\mu.$$

Passing from $\tau_0 = 2\mu$ to $\tau_0 = \mu$, we obtain $\tau_0 \rho = \nu/4 = \delta_F/2$ and so (10.15) holds. Moreover, setting $\epsilon_E := \mu/2$ and $\epsilon_F := \nu/8$, we see that (10.16) is also satisfied. The proof of Theorem 10.3.3 therefore shows that we have $\mathrm{d}(x, \Phi^{-1}(y)) \leq \kappa\, \mathrm{d}(y, \Phi(x))$, where $\kappa = \rho^{-1} = 4\mu\nu^{-1}$. □

Multifunctions of the Form $x \mapsto h(x) - Q$

Recall that given a mapping $h : E \to F$ and a nonempty subset Q of F, we denote by $\Phi_h : E \rightrightarrows F$ the multifunction defined by $\Phi_h(x) := h(x) - Q$, $x \in E$. For multifunctions of this type, we now show that metric regularity is stable under small locally Lipschitz continuous perturbations. More precisely, we pass from Φ_h to $\Phi_{\widehat{h}}$, where the Lipschitz constant of $h - \widehat{h}$ is sufficiently small.

Theorem 10.3.6 (Perturbation Theorem) *Assume that*

> $h, \widehat{h} : E \to F$ *are continuous,* $Q \subseteq F$ *is closed and convex,*
> Φ_h *is metrically regular around* $(\bar{x}, \bar{y}) \in \operatorname{graph} \Phi_h$ *with constant* $\kappa > 0$,
> $h - \widehat{h}$ *is locally L-continuous around* $\bar{x}$ *with Lipschitz constant* $\lambda < \kappa^{-1}$.

Then the multifunction $\Phi_{\widehat{h}}$ *is metrically regular around* $(\bar{x}, \bar{y} - h(\bar{x}) + \widehat{h}(\bar{x}))$ *with constant* $\kappa(\lambda) := \kappa(1 - \kappa\lambda)^{-1}$.

Proof.

(I) Define $g(x) := h(x) - \widehat{h}(x)$ for any $x \in E$. By assumption on Φ_h and g there exist positive constants δ_E and δ_F such that

$$\mathrm{d}(x, \Phi_h^{-1}(y)) \leq \kappa\, \mathrm{d}(y, \Phi_h(x)) \quad \text{whenever } x \in \mathrm{B}(\bar{x}, \delta_E),\ y \in \mathrm{B}(\bar{y}, \delta_F), \tag{10.23}$$

$$\|g(x) - g(x')\| \leq \lambda \|x - x'\| \quad \text{whenever } x, x' \in \mathrm{B}(\bar{x}, \delta_E). \tag{10.24}$$

We shall show that there exist positive constants ϵ_E and ϵ_F such that

$$\mathrm{d}(x, \Phi_{\widehat{h}}^{-1}(y)) \leq \kappa(\lambda)\, \mathrm{d}(y, \Phi_{\widehat{h}}(x)) \tag{10.25}$$

whenever

$$\|x - \bar{x}\| < \epsilon_E, \quad \|y - (\bar{y} - g(\bar{x}))\| < \epsilon_F. \tag{10.26}$$

(II) Without loss of generality we may assume that $g(\bar{x}) = o$. Choose $\beta, \epsilon > 0$ such that

$$\kappa\lambda < \beta < 1, \quad (1 + \epsilon)\kappa\lambda < \beta \tag{10.27}$$

and $\epsilon_E, \epsilon_F > 0$ such that

$$\epsilon_E < \delta_E, \quad \epsilon_F + \lambda \epsilon_E < \delta_F. \tag{10.28}$$

Now let $(x, y) \in E \times F$ satisfy (10.26). We construct recursively a sequence (x_k) in E with the following properties:

$$x_1 := x, \quad x_{k+1} \in \Phi_h^{-1}(y + g(x_k)), \quad k = 1, 2, \ldots, \tag{10.29}$$

$$\|x_{k+1} - x_k\| \le (1 + \epsilon) \, \mathrm{d}\big(x_k, \Phi_h^{-1}(y + g(x_k))\big), \quad k = 1, 2, \ldots \tag{10.30}$$

We have

$$\|x - \bar{x}\| < \delta_E, \quad \|y + g(x) - \bar{y}\| \le \|y - \bar{y}\| + \|g(x)\| \underset{(10.24),(10.28)}{<} \delta_F \tag{10.31}$$

and so (10.23) gives

$$\mathrm{d}\big(x, \Phi_h^{-1}(y + g(x))\big) \le \kappa \, \mathrm{d}\big(h(x) - y - g(x), P\big) = \kappa \, \mathrm{d}\big(y, \Phi_{\widehat{h}}(x)\big).$$

Hence there exists $x_2 \in \Phi_h^{-1}(y + g(x))$ satisfying

$$\|x_2 - x_1\| \le (1 + \epsilon)\kappa \, \mathrm{d}(y, \Phi_{\widehat{h}}(x)) \underset{(10.27)}{<} \lambda^{-1}\beta \, \mathrm{d}(y, \Phi_{\widehat{h}}(x)). \tag{10.32}$$

For $\rho > 0$ let $\alpha(\rho)$ denote the modulus of continuity of h at $\bar{x}$, i.e.,

$$\alpha(\rho) := \sup\{\|h(x) - h(\bar{x})\| \mid x \in \mathrm{B}(\bar{x}, \rho)\}.$$

It follows that

$$\mathrm{d}(y, \Phi_{\widehat{h}}(x)) = \mathrm{d}(h(x) - y - g(x), Q) \le \|(h(x) - y - g(x)) - (h(\bar{x}) - \bar{y})\|$$
$$\le \|h(x) - h(\bar{x})\| + \|y - \bar{y}\| + \|g(x)\| \le \alpha(\epsilon_E) + \epsilon_F + \lambda\epsilon_E. \tag{10.33}$$

Concerning the first inequality, notice that $(\bar{x}, \bar{y}) \in \mathrm{graph}(\Phi_h)$ and so $h(\bar{x}) - \bar{y} \in Q$. In view of (10.33), the estimate (10.32) gives

$$\|x_2 - x_1\| < \lambda^{-1}\beta\big(\alpha(\epsilon_E) + \lambda\epsilon_E + \epsilon_F\big). \tag{10.34}$$

Hence, if ϵ_E and ϵ_F (beside satisfying (10.28)) are small enough, then

$$\|x_2 - \bar{x}\| \le \|x_2 - x_1\| + \|x - \bar{x}\| < \delta_E,$$
$$\|(y + g(x_2)) - \bar{y}\| \le \|(y + g(x)) - \bar{y}\| + \|g(x_2) - g(x)\| < \delta_F; \tag{10.35}$$

here the last inequality follows from (10.24) and (10.31).

(III) We now show by induction that ϵ_E and ϵ_F can be made so small that for each $k \in \mathbb{N}$ we have

$$\|x_k - \bar{x}\| < \delta_E, \quad \|(y + g(x_k)) - \bar{y}\| < \delta_F. \tag{10.36}$$

By (10.26) and (10.35) we know that (10.36) holds for $k < k_0$ if $k_0 = 3$. Suppose now that (10.36) holds for all $k < k_0$, where $k_0 \geq 3$. We have to show that it also holds for k_0. For each $k \in \{2, \ldots, k_0 - 1\}$ we have

$$\|x_{k+1} - x_k\| \underset{(10.30)}{<} \kappa^{-1}\lambda^{-1}\beta\, \mathrm{d}\big(x_k, \Phi_h^{-1}(y + g(x_k))\big). \tag{10.37}$$

We further obtain

$$\begin{aligned} \mathrm{d}\big(x_k, \Phi_h^{-1}(y + g(x_k))\big) &\leq \kappa\, \mathrm{d}\big(y + g(x_k), \Phi_h(x_k)\big) \\ &\leq \kappa\,\|(y + g(x_k)) - (y + g(x_{k-1}))\| \leq \kappa\lambda\|x_k - x_{k-1}\|; \end{aligned} \tag{10.38}$$

here the second inequality holds since $x_k \in \Phi_h^{-1}(y+g(x_{k-1}))$ (see (10.29)) and so $y+g(x_{k-1}) \in \Phi_h(x_k)$. Inserting (10.38) into (10.37) yields $\|x_{k+1} - x_k\| < \beta\|x_k - x_{k-1}\|$ and so

$$\|x_{k+1} - x_k\| < \beta^{k-1}\|x_2 - x_1\|. \tag{10.39}$$

From this we get

$$\|x_{k_0} - x_1\| \leq \|x_{k_0} - x_{k_0-1}\| + \cdots + \|x_2 - x_1\| \leq \frac{1}{1-\beta}\|x_2 - x_1\|$$

$$\underset{(10.32)}{\leq} \frac{\beta}{\lambda(1-\beta)}\, \mathrm{d}(y, \Phi_h(x)), \tag{10.40}$$

which implies

$$\|x_{k_0} - \bar{x}\| \leq \|x_{k_0} - x_1\| + \|x_1 - \bar{x}\| < \frac{\beta}{\lambda(1-\beta)}\, \mathrm{d}(y, \Phi_h(x)) + \epsilon_E. \tag{10.41}$$

Moreover, we have

$$\begin{aligned} \|y + g(x_{k_0}) - \bar{y}\| &\leq \|y - \bar{y}\| + \|g(x_{k_0}) - g(\bar{x})\| \\ &\leq \|y - \bar{y}\| + \lambda\|x_{k_0} - \bar{x}\| \leq \epsilon_F + \lambda\epsilon_E + \beta(1-\beta)^{-1}\, \mathrm{d}(y, \Phi_h(x)). \end{aligned} \tag{10.42}$$

The inequalities (10.41) and (10.42) together with (10.33) show that there exists $\delta > 0$, independent of k, such that (cf. (10.36))

$$x_k \in \mathring{\mathrm{B}}(\bar{x}, \delta), \quad y + g(x_k) \in \mathring{\mathrm{B}}(\bar{y}, \delta) \quad \text{for } k = 1, 2, \ldots \tag{10.43}$$

(IV) From (10.39) and $0 < \beta < 1$ we conclude that (x_k) is a Cauchy sequence and so, since E is complete, is convergent to some $x' \in B(\bar{x}, \delta)$. Since g is continuous, we have $g(x_k) \to g(x')$ as $k \to \infty$. This, (10.29), and the closedness of graph Φ_h imply $(x', y + g(x')) \in \text{graph}\,\Phi_h$. Since $\Phi_h^{-1}(y + g(x')) = \Phi_{\widehat{h}}^{-1}(y)$, it follows that $x' \in \Phi_{\widehat{h}}^{-1}(y)$. We thus obtain

$$d\big(x, \Phi_{\widehat{h}}^{-1}(y)\big) \le \|x_1 - x'\| \underset{(10.40)}{\le} \lambda^{-1}(1 - \beta)^{-1}\beta\, d(y, \Phi_{\widehat{h}}(x)).$$

Letting $\beta \downarrow \kappa\lambda$ (notice (10.27)), we finally see that (10.25) holds for all $x \in \mathring{B}(\bar{x}, \delta)$ and all $y \in \mathring{B}(\bar{y} - g(\bar{x}), \delta)$. $\square$

Now assume that, in addition,

$$h : E \to F \text{ is continuously differentiable at } \bar{x} \in h^{-1}(Q).$$

Then h is F-differentiable on a nonempty open subset U of E containing $\bar{x}$. Let $\rho > 0$ be such that $B(\bar{x}, \rho) \subseteq U$. Moreover, let $\widehat{h} : E \to F$ be the linearization of h at $\bar{x}$, i.e.,

$$\widehat{h}(x) := h(\bar{x}) + h'(\bar{x})(x - \bar{x}), \quad x \in E.$$

Parallel to

$$\Phi_h(x) := h(x) - Q, \quad x \in E, \tag{10.44}$$

we consider the linearized multifunction

$$\Phi_{\widehat{h}}(x) := \widehat{h}(x) - Q = h(\bar{x}) + h'(\bar{x})(x - \bar{x}) - Q, \quad x \in E. \tag{10.45}$$

Since the function $\widehat{h}$ is continuous and affine, the linearized multifunction $\Phi_{\widehat{h}}$ is closed and convex. Hence by Theorem 10.3.5, $\Phi_{\widehat{h}}$ is metrically regular at $(\bar{x}, o)$ if and only if $o \in \text{int range}\,\Phi_{\widehat{h}}$ which is equivalent to

$$o \in \text{int}\big(h(\bar{x}) + h'(\bar{x})(E) - Q\big). \tag{10.46}$$

By the mean value theorem (Proposition 3.3.4), we have

$$\big\|\big(h(x) - \widehat{h}(x)\big) - \big(h(x') - \widehat{h}(x')\big)\big\| = \big\|\big(h(x) - h(x')\big) - h'(\bar{x})(x - x')\big\|$$
$$\le \lambda\|x - \bar{x}\| \quad \text{for all } x, x' \in B(\bar{x}, \rho),$$

where

$$\lambda := \max\{\|h'(x) - h'(\bar{x})\| \mid x \in B(\bar{x}, \rho)\}.$$

Hence the function $h - \widehat{h}$ is locally Lipschitz continuous at $\bar{x}$, where the Lipschitz constant can be made sufficiently small by making ρ small enough. Therefore Theorem 10.3.6 shows that Φ_h is also metrically regular. Conversely, if Φ_h is metrically regular at $(\bar{x}, o)$, then by Theorem 10.3.6 the perturbed multifunction $\Phi_{\widehat{h}}$ is also metrically regular at $(\bar{x}, o)$, which in turn is equivalent to (10.46) by Theorem 10.3.5. Thus we have proved the following.

Proposition 10.3.7 *Let $Q \subseteq F$ be closed and convex, assume that $h : E \to F$ is continuous on E and continuously differentiable at $\bar{x} \in h^{-1}(Q)$. Then the following assertions are equivalent:*

(a) Φ_h *(see (10.44)) is metrically regular at $(\bar{x}, o)$.*
(b) $\Phi_{\hat{h}}$ *(see (10.45)) is metrically regular at $(\bar{x}, o)$.*
(c) *The condition (10.46) holds.*

We want to establish conditions that are at least sufficient for (10.46) or, in view of later applications, for a slightly more general condition. Recall that cr M denotes the core of the set $M \subseteq F$.

Let A be a nonempty subset of E. We consider the *Robinson condition* (see [176])

$$o \in \text{int}\big(h(\bar{x}) + h'(\bar{x})(A - \bar{x}) - Q\big), \tag{10.47}$$

the *core condition*

$$o \in \text{cr}\big(h(\bar{x}) + h'(\bar{x})(A - \bar{x}) - Q\big), \tag{10.48}$$

and the *Zowe–Kurcyusz condition* (see [229])

$$h'(\bar{x})\big(\mathbb{R}_+(A - \bar{x})\big) - \mathbb{R}_+\big(Q - h(\bar{x})\big) = F. \tag{10.49}$$

Proposition 10.3.8 *Assume that h, Q, and $\bar{x}$ are as in Proposition 10.3.7. Assume further that A is a nonempty closed convex subset of E. Then the Robinson condition (10.47), the core condition (10.48), and the Zowe–Kurcyusz condition (10.49) are mutually equivalent.*

Proof. The implication (10.47) $\Longrightarrow$ (10.48) is obvious and the implication (10.48) $\Longrightarrow$ (10.49) is immediately verified.
(10.49) $\Longrightarrow$ (10.47): Define $\Psi : E \times \mathbb{R} \times \mathbb{R} \rightrightarrows F$ by

$$\Psi(x, \sigma, \tau) := \begin{cases} h'(\bar{x})\big(\sigma(x - \bar{x})\big) - \tau(Q - h(\bar{x})) & \text{if } \sigma \geq 0, \ \tau \geq 0, \ x \in A, \\ \emptyset & \text{otherwise.} \end{cases}$$

Then the multifunction Ψ is closed and convex and

$$\text{range } \Psi = h'(\bar{x})\big(\mathbb{R}_+(A - \bar{x})\big) - \mathbb{R}_+\big(Q - h(\bar{x})\big).$$

The condition (10.49), therefore, implies $o \in \text{int range } \Psi$. By the generalized open mapping theorem (Theorem 10.1.2) we see that $o \in \text{int } \Psi\big((\mathbb{R}_+(A - \bar{x}) \cap B_E) \times [0, 1] \times [0, 1]\big)$ and so (10.47) is satisfied. $\qquad\square$

10.4 Openness Bounds of Multifunctions

The following concept will turn out to be a suitable tool for the further study of multifunctions.

Definition 10.4.1 Let the multifunction $\Phi : E \rightrightarrows F$ be open at a linear rate around $(\bar{x}, \bar{y}) \in \operatorname{graph} \Phi$. Then
$\operatorname{ope}(\Phi)(\bar{x}, \bar{y}) :=$ supremum of all linear rates of openness of Φ around $(\bar{x}, \bar{y})$
is called *openness bound* of Φ around $(\bar{x}, \bar{y})$.

We shall consider two classes of multifunctions for which the openness bound can easily be computed.

Processes

A multifunction $\Phi : E \rightrightarrows F$ is called *(convex) process* if $\operatorname{graph}(\Phi)$ is a (convex) cone. Notice that if Φ is a process, so is Φ^{-1}. A process is said to be *bounded* if it maps bounded sets into bounded sets. If Φ is a bounded convex process, then the (finite) expression

$$\|\Phi\| := \sup_{x \in \mathrm{B} \cap \operatorname{Dom} \Phi} \ \inf_{y \in \Phi(x)} \|y\| \qquad (10.50)$$

is called *norm* of Φ.

Proposition 10.4.2 Let $\Phi : E \rightrightarrows F$ be a bounded convex process. If Φ is open at a linear rate around (o, o), then $0 < \|\Phi^{-1}\| < +\infty$ and

$$\operatorname{ope}(\Phi)(o, o) = \frac{1}{\|\Phi^{-1}\|}. \qquad (10.51)$$

Proof.

(I) We verify $\|\Phi^{-1}\| < +\infty$. Let $\rho > 0$ and τ_0 be openness parameters of Φ around (o, o). Then it follows that $\rho \tau \mathrm{B}_F \subseteq \Phi(\tau \mathrm{B}_E)$ for all $\tau \in [0, \tau_0]$. Since Φ is a process, we have $\Phi(\tau \mathrm{B}_E) = \tau \Phi(\mathrm{B}_E)$. Therefore

$$\rho \mathrm{B}_F \subseteq \Phi(\mathrm{B}_E). \qquad (10.52)$$

We conclude that $\|\Phi^{-1}\| \leq 1/\rho$.

(II) Next we show that $\|\Phi^{-1}\| > 0$. Assume $\|\Phi^{-1}\| = 0$. Choose some $y_0 \in \mathrm{B}_F \cap \operatorname{Dom}(\Phi^{-1})$ such that $y_0 \neq o$ (which exists by (10.52)). For each $n \in \mathbb{N}$ there exists $x_n \in \Phi^{-1}(y_0)$ such that $\|x_n\| < 1/n$. It follows that $n x_n \in \mathrm{B}_E$ and $n y_0 \in \Phi(n x_n)$. Hence $\Phi(\mathrm{B}_E)$ is not bounded, which contradicts the hypothesis on Φ.

(III) Now we verify (10.51). As shown in step (I), if ρ is a linear rate of openness, then (10.52) holds. Conversely, if (10.52) holds with some $\rho > 0$, then

$$y + \rho \tau \mathrm{B}_F \subseteq \Phi(x) + \Phi(\tau \mathrm{B}_E) \subseteq \Phi(x + \tau \mathrm{B}_E) \quad \forall (x, y) \in \operatorname{graph} \Phi \ \forall \tau \geq 0.$$

Hence we obtain

$$\text{ope}(\varPhi)(o, o) = \sup\{\rho > 0 \mid \rho B_F \subseteq \varPhi(B_E)\}. \tag{10.53}$$

It is evident that (10.52) is equivalent to $\inf_{x \in \varPhi^{-1}(y)} \|x\| \leq \frac{1}{\rho}$ for any $y \in B_F$. From this and (10.53) we deduce (10.51). $\square$

Single-Valued Multifunctions

Recall that a functional $f : E \to \mathbb{R}$ is identified with the single-valued multifunction $\tilde{f} : E \rightrightarrows \mathbb{R}$ defined by $\tilde{f}(x) := \{f(x)\}$, $x \in E$. In the following we shall say that f is open at a linear rate around $(\bar{x}, f(\bar{x}))$ if $\tilde{f}$ has this property, and we write $\text{ope}(f)(\bar{x})$ instead of $\text{ope}(\tilde{f})(\bar{x}, \tilde{f}(\bar{x}))$.

Proposition 10.4.3 *Let $f : E \to \mathbb{R}$ be locally L-continuous around $\bar{x} \in E$ and assume that $o \notin \partial_o f(\bar{x})$. Then f is open at a linear rate around $(\bar{x}, f(\bar{x}))$ and*

$$\text{ope}(f)(\bar{x}) = - \inf_{\|y\|=1} f^\circ(\bar{x}, y). \tag{10.54}$$

Proof.

(I) First notice that $p := -\inf_{\|y\|=1} f^\circ(\bar{x}, y)$ is a positive real number. In fact, since $o \notin \partial_o f(\bar{x})$, we have $f^\circ(\bar{x}, y_0) < 0$ for some $y_0 \in E$, where we may assume that $\|y_0\| = 1$. Hence $p > 0$. Since $f^\circ(\bar{x}, \cdot)$ is globally L-continuous (Theorem 7.3.2), we have in particular $f^\circ(\bar{x}, y) \geq -\lambda \|y\|$ for each $y \in E$, where $\lambda > 0$ is a local Lipschitz constant of f at $\bar{x}$. Therefore $p \leq \lambda < +\infty$.

(II) We show that $\text{ope}(f)(\bar{x}) \geq p$. Let $\epsilon \in (0, p/4]$ be given. Then there exists $y_\epsilon \in E$ such that $\|y_\epsilon\| = 1$ and $f^\circ(\bar{x}, y_\epsilon) \leq -p + \frac{\epsilon}{2}$. Moreover, by the definition of $f^\circ(\bar{x}, y_\epsilon)$, there exists $\tau_1 \in (0, 1]$ such that for all $x \in \bar{x} + \tau_1 B_E$ and all $\tau \in (0, \tau_1]$,

$$\frac{f(x + \tau y_\epsilon) - f(x)}{\tau} \leq -p + \epsilon,$$

and so $f(x + \tau y_\epsilon) \leq f(x) - \tau(p - \epsilon) < f(x)$. Applying the intermediate value theorem to the restriction of f to $x + \tau B_E$, we obtain

$$[f(x) - \tau(p - \epsilon), \ f(x)] \subseteq f(x + \tau B_E). \tag{10.55}$$

Recall that $(-f)^\circ(\bar{x}, -y_\epsilon) = f^\circ(\bar{x}, y_\epsilon)$ (Theorem 7.3.2). Therefore, a similar argument with f and y_ϵ replaced by $-f$ and $-y_\epsilon$, respectively, shows that there exists $\tau_2 \in (0, 1]$ such that for all $x \in \bar{x} + \tau_2 B_E$ and all $\tau \in (0, \tau_2]$,

$$[f(x), \ f(x) + \tau(p - \epsilon)] \subseteq f(x + \tau B_E). \tag{10.56}$$

Set $\tau_0 := \min\{\tau_1, \tau_2\}$. Combining (10.55) and (10.56), we obtain

$$f(x) + (p - \epsilon) \cdot \tau B_E \subseteq f(x + \tau B_E) \quad \forall\, x \in \bar{x} + \tau_0 B_E \quad \forall\, \tau \in (0, \tau_0],$$

which shows that $p - \epsilon$ is an openness rate of f around $(\bar{x}, f(\bar{x}))$. Letting $\epsilon \downarrow 0$, we conclude that $\mathrm{ope}(f)(\bar{x}) \geq p$. In particular, we see that f is open with a linear rate around $(\bar{x}, f(\bar{x}))$.

(III) Now we show that $\mathrm{ope}(f)(\bar{x}) \leq p$. This together with step (II) verifies (10.54). Let $\rho > 0$ be a linear rate of openness of f around $(\bar{x}, f(\bar{x}))$. Then there exist a neighborhood W of $(\bar{x}, f(\bar{x}))$ and $\tau_0 > 0$ such that

$$f(x) + \rho \tau \eta \in f(x + \tau B_E) \quad \forall\, (x, f(x)) \in W \quad \forall\, \tau \in [0, \tau_0] \quad \forall\, \eta \in [-1, 1].$$

In particular, there exists $y_1 \in B_E$ such that

$$f(x) - \rho \tau = f(x + \tau y_1) \quad \forall\, (x, f(x)) \in W \quad \forall\, \tau \in [0, \tau_0].$$

It is clear that $y_1 \neq o$. By adjusting τ_0 if necessary, we may assume that $\|y_1\| = 1$. It follows that $-\rho = f^\circ(\bar{x}, y_1) \geq -p$ and so $\mathrm{ope}(f)(\bar{x}) \leq p$. $\square$

10.5 Weak Metric Regularity and Pseudo-Lipschitz Continuity

We introduce and study two concepts that are closely related to metric regularity and linear openness.

Definition 10.5.1

(a) The multifunction $\Phi : E \rightrightarrows F$ is said to be *weakly metrically regular* around $(\bar{x}, \bar{y}) \in \mathrm{graph}\, \Phi$, if there exist a neighborhood U of $\bar{x}$, a neighborhood V of $\bar{y}$, and a constant $\kappa > 0$ such that for all $x \in U$ with $\Phi(x) \cap V \neq \emptyset$ and all $y \in V$ one has

$$d(x, \Phi^{-1}(y)) \leq \kappa\, d(y, \Phi(x)). \tag{10.57}$$

(b) The multifunction $\Psi : F \rightrightarrows E$ is said to be *pseudo-Lipschitz* (or to have the *Aubin property*) around $(\bar{y}, \bar{x}) \in \mathrm{graph}\, \Psi$, if there exist a neighborhood U of $\bar{x}$, a neighborhood V of $\bar{y}$, and a (Lipschitz) constant $\lambda > 0$ such that

$$\Psi(y) \cap U \neq \emptyset \quad \forall\, y \in V \quad \text{and} \tag{10.58}$$
$$\Psi(y) \cap U \subseteq \Psi(y') + \lambda \|y - y'\| B_E \quad \forall\, y, y' \in V. \tag{10.59}$$

It is obvious that a metrically regular multifunction is weakly metrically regular. A mapping $T : F \to E$ which is pseudo-Lipschitz around $(\bar{y}, T(\bar{y}))$, where U can be chosen to be E, is locally Lipschitz continuous around $\bar{y}$.

Theorem 10.5.2 *If $\Phi : E \rightrightarrows F$ is a closed-valued multifunction and $(\bar{x}, \bar{y}) \in \mathrm{graph}\, \Phi$, then the following assertions are equivalent:*

(a) Φ is weakly metrically regular around $(\bar{x}, \bar{y})$.

(b) Φ^{-1} is pseudo-Lipschitz around $(\bar{y}, \bar{x})$.

Proof. (a) $\Longrightarrow$ (b): Let the condition of Definition 10.5.1(a) be satisfied, where we may assume that $U := \bar{x} + \epsilon_1 B_E$ and $V := \bar{y} + \epsilon_2 B_F$ with $\epsilon_1 > 0$ and $\epsilon_2 > 0$. Since $\bar{x} \in U$ and $\bar{y} \in \Phi(\bar{x}) \cap V$, we have by (10.57) that

$$d(\bar{x}, \Phi^{-1}(y)) \le \kappa \, d(y, \Phi(\bar{x})) \le \kappa \|y - \bar{y}\| \quad \forall y \in V. \qquad (10.60)$$

Choose a positive number $\tilde{\epsilon}_2 < \epsilon_2$ so small that $\kappa \tilde{\epsilon}_2 \le \epsilon_1$. Set $\tilde{U} := \bar{x} + \kappa \epsilon_2 B_E$ and $\tilde{V} := \bar{y} + \tilde{\epsilon}_2 B_F$. If $y \in \tilde{V}$, then $y \in V$ and (10.60) gives $d(\bar{x}, \Phi^{-1}(y)) \le \kappa \tilde{\epsilon}_2 < \kappa \epsilon_2$. Hence there exists $x \in \Phi^{-1}(y)$ such that $\|x - \bar{x}\| < \kappa \epsilon_2$. This shows that $\Phi^{-1}(y) \cap \tilde{U} \ne \emptyset$ for any $y \in \tilde{V}$. Now let $y, y' \in \tilde{V}$ be given. For any $x \in \Phi^{-1}(y) \cap \tilde{U}$ it follows from (10.57) that

$$d(x, \Phi^{-1}(y')) \le \kappa \, d(y', \Phi(x)).$$

If $\lambda > \kappa$, we thus find some $x' \in \Phi^{-1}(y') \cap \tilde{U}$ satisfying

$$\|x - x'\| \le \lambda \, d(y', \Phi(x)) \le \lambda \|y' - y\|.$$

In this connection notice that if $d(y', \Phi(x)) = 0$, then $y' \in \Phi(x)$ (as $\Phi(x)$ is closed) and so we may take $x' := x$. We have thus shown that (b) holds.

(b) $\Longrightarrow$ (a): Let $\epsilon > 0$ be such that $U := \bar{x} + \epsilon B_E$ and $V := \bar{y} + \epsilon B_F$ satisfy the condition of Definition 10.5.1(b). Set $\tilde{V} := \bar{y} + (\epsilon/3) B_F$. By (10.58) there exists a neighborhood $\tilde{U}$ of $\bar{x}$ such that $\tilde{U} \subseteq U$ and $\Phi^{-1}(y) \cap \tilde{U} \ne \emptyset$ for all $y \in \tilde{V}$. We also have

$$d(y, \Phi(x)) = d(y, \Phi(x) \cap \tilde{V}) \quad \forall x \in \tilde{U} \quad \forall y \in \tilde{V}. \qquad (10.61)$$

To verify this equation, let $y' \in \Phi(x) \setminus \tilde{V}$ be given. Then

$$d(y, y') \ge d(y', \bar{y}) - d(\bar{y}, y) > \epsilon - \frac{\epsilon}{3} = \frac{2}{3}\epsilon.$$

On the other hand, for any $z \in \Phi(x) \cap \tilde{V}$ we obtain

$$d(y, z) \le d(y, \bar{y} + d(\bar{y}, z) \le \frac{\epsilon}{3} + \frac{\epsilon}{3} = \frac{2}{3}\epsilon.$$

Now let $x \in \tilde{U}$ be such that $\Phi(x) \cap \tilde{V} \ne \emptyset$ and let $y \in \tilde{V}$. By (10.59) we have

$$d(x, \Phi^{-1}(y)) \le \lambda \, d(y, \Phi^{-1}(x) \cap U) \le \lambda \|y - z\| \quad \forall z \in \Phi(x) \cap \tilde{V}$$

and so

$$d(x, \Phi^{-1}(y)) \le \lambda \, d(y, \Phi(x) \cap \tilde{V}) \underset{(10.61)}{=} \lambda \, d(y, \Phi(x)),$$

which completes the proof. $\qquad \qquad \square$

10.6 Linear Semiopenness and Related Properties

We modify the concepts introduced in the preceding sections. Assume again that E and F are Banach spaces, $\Phi : E \rightrightarrows F$ is a multifunction and $(\bar{x}, \bar{y}) \in \operatorname{graph} \Phi$.

Definition 10.6.1

(a) The multifunction Φ is said to be *linearly semiopen* around $(\bar{x}, \bar{y})$ if there exist numbers $\rho > 0$ and $\tau_0 > 0$ such that for all $(x, y) \in \operatorname{graph} \Phi$ satisfying $x \in \bar{x} + \tau_0 B_E$ and $y \in \bar{y} + \tau_0 B_F$ and all $\tau \in [0, \tau_0]$ one has

$$y + \rho \tau \|x - \bar{x}\| B_F \subseteq \Phi(x + \tau \|x - \bar{x}\| B_E). \tag{10.62}$$

(b) The multifunction Φ is said to be *metrically semiregular* around $(\bar{x}, \bar{y})$ if there exist numbers $\kappa > 0$ and $\tau_1 > 0$ such that for all $(x, y) \in E \times F$ satisfying $x \in \bar{x} + \tau_1 B_E$, $y \in \bar{y} + \tau_1 B_F$, and $\operatorname{d}(y, \Phi(x)) \le \tau_1 \|x - \bar{x}\|$ one has

$$\operatorname{d}(x, \Phi^{-1}(y)) \le \kappa \operatorname{d}(y, \Phi(x)). \tag{10.63}$$

(c) The multifunction Φ is said to be *semi-pseudo-Lipschitz* around $(\bar{x}, \bar{y})$ if there exist numbers $\lambda > 0$ and $\tau_2 > 0$ such that for all $(x, y) \in \operatorname{graph} \Phi$ satisfying $x \in \bar{x} + \tau_2 B_E$ and $y \in \bar{y} + \tau_2 B_F$ there is a neighborhood $V(y)$ of y such that

$$\Phi(x) \cap V(y) \subseteq \Phi(x') + \lambda \|x - x'\| B_F \quad \forall x' \in x + \tau_2 \|y - \bar{y}\| B_E. \tag{10.64}$$

The numbers $\rho > 0$ and $\tau_0 > 0$ in Definition 10.6.1(a) will be referred to as *semiopenness parameters*. An analogous terminology will be applied for the numbers associated with metric semiregularity and semi-pseudo-Lipschitz continuity.

Remark 10.6.2 It is obvious that linear openness implies linear semiopenness. In fact, if Φ is open around $(\bar{x}, \bar{y})$ with the linear rate ρ' and the parameter τ_0', then Φ is linearly semiopen around $(\bar{x}, \bar{y})$ with the semiopenness parameters $\rho = \rho'$ and $\tau_0 = \min\{\tau_0', \sqrt{\tau_0'}\}$. The norm functional on E, interpreted as a multifunction from E to $\mathbb{R}$, is not open at a linear rate around $(0,0)$ (it is not even an open mapping) but is linearly semiopen around $(0,0)$. Observe that, in contrast to openness at a linear rate, linear semiopenness does not impose a condition on the multifunction at the reference point $(\bar{x}, \bar{y}) \in \operatorname{graph} \Phi$.

Theorem 10.6.3 *If* $\Phi : E \rightrightarrows F$ *is a closed-valued multifunction and* $(\bar{x}, \bar{y}) \in \operatorname{graph} \Phi$, *then the following assertions are mutually equivalent:*

(a) Φ *is linearly semiopen around* $(\bar{x}, \bar{y})$.
(b) Φ *is metrically semiregular around* $(\bar{x}, \bar{y})$.
(c) Φ^{-1} *is semi-pseudo-Lipschitz around* $(\bar{y}, \bar{x})$.

Proof. As in the proof of Theorem 10.3.3 we may assume that $(\bar{x}, \bar{y}) = (o, o)$. The parameters used in the following refer to the respective definition.
(a) $\Longrightarrow$ (b): Choose $\tau_1 > 0$ so small that

$$\tau_1 < \min\left\{\frac{\tau_0}{1 + \rho\tau_0}, \rho\tau_0\right\}.$$

Let $(x, y) \in \tau_1 B_E \times \tau_1 B_F$ be such that $d(y, \Phi(x)) \leq \tau_1\|x\|$. We distinguish two cases:

Case 1. First assume that $d(y, \Phi(x)) > 0$. Then $x \neq o$. Define τ by $d(y, \Phi(x)) = \rho\tau\|x\|$. Then for any $\delta > 0$ there exists $y' \in \Phi(x)$ such that $\|y - y'\| \leq \rho(\tau + \delta)\|x\|$ and so

$$\|y'\| \leq \|y\| + \|y - y'\| \leq \tau_1 + \rho(\tau + \delta)\tau_1.$$

Since $\rho\tau \leq \tau_1$ the choice of τ_1 shows that $\tau < \tau_0$. Hence we find $\delta = \delta(\tau) > 0$ such that $\tau + \delta(\tau) < \tau_0$. We thus obtain

$$x \in \tau_0 B_E, \quad y' \in \tau_0 B_F, \quad y \in y' + \rho\tau_0\|x\|B_F.$$

Since (a) holds, we conclude from (10.62), with y replaced by y', that there exists $x' \in x + (\tau + \delta(\tau))\|x\|B_E$ such that $y \in \Phi(x')$. Consequently,

$$d(x, \Phi^{-1}(y)) \leq \|x - x'\| \leq (\tau + \delta(\tau))\|x\| \leq \frac{\tau + \delta(\tau)}{\rho\tau} d(y, \Phi(x)).$$

Since this holds for all sufficiently small $\delta(\tau)$, we conclude that any $\kappa > \rho^{-1}$ satisfies (10.63).

Case 2. Assume now that $d(y, \Phi(x)) = 0$. Then $y \in \Phi(x)$ as $\Phi(x)$ is closed. Therefore we have $d(x, \Phi^{-1}(y) = 0 \leq \kappa d(y, \Phi(x))$ with the same κ as above.

(b) $\Longrightarrow$ (c): Choose $\tau_2 > 0$ such that $\tau_2 + \tau_2^2 < \tau_1/2$ and take any $\lambda > \kappa$. Fix any $(y, x) \in \operatorname{graph}(\Phi^{-1})$ with $x \in \tau_2 B_E$ and $y \in \tau_2 B_F$. Further let $y' \in y + \tau_2\|x\|B_F$ be given. We are going to show that with some neighborhood $U(x)$ of x, we have

$$\Phi^{-1}(y) \cap U(x) \subseteq \Phi^{-1}(y') + \lambda\|y - y'\|B_E. \tag{10.65}$$

If $x \neq o$, define $U(x) := x + \frac{1}{2}\|x\|B_E$. Now let $x' \in \Phi^{-1}(y) \cap U(x)$ be given. Then we have the estimates

$$\|x'\| \leq \|x\| + \|x - x'\| \leq \tau_2 + \tfrac{1}{2}\tau_2 \leq \frac{3}{4}\tau_1 < \tau_1,$$

$$\|y'\| \leq \|y\| + \tau_2\|x\| \leq (1 + \tau_2)\tau_2 < \tau_1,$$

$$d(y', \Phi(x')) \leq \|y' - y\| \leq \tau_2\|x\| \leq 2\tau_2\|x'\| \leq \tau_1\|x'\|.$$

Since (b) holds, we can apply (10.57) with x replaced by x' to deduce

$$\mathrm{d}(x', \Phi^{-1}(y')) \leq \kappa \, \mathrm{d}(y', \Phi(x')).$$

Since $\lambda > \kappa$, there exists $x'' \in \Phi^{-1}(y')$ satisfying

$$\|x' - x''\| \leq \lambda \, \mathrm{d}(y', \Phi(x')) \leq \lambda \, \|y - y'\|.$$

It follows that $x' \in \Phi^{-1}(y') + \lambda \, \|y - y'\| B_E$. If $x = o$, then $y + \tau_2 \|x\| B_F = \{y\}$, and (10.65) holds with $U(x) := E$.

(c) $\implies$ (a): Set $\rho := 1/\lambda$ and $\tau_0 := \tau_2 \lambda$. Fix any $(x, y) \in \operatorname{graph} \Phi$ with $x \in \tau_0 B_E$ and $y \in \tau_0 B_F$. If $x = o$, there is nothing to prove. Assume now that $x \neq o$. It suffices to show that for any $\tau \in (0, \tau_0]$ the implication

$$\|y' - y\| = \rho \tau \|x\| \implies y' \in \Phi(x + \tau \|x\| B_E)$$

holds true. From $\|y' - y\| = \rho \tau \|x\|$ we conclude $y' \in y + \tau_2 \|x\| B_F$. Now (c) gives

$$x \in \Phi^{-1}(y) \cap U(x) \subseteq \Phi^{-1}(y') + \lambda \, \|y - y'\| B_E.$$

Hence there exists $x' \in \Phi^{-1}(y')$ such that $x \in x' + \lambda \, \|y - y'\| B_E$ and so $x' \in x + \lambda \, \|y - y'\| B_E$. It follows that

$$y' \in \Phi(x + \lambda \, \|y - y'\| B_E) = \Phi(x + \tau \|x\| B_E). \qquad \square$$

Next we show that linear semiopenness is stable under local Lipschitz perturbations (cf. Theorem 10.3.6).

Theorem 10.6.4 *Assume that $\Phi : E \rightrightarrows F$ is closed and linearly semiopen around $(\bar{x}, \bar{y}) \in \operatorname{graph} \Phi$ with semiopenness rate $\rho > 0$ and that $\Psi : E \to F$ is locally L-continuous around $\bar{x}$ with constant $\lambda < \rho$. Then $\Phi + \Psi$ is linearly semiopen around $(\bar{x}, \bar{y} + \Psi(\bar{x}))$ with semiopenness rate $\rho - \lambda$.*

Proof. Without loss of generality we can again assume that $(\bar{x}, \bar{y}) = (o, o)$ and that $\Psi(o) = o$. Beside ρ let τ_0 be a semiopenness parameter of Φ. Then for all $(x, y) \in \operatorname{graph} \Phi$ with $x \in \tau_0 B_E$, $y \in \tau_0 B_F$ and all $\tau \in (0, \tau_0]$ we have

$$y + \rho \tau \|x\| B_F \subseteq \Phi(x + \tau \|x\| B_E). \tag{10.66}$$

Choose $\tau_1 > 0$ such that

$$\tau_1 \leq \min \left\{ \frac{2\tau_0}{3(1 + \lambda)}, \, \frac{1}{2}, \, \frac{\tau_0}{2}, \, \frac{1}{2(\rho - \lambda)} \right\}$$

and that Ψ is Lipschitz continuous on $\tau_1 B_E$ with Lipschitz constant λ. Now let $(x_0, y_0) \in \operatorname{graph}(\Phi + \Psi)$ with $x_0 \in \tau_1 B_E$ and $y_0 \in \tau_1 B_F$ be given. We shall show that for any $\tau \in (0, \tau_1]$ we have

$$y_0 + (\rho - \lambda)\tau \|x_0\| B_F \subseteq (\Phi + \Psi)(x_0 + \tau \|x_0\| B_E). \tag{10.67}$$

This is clear if $x = o$. Now let $x \neq o$. Take any $\tau \in (0, \tau_1]$ and any $y' \in F$ satisfying

$$\|y' - y_0\| = (\rho - \lambda)\tau\|x_0\|.$$

We are going to show the existence of $x' \in E$ such that

$$x' \in x_0 + \tau\|x_0\|\mathbf{B}_E \quad \text{and} \quad y' \in (\Phi + \Psi)(x'). \tag{10.68}$$

By assumption on (x_0, y_0) we have

$$(x_0, y_0 - \Psi(x_0)) \in \text{graph}\,\Phi, \quad \|x_0\| \leq \tau_1 \leq \tau_0,$$

$$\|y_0 - \Psi(x_0)\| \leq \|y_0\| + \|\Psi(x_0)\| \leq \tau_1 + \lambda\|x_0\| \leq (1 + \lambda)\tau_1 \leq \tau_0,$$

$$\|(y' - \Psi(x_0)) - (y_0 - \Psi(x_0))\| \leq (\rho - \lambda)t\|x_0\| \leq \rho\tau_0|x_0|.$$

The last line shows that

$$y' - \Psi(x_0) \in (y_0 - \Psi(x_0)) + (\rho - \lambda)\tau\|x_0\|\mathbf{B}_F.$$

Hence by (10.66) with $y := y_0 - \Psi(x_0)$, there exists $x_1 \in E$ satisfying

$$y' - \Psi(x_0) \in \Phi(x_1), \quad \|x_1 - x_0\| \leq \frac{\rho - \lambda}{\rho}\tau\|x_0\|.$$

Now we proceed by induction. Suppose that for $i = 1, \ldots, n$ we have obtained $x_i \in E$ such that

$$y' - \Psi(x_{i-1}) \in \Phi(x_i), \quad \|x_i - x_{i-1}\| \leq \frac{\rho - \lambda}{\rho}\left(\frac{\lambda}{\rho}\right)^{i-1}\tau\|x_0\|. \tag{10.69}$$

It follows that

$$\|x_n - x_0\| \leq \sum_{i=1}^{n}\|x_i - x_{i-1}\| \leq \frac{\rho - \lambda}{\rho}\tau\|x_0\|\sum_{i=1}^{n}\left(\frac{\lambda}{\rho}\right)^{i-1}$$

$$\leq \frac{\rho - \lambda}{\rho}\tau\|x_0\|\sum_{i=0}^{\infty}\left(\frac{\lambda}{\rho}\right)^{i} = \tau\|x_0\| \leq \frac{1}{2}\|x_0\|. \tag{10.70}$$

On the one hand, this implies $\|x_n\| \leq \frac{3}{4}\|x_0\| \leq \tau_0$. On the other hand, using $\|x_n\| \geq \|x_0\| - \|x_n - x_0\|$ we also obtain $\|x_n\| \geq \frac{1}{2}\|x_0\|$ and so $x_n \neq o$. Furthermore we have

$$\|y' - \Psi(x_{n-1})\| \leq \|y' - y_0\| + \|\Psi(x_{n-1})\| + \|y_0\|$$

$$\leq (\rho - \lambda)\tau_1\|x_0\| + \frac{3}{2}\lambda\|x_0\| + \|y_0\| \leq \tau_1\left(\frac{1}{2}\lambda + \frac{3}{2}\lambda + 1\right) \leq \tau_0$$

and also

$$\left\|\left(y' - \Psi(x_n)\right) - \left(y' - \Psi(x_{n-1})\right)\right\| \leq \lambda \|x_n - x_{n-1}\|$$

$$\leq (\rho - \lambda)\left(\frac{\lambda}{\rho}\right)^i \tau\|x_0\| \leq (\rho - \lambda)\left(\frac{\lambda}{\rho}\right)^i 2\tau_1\|x_n\| \leq \rho\tau_0\|x_n\|.$$

Hence we can apply (10.66) with $x := x_n$ and $y := y' - \Psi(x_{n-1})$ to find $x_{n+1} \in E$ such that $y' - \Psi(x_n) \in \Phi(x_{n+1})$ and

$$\|x_{n+1} - x_n\| \leq \frac{1}{\rho}\|\left(y' - \Psi(x_n)\right) - \left(y' - \Psi(x_{n-1})\right)\|$$

$$\leq \frac{\lambda}{\rho}\|x_n - x_{n-1}\| \leq \frac{\rho - \lambda}{\rho}\left(\frac{\lambda}{\rho}\right)^i \tau\|x_0\|. \tag{10.71}$$

Since these estimates correspond to (10.69), we conclude that a sequence (x_n) exists in E satisfying (10.71) for all $n \in \mathbb{N}$. Since (x_n) is a Cauchy sequence in the Banach space E, it is convergent to some $x' \in E$. The continuity of Ψ gives $\lim_{n\to\infty} \Psi(x_n) = \Psi(x')$. Since $(x_{n+1}, y' - \Psi(x_n)) \in \operatorname{graph} \Phi$ for all n and $\operatorname{graph} \Phi$ is closed, we see that $(x', y' - \Psi(x')) \in \operatorname{graph} \Phi$ and so $y' \in (\Phi + \Psi)(x')$. Finally, $\|x_n - x_0\| \leq \tau\|x_0\|$ for any n implies $\|x' - x_0\| \leq \tau\|x_0\|$, which completes the proof. □

10.7 Linearly Semiopen Processes

For processes there is a close relationship between linear openness and linear semiopenness. This will be elaborated in this section. We assume that

$$E \text{ and } F \text{ are Banach spaces and } \Phi : E \rightrightarrows F \text{ is a process.}$$

The following notions will be helpful.

Definition 10.7.1

(a) A set $S \subseteq E \times F$ is said to be *generating* for the process Φ if $\operatorname{graph} \Phi = \mathbb{R}_+ S$, i.e., $\operatorname{graph} \Phi$ is the cone generated by S.

(b) A set $S \subseteq E \times F$ is said to be *bounded in E* by r (where $r > 0$) if $\|x\| \leq r$ whenever $(x, y) \in S$.

Example 10.7.2 Let Φ be such that $\Phi(o) = \{o\}$. Then the set

$$S := \{(x, y) \in \operatorname{graph} \Phi \mid \|x\| = 1\}$$

is generating for Φ and is bounded in E by $r = 1$.

Lemma 10.7.3 *Let S be generating for Φ and bounded in E by $r > 0$. Assume that Φ is open at a linear rate around each $(x, y) \in S$ with the openness parameters ρ and τ_0 being independent of (x, y). Then Φ is linearly semiopen around (o, o) with the openness parameters ρ and $\tilde{\tau}_0 := \min\{\tau_0, \tau_0/r\}$.*

Proof. Let $(x, y) \in \operatorname{graph} \Phi$, $(x, y) \neq (o, o)$, where $x \in \tilde{\tau}_0 B_E$ and $y \in \tilde{\tau}_0 B_F$. Then $(\lambda x, \lambda y) \in S$ for some $\lambda > 0$. By assumption we have $\lambda y + \rho \tau B_F \subseteq \Phi(\lambda x + \tau B_E)$ for any $\tau \in (0, \tau_0]$. Since $\|\lambda x\| \leq r$, it follows that

$$\lambda y + \rho \tau \lambda \|x\| B_F \subseteq \Phi(\lambda x + \tau \lambda \|x\| B_E) \quad \forall \tau \in (0, \tilde{\tau}_0]$$

and so since Φ is a process,

$$\lambda(y + \rho \tau \|x\| B_F) \subseteq \lambda \Phi(x + \tau \|x\| B_E) \quad \forall \tau \in (0, \tilde{\tau}_0].$$

Dividing by λ results in the assertion. $\qquad\square$

Lemma 10.7.4 *Let Φ be open at a linear rate around $(\bar{x}, \bar{y}) \in \operatorname{graph} \Phi$. Then there are neighborhoods U of $\bar{x}$ and V of $\bar{y}$ as well as positive numbers ρ and τ_0 such that Φ is open at a linear rate around each $(x, y) \in \operatorname{graph}(\Phi) \cap (U \times V)$ with the openness parameters ρ and τ_0.*

Proof. Let $\bar{\rho}$ and $\bar{\tau}_0$ be openness parameters of Φ around $(\bar{x}, \bar{y})$. Then

$$y' + \bar{\rho} \tau B_F \subseteq \Phi(x' + \tau B_E) \tag{10.72}$$

whenever $(x', y') \in \operatorname{graph} \Phi$, $x' \in \bar{x} + \bar{\tau}_0 B_E$, $y' \in \bar{y} + \bar{\tau}_0 B_F$, and $\tau \in (0, \bar{\tau}_0]$. Define

$$\rho := \bar{\rho}, \quad \tau_0 := \bar{\tau}_0/2, \quad U := \bar{x} + \tau_0 B_E, \quad V := \bar{y} + \tau_0 B_F.$$

Now let any $(x, y) \in \operatorname{graph}(\Phi) \cap (U \times V)$ be given. Choose $(x', y') \in \operatorname{graph} \Phi$ such that $x' \in x + \tau_0 B_E$ and $y' \in y + \tau_0 B_F$. Then $x' \in \bar{x} + \bar{\tau}_0 B_E$ and $y' \in \bar{y} + \bar{\tau}_0 B_F$. In view of (10.72) we obtain $y' + \rho \tau B_F \subseteq \Phi(x' + \tau B_E)$ for any $\tau \in (0, \tau_0]$, and the proof is complete. $\qquad\square$

Proposition 10.7.5 *Let $S \subseteq E \times F$ be a generating set for Φ and assume that one of the following conditions is satisfied:*

(a) *S is compact.*
(b) *E is finite dimensional, $\Phi : E \to F$ is single-valued and locally L-continuous around o, and $S = \{(x, \Phi(x)) \mid \|x\| = 1\}$.*

If Φ is open at a linear rate around each point of S, then Φ is linearly semiopen around (o, o).

Proof.

(I) First notice that in either case, S is bounded in E.
(II) By Lemma 10.7.4 we can assign to each $(x, y) \in S$ an open neighborhood $U(x) \times V(y)$ as well as positive numbers $\rho(x, y)$ and $\tau_0(x, y)$ such that Φ is open at a linear rate around each $(x', y') \in U(x) \times V(y)$ with the openness parameters $\rho(x, y)$ and $\tau_0(x, y)$.

(IIIa) If condition (a) is satisfied, then the open covering $\big(U(x) \times V(y)\big)_{(x,y)\in S}$ of S contains a finite subcovering $\big(U(x_i) \times V(y_i)\big)_{i=1,\dots,n}$. Define

$$\rho := \min\{\rho(x_i, y_i) \mid i = 1, \dots, n\}, \quad \tau_0 := \min\{\tau_0(x_i, y_i) \mid i = 1, \dots, n\}$$

and apply Lemma 10.7.3 to see that the assertion is true.

(IIIb) Now let condition (b) be satisfied. Since Φ is positively homogeneous and locally L-continuous around o in particular, Φ is (globally) L-continuous. Hence for each $x \in E$ satisfying $\|x\| = 1$ we can find an open neighborhood $U'(x) \subseteq U(x)$ such that $x' \in U'(x)$ implies $\Phi(x') \in V(\Phi(x))$. Thus Φ is open at a linear rate around each $(x', \Phi(x'))$, where $x' \in U'(x)$, with openness parameters $\rho(x)$ and $\tau_0(x)$. The open covering $\big(U'(x)\big)_{\|x\|=1}$ of the compact set $\{x \in E \mid \|x\| = 1\}$ contains a finite subcovering $\big(U'(x_i)\big)_{i=1,\dots,n}$. Setting

$$\rho := \min\{\rho(x_i) \mid i = 1, \dots, n\}, \quad \tau_0 := \min\{\tau_0(x_i) \mid i = 1, \dots, n\}$$

and applying Lemma 10.7.3 concludes the proof. $\qquad\square$

10.8 Maximal Monotone Multifunctions

The aim of this section is to establish conditions ensuring that a multifunction $\Phi : E \rightrightarrows E^*$ satisfies range $\Phi = E^*$, which means that for any $x^* \in E^*$ the *generalized equation* $x^* \in \Phi(x)$ has a solution $x \in E$. In this connection, the following concept is crucial.

Definition 10.8.1 The multifunction $\Phi : E \rightrightarrows E^*$ is said to be *maximal monotone* if Φ is monotone and graph Φ is not properly contained in the graph of any other monotone multifunction.

Maximal monotone multifunctions play a prominent role in treating parabolic differential equations as evolution equations in Sobolev spaces of Banach space-valued functions. For instance, the generalized time derivative in the time-periodic quasilinear parabolic problem turns out to be a (single-valued) maximal monotone multifunction. A technical remark will be useful.

Remark 10.8.2 The monotone multifunction $\Phi : E \rightrightarrows E^*$ is maximal monotone if and only if the following holds:

$$[(y, y^*) \in E \times E^* \text{ and } \langle y^* - x^*, y - x \rangle \geq 0 \quad \forall\, (x, x^*) \in \text{graph}\,\Phi]$$
$$\Longrightarrow \quad (y, y^*) \in \text{graph}\,\Phi.$$

We show the maximal monotonicity of the subdifferential mapping $\partial_F f$.

Theorem 10.8.3 Let E be a Fréchet smooth Banach space, let $f : E \to \overline{\mathbb{R}}$ be proper and l.s.c. If the multifunction $\partial_F f : E \rightrightarrows E^*$ is monotone, then it is maximal monotone.

Proof. Let $(b, b^*) \in E \times E^*$ be such that $b^* \notin \partial_F f(b)$. In view of Remark 10.8.2, we have to show that there exist $x \in E$ and $x^* \in \partial_F f(x)$ satisfying $\langle b^* - x^*, b - x \rangle < 0$. Since $o \notin \partial_F(f - b^*)(b)$, the point b is not a minimizer of $f - b^*$. Hence there exists $a \in E$ such that $(f - b^*)(a) < (f - b^*)(b) := \rho$. By Zagrodny's approximate mean value theorem (Theorem 9.4.1), there exist a point $c \in [a, b)$ as well as sequences (x_n) in E and (x_n^*) in E^* such that

$$\lim_{n \to \infty} x_n = c, \quad y_n^* := x_n^* - b^* \in \partial_F(f - b^*)(x_n) \quad \forall n,$$

$$\liminf_{n \to \infty} \langle y_n^*, c - x_n \rangle \geq 0, \quad \liminf_{n \to \infty} \langle y_n^*, b - a \rangle > 0.$$

Noting that $b - c = \lambda(b - a)$ with some $\lambda \in (0, 1]$, we conclude that

$$\liminf_{n \to \infty} \langle x_n^* - b^*, b - x_n \rangle \geq \liminf_{n \to \infty} \langle y_n^*, b - c \rangle + \liminf_{n \to \infty} \langle y_n^*, c - x_n \rangle$$

$$\geq \frac{\|b - c\|}{\|b - a\|} \liminf_{n \to \infty} \langle y_n^*, b - a \rangle + \liminf_{n \to \infty} \langle y_n^*, c - x_n \rangle > 0.$$

Since we obviously have $x_n^* \in \partial_F f(x_n)$ for any n, it suffices to set $x := x_n$ and $x^* := x_n^*$ for n sufficiently large. $\qquad\square$

Corollary 10.8.4 *Let E be a Fréchet smooth Banach space. If $f : E \to \overline{\mathbb{R}}$ is proper, convex, and l.s.c., then the subdifferential mapping ∂f (of convex analysis) is maximal monotone.*

Proof. The subdifferential mapping ∂f is monotone (Proposition 4.3.7) and coincides with $\partial_F f$ (Proposition 9.1.9). Hence the assertion follows from Theorem 10.8.3. $\qquad\square$

To describe a class of single-valued maximal monotone multifunctions, we need the following notion. The mapping $T : E \to E^*$ is said to be *hemicontinuous* if for all $x, y, z \in E$ the real function $\tau \mapsto \langle T(x + \tau y), z \rangle$ is continuous on $[0, 1]$. Notice that each hemicontinuous mapping is radially continuous.

Proposition 10.8.5 *If $T : E \to E^*$ is monotone and hemicontinuous, then T is maximal monotone.*

Proof. See Exercise 10.10.3. $\qquad\square$

Before we can establish the announced surjectivity statement, we derive an auxiliary result that is a distinguished relative of the sandwich theorem.

Lemma 10.8.6 *Let E and F be Banach spaces, let $f : E \to \mathbb{R}$ and $g : F \to \mathbb{R}$ be convex functionals and let $A : E \to F$ be a continuous linear mapping. Assume further that:*

(C1) *f and g are l.s.c. and $o \in \mathrm{cr}\,(\mathrm{dom}\, g - A(\mathrm{dom}\, f)$ or*
(C2) *g is continuous at some point of $A(\mathrm{dom}\, f)$.*

Then there exists $y^* \in F^*$ such that for any $x \in E$ and $y \in F$, one has

$$\inf_{x \in E} \left(f(x) + g(Ax) \right) \leq \left(f(x) - \langle y^*, Ax \rangle \right) + \left(g(y) + \langle y^*, y \rangle \right). \tag{10.73}$$

Proof.

(I) We define the value functional $h : F \to \overline{\mathbb{R}}$ by

$$h(u) := \inf_{x \in E} \left(f(x) + g(Ax + u) \right), \quad u \in F.$$

The functional h is convex and satisfies $\operatorname{dom} h = \operatorname{dom} g - A(\operatorname{dom} f)$. We now show that $o \in \operatorname{int} \operatorname{dom} h$.

(II) First we assume that condition (C1) is satisfied. Passing to suitable translations of f and g if necessary, we may assume that $f(o) = g(o) = 0$. Let

$$M := \bigcup_{x \in B_E} \{ u \in F \mid f(x) + g(Ax + u) \leq 1 \}.$$

Obviously M is convex.

(IIa) We show that M is absorbing. Let $y \in F$ be given. By (C1) there exists $\tau_0 > 0$ such that $\tau y \in \operatorname{dom} g - A(\operatorname{dom} f)$ whenever $|\tau| \leq \tau_0$. For each such τ let $x \in \operatorname{dom} f$ be such that $Ax + \tau y \in \operatorname{dom} g$. Then $f(x) + g(Ax + \tau y) =: r < +\infty$. Let $s \geq \max\{\|x\|, |r|, 1\}$. Since f and g are convex and $f(o) = g(o) = 0$, we deduce that

$$f\left(\frac{1}{s}x\right) + g\left(A\left(\frac{1}{s}x\right) + \frac{\tau}{s}y\right) \leq 1$$

and so $(\tau/s)y \in M$. Hence M is absorbing, i.e., $o \in \operatorname{cr} M$.

(IIb) Next we show that M is cs-closed. Assume that $\lambda_i \geq 0$, $\sum_{i=1}^{\infty} \lambda_i = 1$, $y_i \in M$ and $y := \sum_{i=1}^{\infty} \lambda_i y_i \in F$. Then for each i there is an $x_i \in B_E$ satisfying

$$f(x_i) + g(Ax_i + y_i) \leq 1. \tag{10.74}$$

Let $\epsilon > 0$ be given. Then there exists i_0 such that $\sum_{i=m}^{n} \lambda_i < \epsilon$ whenever $n > m \geq i_0$. It follows that

$$\left\| \sum_{i=m}^{n} \lambda_i x_i \right\| \leq \sum_{i=m}^{n} \lambda_i \|x_i\| < \epsilon \quad \text{whenever } n > m \geq i_0.$$

Hence $\sum_{i=1}^{\infty} \lambda_i x_i$ is convergent to some x in the Banach space E. The sequence (x_i) is contained in the ball B_E which is cs-closed (Lemma 1.2.2), therefore $x \in B_E$. Since f and g are convex and l.s.c. and A is linear and continuous, we deduce from (10.74) that

$$f(x) + g(Ax + y) = f\left(\sum_{i=1}^{\infty} \lambda_i x_i\right) + g\left(\sum_{i=1}^{\infty} \lambda_i (Ax_i + y_i)\right)$$

$$\leq \sum_{i=1}^{\infty} \lambda_i f(x_i) + \sum_{i=1}^{\infty} \lambda_i g(Ax_i + y_i) \leq 1$$

and so $y \in M$. Thus M is cs-closed.

(IIc) Now Proposition 1.2.3 shows that $\operatorname{cr} M = \operatorname{int} M$. This together with step (IIa) yields $o \in \operatorname{int} M$.

(IId) Since h is convex and bounded above on M and $o \in \operatorname{int} M$, the functional h is continuous at o.

(III) Now we assume that condition (C2) is satisfied. Let $\bar{y} \in A(\operatorname{dom} f)$ be such that g is continuous at $\bar{y}$. Then there exists $r > 0$ such that $g(\bar{y} + u) \leq g(\bar{y} + 1$ for any $u \in B_F(o, r)$. Let $\bar{x} \in \operatorname{dom} f$ be such that $\bar{y} = A\bar{x}$. Then we obtain

$$h(u) \leq f(\bar{x}) + g(A\bar{x} + u) \leq f(\bar{x}) + g(\bar{y}) + 1 \quad \forall u \in B_F(o, r).$$

As in step (IId) it follows that h is continuous at o.

(IV) Since h is convex and continuous at o, Proposition 4.1.6 implies that there exists $-y^* \in \partial h(o)$. We thus have

$$\inf_{x \in E} \left(f(x) + g(Ax) \right) = h(o) \leq h(u) + \langle y^*, u \rangle$$

$$\leq f(x) + g(Ax + u) + \langle y^*, u \rangle \quad \forall x \in E \, \forall u \in F. \tag{10.75}$$

Now, if $x \in E$ and $y \in F$ are given, then inserting $u := y - Ax$ in (10.75), we obtain (10.73). $\square$

Recall that the duality mapping $J : E \rightrightarrows E^*$ is defined by $J = \partial j$, where $j(z) := \frac{1}{2}\|z\|^2$. By Remark 4.6.3, J can be written as

$$J(z) = \{z^* \in E^* \mid \|z^*\| = \|z\|, \ \langle z^*, z \rangle = \|z\|^2\}, \quad z \in E. \tag{10.76}$$

If E is reflexive, then on identifying E^{**} with E, the duality mapping $J : E^* \rightrightarrows E$ is defined analogously.

Theorem 10.8.7 *Let E be a reflexive Banach space. If $\Phi : E \rightrightarrows E^*$ is maximal monotone, then* $\operatorname{range}(\Phi + \lambda J) = E^*$ *for any $\lambda > 0$.*

Proof.

(I) Suppose for the moment we had already verified the existence of $z \in E$ satisfying $o \in (\Phi + J)(z)$. Now let $\lambda > 0$ and $z^* \in E^*$ be given. Since $\lambda^{-1}(\Phi - z^*)$ inherits the maximal monotonicity from Φ, we can conclude that there exists $z \in E$ satisfying $o \in (\lambda^{-1}(\Phi - z^*) + J)(z)$. It follows that $z^* \in (\Phi + \lambda J)(z)$. Hence it remains to show that the generalized equation $o \in (\Phi + J)(z)$ has a solution.

(II) Define $f_\Phi : E \times E^* \to \overline{\mathbb{R}}$ by

$$f_\Phi(x, x^*) :=$$

$$\sup_{y^* \in \Phi(y)} \left(\langle y^*, x \rangle + \langle x^*, y \rangle - \langle y^*, y \rangle \right) = \langle x^*, x \rangle + \sup_{y^* \in \Phi(y)} \langle x^* - y^*, y - x \rangle.$$

The function f_Φ is proper, convex, and l.s.c. Since Φ is maximal monotone, it follows that

$$f_\Phi(x, x^*) \geq \langle x^*, x \rangle, \tag{10.77}$$

and by Remark 10.8.2 equality holds if and only if $x^* \in \Phi(x)$.

(III) For any $x \in E$ and $x^* \in E^*$ we have

$$0 \leq \frac{1}{2}(\|x\|^2 + \|x^*\|^2) - \|x\| \cdot \|x^*\| \leq \frac{1}{2}(\|x\|^2 + \|x^*\|^2) + \langle x^*, x \rangle \tag{10.78}$$

and so (10.77) passes into

$$0 \leq f_\Phi(x, x^*) + \frac{1}{2}(\|x\|^2 + \|x^*\|^2). \tag{10.79}$$

Lemma 10.8.6 (with E and F replaced by $E \times E^*$) now ensures the existence of $(z, z^*) \in E \times E^*$ such that for all $(y, y^*) \in E \times E^*$,

$$0 \leq f_\Phi(x, x^*) - \langle z^*, x \rangle - \langle x^*, z \rangle + \frac{1}{2}(\|y\|^2 + \|y^*\|^2) + \langle z^*, y \rangle + \langle y^*, z \rangle.$$

Choose $y^* \in -J(z)$ and apply (10.76), analogously choose $y \in -J(z^*)$. Then (10.79) gives

$$f_\Phi(x, x^*) - \langle z^*, x \rangle - \langle x^*, z \rangle \geq \frac{1}{2}(\|z\|^2 + \|z^*\|^2). \tag{10.80}$$

Let $x^* \in \Phi(x)$. Then $f_\Phi(x, x^*) = \langle x^*, x \rangle$ (cf. step (II)). Hence (10.80) can be written as

$$\langle x^* - z^*, x - z \rangle \geq \frac{1}{2}(\|z\|^2 + \|z^*\|^2) + \langle z^*, z \rangle \geq 0, \tag{10.81}$$

where the last inequality follows from (10.78). Since (10.81) holds for any x, x^* satisfying $x^* \in \Phi(x)$ and Φ is maximal monotone, we can conclude that $z^* \in \Phi(z)$. Setting $x := z$ and $x^* := z^*$ in (10.81), we obtain $\frac{1}{2}(\|z\|^2 + \|z^*\|^2) + \langle z^*, z \rangle = 0$, which by (10.76) implies $-z^* \in J(z)$ and so $o \in (\Phi + J)(z)$. $\quad\square$

With the aid of Theorem 10.8.7 we shall establish a result on the surjectivity of Φ. For this, we need the following notion.

The multifunction $\Phi : E \rightrightarrows E^*$ is said to be *coercive* if $\mathrm{Dom}\,\Phi$ is bounded, or $\mathrm{Dom}\,\Phi$ is unbounded and

$$\frac{\inf\{\langle x^*, x \rangle \mid x^* \in \Phi(x)\}}{\|x\|} \to +\infty \quad \text{as } x \in \mathrm{Dom}\,\Phi, \|x\| \to +\infty.$$

Theorem 10.8.8 (Surjectivity Theorem) *Let E be a reflexive Banach space. If $\Phi : E \rightrightarrows E^*$ is maximal monotone and coercive, then* $\mathrm{range}\,\Phi = E^*$.

Proof.

(I) By Proposition 4.7.14, E admits an equivalent norm that is F-differentiable on $E \setminus \{o\}$ and so the duality mapping J with respect to this norm is single-valued. Notice that Φ is also coercive with respect to the equivalent norm.

(II) Let $z^* \in E^*$ be given. Choose a sequence (λ_k) of positive numbers tending to zero. By Theorem 10.8.7, for each k there exists $x_k \in \operatorname{Dom} \Phi$ such that $z^* \in (\Phi + \lambda_k J)(x_k)$ and so there exists $x_k^* \in \Phi(x_k)$ satisfying

$$z^* = x_k^* + \lambda_k J(x_k). \tag{10.82}$$

(III) If $\operatorname{Dom} \Phi$ is bounded, then the sequence (x_k) is also bounded. Assume now that $\operatorname{Dom} \Phi$ is unbounded. Since Φ is coercive, there exists $\rho > 0$ such that

$$\frac{\langle x^*, x \rangle}{\|x\|} > \|z^*\| \quad \text{whenever } x \in \operatorname{Dom} \Phi, \ \|x\| > \rho, \ x^* \in \Phi(x).$$

For these x we obtain

$$\langle x^* - z^*, x \rangle = \langle x^*, x \rangle - \langle z^*, x \rangle > \|x\| \cdot \|z^*\| - \langle z^*, x \rangle \geq 0.$$

On the other hand, in view of (10.76) and (10.82) we have

$$\langle x_k^* - z^*, x_k \rangle = -\lambda_k \langle J(x_k), x_k \rangle = -\lambda_k \|x_k\|^2 \leq 0.$$

Therefore we must conclude that $\|x_k\| \leq \rho$ for any k. This further implies

$$\|x_k^* - z^*\| = \lambda_k \|J(x_k)\| = \lambda_k \|x_k\| \leq \lambda_k \rho \to 0 \quad \text{as } k \to \infty.$$

(IV) Since E is reflexive, the bounded sequence (x_k) contains a subsequence (z_k) that is weakly convergent to some $z \in E$. We show that z satisfies $z^* \in \Phi(z)$. Since Φ is monotone, we obtain for any $y \in \operatorname{Dom} \Phi$ and $y^* \in \Phi(y)$ that $\langle z_k^* - y^*, z_k - y \rangle \geq 0$ for any k. In this context, (z_k^*) denotes the corresponding subsequence of (x_k^*). Since by step (III) the sequence (z_k^*) is norm convergent to z^*, the last inequality implies $\langle z^* - y^*, z - y \rangle \geq 0$ (Exercise 10.10.1). Since Φ is maximal monotone, we conclude that $z \in \operatorname{Dom} \Phi$ and $z^* \in \Phi(z)$. $\qquad\square$

As an immediate consequence of Theorem 10.8.8 and Proposition 10.8.5 we obtain:

Corollary 10.8.9 *If E is a reflexive Banach space and $T : E \to E^*$ is monotone, hemicontinuous, and coercive, then* $\operatorname{range} T = E^*$.

10.9 Convergence of Sets

Convention. Throughout this section, unless otherwise specified, E is a normed vector space and F is a locally convex (not necessarily normed) vector space.

Here we consider this more general setting because we will later apply the following concepts to multifunctions of the form $\Phi : E \rightrightarrows E^*$, where E^* is equipped with the weak* topology $\sigma(E^*, E)$.

Definition 10.9.1 Let $(S_\alpha)_{\alpha \in A}$, where A is a directed set, be a *generalized* sequence of subsets of F.

(a) The *Painlevé–Kuratowski upper limit* $\operatorname{Lim\,sup}_{\alpha \in A} S_\alpha$ is the set of cluster points of generalized sequences $(v_\alpha)_{\alpha \in A}$, where $v_\alpha \in S_\alpha$ for any $\alpha \in A$.

(b) The *Painlevé–Kuratowski lower limit* $\operatorname{Lim\,inf}_{\alpha \in A} S_\alpha$ is the set of limits of generalized sequences $(v_\alpha)_{\alpha \in A}$, where $v_\alpha \in S_\alpha$ for any $\alpha \in A$.

The definition applies in particular to a sequence $(S_k)_{k \in \mathbb{N}}$ in F, in which case we write $\operatorname{Lim\,sup}_{k \to \infty}, S_k$ and $\operatorname{Lim\,inf}_{k \to \infty}, S_k$, respectively.

Now let $\Phi : E \rightrightarrows F$ be a multifunction and $\bar{x} \in \operatorname{Dom} \Phi$. We consider Φ as a generalized sequence $(\Phi(x))_{x \in E}$ in F, where E is directed by the preorder $y \succeq x$ if and only if $\|y - \bar{x}\| \leq \|x - \bar{x}\|$. The resulting Painlevé–Kuratowski upper (lower) limit is then written

$$\operatorname{Lim\,sup}_{x \to \bar{x}} \Phi(x) \quad \text{and} \quad \operatorname{Lim\,inf}_{x \to \bar{x}} \Phi(x),$$

respectively. Lemma 10.9.2 gives the explicit characterization of these limits.

Lemma 10.9.2 *If $\Phi : E \rightrightarrows F$ is a multifunction and $\bar{x} \in \operatorname{Dom} \Phi$, then:*

(a) $\operatorname{Lim\,sup}_{x \to \bar{x}} \Phi(x)$ *is the set of cluster points of generalized sequences $(v_\alpha)_{\alpha \in A}$, where $v_\alpha \in \Phi(x_\alpha)$ for any α in the directed set A and $(x_\alpha)_{\alpha \in A}$ is convergent in E to $\bar{x}$.*

(b) $\operatorname{Lim\,inf}_{x \to \bar{x}} \Phi(x)$ *is the set of limits of generalized sequences $(v_\alpha)_{\alpha \in A}$, where $v_\alpha \in \Phi(x_\alpha)$ for any $\alpha \in A$ and $(x_\alpha)_{\alpha \in A}$ is convergent in E to $\bar{x}$.*

Proof. See Exercise 10.10.4. □

The following result is also easily verified.

Lemma 10.9.3 *One always has*

$$\operatorname{Lim\,inf}_{x \to \bar{x}} \Phi(x) \subseteq \operatorname{cl} \Phi(\bar{x}) \subseteq \operatorname{Lim\,sup}_{x \to \bar{x}} \Phi(x).$$

Proof. See Exercise 10.10.5. □

If, in addition, F is a normed vector space, we have a simple characterization of these concepts. We write $x \to^{\Phi} \bar{x}$ if $x \in \text{Dom}\,\Phi$ and $x \to \bar{x}$.

Lemma 10.9.4 *If E and F are normed vector spaces and $\Phi : E \rightrightarrows F$ is a multifunction, then*

$$\text{Lim}\sup_{x \to \bar{x}} \Phi(x) = \left\{ v \in F \mid \liminf_{x \to^{\Phi} \bar{x}} d(\Phi(x), v) = 0 \right\},$$

$$\text{Lim}\inf_{x \to \bar{x}} \Phi(x) = \left\{ v \in F \mid \lim_{x \to^{\Phi} \bar{x}} d(\Phi(x), v) = 0 \right\},$$

and these limits are closed sets.

Proof. See Exercise 10.10.6.   □

We shall also make use of a sequential variant of the above concepts.

Definition 10.9.5 Let E be a normed vector space, $\Phi : E \rightrightarrows E^*$ be a multifunction, and $\bar{x} \in \text{Dom}\,\Phi$. The *sequential Painlevé–Kuratowski upper limit* $\text{sLim}\sup_{x \to \bar{x}} \Phi(x)$ of Φ is defined to be the set of all $x^* \in E^*$ for which there exist a sequence (x_k) in $\text{Dom}\,\Phi$ that is norm convergent to $\bar{x}$ and a sequence (x_k^*) in E^* that is $\sigma(E^*, E)$ convergent to x^* such that $x_k^* \in \Phi(x_k)$ for all $k \in \mathbb{N}$.

Lemma 10.9.6 is an immediate consequence of the definitions.

Lemma 10.9.6 *Let E be a normed vector space and equip E^* with the topology $\sigma(E^*, E)$. Further let $\Phi : E \rightrightarrows E^*$. Then*

$$\text{sLim}\sup_{x \to \bar{x}} \Phi(x) \subseteq \text{Lim}\sup_{x \to \bar{x}} \Phi(x).$$

Now we introduce a convergence concept for a generalized sequence of functions.

Definition 10.9.7 Let $(\varphi_\alpha)_{\alpha \in A}$ be a generalized sequence of functions $\varphi_\alpha : F \to \overline{\mathbb{R}}$:

(a) The *upper epi-limit* of $(\varphi_\alpha)_{\alpha \in A}$ is the function $\text{eLim}\sup_{\alpha \in A} \varphi_\alpha$ whose epigraph is the Painlevé–Kuratowski lower limit of $\text{epi}\,\varphi_\alpha$:

$$\text{epi}\left(\text{eLim}\sup_{\alpha \in A} \varphi_\alpha\right) := \text{Lim}\inf_{\alpha \in A}(\text{epi}\,\varphi_\alpha). \tag{10.83}$$

(b) The *lower epi-limit* of $(\varphi_\alpha)_{\alpha \in A}$ is the function $\text{eLim}\inf_{\alpha \in A} \varphi_\alpha$ whose epigraph is the Painlevé–Kuratowski upper limit of $\text{epi}\,\varphi_\alpha$:

$$\text{epi}\left(\text{eLim}\inf_{\alpha \in A} \varphi_\alpha\right) := \text{Lim}\sup_{\alpha \in A}(\text{epi}\,\varphi_\alpha). \tag{10.84}$$

(c) If the upper and the lower epi-limit of $(\varphi_\alpha)_{\alpha \in A}$ coincide, then this function is called *epi-limit* of $(\varphi_\alpha)_{\alpha \in A}$ and is denoted $\text{eLim}_{\alpha \in A} \varphi_\alpha$.

It is left as Exercise 10.10.7 to show that the right-hand sides of (10.83) and (10.84) in fact are the epigraphs of functions. If, in particular, (φ_k) is a sequence of functions, then the corresponding epi-limit functions are defined, respectively, by

$$\text{epi}\big(\text{eLim sup}_{k\to\infty} \varphi_k\big) := \text{Lim inf}_{k\to\infty}(\text{epi}\,\varphi_k),$$

$$\text{epi}\big(\text{eLim inf}_{k\to\infty} \varphi_k\big) := \text{Lim sup}_{k\to\infty}(\text{epi}\,\varphi_k),$$

$$\text{epi}\big(\text{eLim}_{k\to\infty} \varphi_k\big) := \text{epi}\big(\text{eLim sup}_{k\to\infty} \varphi_k\big) = \text{epi}\big(\text{eLim inf}_{k\to\infty} \varphi_k\big).$$

In a normed vector space we have a simple characterization of the epi-limit.

Lemma 10.9.8 *Let F be a normed vector space and $\varphi_k : F \to \overline{\mathbb{R}}$ for all $k \in \mathbb{N}$. Then $\varphi = \text{eLim}_{k\to\infty} \varphi_k$ if and only if for each $x \in F$ one has*

$$\varphi(x) \leq \liminf_{k\to\infty} \varphi_k(x_k) \quad \textit{for any sequence } x_k \to x \quad \textit{and}$$

$$\varphi(x) \geq \limsup_{k\to\infty} \varphi_k(x_k) \quad \textit{for some sequence } x_k \to x.$$

Proof. See Exercise 10.10.8. □

10.10 Bibliographical Notes and Exercises

The presentation of Sects. 10.1–10.5 was strongly influenced by Bonnans and Shapiro [16]. Lemma 10.1.1 is due to Robinson [176]. The generalized open mapping theorem (Theorem 10.1.2) is due, independently, to Robinson [176] and Ursescu [211]. Proposition 10.2.1 is a result of Tuy [210], Proposition 10.2.3 appears in this context in [196].

The concept of openness at a linear rate is inherent in Lyusternik's original proof (of 1934) of the tangent space theorem (see Theorem 11.4.2). The concept was developed for single-valued mappings, under different names, by Dmitruk et al. [54], Dolecki [55], Warga [214], and others.

Ioffe [94] obtained Theorem 10.3.3, the proof given here follows Bonnans and Shapiro [16]. The stability theorem (Theorem 10.3.5) is due, independently, to Robinson [175] and Ursescu [211]. The perturbation theorem (Theorem 10.3.6) can be traced back to Lyusternik [128] and Graves [79]. For the estimate in the theorem see Ioffe [93].

Proposition 10.4.3 is a result of Pühl [172]. Jourani and Thibault [107] have established a corresponding sufficient condition for linear openness of certain classes of multifunctions but without a representation of the openness bound. Pühl [171] gives a nice geometric proof for the fact that a continuous convex functional f is open at a linear rate around $\bar{x}$ if and only if $\bar{x}$

is not a minimizer of f. This paper also contains sufficient conditions for cone-convex mappings to be open at a linear rate. Theorem 10.5.2 and its proof are taken from Borwein and Zhuang [25]. Penot [164] showed that the assumption that Φ be closed-valued can be dropped; see also Mordukhovich and Shao [139].

Concerning the results of Sects. 10.6 and 10.7 we refer to Pühl and Schirotzek [173], see also the doctoral thesis of Heidrun Pühl [172].

The elegant proof of Theorem 10.8.3 is taken from Borwein and Zhu [24]. In this connection, we had to assume that E be a Fréchet smooth Banach space so that we could apply Theorem 10.8.3 which in turn was deduced with the aid of Theorem 9.4.1. By a quite different proof, Rockafellar [181] showed that the conclusion of Corollary 10.8.4 holds in any Banach space E. A substantially simpler proof of Rockafellar's result is due to Simons [198] (cf. Phelps [165]). Groh [80] considers monotone operators via certain scalarizations called *forms*. Also see Exercise 10.10.2.

Theorem 10.8.7 is due to Rockafellar [182]. Lemma 10.8.6 and the astoundingly easy proof of Theorem 10.8.7 are taken from Borwein and Zhu [24]. In this context, the function f_Φ is called *Fitzpatrick function* by Borwein and Zhu. Theorem 10.8.8 together with results on the maximal monotonicity of the sum of two multifunctions admits various important applications. For readers interested in this subject, we recommend the comprehensive two-volume monograph by Zeidler [223, 224].

As standard references to multifunction theory we recommend Aubin and Frankowska [8] (infinite-dimensional spaces) and Rockafellar and Wets [189] (finite-dimensional spaces).

Exercise 10.10.1 Assume that E is a Banach space, (x_k) is a sequence in E and (x_k^*) is a sequence in E^*. Show that $\lim_{k\to\infty}\langle x_k^*, x_k\rangle = \langle x^*, x\rangle$ if

(a) (x_k) is norm convergent to x and (x_k^*) is weak* convergent to x^* or
(b) (x_k) is weakly convergent to x and (x_k^*) is norm convergent to x^*.

Exercise 10.10.2 Prove the following assertion:
If E is a Banach space and $f : E \to \mathbb{R}$ is convex and continuous, then $\partial f : E \rightrightarrows E^*$ is maximal monotone.
Hint: Follow the proof of Theorem 10.8.3 but apply the mean value theorem of Proposition 3.3.1 instead of Theorem 9.4.1 (cf. Phelps [165]).

Exercise 10.10.3 Verify Proposition 10.8.5.

Exercise 10.10.4 Prove Lemma 10.9.2.

Exercise 10.10.5 Verify Lemma 10.9.3.

Exercise 10.10.6 Prove Lemma 10.9.4.

Exercise 10.10.7 Prove that the right-hand sides of (10.83) and (10.84) in fact are the epigraphs of functions.

Hint: Show that a set $S \subseteq F \times \mathbb{R}$ is the epigraph of a function $\varphi : F \to \overline{\mathbb{R}}$ if and only if it has the following two properties:

(a) $(x, t) \in S$ and $t' > t$ imply $(x, t') \in S$ *and*
(b) $x \in F$ and $t^* := \inf\{t \in \mathbb{R} \mid (x, t) \in S\}$ imply $(x, t^*) \in S$.

Exercise 10.10.8 Prove Lemma 10.9.8.

11

Tangent and Normal Cones

11.1 Tangent Cones: First Properties

In this section, unless otherwise specified, we assume that E is a normed vector space, A is a nonempty subset of E, and $\bar{x} \in A$. Resuming the discussion started in Sect. 7.1, we define various *tangent cones* as local approximations of A near $\bar{x}$. By $x \to_A \bar{x}$ we mean that $x \in A$ and $x \to \bar{x}$.

Definition 11.1.1 One defines

$$T_r(A, \bar{x}) := \{y \in E \mid \exists \tau_k \downarrow 0 \, \forall k : \bar{x} + \tau_k y \in A\},$$
$$\text{cone of radial directions to } A \text{ at } \bar{x},$$

$$T(A, \bar{x}) := \{y \in E \mid \exists \tau_k \downarrow 0 \, \exists y_k \to y \, \forall k : \bar{x} + \tau_k y_k \in A\},$$
$$\text{contingent cone to } A \text{ at } \bar{x},$$

$$T_C(A, \bar{x}) := \{y \in E \mid \forall x_k \to_A \bar{x} \, \forall \tau_k \downarrow 0 \, \exists y_k \to y \, \forall k : x_k + \tau_k y_k \in A\},$$
$$\text{Clarke tangent cone to } A \text{ at } \bar{x},$$

$$I_r(A, \bar{x}) := \{y \in E \mid \exists \epsilon > 0 \, \forall \tau \in (0, \epsilon) : \bar{x} + \tau y \in A\},$$
$$\text{cone of radial inner directions, or}$$
$$\text{cone of feasible directions, to } A \text{ at } \bar{x},$$

$$I(A, \bar{x}) := \{y \in E \mid \exists \epsilon > 0 \, \forall \tau \in (0, \epsilon) \, \forall z \in B(y, \epsilon) : \bar{x} + \tau z \in A\},$$
$$\text{cone of inner directions to } A \text{ at } \bar{x},$$

$$H(A, \bar{x}) := \{y \in E \mid \forall x_k \to_A \bar{x} \, \forall \tau_k \downarrow 0 \, \forall y_k \to y \, \forall k : x_k + \tau_k y_k \in A\},$$
$$\text{cone of hypertangents to } A \text{ at } \bar{x}.$$

Proposition 11.1.2

(a) *One has*

$$I(A, \bar{x}) \subseteq I_r(A, \bar{x}) \subseteq T_r(A, \bar{x}) \subseteq T(A, \bar{x}),$$
$$H(A, \bar{x}) \subseteq T_C(A, \bar{x}) \subseteq T(A, \bar{x}).$$

Each of the sets is a cone, $I(A, \bar{x})$ and $H(A, \bar{x})$ may be empty, the other cones contain the zero element.

(b) *$T(A, \bar{x})$ and $T_C(A, \bar{x})$ are closed, $T_C(A, \bar{x})$ is also convex.*
(c) *If U is a neighborhood of $\bar{x}$, then $T(A, \bar{x}) = T(A \cap U, \bar{x})$, analogously for the other cones considered in (a).*
(d) *If A is convex, then*

$$I_r(A, \bar{x}) = T_r(A, \bar{x}) = \mathbb{R}_+(A - \bar{x}),$$
$$T(A, \bar{x}) = \mathrm{cl}\big(\mathbb{R}_+(A - \bar{x})\big),$$
$$I(A, \bar{x}) = \{\rho(x - \bar{x}) \mid \rho > 0, \ x \in \mathrm{int}\, A\}.$$

Proof.

(I) We show that $T(A, \bar{x})$ is closed. Let (z_n) be a sequence in $T(A, \bar{x})$ converging to some $y \in E$. For each $n \in \mathbb{N}$ there exist sequences $(\tau_k^{(n)})$ in $(0, +\infty)$ and $(y_k^{(n)})$ in E satisfying

$$\tau_k^{(n)} \downarrow 0, \quad y_k^{(n)} \to z_n \text{ as } k \to \infty \quad \text{and} \quad \bar{x} + \tau_k^{(n)} y_k^{(n)} \in A \ \forall k \in \mathbb{N}.$$

Hence for each $n \in \mathbb{N}$ there exists $k(n)$ such that

$$\tau_k^{(n)} < \tfrac{1}{n} \quad \text{and} \quad \|y_k^{(n)} - z_n\| < \tfrac{1}{n} \quad \forall k \geq k(n).$$

Setting $\tau_n := \tau_{k(n)}^{(n)}$ and $y_n := y_{k(n)}^{(n)}$, we obtain $\tau_n \downarrow 0$ and $y_n \to y$ as $n \to \infty$ as well as $\bar{x} + \tau_n y_n \in A$ for each n. Therefore $y \in T(A, \bar{x})$.

(II) We now verify that $T_C(A, \bar{x})$ is convex. Let $y_1, y_2 \in T_C(A, \bar{x})$. Since $T_C(A, \bar{x})$ is a cone, we only have to show that $y_1 + y_2 \in T_C(A, \bar{x})$. Assume that $\tau_k \downarrow 0$ and $x_k \to_A \bar{x}$ as $k \to \infty$. Then there exists $(y_k^{(1)})$ in E such that $y_k^{(1)} \to y_1$ as $k \to \infty$ and $x_k^{(1)} := x_k + \tau_k y_k^{(1)} \in A$ for each k. Since $x_k^{(1)} \to \bar{x}$, there also exists $(y_k^{(2)})$ in E satisfying

$$y_k^{(2)} \to y_2 \quad \text{and} \quad x_k + \tau_k\big(y_k^{(1)} + y_k^{(2)}\big) = x_k^{(1)} + \tau_k y_k^{(2)} \in A \ \ \forall k \in \mathbb{N}.$$

Since we also have $y_k^{(1)} + y_k^{(2)} \to y_1 + y_2$, we conclude that $y_1 + y_2 \in T_C(A, \bar{x})$.

The verification of the remaining assertions is left as Exercise 11.7.1. $\square$

Statement (c) of the proposition means that the approximating cones depend on the local properties of A near $\bar{x}$ only.

Figure 11.1 supplements the figures in Sect. 7.1.

Proposition 11.1.3 *If $A, B \subseteq E$ and $\bar{x} \in A \cap B$, then*

$$T_r(A, \bar{x}) \cap I_r(B, \bar{x}) \subseteq T_r(A \cap B, \bar{x}), \tag{11.1}$$
$$T(A, \bar{x}) \cap I(B, \bar{x}) \subseteq T(A \cap B, \bar{x}). \tag{11.2}$$

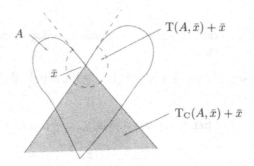

Fig. 11.1

Proof. We verify (11.2), leaving (11.1) as Exercise 11.7.2. Let $y \in T(A, \bar{x}) \cap I(B, \bar{x})$. Then there exist sequences $\tau_k \downarrow 0$ and $y_k \to y$ such that $\bar{x} + \tau_k y_k \in A$ for any $k \in \mathbb{N}$. Further there exists $\epsilon > 0$ such that $\bar{x} + \tau B(y, \epsilon) \in B$ for any $\tau \in (0, \epsilon)$. For all sufficiently large k we therefore obtain $\bar{x} + \tau_k y_k \in A \cap B$. Hence $y \in T(A \cap B, \bar{x})$. $\qquad \square$

The following example shows that in the above formulas, $I(B, \bar{x})$ cannot be replaced by $T(B, \bar{x})$.

Example 11.1.4 In $E := \mathbb{R}^2$ consider the sets $A := \{(x, y) \in \mathbb{R}^2 \mid y \geq x^2\}$ and $B := \{(x, y) \in \mathbb{R}^2 \mid y \leq -x^2\}$. Then with $\bar{x} := (0, 0)$ we have

$$T(A, \bar{x}) \cap T(B, \bar{x}) = \mathbb{R} \times \{0\} \quad \text{but} \quad T(A \cap B, \bar{x}) = \{(0, 0)\}.$$

Proposition 11.1.5 *There always holds*

$$T(A, \bar{x}) = \{y \in E \mid \liminf_{\tau \downarrow 0} \tau^{-1} d_A(\bar{x} + \tau y) = 0\},$$
$$T_C(A, \bar{x}) = \{y \in E \mid \liminf_{\substack{\tau \downarrow 0 \\ x \to_A \bar{x}}} \tau^{-1} d_A(x + \tau y) = 0\}.$$

Proof. See Exercise 11.7.3. $\qquad \square$

Now we establish a representation of the Clarke tangent cone in terms of the Clarke directional derivative. It is easy to see that

$$\left| d_A(x) - d_A(y) \right| \leq \|x - y\| \quad \forall x, y \in E, \tag{11.3}$$

i.e., the distance functional $d_A(\cdot)$ is (globally) L-continuous and so $d_A^\circ(x, y)$ exists for all $x, y \in E$.

Proposition 11.1.6 *One has*

$$T_C(A, \bar{x}) = \{y \in E \mid d_A^\circ(\bar{x}, y) = 0\}.$$

Proof.

(I) Let $y \in E$ satisfy $d_A^\circ(\bar{x}, y) = 0$. Further let sequences $\tau_k \downarrow 0$ and $x_k \to_A \bar{x}$ be given. Then

$$0 = \limsup_{\substack{\tau \downarrow 0 \\ x \to \bar{x}}} \tfrac{1}{\tau}\big(d_A(x + \tau y) - d_A(x)\big) \geq \limsup_{k \to \infty} \tfrac{1}{\tau_k}\big(d_A(x_k + \tau_k y) - \underbrace{d_A(x_k)}_{=0}\big).$$

On the other hand, we have $\liminf_{k \to \infty} \tfrac{1}{\tau_k} d_A(x_k + \tau_k y) \geq 0$ and so

$$\lim_{k \to \infty} \tfrac{1}{\tau_k} d_A(x_k + \tau_k y) = 0.$$

By the definition of d_A, for each $k \in \mathbb{N}$ there exists $z_k \in A$ satisfying

$$\|z_k - (x_k + \tau_k y)\| \leq d_A(x_k + \tau_k y) + \tfrac{\tau_k}{k}.$$

Setting $y_k := \tfrac{1}{\tau_k}(z_k - x_k)$, we obtain

$$\|y_k - y\| = \tfrac{1}{\tau_k}\|z_k - (x_k + \tau_k y)\| \leq \tfrac{1}{\tau_k} d_A(x_k + \tau_k y) + \tfrac{1}{k} \to 0 \quad \text{as } k \to \infty.$$

Further we have $x_k + \tau_k y_k = z_k \in A$ for each k. Hence $y \in T_C(A, \bar{x})$.

(II) Now let $y \in T_C(A, \bar{x})$. By the definition of d_A°, there exist $\tau_k \downarrow 0$ and $x_k' \to \bar{x}$ satisfying

$$d_A^\circ(\bar{x}, y) = \lim_{k \to \infty} \tfrac{1}{\tau_k}\big(d_A(x_k' + \tau_k y) - d_A(x_k')\big). \tag{11.4}$$

Notice that the sequence (x_k') need not belong to A. But since there does exist a sequence (x_k) in A converging to $\bar{x}$ (set, for example, $x_k := \bar{x}$ for each k), we have

$$d_A^\circ(\bar{x}, y) \geq \limsup_{k \to \infty} \tfrac{1}{\tau_k}\big(d_A(x_k + \tau_k y) - d_A(x_k)\big) \geq 0.$$

Therefore it suffices to show that the right-hand side of (11.4) is not greater than zero. Let $z_k \in A$ be such that

$$\|z_k - x_k'\| \leq d_A(x_k') + \tfrac{\tau_k}{k}. \tag{11.5}$$

Since $x_k' \to \bar{x}$ as $k \to \infty$, it follows that

$$\|z_k - \bar{x}\| \leq \|z_k - x_k'\| + \|x_k' - \bar{x}\| \leq d_A(x_k') + \tfrac{\tau_k}{k} + \|x_k' - \bar{x}\| \to 0$$

and so $z_k \to \bar{x}$. Since $y \in T_C(A, \bar{x})$, there exists $y_k \to y$ satisfying $z_k + \tau_k y_k \in A$ for each $k \in \mathbb{N}$. Moreover, since d_A is L-continuous with L-constant 1 (see (11.3)), we obtain

$$d_A(x_k' + \tau_k y) \leq d_A(z_k + \tau_k y_k) + \|z_k - x_k'\| + \tau_k \|y_k - y\|$$
$$\underset{(11.5)}{\leq} d_A(x_k') + \tau_k\big(\tfrac{1}{k} + \|y_k - y\|\big).$$

Hence the right-hand side of (11.4) is in fact at most equal to zero. $\qquad\square$

Corollary 11.1.7 *If A is convex, then $T_C(A, \bar{x}) = T(A, \bar{x}) = \mathrm{cl}\,\mathbb{R}_+(A - \bar{x})$.*

Proof. By Proposition 11.1.2(d) we know that the second equation holds true. We now show that $T_C(A, \bar{x}) = \mathrm{cl}\,\mathbb{R}_+(A - \bar{x})$. Since A is convex, the functional d_A is convex, and it is also L-continuous. Hence d_A is regular (Remark 7.4.2) and so $d_A^\circ(\bar{x}, \cdot) = d_{A,G}(\bar{x}, \cdot)$. By Proposition 11.1.6, $u \in T_C(A, \bar{x})$ is equivalent to $d_{A,G}(\bar{x}, u) = 0$ and so to $\lim_{\tau \downarrow 0} \tau^{-1} d_A(\bar{x} + \tau u) = 0$. The latter relation holds if and only if for each $k \in \mathbb{N}$ there exist $\tau_k \in (0, \frac{1}{k})$ and $x_k \in A$ such that

$$u_k := \frac{1}{\tau_k}\left(\bar{x} + \tau_k u - x_k\right) \to 0 \quad \text{as } k \to \infty.$$

Noting that $u = \frac{1}{\tau_k}\left(x_k - \bar{x}\right) + u_k$ completes the proof. $\qquad\square$

The following "ball characterizations" of the Clarke tangent cone and the hypertangent cone will be useful in the sequel.

Lemma 11.1.8

(a) *One has $y \in T_C(A, \bar{x})$ if and only if for any $\epsilon > 0$ there exists $\delta > 0$ such that*
$$A \cap B(\bar{x}, \delta) + \tau y \subseteq A + \tau B(o, \epsilon) \quad \forall \tau \in (0, \delta).$$

(b) *One has $y \in H(A, \bar{x})$ if and only if there exists $\epsilon > 0$ such that*
$$u + \tau v \in A \quad \text{whenever } u \in A \cap B(\bar{x}, \epsilon), \ v \in B(y, \epsilon), \ \tau \in (0, \epsilon).$$

Proof. See Exercise 11.7.4. $\qquad\square$

The next result will be applied in Sect. 12.3 to derive a multiplier rule.

Proposition 11.1.9 $H(A, \bar{x})$ *is always open. If $H(A, \bar{x})$ is nonempty, then* $\mathrm{int}\,T_C(A, \bar{x}) = H(A, \bar{x})$.

Proof.

(I) It follows easily from Lemma 11.1.8 that the cone $H(A, \bar{x})$ is open. Since it is a subset of $T_C(A, \bar{x})$, we always have $H(A, \bar{x}) \subseteq \mathrm{int}\,T_C(A, \bar{x})$.

(IIa) Assuming now that $H(A, \bar{x})$ is nonempty, we have to show that

$$\mathrm{int}\,T_C(A, \bar{x}) \subseteq H(A, \bar{x}). \tag{11.6}$$

This will be done when we have verified the relation

$$H(A, \bar{x}) + T_C(A, \bar{x}) \subseteq H(A, \bar{x}). \tag{11.7}$$

In fact, let $y \in \mathrm{int}\,T_C(A, \bar{x})$ be given. Choose $z \in H(A, \bar{x})$. Then $y - \eta z \in T_C(A, \bar{x})$ for some sufficiently small $\eta > 0$. Since we also have $\eta z \in H(A, \bar{x})$, we see that

$$y = \eta z + (y - \eta z) \underset{(11.7)}{\in} H(A, \bar{x}).$$

(IIb) Thus it remains to verify (11.7). Let $y_1 \in \mathrm{H}(A, \bar{x})$ and $y_2 \in \mathrm{T}_C(A, \bar{x})$ be given. We have to show that for some $\epsilon > 0$,

$$A \cap \mathrm{B}(\bar{x}, \epsilon) + \tau \mathrm{B}(y_1 + y_2, \epsilon) \subseteq A \quad \forall \tau \in (0, \epsilon). \tag{11.8}$$

Since $y_1 \in \mathrm{H}(A, \bar{x})$, there exists $\epsilon_1 > 0$ such that

$$A \cap \mathrm{B}(\bar{x}, \epsilon_1) + \tau \mathrm{B}(y_1, \epsilon_1) \subseteq A \quad \forall \tau \in (0, \epsilon_1). \tag{11.9}$$

Furthermore, $y_2 \in \mathrm{T}_C(A, \bar{x})$ implies that for some $\epsilon_2 > 0$ we have

$$A \cap \mathrm{B}(\bar{x}, \epsilon_2) + \tau y_2 \subseteq A + \mathrm{B}\left(o, \tau \frac{\epsilon_1}{2}\right) \quad \forall \tau \in (0, \epsilon_2). \tag{11.10}$$

Now let ϵ be such that

$$0 < \epsilon < \min\left\{\epsilon_2, \frac{\epsilon_1}{2}, \frac{\epsilon_1}{1 + \epsilon_1 + \|y_2\|}\right\}.$$

Let y be an element of the left-hand side of (11.7). It follows that $y = z + \tau(y_1 + y_2 + \epsilon z')$, where $z \in A \cap \mathrm{B}(\bar{x}, \epsilon)$ and $z' \in \mathrm{B}(o, 1)$. Since $\epsilon \leq \epsilon_2$, by (11.10) we see that for some $z'' \in \mathrm{B}(o, 1)$ we have $z + \tau y_2 - \tau \frac{\epsilon_1}{2} z'' \in A$. Moreover, we obtain

$$\left\|\bar{x} - \left(z + \tau y_2 - \tau \frac{\epsilon_1}{2} z''\right)\right\| \leq \|\bar{x} - z\| + \tau \left\|y_2 - \frac{\epsilon_1}{2} z''\right\|$$
$$< \epsilon + \epsilon\left(\|y_2\| + \frac{\epsilon_1}{2}\right) < \epsilon_1$$

and so $z + \tau y_2 - \tau \frac{\epsilon_1}{2} z'' \in A \cap \mathrm{B}(\bar{x}, \epsilon_1)$. In view of (11.9), it follows that

$$z + \tau y_2 - \tau \frac{\epsilon_1}{2} z'' + \tau \mathrm{B}(y_1, \epsilon_1) \subseteq A. \tag{11.11}$$

Notice that

$$y = z + \tau\left(y_2 - \frac{\epsilon_1}{2} z''\right) + \tau\left(y_1 + \left(\epsilon z' + \frac{\epsilon_1}{2} z''\right)\right),$$

where $\left\|\epsilon z' + \frac{\epsilon_1}{2} z''\right\| \leq \epsilon + \frac{\epsilon_1}{2} < \epsilon_1$. Hence (11.11) implies $y \in A$. We have thus verified (11.7) and so (11.6).   □

In view of Proposition 11.1.9 we give the following:

Definition 11.1.10 The set A is said to be *epi-Lipschitzian* at $\bar{x}$ if $\mathrm{H}(A, \bar{x})$ is nonempty. If $\mathrm{H}(A, x)$ is nonempty for all $x \in A$, then A is said to be *epi-Lipschitzian*.

Remark 11.1.11 Rockafellar [184] showed that if E is finite dimensional, then A is epi-Lipschitzian at $\bar{x}$ if and only if $\operatorname{int} \mathrm{T}_C(A, \bar{x})$ is nonempty. He also gave an example of a *convex* subset A of an infinite-dimensional normed vector space such that $\operatorname{int} \mathrm{T}_C(A, \bar{x})$ is nonempty but A is not epi-Lipschitzian.

The following result provides an important class of epi-Lipschitzian sets.

Proposition 11.1.12 *If A is convex and $\operatorname{int} A$ is nonempty, then A is epi-Lipschitzian.*

Proof.

(I) Let $x' \in \operatorname{int} A$ and choose $\epsilon > 0$ such that $B(x', 2\epsilon) \subseteq A$. Now let $\bar{x}$ be any element of A. We show that $y := x' - \bar{x} \in H(A, \bar{x})$. First notice that

$$\bar{x} + B(y, 2\epsilon) \subseteq A. \tag{11.12}$$

In fact, if $v \in B(y, 2\epsilon)$ and $v' := \bar{x} + v$, then $\|v' - x'\| = \|v - y\| \leq 2\epsilon$ and so $v' \in B(x', 2\epsilon) \subseteq A$.

(II) Now let $u \in A \cap B(\bar{x}, \epsilon)$ and $v \in B(y, \epsilon)$ be given. We then have

$$\|(v + u - \bar{x}) - y\| \leq \|v - y\| + \|u - \bar{x}\| \leq 2\epsilon$$

and so (by (11.12)) $u + v = \bar{x} + (v + u - \bar{x}) \in A$. Since A is convex, we obtain $u + \tau v = \tau(u+v) + (1-\tau)u \in A$ for any $\tau \in [0, 1]$. By Lemma 11.1.8 we conclude that $y \in H(A, \bar{x})$. $\qquad\square$

In analogy to Proposition 11.1.3 we have the following intersection result.

Proposition 11.1.13 *If $A, B \subseteq E$ and $\bar{x} \in A \cap B$, then*

$$T_C(A, \bar{x}) \cap H(B, \bar{x}) \subseteq T_C(A \cap B, \bar{x}).$$

Proof. See Exercise 11.7.5. $\qquad\square$

We formulate without proof a stronger intersection result due to Rockafellar [183].

Proposition 11.1.14 *Let $A, B \subseteq E$, let $\bar{x} \in A \cap B$, and assume that $T_C(A, \bar{x}) \cap H(B, \bar{x})$ is nonempty. Then*

$$T_C(A, \bar{x}) \cap T_C(B, \bar{x}) \subseteq T_C(A \cap B, \bar{x}).$$

11.2 Normal Cones: First Properties

Let again A be a nonempty subset of the normed vector space E and let $\bar{x} \in A$. We now define several normal cones to A at $\bar{x}$.

Definition 11.2.1 If A is convex, then

$$N(A, \bar{x}) := (A - \bar{x})^\circ = \{v \in E^* \mid \langle v, x - \bar{x} \rangle \leq 0 \quad \forall x \in A\}$$

is called *normal cone* to A at $\bar{x}$ in the sense of convex analysis.

Lemma 11.2.2 *If A is convex, then*

$$N(A, \bar{x}) = T(A, \bar{x})^\circ \quad and \quad N(A, \bar{x}) = \partial \delta_A(\bar{x}).$$

Proof. The first equation follows by Proposition 11.1.2(d) and the second is an immediate consequence of the definition of $N(A, \bar{x})$. $\qquad\square$

In the nonconvex case, the definition of normal cone is modeled on one or the other of the preceding equations.

Definition 11.2.3 The cone $N_C(A, \bar{x}) := T_C(A, \bar{x})^\circ$ is called *Clarke normal cone to A at $\bar{x}$*.

Recall that $\mathrm{cl}^* M$ denotes the $\sigma(E^*, E)$-closure of $M \subseteq E^*$.

Proposition 11.2.4 *One has $N_C(A, \bar{x}) = \mathrm{cl}^* \big(\mathbb{R}_+ \partial_\circ \mathrm{d}_A(\bar{x}) \big)$.*

Proof. By Proposition 7.3.7, we have $\mathrm{d}_A^\circ(\bar{x}, y) = \max\{\langle v, y \rangle \mid v \in \partial_\circ \mathrm{d}_A(\bar{x})\}$. Using this and Proposition 11.1.6, we obtain

$$y \in T_C(A, \bar{x}) \iff \mathrm{d}_A^\circ(\bar{x}, y) = 0 \iff \langle v, y \rangle \leq 0 \, \forall v \in \partial_\circ \mathrm{d}_A(\bar{x})$$
$$\iff y \in \big(\partial_\circ \mathrm{d}_A(\bar{x}) \big)^\circ,$$

which implies $N_C(A, \bar{x}) = \big(\partial_\circ \mathrm{d}_A(\bar{x}) \big)^{\circ\circ}$. The bipolar theorem (Proposition 2.3.3) finally yields the assertion. In this connection, recall that the Clarke subdifferential is convex. $\qquad\square$

Definition 11.2.5 Let A be a nonempty subset of E and $\bar{x} \in A$. Then we call

$$\begin{aligned}
N_F(A, \bar{x}) &:= \partial_F \delta_A(\bar{x}) \quad \textit{Fréchet normal cone to A at $\bar{x}$,} \\
N_V(A, \bar{x}) &:= \partial_V \delta_A(\bar{x}) \quad \textit{viscosity normal cone to A at $\bar{x}$,} \\
N_P(A, \bar{x}) &:= \partial_P \delta_A(\bar{x}) \quad \textit{proximal normal cone to A at $\bar{x}$.}
\end{aligned}$$

Each $u \in N_F(A, \bar{x})$ is said to be a *Fréchet normal* to A at $\bar{x}$, analogously we use *viscosity normal* and *proximal normal*.

We first give a simple but useful characterization of Fréchet normal cones.

Proposition 11.2.6 *Let A be a nonempty subset of E and $\bar{x} \in A$. Then for any $x^* \in E^*$, the following assertions are mutually equivalent:*

(a) $x^* \in N_F(A, \bar{x})$.
(b) *For every $\epsilon > 0$ there exists $\delta > 0$ such that*

$$\langle x^*, x - \bar{x} \rangle \leq \epsilon \|x - \bar{x}\| \quad \forall x \in A \cap B(\bar{x}, \delta).$$

(c) *There exists a function $\varphi : E \to \mathbb{R}$ that is F-differentiable at $\bar{x}$ with $\varphi'(\bar{x}) = x^*$ and attains a maximum over A at $\bar{x}$.*

Proof.

(a) $\Longrightarrow$ (b): This follows immediately from the definition of $N_F(A, \bar{x})$.

(b) $\Longrightarrow$ (c): It is easy to check that the function $\varphi : E \to \mathbb{R}$ defined by

$$\varphi(x) := \begin{cases} \min\{0, \langle x^*, x - \bar{x} \rangle\} & \text{if } x \in A, \\ \langle x^*, x - \bar{x} \rangle & \text{otherwise} \end{cases}$$

has the required properties.

(c) $\Longrightarrow$ (a): According to (c) we have

$$\varphi(x) = \varphi(\bar{x}) + \langle x^*, x - \bar{x} \rangle + r(x) \quad \text{where} \quad r(x)/\|x - \bar{x}\| \to 0 \quad \text{as } x \to \bar{x}.$$

Since $\varphi(x) \le \varphi(\bar{x})$ for any $x \in A$, statement (a) follows. $\qquad\square$

In a Hilbert space, proximal normals can be characterized in various ways.

Proposition 11.2.7 *Let A be a nonempty subset of the Hilbert space E and let $\bar{x} \in A$. Then for any $u \in E$, the following assertions are mutually equivalent:*

(a) *$u \in N_P(A, \bar{x})$.*

(b) *Either $u = o$ or there exist $\lambda > 0$ and $z \in E \setminus A$ such that $u = \lambda(z - \bar{x})$ and $\bar{x} \in \operatorname{proj}_A(z)$.*

(c) *There exists $\rho \ge 0$ such that $(u \mid x - \bar{x}) \le \rho\|x - \bar{x}\|^2$ for any $x \in A$.*

(d) *There exist $\sigma \ge 0$ and $\epsilon > 0$ such that $(u \mid x - \bar{x}) \le \sigma\|x - \bar{x}\|^2$ for any $x \in A \cap B(\bar{x}, \epsilon)$.*

(e) *There exists $\tau > 0$ such that $d_A(\bar{x} + \tau u) = \tau\|u\|$.*

Proof. We prepare the proof with two observations. First, we have

$$\bar{x} \in \operatorname{proj}_A(z) \quad \Longleftrightarrow \quad \|z - \bar{x}\| \le \|z - x\| \; \forall x \in A$$
$$\Longleftrightarrow \quad (z - \bar{x} \mid x - \bar{x}) \le (z - x \mid z - x) \; \forall x \in A.$$

Simplifying the inner products in the last inequality leads to

$$\bar{x} \in \operatorname{proj}_A(z) \quad \Longleftrightarrow \quad (z - \bar{x} \mid x - \bar{x}) \le \frac{1}{2}\|x - \bar{x}\|^2 \quad \forall x \in A. \qquad (11.13)$$

Second, for any $\tau > 0$ we have

$$\|x - (\bar{x} + \tau u)\|^2 = \|x - \bar{x}\|^2 - 2\tau(u \mid x - \bar{x}) + \tau^2\|u\|^2. \qquad (11.14)$$

(a) $\Longrightarrow$ (b): This is obvious for $u = o$. If $u \in \partial_P \delta_A(\bar{x})$ and $u \ne o$, then there exists $\sigma > 0$ satisfying

$$(u \mid x - \bar{x}) - \sigma\|x - \bar{x}\|^2 \le 0 \quad \forall x \in A.$$

By (11.13) we obtain $\bar{x} \in \operatorname{proj}_A(z)$, where $z := \bar{x} + \frac{1}{2\sigma}u$. Observe that $z \notin A$ and $u = 2\sigma(z - \bar{x})$.

(b) $\implies$ (c): Let u be as in (b). In view of (11.13) it follows that $(u \mid x - \bar{x}) \leq \frac{\lambda}{2}\|x - \bar{x}\|^2$ for each $x \in A$.

(c) $\implies$ (d) is obvious.

(d) $\implies$ (e): Assume that (d) holds and choose any $\tau > 0$. Using (11.14), we obtain

$$\|x - (\bar{x} + \tau u)\|^2 \geq (1 - 2\tau\sigma)\|x - \bar{x}\|^2 + \tau^2\|u\|^2 \quad \forall x \in A \cap \mathrm{B}(\bar{x}, \epsilon).$$

Since for $x := \bar{x}$ we have $\|x - (\bar{x} + \tau u)\|^2 = \tau^2\|u\|^2$, we conclude that $d_A(\bar{x} + \tau u) = \tau\|u\|$.

(e) $\implies$ (a): The condition (e) implies

$$\|x - (\bar{x} + \tau u)\|^2 \geq \tau^2\|u\|^2 \quad \forall x \in A, \tag{11.15}$$

which by (11.14) entails

$$\delta_A(x) - \delta_A(\bar{x}) \geq (u \mid x - \bar{x}) - \frac{1}{2\tau}\|x - \bar{x}\|^2 \quad \forall x \in E \tag{11.16}$$

and so $u \in \mathrm{N}_P(A, \bar{x})$. $\qquad\square$

Geometric Interpretation

The equivalence of (a) and (b) means that $\mathrm{N}_P(A, \bar{x})$ collects all points u on rays emanating from $\bar{x}$ and meeting some point $z \in E \setminus A$ for which $\bar{x}$ is the best approximation with respect to A (Fig. 11.2).

It is clear that generally we have $\mathrm{N}_P(A, \bar{x}) \subseteq \mathrm{N}_F(A, \bar{x})$.

Proposition 11.2.8 *Assume that E is a Fréchet smooth Banach space, A is convex and closed, and $\bar{x} \in A$. Then*

$$\mathrm{N}_P(A, \bar{x}) = \mathrm{N}_F(A, \bar{x}) = \mathrm{N}_C(A, \bar{x}) = \mathrm{N}(A, \bar{x}).$$

Proof. See Exercise 11.7.6. $\qquad\square$

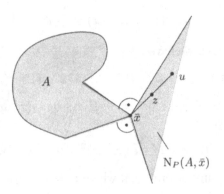

Fig. 11.2

Observe that the closedness of A is assumed only to ensure that δ_A is l.s.c. which enters the definition of the proximal subdifferential. If A is convex (and closed), then $u \in N_P(A, \bar{x})$ if and only if (b) of Proposition 11.2.7 holds, which is geometrically interpreted in Remark 5.3.2 and Fig. 5.1.

11.3 Tangent and Normal Cones to Epigraphs

Let $f : E \to \overline{\mathbb{R}}$ be proper and let $\bar{x} \in \operatorname{dom} f$. Our aim now is to give representations of approximating cones to $\operatorname{epi} f$ at $\bar{x}$. Recall the lower directional H-derivative

$$\underline{f}_H(\bar{x}, y) = \liminf_{\substack{\tau \downarrow 0 \\ z \to y}} \tfrac{1}{\tau}\big(f(\bar{x} + \tau z) - f(\bar{x})\big), \quad y \in E.$$

Proposition 11.3.1 *Let $f : E \to \overline{\mathbb{R}}$ be proper and $\bar{x} \in \operatorname{dom} f$.*

(a) *There always holds*

$$\mathrm{T}\big(\operatorname{epi} f, (\bar{x}, f(\bar{x}))\big) = \operatorname{epi} \underline{f}_H(\bar{x}, \cdot).$$

(b) *If f is locally L-continuous around $\bar{x}$, then*

$$\mathrm{T}_C\big(\operatorname{epi} f, (\bar{x}, f(\bar{x}))\big) = \operatorname{epi} f^\circ(\bar{x}, \cdot).$$

Proof.

(a) (I) Let $(y, \rho) \in \mathrm{T}\big(\operatorname{epi} f, (\bar{x}, f(\bar{x}))\big)$ be given. Then there exist sequences (τ_k) in $(0, +\infty)$ and (z_k, ρ_k) in $E \times \mathbb{R}$ satisfying $\tau_k \downarrow 0$, $z_k \to y$, and $\rho_k \to \rho$ as $k \to \infty$ such that $(\bar{x}, f(\bar{x})) + \tau_k(z_k, \rho_k) \in \operatorname{epi} f$ for any $k \in \mathbb{N}$. It follows that

$$\frac{1}{\tau_k}\big(f(\bar{x} + \tau_k z_k) - f(\bar{x})\big) \le \rho_k \quad \forall k \in \mathbb{N}$$

and so $\underline{f}_H^+(\bar{x}, y) \le \rho$ which means $(y, \rho) \in \operatorname{epi} \underline{f}_H^+(\bar{x}, \cdot)$.

(II) Now let $(y, \rho) \in \operatorname{epi} \underline{f}_H^+(\bar{x}, \cdot)$ be given. We then have

$$\inf_{\substack{0 < \tau < \epsilon \\ \|z - y\| < \epsilon}} \tfrac{1}{\tau}\big(f(\bar{x} + \tau z) - f(\bar{x})\big) \le \rho \quad \forall \epsilon > 0.$$

Hence for any $k \in \mathbb{N}$ there exists $\tau_k \in (0, \tfrac{1}{k})$ and $z_k \in B(y, \tfrac{1}{k})$ such that

$$\frac{1}{\tau_k}\big(f(\bar{x} + \tau_k z_k) - f(\bar{x})\big) < \rho + \frac{1}{k} \quad \forall k \in \mathbb{N}.$$

We thus see that $\tau_k \downarrow 0$, $(z_k, \rho + \tfrac{1}{k}) \to (y, \rho)$, and $(\bar{x}, f(\bar{x})) + \tau_k(z_k, \rho + \tfrac{1}{k}) \in \operatorname{epi} f$. Hence $(y, \rho) \in \mathrm{T}\big(\operatorname{epi} f, (\bar{x}, f(\bar{x}))\big)$.

(b) This is verified analogously. $\qquad\square$

Corollary 11.3.2 *If $f : E \to \overline{\mathbb{R}}$ is locally L-continuous around $\bar{x} \in \operatorname{dom} f$, then for any $x^* \in E^*$ one has*

$$x^* \in \partial_o f(\bar{x}) \quad \Longleftrightarrow \quad (x^*, -1) \in N_C(\operatorname{epi} f, (\bar{x}, f(\bar{x}))).$$

Proof. By Proposition 7.3.7 we have $x^* \in \partial_o f(\bar{x})$ if and only if $f^\circ(\bar{x}, y) \geq \langle x^*, y \rangle$ for any $y \in E$. Hence using Proposition 11.3.1 and the definition of the Clarke normal cone, we obtain

$$
\begin{aligned}
x^* \in \partial_o f(\bar{x}) \quad &\Longleftrightarrow \quad (y, \rho) \in \operatorname{epi} f^\circ(\bar{x}, \cdot) \quad \forall y \in E \quad \forall \rho \geq f^\circ(\bar{x}, y) \\
&\Longleftrightarrow \quad \langle (x^*, -1), (y, \rho) \rangle = \langle x^*, y \rangle - \rho \leq 0 \quad \forall (y, \rho) \in \operatorname{epi} f^\circ(\bar{x}, \cdot) \\
&\Longleftrightarrow \quad (x^*, -1) \in T_C\big(\operatorname{epi} f, (\bar{x}, , f(\bar{x}))\big)^\circ = N_C\big(\operatorname{epi} f, (\bar{x}, f(\bar{x}))\big). \quad \square
\end{aligned}
$$

If f is strictly H-differentiable at $\bar{x}$, then by Proposition 7.3.9 the assertion of the corollary reduces to $(f'(\bar{x}), -1) \in N_C(\operatorname{epi} f, (\bar{x}, f(\bar{x})))$. In the language of differential geometry in the plane this means that $(f'(\bar{x}), -1)$ is a normal vector to graph f at the point $(\bar{x}, f(\bar{x}))$.

Remark 11.3.3 By Proposition 11.3.1 we have

$$\underline{f}_H(\bar{x}, y) = \inf\{\rho \in \mathbb{R} \mid (y, \rho) \in T(\operatorname{epi} f, (\bar{x}, f(\bar{x})))\} \quad \forall y \in E,$$

$$f^\circ(\bar{x}, y) = \inf\{\rho \in \mathbb{R} \mid (y, \rho) \in T_C(\operatorname{epi} f, (\bar{x}, f(\bar{x})))\} \quad \forall y \in E.$$

This shows that directional derivatives can also be defined with the aid of approximating cones. Furthermore, Corollary 11.3.2 and Proposition 11.3.4 (which is the proximal subdifferential analogue to Corollary 11.3.2) indicate that subdifferentials can be defined via normal cones.

Proposition 11.3.4 *Let E be a Hilbert space, let $f : E \to \overline{\mathbb{R}}$ be proper and l.s.c., and let $\bar{x} \in \operatorname{dom} f$. Then for each $x^* \in E$ one has*

$$x^* \in \partial_P f(\bar{x}) \quad \Longleftrightarrow \quad (x^*, -1) \in N_P(\operatorname{epi} f, (\bar{x}, f(\bar{x}))).$$

Proof. $\Longrightarrow$: Let $x^* \in \partial_P f(\bar{x})$ be given. By the definition of the proximal subdifferential there exist $\sigma > 0$ and $\delta > 0$ such that for all $x \in B(\bar{x}, \delta)$ and all $\alpha \geq f(x)$ we have

$$\alpha - f(\bar{x}) + \sigma\big[\|x - \bar{x}\|^2 + (\alpha - f(\bar{x}))^2\big] \geq (x^* \mid x - \bar{x}). \tag{11.17}$$

In other words, if $x \in B(\bar{x}, \delta)$ and $(x, \alpha) \in \operatorname{epi} f$, then

$$
\begin{aligned}
\big((x^*, -1) \mid (x, \alpha) - (\bar{x}, f(\bar{x}))\big) &= \big((x^*, -1) \mid (x - \bar{x}, \alpha - f(\bar{x}))\big) \\
&= (x^* \mid x - \bar{x}) - \alpha + f(\bar{x}) \\
&\underset{(11.17)}{\leq} \sigma\|(x, \alpha) - (\bar{x}, f(\bar{x}))\|^2.
\end{aligned}
$$

By Proposition 11.2.7 we conclude that $(x^*, -1) \in N_P\big(\text{epi } f, (x, f(x))\big)$.

$\Longleftarrow$: Assume now that $(x^*, -1) \in N_P\big(\text{epi } f, (\bar{x}, f(\bar{x}))\big)$. By Proposition 11.2.7 there exists $\eta > 0$ such that for all $(x, \alpha) \in \text{epi } f$ we have

$$\big(\eta(x^*, -1) \,|\, (x - \bar{x}, \alpha - f(\bar{x}))\big) \leq \tfrac{1}{2}\|(x - \bar{x}, \alpha - f(\bar{x}))\|^2,$$

which in view of (11.13) implies that

$$(\bar{x}, f(\bar{x})) \in \text{proj}_{\text{epi} f}(p), \quad \text{where } p := (\bar{x}, f(\bar{x})) + \eta(x^*, -1). \tag{11.18}$$

The definition of the projection now shows (cf. Fig. 11.3, where $\delta' := \|\eta(x^*, -1)\|$) that

$$\|\eta(x^*, -1)\|^2 \leq \|p - (x, \alpha)\|^2 \quad \forall\, (x, \alpha) \in \text{epi } f.$$

In particular, choosing $\alpha = f(x)$ we obtain

$$\delta^2\|x^*\|^2 + \delta^2 \leq \|\bar{x} - y + \delta x^*\|^2 \big(f(\bar{x}) - f(x) - \delta\big)^2.$$

Evaluating $\|\cdot\|^2$ via the inner product, the latter inequality passes into

$$\delta^2 + 2\delta\,(x^* \,|\, x - \bar{x}) - \|x - \bar{x}\|^2 \leq \big(f(x) - f(\bar{x}) + \delta\big)^2. \tag{11.19}$$

Now choose $\epsilon > 0$ so small that for each $x \in B(\bar{x}, \epsilon)$ the left-hand side of (11.19) is positive and at the same time $f(x) > f(\bar{x}) - \delta$ (the latter being possible since f is l.s.c. at $\bar{x}$). For each such x we obtain from (11.19) that $f(x) \geq g(x)$, where

$$g(x) := f(\bar{x}) - \delta + \big(\delta^2 + 2\delta\,(x^* \,|\, x - \bar{x}) - \|x - \bar{x}\|^2\big)^{1/2}.$$

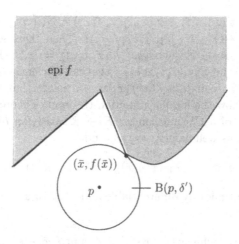

Fig. 11.3

By Example 3.6.2 the function g is twice continuously differentiable on $B(\bar{x}, \epsilon)$ (upon diminishing ϵ if necessary) and $g'(\bar{x}) = x^*$. Now Proposition 3.5.1 implies that with some $\sigma > 0$ we have

$$g(x) \geq g(\bar{x}) + (x^* \,|\, x - \bar{x}) - \sigma \|x - \bar{x}\|^2 \quad \forall\, x \in B(\bar{x}, \epsilon).$$

This together with $f(x) \geq g(x)$ and $f(\bar{x}) = g(\bar{x})$ gives

$$f(x) \geq f(\bar{x}) + (x^* \,|\, x - \bar{x}) - \sigma |x - \bar{x}|^2 \quad \forall\, x \in B(\bar{x}, \epsilon)$$

and so $x^* \in \partial_P f(\bar{x})$. $\square$

Now we derive a result analogous to Proposition 11.3.4 for the viscosity (Fréchet) subdifferential and the viscosity (Fréchet) normal cone.

Proposition 11.3.5 *Let $f : E \to \bar{\mathbb{R}}$ be proper and l.s.c., and let $\bar{x} \in \mathrm{dom}\, f$. Then for each $x^* \in E$ one has*

$$x^* \in \partial_V f(\bar{x}) \quad \Longleftrightarrow \quad (x^*, -1) \in N_V\big(\mathrm{epi} f, (\bar{x}, f(\bar{x}))\big). \tag{11.20}$$

If, in addition, E is a Fréchet smooth Banach space, then

$$x^* \in \partial_F f(\bar{x}) \quad \Longleftrightarrow \quad (x^*, -1) \in N_F\big(\mathrm{epi} f, (\bar{x}, f(\bar{x}))\big). \tag{11.21}$$

Proof.

(I) Let $x^* \in \partial_V f(\bar{x})$ be given. By definition there exists a C^1 function g such that $g'(\bar{x}) = x^*$ and $f - g$ attains a local minimum at $\bar{x}$. Define $h(y, r) := g(y) - r$ for y near $\bar{x}$ and $r \in \mathbb{R}$. Then h is a C^1 function satisfying $h'(\bar{x}, f(\bar{x})) = (x^*, -1)$ and

$$\delta_{\mathrm{epi}\, f}(y, r) - h(y, r) \geq \delta_{\mathrm{epi}\, f}(\bar{x}, f(\bar{x})) - h(\bar{x}, f(\bar{x})).$$

It follows that $(x^*, -1) \in \partial_V \delta_{\mathrm{epi}\, f}(\bar{x}, f(\bar{x})) = N_V\big(\mathrm{epi}\, f, (\bar{x}, f(\bar{x}))\big)$.

(II) Now let $(x^*, -1) \in N_V\big(\mathrm{epi}\, f, (\bar{x}, f(\bar{x}))\big)$. Then there exists a C^1 function $h : E \times \mathbb{R} \to \mathbb{R}$ such that $h'(\bar{x}, f(\bar{x})) = (x^*, -1)$ and $h(x, t) \leq h(\bar{x}, f(\bar{x})) = 0$ for any $(x, t) \in \mathrm{epi}\, f$ that is sufficiently close to $(\bar{x}, f(\bar{x}))$. Concerning the equation $h(\bar{x}, f(\bar{x})) = 0$, notice that h can be chosen in this way. Now the implicit function theorem (Theorem 3.7.2) ensures the existence of a C^1 function $g : E \to \mathbb{R}$ satisfying $h(x, g(x)) = 0$ for any x near $\bar{x}$ as well as $g(\bar{x}) = f(\bar{x})$ and

$$g'(\bar{x}) = -h_{|2}(\bar{x}, g(\bar{x}))^{-1} \circ h_{|1}(\bar{x}, g(\bar{x})) = x^*.$$

Since h is continuously differentiable and $h_{|2}(\bar{x}, g(\bar{x})) = -1$, there exists $\epsilon > 0$ such that

$$h(x, t) < h(x, s) \quad \text{whenever} \quad x \in B(\bar{x}, \epsilon) \text{ and } f(\bar{x}) - \epsilon < s < t < f(\bar{x}) + \epsilon. \tag{11.22}$$

Since g is continuous at $\bar{x}$ and f is l.s.c. at $\bar{x}$, there exists $\delta \in (0, \epsilon)$ such that

$$f(\bar{x}) - \epsilon \le g(x) \le f(\bar{x}) + \epsilon \quad \text{and} \quad f(x) > f(\bar{x}) - \epsilon \quad \forall x \in \mathrm{B}(\bar{x}, \delta).$$

Now let $x \in \mathrm{B}(\bar{x}, \delta)$. If $f(x) \ge f(\bar{x}) + \epsilon$, then we immediately have $f(x) - g(x) \ge 0 = f(\bar{x}) - g(\bar{x})$. But the latter inequality also holds if $f(x) < f(\bar{x}) + \epsilon$ because in this case $h(x, f(x)) \le 0 = h(x, g(x))$ by (11.22). Thus, we have shown that $x^* \in \partial_V f(\bar{x})$.

(III) The additional assertion follows from the above by virtue of Theorem 9.1.7 and the definition of the Fréchet normal cone. In this connection notice that if E is Fréchet smooth, so is $E \times \mathbb{R}$. □

Remark 11.3.6 Here we obtained the characterization (11.21) as a by-product of (11.20). In a more general setting we shall later see that (11.21) holds without the hypothesis that E be Fréchet smooth.

11.4 Representation of Tangent Cones

Our aim in this section is to characterize (subsets of) approximating cones of sets given by inequalities and/or equations.

Approximating $h^{-1}(Q)$

We make the following assumptions:

(A) E and F are Banach spaces, Q is a nonempty closed convex subset of F, $h : E \to F$ is continuous on E and continuously differentiable at $\bar{x} \in h^{-1}(Q)$.

Theorem 11.4.1 *Let the assumptions* (A) *and the Robinson condition*

$$o \in \mathrm{int}\big(h(\bar{x}) + h'(\bar{x})(E) - Q\big)$$

be satisfied. Then

$$\mathrm{T}(h^{-1}(Q), \bar{x}) = h'(\bar{x})^{-1}\big(\mathrm{T}(Q, h(\bar{x}))\big). \tag{11.23}$$

Proof.

(I) We show

$$h'(\bar{x})^{-1}\big(\mathrm{T}(Q, h(\bar{x}))\big) \subseteq \mathrm{T}(h^{-1}(Q), \bar{x}). \tag{11.24}$$

Let $y \in h'(\bar{x})^{-1}\big(\mathrm{T}(Q, h(\bar{x}))\big)$. By Proposition 11.1.5 there exists $\tau_k \downarrow 0$ such that

$$\lim_{k \to \infty} \tau_k^{-1} \mathrm{d}(Q, h(\bar{x}) + \tau_k h'(\bar{x})y) = 0. \tag{11.25}$$

Since h is continuously differentiable at $\bar{x}$, we further have

$$h(\bar{x} + \tau y) = h(\bar{x}) + \tau h'(\bar{x})y + \mathbf{o}(\tau), \quad \tau \downarrow 0. \tag{11.26}$$

Hence $\tau_k^{-1}\mathrm{d}(Q, h(\bar{x} + \tau_k y)) \to 0$ as $k \to \infty$. By Proposition 10.3.7 there exists $\kappa > 0$ such that $\mathrm{d}(h^{-1}(Q), \bar{x} + \tau y) \leq \kappa\,\mathrm{d}(Q, h(\bar{x} + \tau y))$. It follows that

$$\lim_{k \to \infty} \tau_k^{-1}\mathrm{d}(h^{-1}(Q), \bar{x} + \tau_k y) = 0 \tag{11.27}$$

and so $y \in \mathrm{T}(h^{-1}(Q), \bar{x})$.

(II) Now we show the reverse inclusion to (11.24). So let $y \in \mathrm{T}(h^{-1}(Q), \bar{x})$ be given. Then there exists $\tau_k \downarrow 0$ such that (11.27) holds. Since the mapping h is continuously differentiable at $\bar{x}$, it is locally Lipschitz continuous there. Hence there exists $\lambda > 0$ such that $\mathrm{d}(Q, h(x)) \leq \lambda\,\mathrm{d}(h^{-1}(Q), x)$ for any x near $\bar{x}$. This and (11.27) imply that $\tau_k^{-1}\mathrm{d}(h(\bar{x} + \tau_k y), Q) \to 0$ as $k \to \infty$. Using (11.26) again, we see that (11.25) holds and so $h'(\bar{x})y \in \mathrm{T}(Q, h(\bar{x}))$. $\qquad\square$

Notice that we verified the reverse inclusion to (11.24) without making use of the Robinson condition. However, the crucial inclusion for deriving multiplier rules in Chap. 12 will be (11.24), and this inclusion has been verified with the aid of Proposition 10.3.7 which is bound to the Robinson condition or one of the equivalent conditions (see Proposition 10.3.8).

As a special case, Theorem 11.4.1 contains the following classical result (cf. Fig. 7.3).

Theorem 11.4.2 (Tangent Space Theorem) *Let the assumptions* (A) *with* $Q = \{o\}$ *be satisfied. If* $h'(\bar{x})$ *is surjective, then*

$$\mathrm{T}(\ker h, \bar{x}) = \ker h'(\bar{x}). \tag{11.28}$$

Example 11.4.3 Let $E = F := \mathbb{R}^2$, $\bar{x} := (1, 0)$, and $h = (h_1, h_2)$, where

$$h_1(x_1, x_2) := -x_2, \quad h_2(x_1, x_2) := x_2 + (x_1 - 1)^3.$$

Then $\mathrm{T}(\ker h, \bar{x}) = \{(0, 0)\}$ but $\ker h'(\bar{x}) = \mathbb{R} \times \{0\}$, i.e., (11.28) is not valid. Notice that in this case, $h'(\bar{x})(\mathbb{R}^2) = \{\alpha(-1, 1) \mid \alpha \in \mathbb{R}\}$ and so $h'(\bar{x})$ is not surjective.

Approximating $A \cap h^{-1}(Q)$

Now we consider sets of the form

$$A \cap h^{-1}(Q) = \{x \in E \mid x \in A,\ h(x) \in Q\}.$$

Theorem 11.4.4 *In addition to the assumptions* (A), *let now A be a nonempty closed convex subset of E and assume that the Robinson condition*

$$o \in \text{int}\big(h(\bar{x}) + h'(\bar{x})(A - \bar{x}) - Q\big) \qquad (11.29)$$

is satisfied. Then one has

$$T(A \cap h^{-1}(Q), \bar{x}) = T(A, \bar{x}) \cap h'(\bar{x})^{-1}\big(T(Q, h(\bar{x}))\big). \qquad (11.30)$$

Proof. Setting

$$\tilde{F} := E \times F, \quad \tilde{Q} := A \times Q, \quad \tilde{h}(x) := (x, h(x)), \quad x \in E,$$

we have

$$\tilde{h}^{-1}(\tilde{Q}) = A \cap h^{-1}(Q), \quad \tilde{h}'(\bar{x})y = (y, h'(\bar{x})y) \quad \text{for all } y \in E,$$
$$T(\tilde{Q}, \tilde{h}(\bar{x})) = T(A, \bar{x}) \times T(Q, h(\bar{x})).$$

Hence the assertion follows from Theorem 11.4.1 as soon as we have shown that the Robinson condition

$$(o, o) \in \text{int}\big(\tilde{h}(\bar{x}) + \tilde{h}'(\bar{x})(E) - \tilde{Q}\big) \qquad (11.31)$$

is satisfied. According to (11.29) there exists $\epsilon > 0$ such that

$$\epsilon B_F \subseteq h(\bar{x}) + h'(\bar{x})(A - \bar{x}) - Q. \qquad (11.32)$$

Choose $\delta > 0$ such that $\|z - h'(\bar{x})x_1\| < \epsilon$ whenever $|x_1| < \delta$ and $\|z\| < \delta$. Now let $(y, z) \in \delta B_{E \times F}$ be given. Set $x_1 := y$. By (11.32) there exists $x_2 \in A - \bar{x}$ such that

$$z - h'(\bar{x})x_1 \in h(\bar{x}) + h'(\bar{x})x_2 - Q.$$

Defining $x := x_1 + x_2$, we have $(y, z) \in \tilde{h}(\bar{x}) + \tilde{h}'(\bar{x})x - \tilde{Q}$. This verifies (11.31), and the proof is complete. $\qquad \Box$

Approximating Sublevel Sets

Now we want to approximate the set

$$M := \{x \in E \mid x \in A, \; g_i(x) \leq 0 \; (i = 1, \ldots, m), \; h(x) = o\}, \qquad (11.33)$$

where

$$g_1, \ldots, g_m : E \to \mathbb{R}, \quad h : E \to F.$$

Of course, M can be written as $A \cap \hat{h}^{-1}(Q)$ by setting

$$\hat{h} := (g_1, \ldots, g_m, h) : E \to \mathbb{R}^m \times F, \quad Q := -\mathbb{R}_+^m \times \{o\}.$$

However, if we want to apply Theorem 11.4.4, we have to assume that beside the mapping h, the functions $g_1, \ldots, g_m$ are also continuously differentiable at $\bar{x}$. By a somewhat different approach we shall now show that we can do with weaker differentiability hypotheses on the functions g_i. Therefore we first consider the set

$$M_1 := \{x \in E \mid x \in A, \ g_i(x) \le 0 \ (i = 1, \ldots, m)\}. \tag{11.34}$$

We define

$$I := \{1, \ldots, m\}, \quad I(\bar{x}) := \{i \in I \mid g_i(\bar{x}) = 0\}.$$

The set $I(\bar{x})$ is the index set of the constraint functions that are *active* or *binding* at the point $\bar{x}$. In Fig. 11.4 we have $1 \in I(\bar{x})$ and $2 \notin I(\bar{x})$. It will turn out that for $i \notin I(\bar{x})$, the constraint $g_i(x) \le 0$ is not critical provided the function g_i is upper semicontinuous at $\bar{x}$.

Figure 11.5 indicates what we can expect in $\mathbb{R}^n$ if $i \in I(\bar{x})$ and g_i is differentiable at $\bar{x}$. An "admissible" direction y satisfies $\frac{\pi}{2} < \alpha < \pi$ and so $\langle \nabla g_i(\bar{x}), y \rangle \le 0$. If g_i is not differentiable at $\bar{x}$, then we use a (radial) upper convex approximation γ_i. We set

$$\gamma := (\gamma_i)_{i \in I(\bar{x})},$$
$$L^<(\gamma, \bar{x}) := \{y \in E \mid \gamma_i(y) < 0 \ \forall i \in I(\bar{x})\},$$
$$L^\le(\gamma, \bar{x}) := \{y \in E \mid \gamma_i(y) \le 0 \ \forall i \in I(\bar{x})\}.$$

The sets $L^<(\gamma, \bar{x})$ and $L^\le(\gamma, \bar{x})$, which are obviously cones, are called *linearizing cones* of γ at $\bar{x}$.

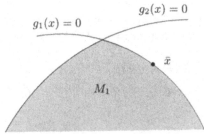

Fig. 11.4

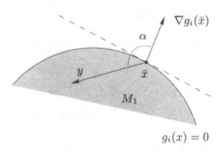

Fig. 11.5

Proposition 11.4.5 *Let E be a normed vector space, A be a nonempty subset of E, and $g_i : E \to \mathbb{R}$ for $i \in I$. Assume that for $i \in I \setminus I(\bar{x})$ the function g_i is upper semicontinuous at $\bar{x}$. Let M_1 be defined by (11.34).*

(a) *If $\gamma_i \in \mathrm{UC_r}(g_i, \bar{x})$ for $i \in I(\bar{x})$, then*

$$L^<(\gamma, \bar{x}) \subseteq \mathrm{I_r}(M_1, \bar{x}), \tag{11.35}$$

$$\mathrm{T_r}(A, \bar{x}) \cap L^<(\gamma, \bar{x}) \subseteq \mathrm{T_r}(A \cap M_1, \bar{x}). \tag{11.36}$$

If, in addition, $K_r \subseteq \mathrm{T_r}(A, \bar{x})$ is a convex cone satisfying

$$K_r \cap L^<(\gamma, \bar{x}) \neq \emptyset, \tag{11.37}$$

then

$$K_r \cap L^{\leq}(\gamma, \bar{x}) \subseteq \mathrm{cl}\, \mathrm{T_r}(A \cap M_1, \bar{x}). \tag{11.38}$$

(b) *If $\gamma_i \in \mathrm{UC}(g_i, \bar{x})$ for $i \in I(\bar{x})$, then*

$$L^<(\gamma, \bar{x}) \subseteq \mathrm{I}(M_1, \bar{x}), \tag{11.39}$$

$$\mathrm{T}(A, \bar{x}) \cap L^<(\gamma, \bar{x}) \subseteq \mathrm{T}(A \cap M_1, \bar{x}). \tag{11.40}$$

If, in addition, $K \subseteq \mathrm{T}(A, \bar{x})$ is a convex cone satisfying

$$K \cap L^<(\gamma, \bar{x}) \neq \emptyset, \tag{11.41}$$

then

$$K \cap L^{\leq}(\gamma, \bar{x}) \subseteq \mathrm{T}(A \cap M_1, \bar{x}). \tag{11.42}$$

Proof. Ad (11.39). Let $y \in L^<(\gamma, \bar{x})$. Then we have $(\bar{g}_i)^+_H(\bar{x}, y) \leq \gamma_i(y) < 0$ for any $i \in I(\bar{x})$. Hence for any such i there exists $\epsilon_i \in (0, 1)$ such that

$$\frac{1}{\tau}\big[g_i(\bar{x} + \tau z) - \underbrace{g_i(\bar{x})}_{=0}\big] < 0 \quad \forall \tau \in (0, \epsilon_i) \,\forall z \in \mathrm{B}(y, \epsilon_i). \tag{11.43}$$

Now let $i \in I \setminus I(\bar{x})$. Then $g_i(\bar{x}) < 0$ and g_i is upper semicontinuous at $\bar{x}$. Hence there exists $\delta_i > 0$ such that

$$g_i(x) < 0 \quad \forall x \in \mathrm{B}(\bar{x}, \delta_i). \tag{11.44}$$

Set

$$\tilde{\epsilon} := \min\{\epsilon_i \,|\, i \in I(\bar{x})\}, \;\, \delta := \min\{\delta_i \,|\, i \in I \setminus I(\bar{x})\}, \;\, \epsilon := \min\Big\{\tilde{\epsilon}, \frac{\delta}{\tilde{\epsilon} + \|y\|}\Big\}.$$

If now $\tau \in (0, \epsilon)$ and $z \in \mathrm{B}(y, \tilde{\epsilon})$, then it follows from (11.43) that $g_i(\bar{x} + \tau z) < 0$ for each $i \in I(\bar{x})$. Furthermore we obtain

$$\|(\bar{x} + \tau z) - \bar{x}\| = \tau\|z\| \leq \tau(\|z - y\| + \|y\|) \leq \epsilon(\tilde{\epsilon} + \|y\|) \leq \delta$$

which, by (11.44), implies $g_i(\bar{x} + \tau z) < 0$ for each $i \in I \setminus I(\bar{x})$. Thus we conclude that $y \in I(M_1, \bar{x})$.

Ad (11.40). This follows from (11.39) and Lemma 11.1.3.

Ad (11.42). Let $y_0 \in K \cap L^<(\gamma, \bar{x})$ and $y \in K \cap L^{\leq}(\gamma, \bar{x})$. For $\lambda \in (0,1)$, set $y_\lambda := \lambda y_0 + (1 - \lambda)y$. Then we have $y_\lambda \in K \subseteq T(A, \bar{x})$ and

$$\gamma_i(y_\lambda) \leq \lambda \gamma_i(y_0) + (1 - \lambda)\gamma_i(y) < 0 \quad \forall i \in I(\bar{x})$$

and so $y_\lambda \in L^<(\gamma\bar{x})$. In view of (11.39) and Proposition 11.1.3, we conclude that

$$y_\lambda \in T(A, \bar{x}) \cap I(M_1, \bar{x}) \subseteq T(A \cap M_1, \bar{x})$$

and so $y = \lim_{\lambda \downarrow 0} y_\lambda \in T(A \cap M_1, \bar{x})$ because the latter set is closed (Proposition 11.1.2). It is left as Exercise 11.7.9 to verify the assertions (11.35), (11.36), and (11.38). □

Remark 11.4.6

(a) The assumption that g_i, for each $i \in I(\bar{x})$, admits a (radial) upper convex approximation γ_i at $\bar{x}$ may be interpreted as a "differentiability" requirement. If g_i is locally L-continuous around $\bar{x}$, then by Theorem 7.3.2, the Clarke directional derivative $g_i^\circ(\bar{x}, \cdot)$ as well as the Michel–Penot directional derivative $g_i^\diamond(\bar{x}, \cdot)$ are possible choices for $\gamma_i \in UC(g_i, \bar{x})$.

(b) The above proof shows that for (11.35) and (11.39) the convexity of γ_i, $i \in I(\bar{x})$, is dispensable. Hence these statements also hold with γ_i replaced by the upper directional G-derivative (or H-derivative) of g_i at $\bar{x}$.

(c) The conditions (11.37) and (11.41) are called *regularity conditions* or *constraint qualifications*. Notice that the set $L^<(\gamma, \bar{x})$ may be empty. In Proposition 11.4.5, the set A is not assumed to be convex. However, if A happens to be convex, then $T_r(A, \bar{x})$ and $T(A, \bar{x})$ are convex cones (Proposition 11.1.2) and so we can choose $K_r := T_r(A, \bar{x})$, $K := T_r(A, \bar{x})$ or $K := T(A, \bar{x})$.

In the preceding results we used the contingent cone for the local approximation of sublevel sets. Now we choose the Clarke tangent cone. We start with a definition.

Definition 11.4.7 The subset M of E is said to be *tangentially regular* at $\bar{x} \in M$ if $T_C(M, \bar{x}) = T(M, \bar{x})$.

Now we consider the sublevel set

$$M := \{x \in E \mid g(x) \leq g(\bar{x})\}. \tag{11.45}$$

Theorem 11.4.8 *Let $g : E \to \mathbb{R}$ be locally L-continuous around $\bar{x} \in E$ and assume that $o \notin \partial_\circ g(\bar{x})$. Then, with M according to (11.45), one has*

$$\{y \in E \mid g^\circ(\bar{x}, y) \leq 0\} \subseteq T_C(M, \bar{x}). \tag{11.46}$$

If g is regular at $\bar{x}$, then (11.46) holds with equality and M is tangentially regular at $\bar{x}$.

Proof.

(I) By the definition of $\partial_o g(\bar{x})$ and since $o \notin \partial_o g(\bar{x})$, there exists $y_0 \in E$ such that $g°(\bar{x}, y_0) < 0$. Now let $\tilde{y}$ be any element of the left-hand side of (11.46). Since $g°(\bar{x}, \cdot)$ is sublinear, it follows that $g°(\bar{x}, \tilde{y} + \epsilon y_0) < 0$ for each $\epsilon > 0$. In step (II) we shall show that every $y \in E$ satisfying $g°(\bar{x}, y) < 0$ belongs to $T_C(M, \bar{x})$. It then follows that $\tilde{y} + \epsilon y_0 \in T_C(M, \bar{x})$ for each $\epsilon > 0$, and since $T_C(M, \bar{x})$ is closed (Proposition 11.1.2), we conclude letting $\epsilon \downarrow 0$ that $\tilde{y} \in T_C(M, \bar{x})$. This verifies (11.46).

(II) Let $y \in E$ be such that $g°(\bar{x}, y) < 0$. By the definition of $g°(\bar{x}, y)$ there exist $\epsilon > 0$ and $\delta > 0$ such that

$$g(x + \tau y) - g(x) \leq -\delta\tau \quad \forall x \in B(\bar{x}, \epsilon) \quad \forall \tau \in (0, \epsilon). \tag{11.47}$$

Now let (x_k) be a sequence in M converging to $\bar{x}$ and $\tau_k \downarrow 0$. In view of (11.47) we obtain for all sufficiently large k,

$$g(x_k + \tau_k y) \leq g(x_k) - \delta\tau_k \leq g(\bar{x}) - \delta\tau_k$$

and so $x_k + \tau_k y \in M$. This shows that $y \in T_C(M, \bar{x})$.

(III) We verify the regularity statement. Assume that g is regular at $\bar{x}$. Since (11.46) is already verified and $T_C(M, \bar{x}) \subseteq T(M, \bar{x})$ always holds, it suffices to show that $T(M, \bar{x})$ is a subset of the left-hand side of (11.46). Thus let $y \in T(M, \bar{x})$ be given. Then Proposition 11.1.5 shows that $\liminf_{\tau \downarrow 0} \tau^{-1} d_M(\bar{x} + \tau y) = 0$. Hence for each $\epsilon > 0$ we find a sequence $\tau_k \downarrow 0$ such that for all sufficiently large k we have $d_M(\bar{x} + \tau_k y) \leq \epsilon\tau_k$. Thus there exists $x_k \in M$ satisfying $\|(\bar{x} + \tau_k y) - x_k\| \leq 2\epsilon\tau_k$. Let λ denote a local Lipschitz constant of g around $\bar{x}$. Then we obtain, again for k sufficiently large,

$$g(\bar{x} + \tau_k y) - g(x_k) \leq \lambda\|(\bar{x} + \tau_k y) - x_k\| \leq 2\epsilon\tau_k\lambda$$

and so

$$\frac{1}{\tau_k}\left(g(\bar{x} + \tau_k y) - g(\bar{x})\right) \leq \frac{1}{\tau_k}\left(g(x_k) - g(\bar{x})\right) + 2\epsilon\lambda \leq 2\epsilon\lambda.$$

Letting $k \to \infty$ and then $\epsilon \downarrow 0$, we conclude that $g°(\bar{x}, y) = g_G(\bar{x}, y) \leq 0$. $\square$

Approximating Level–Sublevel Sets

We return to the contingent cone, now considering the set

$A \cap M$, where
$$M := \{x \in E \mid g_i(x) \leq 0 \; (i = 1, \ldots, m), \; h(x) = o\}, \quad \bar{x} \in A \cap M.$$

We again set $I := \{1, \ldots, m\}$ and $I(\bar{x}) := \{i \in I \mid g_i(\bar{x}) = 0\}$. We make the following hypotheses:

(H) E and F are Banach spaces, $A \subseteq E$ is nonempty closed and convex,
   $g_i : E \to \mathbb{R}$ for $i \in I$, $\gamma_i \in \mathrm{UC}(g_i, \bar{x})$ for $i \in I(\bar{x})$,
   g_i is upper semicontinuous at $\bar{x}$ for $i \in I \setminus I(\bar{x})$,
   $h : E \to F$ is continuous on E and continuously differentiable at $\bar{x}$.

Theorem 11.4.9 *Assume that the hypotheses* (H) *hold and that*

$$h'(\bar{x})\big(\mathbb{R}_+(A - \bar{x})\big) = F. \tag{11.48}$$

Then:

(a) *There always holds*

$$\mathbb{R}_+(A - \bar{x}) \cap \mathrm{L}^<(\gamma, \bar{x}) \cap \ker h'(\bar{x}) \subseteq \mathrm{T}(A \cap M, \bar{x}).$$

(b) *If*

$$\mathbb{R}_+(A - \bar{x}) \cap \mathrm{L}^<(\gamma, \bar{x}) \cap \ker h'(\bar{x}) \neq \emptyset, \tag{11.49}$$

then

$$\mathbb{R}_+(A - \bar{x}) \cap \mathrm{L}^\leq(\gamma, \bar{x}) \cap \ker h'(\bar{x}) \subseteq \mathrm{T}(A \cap M, \bar{x}).$$

Proof.

(a) Let $y \in \mathbb{R}_+(A - \bar{x}) \cap \mathrm{L}^<(\gamma, \bar{x}) \cap \ker h'(\bar{x})$. By Theorem 11.4.4 (with $Q = \{o\}$) we obtain $y \in \mathrm{T}(A \cap \ker h, \bar{x})$, and Proposition 11.4.5(b) implies $y \in \mathrm{I}(M_1, \bar{x})$. Thus the assertion follows with the aid of Proposition 11.1.3.

(b) This is verified in analogy to formula (11.42) in Proposition 11.4.5.   □

11.5 Contingent Derivatives and a Lyusternik Type Theorem

We introduce a derivative-like concept for multifunctions. In Sect. 13.2 we shall study an alternative construction.

Definition 11.5.1 Let $\Phi : E \rightrightarrows F$ be a multifunction and $(\bar{x}, \bar{y}) \in \mathrm{graph}\, \Phi$. The multifunction $\mathrm{D}\Phi(\bar{x}, \bar{y}) : E \rightrightarrows F$ defined by

$$\mathrm{D}\Phi(\bar{x}, \bar{y})(u) := \underset{u' \to u, \tau \downarrow 0}{\mathrm{Lim\,sup}} \frac{1}{\tau}\big(\Phi(\bar{x} + \tau u') - \bar{y}\big), \quad u \in E,$$

is called *contingent derivative* of Φ at $(\bar{x}, \bar{y})$.

The definition implies

$$\text{graph}\big(D\varPhi(\bar{x},\bar{y})\big) = T\big(\text{graph}\,\varPhi,(\bar{x},\bar{y})\big), \tag{11.50}$$

where the right-hand side is the contingent cone to graph $\varPhi$ at $(\bar{x},\bar{y})$. This is why $D\varPhi(\bar{x},\bar{y})$ is called *contingent derivative*. If $\varPhi : E \to F$ is Hadamard differentiable at $\bar{x}$, then

$$D\varPhi(\bar{x},\varPhi(\bar{x}))(u) = \{\varPhi'(\bar{x})u\} \quad \forall u \in E,$$

i.e., $D\varPhi(\bar{x},\varPhi(\bar{x}))$ can be identified with the Hadamard derivative $\varPhi'(\bar{x})$.

The contingent derivative turns out to be an appropriate tool to establish a tangential approximation of $\ker\varPhi$, where $\varPhi$ is a multifunction (cf. the tangential approximations derived in Sect. 11.4).

Theorem 11.5.2 *Let E and F be Banach spaces. If the multifunction $\varPhi :$ $E \rightrightarrows F$ is linearly semiopen around $(\bar{x},o) \in \text{graph}\,\varPhi$, then*

$$T(\ker\varPhi,\bar{x}) = \ker\big(D\varPhi(\bar{x},o)\big). \tag{11.51}$$

Proof.

(I) First let $u \in T(\ker\varPhi,\bar{x})$ be given. Then there exist sequences $u_k \to u$ and $\tau_k \downarrow 0$ such that $o \in \varPhi(\bar{x}+\tau_k u_k)$ for any $k \in \mathbb{N}$. It follows that

$$o \in \limsup_{k\to\infty} \frac{1}{\tau_k}\varPhi(\bar{x}+\tau_k u_k) \subseteq D\varPhi(\bar{x},o)(u).$$

Hence $u \in \ker\big(D\varPhi(\bar{x},o)\big)$.

(II) Now let $u \in \ker\big(D\varPhi(\bar{x},o)\big)$ be given. If $u = o$, then $u \in T(\ker\varPhi,\bar{x})$. Thus assume that $u \neq o$. Denote the semiopenness parameters of $\varPhi$ around $(\bar{x},o)$ by ρ and τ_0. By definition of $D\varPhi(\bar{x},o)$ there exist sequences $u_k \to u$, $\tau_k \downarrow 0$, and $v_k \in \varPhi(\bar{x}+\tau_k u_k)$ such that $\lim_{k\to\infty} v_k/\tau_k = o$. For sufficiently large k, we have $\bar{x}+\tau_k u_k \in \bar{x}+\tau_0 B_E$ and $v_k \in \tau_0 B_F$. Again for sufficiently large k, we further have $\|u_k\| \geq \frac{1}{2}\|u\|$ and so

$$\tau'_k := \frac{\|v_k\|}{\rho\tau_k\|u_k\|} \leq \frac{2}{\rho\|u\|}\cdot\frac{\|v_k\|}{\tau_k} \to 0 \quad \text{as } k \to \infty.$$

Define $z_k := \bar{x} + \tau_k u_k$. Then we obtain

$$o \in v_k + \|v_k\|B_F = v_k + \rho\tau'_k\|z_k - \bar{x}\|B_F \subseteq \varPhi(z_k + \tau'_k\|z_k - \bar{x}\|B_E);$$

here the latter inclusion is a consequence of the linear semiopenness of $\varPhi$. Hence there exists $z'_k \in \ker(\varPhi)$ satisfying $\|z'_k - z_k\| \leq \tau'_k\tau\|u_k\|$. Define

$y_k := \frac{1}{\tau_k}(z'_k - \bar{x})$. Then $\bar{x} + \tau_k y_k \in \ker \Phi$. It remains to show that $y_k \to u$ as $k \to \infty$. This follows from

$$\|y_k - u_k\| = \frac{1}{\tau_k}\|z'_k - z_k\| \le \tau'_k \|u_k\| \to 0 \quad \text{as } k \to \infty$$

and so

$$\|y_k - u\| \le \|y_k - u_k\| + \|u_k - u\| \to 0 \quad \text{as } k \to \infty. \quad \square$$

Remark 11.5.3 We show that the classical tangent space theorem (Theorem 11.4.2) can be regained from Theorem 11.5.2. We use the assumptions and the notation of Theorem 11.4.2. Since the continuous linear mapping $h'(\bar{x})$ is surjective, it is open by the Banach open mapping theorem. Hence $x \mapsto h'(\bar{x})(x - \bar{x})$ is open at a linear rate, say ρ, around $\bar{x}$. Furthermore, since h is continuously differentiable at $\bar{x} \in \ker(h)$, there exists a mapping $r : E \to F$ such that

$$h(x) = h'(\bar{x})(x - \bar{x}) + r(x), \quad \text{where } \lim_{x \to \bar{x}} r(x)/\|x\| = o, \tag{11.52}$$

and we have

$$\frac{1}{\|x_1 - x_2\|}\big(r(x_1) - r(x_2)\big)$$
$$= \frac{1}{\|x_1 - x_2\|}\Big[\big(h(x_1) - h(x_2)\big) - h'(\bar{x})(x_1 - x_2)\Big] \to o \quad \text{as } x_1, x_2 \to \bar{x}.$$

Here the limit relation holds by Proposition 3.2.4(v) because h is strictly differentiable (Proposition 3.4.2). Hence for any $\epsilon > 0$ there exists $\delta > 0$ such that

$$\|r(x_1) - r(x_2)\| \le \epsilon \|x_1 - x_2\| \quad \forall x_1, x_2 \in \mathrm{B}_E(\bar{x}, \delta). \tag{11.53}$$

In particular, we can take $\epsilon \in (0, \rho)$. Then Theorem 10.3.6 shows that the mapping h is linearly semiopen around $\bar{x}$. Therefore Theorem 11.5.2 implies

$$\mathrm{T}(\ker h, \bar{x}) = \ker\big(\mathrm{D}h(\bar{x})\big) = \ker h'(\bar{x}). \tag{11.54}$$

The above argument suggests how to slightly weaken the hypotheses of the Lyusternik theorem. In this connection, we consider the multifunction $h'(\bar{x})^{-1} : F \rightrightarrows E$ which is defined as usual by

$$h'(\bar{x})^{-1}(y) := \{x \in E \mid h'(\bar{x})(x) = y\}, \quad y \in F.$$

Notice that $h'(\bar{x})^{-1}$ is a process and $\|h'(\bar{x})^{-1}\|$ denotes its norm according to (10.50).

Proposition 11.5.4 *Assume that $h : E \to F$ is F-differentiable at $\bar{x} \in \ker(h)$, that $h'(\bar{x})$ is surjective, and that the mapping r in (11.52) is locally Lipschitz continuous at $\bar{x}$ with a Lipschitz constant $\lambda < 1/\|h'(\bar{x})^{-1}\|$. Then (11.54) holds.*

Proof. The mapping $\Phi : x \mapsto h'(\bar{x})(x)$, interpreted as a multifunction, is open at a linear rate around (o, o) (cf. Remark 11.5.3) and is a bounded process. By Proposition 10.4.2, the openness bound of Φ around (o, o) is $\mathrm{ope}(\Phi)(o, o) = 1/\|h'(\bar{x})^{-1}\|$. Since $h'(\bar{x})$ is linear, the mapping $x \mapsto h'(x - \bar{x})$ is open at a linear rate around $(\bar{x}, o)$ with the same openness bound $1/\|h'(\bar{x})^{-1}\|$. Now the assertion follows by Theorems 10.3.6 and 11.5.2. $\square$

Example 11.5.5 Define $h : \mathbb{R}^2 \to \mathbb{R}$ defined by

$$h(x_1, x_2) := \begin{cases} ax_1 + bx_2 + x_1 x_2 & \text{if } x_1 \geq 0, \\ ax_1 + bx_2 - x_1 x_2 & \text{if } x_1 < 0, \end{cases}$$

where $|a| + |b| > 2$. The function h satisfies the above assumptions at $(0, 0)$ while h is not differentiable at $(0, x_2)$ if $x_2 \neq 0$ and so the classical Lyusternik theorem does not apply.

11.6 Representation of Normal Cones

In this section we characterize normal cones of a set $M \subseteq E$ which is defined by an inequality or an equation.

The Clarke Normal Cone to Sublevel Sets

We start with the set

$$M := \{x \in E \mid f(x) \leq f(\bar{x})\}. \tag{11.55}$$

Theorem 11.6.1 *Let* $f : E \to \overline{\mathbb{R}}$ *be proper and locally L-continuous around* $\bar{x}$. *Assume that* $o \notin \partial_\circ f(\bar{x})$. *Then, with* M *as in* (11.55), *one has*

$$\mathrm{N}_C(M, \bar{x}) \subseteq \mathbb{R}_+ \partial_\circ f(\bar{x}). \tag{11.56}$$

If, in addition, f is regular at $\bar{x}$, then (11.56) *holds as an equation.*

Proof. Taking polars in (11.46) (see Theorem 11.4.8), we obtain

$$\mathrm{N}_C(M, \bar{x}) \subseteq \{y \in E \mid f^\circ(\bar{x}, y) \leq 0\}^\circ = \left(\partial_\circ f(\bar{x})\right)^{\circ\circ}. \tag{11.57}$$

Here the equation follows by Proposition 7.3.7(b). Applying the bipolar theorem to the right-hand side and recalling that $\partial_\circ f(\bar{x})$ is weak* compact not containing o, the assertion follows. If f is regular at $\bar{x}$, then by Theorem 11.4.8 the inclusion in (11.57) is an equation and so is the inclusion in (11.56). $\square$

Now let M be defined by

$$M := \{x \in E \mid f_i(x) \leq 0, \quad i = 1, \ldots, n\}. \tag{11.58}$$

Corollary 11.6.2 *For $i = 1, \ldots, n$, let $f_i : E \to \mathbb{R}$ be strictly H-differentiable at $\bar{x}$, where $f_1(\bar{x}) = \cdots = f_n(\bar{x}) = 0$. If the functionals $f_1'(\bar{x}), \ldots, f_n'(\bar{x})$ are positively linearly independent, then the set M in (11.58) is regular at $\bar{x}$ and one has*

$$\mathrm{N}_C(M, \bar{x}) = \Big\{ \sum_{i=1}^{n} \lambda_i f_i'(\bar{x}) \mid \lambda_i \geq 0, \ i = 1, \ldots, n \Big\}. \tag{11.59}$$

Proof. Define $f := \max\{f_1, \ldots, f_n\}$. By Proposition 7.3.9(c), each f_i is locally L-continuous around $\bar{x}$ and so is f. Moreover, by Proposition 7.4.7, f is also regular at $\bar{x}$. Since $M = \{x \in E \mid f(x) \leq 0\}$, Theorem 11.6.1 implies that (11.56) holds with equality. Finally, the maximum rule of Proposition 7.4.7 yields (11.59). □

The Proximal Normal Cone to Sublevel Sets

Next we give the complete, geometrically appealing proof for the representation of $\mathrm{N}_P(M, \bar{x})$ in a Hilbert space and then indicate the more technical proof for $\mathrm{N}_F(M, \bar{x})$ in a Fréchet smooth Banach space. We consider the set

$$M := \{x \in E \mid f(x) \leq 0\}. \tag{11.60}$$

Theorem 11.6.3 *Let E be a Hilbert space and let M be given by (11.60), where $f : E \to \overline{\mathbb{R}}$ is proper and l.s.c. Let $\bar{x} \in M$ and $u \in N_P(M, \bar{x})$. Then either*

(C1) *for any $\epsilon > 0$ and $\eta > 0$ there exists $x \in E$ such that*

$$\|x - \bar{x}\| < \eta, \quad |f(x) - f(\bar{x})| < \eta, \quad \partial_P f(x) \cap \mathrm{B}_{E^*}(o, \epsilon) \neq \emptyset$$

or

(C2) *for any $\epsilon > 0$ there exist $x \in E$, $v \in \partial_P f(x)$, and $\lambda > 0$ such that*

$$\|x - \bar{x}\| < \epsilon, \quad |f(x) - f(\bar{x})| < \epsilon, \quad \|\lambda v - u\| < \epsilon.$$

Proof. Assuming that (C1) does not hold, we shall show that (C2) is valid. Obviously we may suppose that $u \neq o$.

(I) Since $u \in N_P(M, \bar{x})$, Proposition 11.2.7 implies that there exist $\rho > 0$ and $\sigma > 0$ such that

$$0 \geq (u \mid x - \bar{x}) - \sigma \|x - \bar{x}\|^2 \quad \forall x \in M \cap \mathrm{B}(\bar{x}, \rho\|u\|). \tag{11.61}$$

Since f is l.s.c., there exists $m > 0$ such that, on diminishing ρ if necessary, $f(x) \geq -m$ for all $x \in \mathrm{B}(\bar{x}, \rho\|u\|)$. The functional $x \mapsto (u \mid x - \bar{x}) - \sigma \|x - \bar{x}\|^2$ attains a positive value at $x = \bar{x} + 2\eta u$ if $\eta \in (0, \frac{1}{2\sigma})$. By continuity, for $\eta > 0$ sufficiently small we have

$$(u \mid x - \bar{x}) - \sigma \|x - \bar{x}\|^2 > 0 \quad \forall x \in K := \mathrm{B}(\bar{x} + 2\eta u, \, 2\eta\|u\|) \setminus \{\bar{x}\}.$$

Hence if $x \in K$, then $x \notin M$ and so $f(x) > 0$.

(II) For any positive $\alpha < \min\{\eta, 1/m\}$ let

$$h_\alpha(x) := \alpha^{-1}\left(\max\{0, \|x - \bar{x} - \eta u\| - \eta\|u\|\}\right)^2,$$
$$p_\alpha(z) := f(z) + h_\alpha(z) + \delta_{B(\bar{x}, \rho\|u\|)}(z).$$

Since (C1) does not hold, o is not a proximal subderivative of f at $\bar{x}$ and so $\inf_E p_\alpha < 0$. By Theorem 8.3.3 applied to p_α with

$$\lambda := \sqrt{2\epsilon_\alpha/\alpha}, \quad \epsilon := \epsilon_\alpha := \min\{\alpha/2, -\inf_E p_\alpha/2\}$$

there exist $y_\alpha, w_\alpha \in E$ such that

$$\|y_\alpha - w_\alpha\| < \lambda, \quad p_\alpha(y_\alpha) < \inf_E p_\alpha + \epsilon_\alpha < 0,$$

and the functional

$$z \mapsto p_\alpha(z) + \frac{\alpha}{2}\|z - w_\alpha\|^2 \tag{11.62}$$

attains a global minimum at $z = y_\alpha$. Moreover, since $p_\alpha(y_\alpha) < 0$, we have $y_\alpha \in B(\bar{x}, \rho\|u\|)$. We further obtain the following estimate:

$$\begin{aligned}
h_\alpha(y_\alpha) &\le h_\alpha(y_\alpha) + \frac{\alpha}{2}\|y_\alpha - w_\alpha\|^2 \\
&< h_\alpha(y_\alpha) + \epsilon_\alpha \quad \text{(because } \|y_\alpha - w_\alpha\| < \lambda) \\
&= p_\alpha(y_\alpha) + \epsilon_\alpha - f(y_\alpha) < \inf_E p_\alpha + 2\epsilon_\alpha - f(y_\alpha) \\
&\le -f(y_\alpha) \le m \quad \text{(because } 2\epsilon_\alpha \le -\inf_E p_\alpha).
\end{aligned}$$

The definition of h_α shows that

$$\|y_\alpha - \bar{x} - \eta u\| \le \sqrt{m\alpha} + \eta\|u\|, \tag{11.63}$$

i.e.,

$$y_\alpha \in K_\alpha := B(\bar{x} + \eta u, \sqrt{m\alpha} + \eta\|u\|).$$

On the other hand, $p_\alpha(y_\alpha) < 0$ implies $f(y_\alpha) < 0$ and so $y_\alpha \notin K$ (cf. Fig. 11.6), i.e.,

$$\|y_\alpha - \bar{x} - 2\eta u\| > 2\eta\|u\|. \tag{11.64}$$

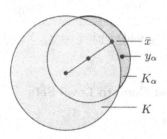

Fig. 11.6

Applying the parallelogram identity to $y_\alpha - \bar{x} - \eta u$ and ηu, we obtain

$$\|y_\alpha - \bar{x}\|^2 + \|y_\alpha - \bar{x} - 2\eta u\|^2 = 2\|y_\alpha - \bar{x} - \eta u\|^2 + 2\eta^2\|u\|^2.$$

The estimates (11.63) and (11.64) now give

$$\|y_\alpha - \bar{x}\|^2 \leq 2(\eta\|u\| + \sqrt{m\alpha})^2 + 2\eta^2\|u\|^2 - 4\eta^2\|u\|^2 = 4\eta\|u\|\sqrt{m\alpha} + m\alpha,$$

showing that $\lim_{\alpha \to 0} y_\alpha = \bar{x}$.

(III) We further have

$$f(y_\alpha) \leq p_\alpha(y_\alpha) < \inf_E p_\alpha + \epsilon_\alpha \leq p_\alpha(\bar{x}) + \epsilon_\alpha = f(\bar{x}) + \epsilon_\alpha$$

and so $\lim_{\alpha \to 0} f(y_\alpha) = f(\bar{x})$ (recall that f is l.s.c. and $\epsilon_\alpha \to 0$ as $\alpha \to 0$).

(IV) Since y_α is a global minimizer of the functional in (11.62) and y_α is in $\mathring{B}(\bar{x}, \rho\|u\|)$ for α sufficiently small, the functional

$$f - g_\alpha, \quad \text{where } g_\alpha(z) := -h_\alpha(z) - \frac{\alpha}{2}\|z - w_\alpha\|^2$$

attains a local minimum at y_α. We have

$$g'_\alpha(y_\alpha) = -h'(y_\alpha) - \alpha(y_\alpha - w_\alpha) = k(\alpha)(\bar{x} + \eta u - y_\alpha) + \alpha(w_\alpha - y_\alpha),$$

where

$$k(\alpha) := \begin{cases} \dfrac{2(\|y_\alpha - \bar{x} - \eta u\| - \eta\|u\|)}{\alpha\|y_\alpha - \bar{x} - \eta u\|} & \text{if } h_\alpha(y_\alpha) > 0, \\ 0 & \text{if } h_\alpha(y_\alpha) = 0. \end{cases}$$

Observe that the F-derivative of h_α is locally L-continuous and that in particular $h'_\alpha(x) = o$ whenever $h_\alpha(x) = 0$. Hence for each α sufficiently small, $g'_\alpha(y_\alpha)$ is a proximal subgradient of f at $\bar{x}$.

(V) Assume that $\liminf_{\alpha \to 0} k(\alpha) = 0$. Then for some sequence $\alpha_i \to 0$ we had $g'(y_{\alpha_i}) \to o$ as $i \to \infty$ and so (C1) would hold: a contradiction. Therefore the lim inf is positive. We have $g'_\alpha(y_\alpha) = \eta k(\alpha)(u + o(1))$ as $\alpha \to 0$. It follows that for α small enough,

$$\|y_\alpha - \bar{x}\| < \epsilon, \quad |f(y_\alpha) - f(\bar{x})| < \epsilon, \quad \left\|\frac{1}{\eta k(\alpha)}g'_\alpha(y_\alpha) - u\right\| < \epsilon.$$

Hence assertion (C2) holds with $x := y_\alpha$, $\lambda := 1/\eta k(\alpha)$, and $v := g'(y_\alpha)$. $\square$

The Proximal Normal Cone to Level Sets

Now we consider the set

$$M := \{x \in E \mid f(x) = 0\}. \tag{11.65}$$

Theorem 11.6.4 *Let E be a Hilbert space and let M be given by (11.65), where $f : E \to \mathbb{R}$ is continuous. Let $\bar{x} \in M$ and $u \in N_P(M, \bar{x})$. Then either*

(D1) *for any $\epsilon > 0$ and $\eta > 0$ there exists $x \in E$ such that*

$$\|x - \bar{x}\| < \eta, \quad |f(x) - f(\bar{x})| < \eta,$$
$$\left(\partial_P f(x) \cup \partial_P(-f)(x)\right) \cap B_{E^*}(o, \epsilon) \neq \emptyset$$

or

(D2) *for any $\epsilon > 0$ there exist $x \in E$, $v \in \partial_P f(x) \cup \partial_P(-f)(x)$, and $\lambda > 0$ such that*

$$\|x - \bar{x}\| < \epsilon, \quad |f(x) - f(\bar{x})| < \epsilon, \quad \|\lambda v - u\| < \epsilon.$$

Sketch of the Proof. As in the proof of Theorem 11.6.3 we may suppose that $u \neq o$. Define ρ, K, and h_α as above, and m as upper bound of $|f|$. Then $f(x) \neq 0$ for all $x \in K$ and so (since K is convex and f is continuous), f does not change sign on K. Define

$$p_\alpha := \begin{cases} f + h_\alpha + \delta_{B(\bar{x}, \rho\|u\|)} & \text{if } f \text{ is positive on } K, \\ -f + h_\alpha + \delta_{B(\bar{x}, \rho\|u\|)} & \text{if } f \text{ is negative on } K. \end{cases}$$

The argument is now analogous to that in the preceding proof. □

The Fréchet Normal Cone to Sublevel Sets

Now we pass to the Fréchet normal cone. In this connection, we shall make use of the following condition:

$$\liminf_{x \to_f \bar{x}} \mathrm{d}\left(o, \partial_F f(x)\right) > 0; \tag{11.66}$$

here, $x \to_f \bar{x}$ means that $x \to \bar{x}$ and $f(x) \to f(\bar{x})$. Condition 11.66 will serve to exclude a case analogous to (C1) in Theorem 11.6.3.

Theorem 11.6.5 *Let E be a Fréchet smooth Banach space and let $M := \{x \in E \mid f(x) \leq 0\}$, where $f : E \to \overline{\mathbb{R}}$ is proper and l.s.c. Assume that (11.66) holds. Let $\bar{x} \in M$ and $u \in N_F(M, \bar{x})$. Then for any $\epsilon > 0$ there exist $x \in E$, $v \in \partial_F f(x)$, and $\lambda > 0$ such that*

$$\|x - \bar{x}\| < \epsilon, \quad |f(x) - f(\bar{x})| < \epsilon, \quad \|\lambda v - u\| < \epsilon.$$

Sketch of the Proof. Let c be such that $0 < c < \liminf_{x \to_f \bar{x}} \mathrm{d}\left(o, \partial_F f(x)\right)$. It is shown that $\eta, \delta \in (0, \epsilon)$ can be chosen such that

$$\left(\bar{x} + K(u, \eta)\right) \cap M \cap B(\bar{x}, \delta) = \{\bar{x}\},$$

where $K(u, \eta)$ denotes the Bishop–Phelps cone. Define $A := \bar{x} + K(u, 2\eta)$ and $g_i(x) := f(x) + i\,\mathrm{d}_A(x)$ for any $x \in E$. We distinguish two cases:

(I) If $\inf\limits_{B(\bar{x}, \delta)} g_i < 0$, then Ekeland's variational principle of Corollary 8.2.6

ensures the existence of some $y_i \in B(\bar{x}, \delta)$ that minimizes the functional

$$h_i(x) := g_i(x) + \frac{1}{i}\|x - y_i\|$$

over $B(\bar{x}, \delta)$ and is such that $g_i(y_i) < 0$. It can be shown that

$$\frac{\eta}{2\eta + 1}\|y_i - \bar{x}\| \le \mathrm{d}_A(y_i) \to 0 \quad \text{as } i \to \infty$$

and so $y_i \to \bar{x}$ as $i \to \infty$. Therefore, $y_i \in \mathring{B}(\bar{x}, \delta)$ for i sufficiently large.

(II) If $\inf\limits_{B(\bar{x}, \delta)} g_i = 0$, then set $y_i := \bar{x}$ for any i.

Hence in both cases, y_i is a local minimizer of h_i for any sufficiently large i. By the approximate sum rule of Theorem 9.2.6, for these i there exist $x_i, z_i \in \mathring{B}(\bar{x}, \delta)$, $x_i^* \in \partial_F f(x_i)$, and $z_i^* \in \partial_F \mathrm{d}_A(z_i)$ such that $|f(x_i) - f(\bar{x})| < \delta$ and

$$\|x_i^* + i z_i^*\| < \eta + 1/i. \tag{11.67}$$

By using the separation theorem it can be shown that

$$\partial_F \mathrm{d}_A(z_i) \subseteq \{\alpha(-u + 2\eta\|u\|B_{E^*}) \mid \alpha > 0\} \cap B_{E^*}.$$

Therefore, $z_i^* = \alpha_i(-u + 2\eta\|u\|)b^*$ for some $\alpha_i > 0$ and some $b^* \in B_{E^*}$. Now it follows from (11.67) that $\|x_i^* - i\alpha_i u\| < 2i\alpha_i\|u\|\eta + \eta + 1/i$. We must conclude that $i\alpha_i > c/(2\|u\|(1 + 2\eta))$ because otherwise we had $\|x_i^*\| < c$ which contradicts the choice of c. Letting $\lambda_i := 1/(i\alpha_i)$, we obtain

$$\|\lambda_i x_i^* - u\| < 2\eta\|u\| + \frac{2\eta\|u\|(1 + 2\eta)}{c} + \frac{2\|u\|(1 + 2\eta)}{ic}.$$

Thus, if $i > 4\|u\|(1 + 2\eta)/c\epsilon$, then setting $\lambda := \lambda_i$, $x := x_i$, and $v := x_i^*$, we have $\|\lambda v - u\| < \epsilon$. $\qquad\square$

The Fréchet Normal Cone to Level Sets

Now we consider the condition

$$\liminf_{x \to \bar{x}} \mathrm{d}\big(o, \partial_F f(x) \cup \partial_F(-f)(x)\big) > 0. \tag{11.68}$$

Theorem 11.6.6 *Let E be a Fréchet smooth Banach space and let $M := \{x \in E \mid f(x) = 0\}$, where $f : E \to \mathbb{R}$ is continuous. Assume that (11.68) holds. Let $\bar{x} \in M$ and $u \in N_F(M, \bar{x})$. Then for any $\epsilon > 0$ there exist $x \in E$, $v \in \partial_F f(x) \cup \partial_F(-f)(x)$, and $\lambda > 0$ such that*

$$\|x - \bar{x}\| < \epsilon, \quad |f(x) - f(\bar{x})| < \epsilon, \quad \|\lambda v - u\| < \epsilon.$$

Sketch of the Proof. Let c be such that

$$0 < c < \liminf_{x \to \bar{x}} \mathrm{d}\big(o, \partial_F f(x) \cup \partial_F(-f)(x)\big).$$

Again it is shown that $\eta, \delta \in (0, \epsilon)$ can be chosen such that

$$\big(\bar{x} + K(u, \eta)\big) \cap M \cap \mathrm{B}(\bar{x}, \delta) = \{\bar{x}\}.$$

Since f is continuous, it follows that:

(a) $f(x) \geq 0$ for all $x \in \big(\bar{x} + K(u, \eta)\big) \cap \mathrm{B}(\bar{x}, \delta)$ or
(b) $f(x) \leq 0$ for all $x \in \big(\bar{x} + K(u, \eta)\big) \cap \mathrm{B}(\bar{x}, \delta)$.

Define $A := \bar{x} + K(u, 2\eta)$ and for each i,

$$g_i := \begin{cases} f + i\,\mathrm{d}_A & \text{in case (a),} \\ -f + i\,\mathrm{d}_A & \text{in case (b).} \end{cases}$$

The remainder of the proof is analogous to that of Theorem 11.6.5.   □

11.7 Bibliographical Notes and Exercises

Normal cones to convex sets can be traced back to Minkowski [131] and were studied systematically by Fenchel [66]. The contingent cone goes back to Bouligand [26]. Clarke [34] introduced (in finite-dimensional spaces) the cone $\mathrm{T}_C(A, \bar{x})$ according to the formula in Proposition 11.1.6. The sequential characterization of $\mathrm{T}_C(A, \bar{x})$, taken here as definition, is due to Hiriart-Urruty [85]. Proposition 11.1.9 was established by Rockafellar [184].

Proximal normals already appear, under the name *perpendicular vectors*, with Clarke [33]. The results on proximal normals presented above are essentially taken from Clarke et al. [39]. Theorem 11.4.8 is due to Rockafellar [183] (cf. Clarke [36]).

Theorem 11.4.1 was established by Robinson [175] and Zowe and Kurcyusz [228], while Theorem 11.4.2 is a classical result of Lyusternik [128] (see also Graves [79]). The concept of contingent cone was introduced by Aubin [5]. Theorem 11.5.2 is due to Pühl and Schirotzek [173] (see also Pühl [172]). Proposition 11.5.4 and its proof are taken from Pühl [172]. In a different way, Ledzevicz and Walczak [122] deduced the result under the additional hypothesis that $\ker h'(\bar{x})$ has a topological complement in E. These authors also constructed Example 11.5.5. Lyusternik type results related to Theorem 11.5.2 were established, among others, by Cominetti [40], Klatte and Kummer [110], and Penot [161].

Theorems 11.6.3–11.6.6 are taken from Borwein et al. [21] (see also Borwein and Zhu [24]).

For more results on tangents and normals, we refer to Aubin and Frankowska [8] and Clarke et al. [39]. By now, a lot of further tangent and normal sets, not all of them cones and most of them not convex, have been introduced in the literature (see the detailed discussion in Rockafellar and Wets [189]).

Exercise 11.7.1 Fill the gaps in the proof of Proposition 11.1.2.

Exercise 11.7.2 Prove the inclusion (11.1) in Proposition 11.1.3.

Exercise 11.7.3 Verify Proposition 11.1.5.

Exercise 11.7.4 Prove Lemma 11.1.8.

Exercise 11.7.5 Prove Proposition 11.1.13.

Exercise 11.7.6 Prove Proposition 11.2.8.

Exercise 11.7.7 Let $A \subseteq E$ be closed and $\bar{x} \in A$. Show that $N_F(A, \bar{x})$ is closed and convex.

Exercise 11.7.8 Let A be a closed subset of a Fréchet smooth Banach space and $\bar{x} \in A$. Formulate and verify a representation of $N_C(A, \bar{x})$ in terms of F-normals to A at $\bar{x}$. *Hint*: Recall Proposition 9.5.1.

Exercise 11.7.9 Verify the assertions (11.35), (11.36), and (11.38) of Proposition 11.4.5.

Exercise 11.7.10 Define for any $(x_1, x_2) \in \mathbb{R}^2$,

$$g_1(x_1, x_2) := x_2 - x_1^3, \ g_2(x_1, x_2) := -x_2, \ \tilde{g}_1 = g_1, \ \tilde{g}_2 := g_2, \ \tilde{g}_3(x_1, x_2) := -x_1.$$

Show that

$$M := \{(x_1, x_2) \in \mathbb{R}^2 \mid g_i(x_1, x_2) \leq 0 \ (i = 1, 2)\}$$
$$= \{(x_1, x_2) \in \mathbb{R}^2 \mid \tilde{g}_k(x_1, x_2) \leq 0 \ (k = 1, 2, 3)\}$$

but for $\bar{x} := (0, 0)$ one has $L^{\leq}((\gamma_1, \gamma_2), \bar{x}) \neq L^{\leq}((\tilde{\gamma}_1, \tilde{\gamma}_2, \tilde{\gamma}_3), \bar{x})$. Also calculate $T(M, \bar{x})$ (cf. Elster et al. [59]).

Exercise 11.7.11 The aim of this exercise is to generalize Theorem 11.4.9 (cf. Schirotzek [196]). Assume that E and H are Banach spaces, G is a normed vector space, $A \subseteq E$ is nonempty closed and convex, $P \subseteq G$ is a convex cone with nonempty interior, and $Q \subseteq H$ is a closed convex cone. Further let $g : E \to G$ and $h : E \to H$ be given, define

$$M := \{x \in E \mid g(x) \in -P, \ h(x) \in -Q\}$$

and let $\bar{x} \in A \cap M$. Assume that the directional H-derivative $d_H g(\bar{x}, \cdot)$ exists and is P-convex on E, and that h is continuous on E and continuously differentiable at $\bar{x}$. Finally define

$$L^i(g, \bar{x}) := \{y \in E \mid d_H(\bar{x}, y) \in -\operatorname{int} P + \mathbb{R} \, g(\bar{x})\},$$
$$L(g, \bar{x}) := \{y \in E \mid d_H(\bar{x}, y) \in -P + \mathbb{R} \, g(\bar{x})\},$$
$$L(h, \bar{x}) := \{y \in E \mid h'(\bar{x})y \in -Q + \mathbb{R} \, h(\bar{x})\}.$$

(a) Modeling the proof of Theorem 11.4.9(b), prove the following:

Theorem 11.7.12 *Assume that*

$$\mathbb{R}_+(A - \bar{x}) \cap L^i(g, \bar{x}) \cap L(h, \bar{x}) \neq \emptyset \quad and$$
$$h'(\bar{x})(\mathbb{R}_+(A - \bar{x})) + \mathbb{R}_+(Q + h(\bar{x})) = H.$$

Then one has

$$\mathbb{R}_+(A - \bar{x}) \cap L(g, \bar{x}) \cap L(h, \bar{x}) \subseteq T(A \cap M, \bar{x}).$$

(b) Formulate and verify a corresponding result that is analogous to Theorem 11.4.9(a).

Optimality Conditions for Nonconvex Problems

12.1 Basic Optimality Conditions

Let $f : E \to \overline{\mathbb{R}}$ be proper, let $A \subseteq E$, and let $\bar{x} \in A \cap \operatorname{dom} f$. As shown in Sect. 7.1, the method of tangent directions leads to the following result.

Proposition 12.1.1 *Let $\bar{x}$ be a local minimizer of f on A.*

(a) *One has $\overline{f}_G(\bar{x}, y) \geq 0$ for any $y \in \mathrm{T_r}(A, \bar{x})$ and $\overline{f}_H(\bar{x}, y) \geq 0$ for any $y \in \mathrm{T}(A, \bar{x})$.*

(b) *If f is G-differentiable at $\bar{x}$, then $\langle f'(\bar{x}), y \rangle \geq 0$ for any $y \in \mathrm{T_r}(A, \bar{x})$.*

(c) *If f is H-differentiable at $\bar{x}$, then $\langle f'(\bar{x}), y \rangle \geq 0$ for any $y \in \mathrm{T}(A, \bar{x})$.*

We apply the method of penalization (cf. Sect. 7.1). In the case of a locally L-continuous functional, a penalty term of the form $\lambda \mathrm{d}_A$ is adequate.

Proposition 12.1.2 *Let f be locally L-continuous with L-constant $\hat{\lambda}$ on an open set U containing A. If $\bar{x}$ is a minimizer of f on A, then for all $\lambda \geq \hat{\lambda}$, $\bar{x}$ is a minimizer of $f + \lambda \mathrm{d}_A$ on U (hence a local minimizer of $f + \lambda \mathrm{d}_A$ on E).*

Proof. By assumption we have

$$f(z) - \hat{\lambda}\|y - z\| \leq f(y) \leq f(z) + \hat{\lambda}\|y - z\| \quad \forall y, z \in U, \quad f(\bar{x}) \leq f(y) \quad \forall y \in A.$$

Let $z \in U$ and $\epsilon > 0$. Then there exists $y \in A$ such that $\|y - z\| \leq \mathrm{d}_A(z) + \epsilon$. It follows that

$$f(\bar{x}) + \lambda \mathrm{d}_A(\bar{x}) = f(\bar{x}) \leq f(y) \leq f(z) + \hat{\lambda}\|y - z\| \leq f(z) + \lambda \mathrm{d}_A(z) + \lambda\epsilon.$$

Letting $\epsilon \downarrow 0$ proves the assertion. $\qquad\square$

Now we can supplement the optimality criteria in Proposition 12.1.1.

Proposition 12.1.3 *Let f be locally L-continuous on an open set containing A. If $\bar{x}$ is a local minimizer of f on A, then*

$$f^\circ(\bar{x}, y) \geq 0 \quad \forall y \in T_C(A, \bar{x}) \quad \text{and} \quad o \in \partial_o f(\bar{x}) + N_C(A, \bar{x}).$$

Proof.

(I) Let $\eta > 0$ be such that $\bar{x}$ minimizes f on $A_\eta := A \cap B(\bar{x}, \eta)$. Then

$$0 \leq (f + \lambda d_{A_\eta})^\circ(\bar{x}, y) \leq f^\circ(\bar{x}, y) + \lambda d_{A_\eta}^\circ(\bar{x}, y) \quad \forall y \in E; \qquad (12.1)$$

the inequalities are a consequence of Proposition 12.1.2 and the definition of Clarke's directional derivative. By Proposition 11.1.6 we obtain $f^\circ(\bar{x}, y) \geq 0$ for all $y \in T_C(A_\eta, \bar{x})$. By Proposition 11.1.2(c), $T_C(A_\eta, \bar{x})$ is equal to $T_C(A, \bar{x})$.

(II) From (12.1) we deduce $o \in \partial_o(f + \lambda d_{A_\eta})(\bar{x})$. The sum rule (Proposition 7.4.3) and Proposition 11.2.4 show that $o \in \partial_o f(\bar{x}) + N_C(A_\eta, \bar{x})$. As in step (I) we have $N_C(A_\eta, \bar{x}) = N_C(A, \bar{x})$. $\square$

Next we assume that f is l.s.c. only. Again we apply the method of penalization, now with the penalty term δ_A. Recall that if $\bar{x}$ is a local minimizer of f on A, then $\bar{x}$ is a local minimizer of $f + \delta_A$ on E. The generalized Fermat rule of Proposition 9.1.5 thus yields

$$o \in \partial_F(f + \delta_A)(\bar{x}). \qquad (12.2)$$

It remains to apply a sum rule. However, in general we only have *approximate* sum rules for F-subdifferentials unless f is F-differentiable at $\bar{x}$. Accordingly we obtain approximate optimality conditions only. In particular, applying the weak approximate sum rule of Theorem 9.2.7 to (12.2), we obtain:

Proposition 12.1.4 *Assume that E is a Fréchet smooth Banach space, $f : E \to \overline{\mathbb{R}}$ is proper and l.s.c., and A is a closed subset of E. Let $\bar{x} \in A$ be a local minimizer of f on A. Then for any $\epsilon > 0$ and any weak* neighborhood V of zero in E^*, there exist $x_0, x_1 \in B(\bar{x}, \epsilon)$ such that $x_1 \in A$, $|f(x_0) - f(x_1)| < \epsilon$, and*

$$o \in \partial_F f(x_0) + N_F(A, x_1) + V.$$

Proof. See Exercise 12.6.1. $\square$

In Sect. 13.7 we shall derive, in Fréchet smooth Banach spaces, exact necessary optimality conditions in terms of another subdifferential.

To conclude, recall that $\partial^F f(\bar{x}) := -\partial_F(-f)(\bar{x})$ denotes the F-superdifferential of f at $\bar{x}$. For certain classes of nondifferentiable functionals this is an adequate derivative-like object, e.g., for concave continuous functionals. In this case, we have a quite strong optimality condition.

Proposition 12.1.5 *Assume that E is a Fréchet smooth Banach space and $f : E \to \overline{\mathbb{R}}$ is proper and l.s.c. If $\bar{x}$ is a local minimizer of f on A, then*

$$-\partial^F f(\bar{x}) \subseteq N_F(A, \bar{x}).$$

Proof. There is nothing to prove if $\partial^F f(\bar{x})$ is empty. Now let $x^* \in -\partial^F f(\bar{x})$ be given. Then there exists a function $g : E \to \mathbb{R}$ that is F-differentiable at $\bar{x}$ with $g'(\bar{x}) = x^*$ and

$$0 = (-f - g)(\bar{x}) \leq (-f - g)(x) \quad \text{for any } x \text{ near } \bar{x}.$$

By assumption on $\bar{x}$ we further have $f(\bar{x}) \leq f(x)$ for any $x \in A$ near $\bar{x}$. It follows that $\bar{x}$ is a local minimizer of $-g$ on A. From Proposition 9.2.2 we obtain that

$$x^* = -(-g)'(\bar{x}) \in \partial_F \delta_A(\bar{x}) = N_F(A, \bar{x}),$$

which completes the proof. □

12.2 Application to the Calculus of Variations

We consider the classical *fixed end point problem* in the calculus of variations:

$$\text{Minimize } f(x) := \int_a^b \varphi\big(t, x(t), \dot{x}(t)\big) \, dt, \quad x \in A,$$

where

$$A := \{x \in E \mid x(a) = \alpha, \ x(b) = \beta\}, \quad E := \mathrm{AC}^\infty[a, b]. \tag{12.3}$$

Recall that E is a Banach space with respect to the norm $\|x\|_{1,\infty}$ (see Example 3.6.3). The following results hold analogously for absolutely continuous functions on $[a, b]$ with values in $\mathbb{R}^n$.

The Smooth Case

First we repeat a classical result, making the following assumptions:

(A) The real-valued function $(t, x, v) \mapsto \varphi(t, x, v)$ is continuous on $[a, b] \times \mathbb{R} \times \mathbb{R}$ and has continuous first-order partial derivatives with respect to x and v there; a, b, α, β are given real numbers with $a < b$.

If $\bar{x} \in E$, we write

$$\overline{\varphi}(t) := \varphi\big(t, \bar{x}(t), \dot{\bar{x}}(t)\big), \quad t \in [a, b].$$

As shown in Example 3.6.3, the functional f is G-differentiable (even continuously differentiable) at each $\bar{x} \in E$ and

$$\langle f'(\bar{x}), y \rangle = \int_a^b \big(\overline{\varphi}_x(t) \cdot y(t) + \overline{\varphi}_v(t) \cdot \dot{y}(t)\big) \, dt, \quad y \in E.$$

It is immediate that

$$T(A, \bar{x}) = T_r(A, \bar{x}) = \{x \in E \mid x(a) = x(b) = 0\} =: E_0. \qquad (12.4)$$

Assume now that $\bar{x} \in A$ is a local minimizer of f on A. Then Proposition 12.1.1 gives

$$\int_a^b \left(\overline{\varphi}_x(t) \cdot y(t) + \overline{\varphi}_v(t) \cdot \dot{y}(t) \right) dt = 0 \quad \forall y \in E_0; \qquad (12.5)$$

notice that we have equality here since E_0 is a linear subspace of E. Let

$$q(t) := \int_a^t \overline{\varphi}_x(s) \, ds, \quad t \in [a, b].$$

Then q is absolutely continuous and

$$\dot{q}(t) = \overline{\varphi}_x(t) \quad \text{for almost all } t \in [a, b]. \qquad (12.6)$$

Using this and applying partial integration to the first term on the left-hand side of (12.5), we obtain

$$\int_a^b \left(\overline{\varphi}_v(t) - q(t) \right) \cdot \dot{y}(t) \, dt = 0 \quad \forall y \in E_0. \qquad (12.7)$$

Now we need the following *fundamental lemma of the calculus of variations*.

Lemma 12.2.1 (Du Bois–Reymond) *Assume that $g, h \in \mathcal{L}^1[a, b]$ and*

$$\int_a^b \left(h(t) \cdot y(t) + g(t) \cdot \dot{y}(t) \right) dt = 0 \quad \forall y \in E_0.$$

Then the function g is absolutely continuous and satisfies $\dot{g}(t) = h(t)$ for almost all $t \in [a, b]$.

The *proof* of this lemma or one of its various modifications can be found in any standard book on the calculus of variations (see, for instance, Cesari [30] or Giaquinta and Hildebrandt [71], see also Loewen [123]).

Applying Lemma 12.2.1 with $h = o$ to (12.7), we obtain the following result.

Proposition 12.2.2 *Let the assumptions (A) be satisfied. If $\bar{x}$ is a local solution of (12.3), then there exists an absolutely continuous function $p : [a, b] \to \mathbb{R}$ such that*

$$\begin{pmatrix} \dot{p}(t) \\ p(t) \end{pmatrix} = \begin{pmatrix} \overline{\varphi}_x(t) \\ \overline{\varphi}_v(t) \end{pmatrix} \quad \text{for almost all } t \in [a, b]. \qquad (12.8)$$

Eliminating the function p, we see that under the assumptions of Proposition 12.2.2 the function $t \mapsto \overline{\varphi}_v(t)$ is absolutely continuous on $[a, b]$ and satisfies

$$\frac{d}{dt} \overline{\varphi}_v(t) = \overline{\varphi}_x(t) \quad \text{for almost all } t \in [a, b]. \qquad (12.9)$$

This is the *Euler–Lagrange equation* for Problem (12.3).

The Nonsmooth Case

Our purpose now is to weaken the differentiability hypotheses of (A).

Denote by $\mathfrak{L}_1$ the σ-algebra of all Lebesgue measurable subsets of $[a, b]$, by $\mathfrak{B}_n$ the σ-algebra of all Borel subsets of $\mathbb{R}^n$, and by $\mathfrak{L}_1 \times \mathfrak{B}_n$ the corresponding product σ-algebra. Let $\bar{x} \in M$ be given. We make the following assumptions:

(Â) The function $\varphi : [a, b] \times \mathbb{R} \times \mathbb{R} \to \mathbb{R} \cup \{+\infty\}$ is $\mathfrak{L}_1 \times \mathfrak{B}_2$-measurable. There exist $\epsilon > 0$ and a positive function $g \in \mathcal{L}^1[a, b]$ such that for almost all $t \in [a, b]$ the function $(x, v) \mapsto \varphi(t, x, v)$ is real-valued and Lipschitz continuous on $(\bar{x}(t), \dot{\bar{x}}(t)) + \epsilon B$ with Lipschitz constant $g(t)$.

We establish a generalization of Proposition 12.2.2.

Theorem 12.2.3 *Let the assumptions* (Â) *be satisfied. If $\bar{x}$ is a local solution of* (12.3), *then there exists an absolutely continuous function $p : [a, b] \to \mathbb{R}$ such that*

$$\begin{pmatrix} \dot{p}(t) \\ p(t) \end{pmatrix} \in \partial_\circ \varphi(t, \bar{x}(t), \dot{\bar{x}}(t)) \quad \text{for almost all } t \in [a, b]. \tag{12.10}$$

Here, $\partial_\circ \varphi(t, \bar{x}(t), \dot{\bar{x}}(t))$ denotes the Clarke subdifferential of the function $(x, v) \mapsto \varphi(t, x, v)$ at the point $((x(t), \dot{x}(t))$.

Notice that under the assumptions (A), Proposition 7.3.9 implies

$$\partial_\circ \varphi(t, \bar{x}(t), \dot{\bar{x}}(t)) = \begin{pmatrix} \overline{\varphi}_x(t) \\ \overline{\varphi}_v(t) \end{pmatrix}$$

so that (12.10) passes into (12.8).

To prepare the proof of Theorem 12.2.3, we quote two propositions. The first is a measurable selection statement which we take from Loewen [123].

Proposition 12.2.4 *Let $g : [a, b] \times \mathbb{R} \to \mathbb{R} \cup \{+\infty\}$ be $\mathfrak{L}_1 \times \mathfrak{B}_1$-measurable and let $p \in [1, +\infty)$. Assume that:*

- *For each $t \in [a, b]$ the function $v \mapsto g(t, v)$ is lower semicontinuous and real-valued at some point of $\mathbb{R}$.*
- *There exists $\tilde{u} \in \mathcal{L}^p[a, b]$ such that the function $t \mapsto g(t, \tilde{u}(t))$ is integrable over $[a, b]$.*

Then the integral of the function $t \mapsto \inf_{v \in \mathbb{R}} g(t, v)$ over $[a, b]$ is defined (perhaps equal to $-\infty$), and one has

$$\int_a^b \inf_{v \in \mathbb{R}} g(t, v) \, dt = \inf_{u \in \mathcal{L}^p[a, b]} \int_a^b g(t, u(t)) \, dt.$$

The next result is Aubin's nonsymmetric minimax theorem (see Aubin [6] or Aubin and Ekeland [7]).

Proposition 12.2.5 *Let K be a compact convex subset of a topological vector space, let U be a convex subset of a vector space, and let $h : K \times U \to \mathbb{R}$ be a function such that:*

- *For each $u \in U$, the function $x \mapsto h(x, u)$ is convex and l.s.c.*
- *For each $x \in K$, the function $u \mapsto -h(x, u)$ is convex.*

Then there exists $\hat{x} \in K$ such that

$$\sup_{u \in U} h(\hat{x}, u) = \sup_{u \in U} \inf_{x \in K} h(x, u).$$

In particular,

$$\inf_{x \in K} \sup_{u \in U} h(x, u) = \sup_{u \in U} \inf_{x \in K} h(x, u).$$

Proof of Theorem 12.2.3. By Proposition 12.1.1 and (12.4) we have for each $y \in E_0$,

$$0 \le \overline{f}_G(\bar{x}, y)$$
$$= \limsup_{\tau \downarrow 0} \int_a^b \tau^{-1} \Big(\varphi\big(t, \bar{x}(t) + \tau y(t), \dot{\bar{x}}(t) + \tau \dot{y}(t)\big) - \varphi\big(t, \bar{x}(t), \dot{\bar{x}}(t)\big) \Big) \, dt.$$

By assumption ($\hat{A}$), the integrand on the right-hand side is a measurable function of t and, for almost all $t \in [a, b]$, is majorized by the integrable function $g(t) \|(y(t), \dot{y}(t))\|$. Hence by a variant of the Fatou lemma, we have $\limsup \int_a^b \cdots \le \int_a^b \limsup \cdots$, and it follows that

$$0 \le \int_a^b \varphi^\circ\big(t, \bar{x}(t), \dot{\bar{x}}(t); y(t), \dot{y}(t)\big) \, dt; \tag{12.11}$$

here $\varphi^\circ(\cdots)$ denotes the Clarke directional derivative of the function $(x, v) \mapsto \varphi(t, x, v)$ at the point $(\bar{x}(t), \dot{\bar{x}}(t))$ in the direction $(y(t), \dot{y}(t))$. The relation (12.11) holds for any $y \in E_0$, in particular for $y = o$. Therefore we have

$$0 = \inf_{y \in E_0} \int_a^b \varphi^\circ\big(t, \bar{x}(t), \dot{\bar{x}}(t); y(t), \dot{y}(t)\big) \, dt.$$

Applying Proposition 7.3.7 we further obtain

$$0 = \inf_{y \in E_0} \int_a^b \sup\{\langle (u(t), p(t)), (y(t), \dot{y}(t)) \rangle \mid (u(t), p(t)) \in \partial_\circ \varphi(t, \bar{x}(t), \dot{\bar{x}}(t))\} \, dt,$$

which by Proposition 12.2.4 passes into

$$0 = \inf_{y \in E_0} \sup_{(u,p) \in K} \int_a^b \big(u(t)\, y(t) + p(t)\, \dot{y}(t)\big) \, dt, \tag{12.12}$$

where

$$K := \{(u, p) \in \mathcal{L}^1 \times \mathcal{L}^1 \mid (u(t), p(t)) \in \partial_\circ \varphi(t, \bar{x}(t), \dot{\bar{x}}(t)), \ t \in [a, b] \ (\text{a.e.})\};$$

here, $\mathcal{L}^1$ stands for $\mathcal{L}^1[a, b]$. The set K is convex and weakly compact and the integral in (12.12) depends bilinearly on u, p. Hence by Proposition 12.2.5, the infimum and the supremum in (12.12) can be interchanged and the supremum is attained at some $(u, p) \in K$, i.e., we have

$$0 = \inf_{y \in E_0} \int_a^b \big(u(t)\, y(t) + p(t)\, \dot{y}(t) \big) \, \mathrm{d}t.$$

Since the integral depends linearly on y, we can conclude that

$$0 = \int_a^b \big(u(t)\, y(t) + p(t)\, \dot{y}(t) \big) \, \mathrm{d}t \quad \forall y \in E_0.$$

Applying Lemma 12.2.1, we see that p is absolutely continuous and $\dot{p}(t) = u(t)$ for almost all $t \in [a, b]$. In view of the definition of K, the proof is complete.
□

Finally we indicate how to extend the class of problems that can be treated using the tools of nonsmooth analysis. Recall that $E := AC^\infty[a, b]$. Consider the so-called *Bolza problem*

$$f(x) := \gamma\big(x(a), x(b)\big) + \int_a^b \varphi\big(t, x(t), \dot{x}(t)\big) \, \mathrm{d}t \longrightarrow \min, \ x \in E. \qquad (12.13)$$

This is a classical problem if the function φ satisfies the assumptions (A) and the function $(\xi, \eta) \mapsto \gamma(\xi, \eta)$ mapping $\mathbb{R}^2$ into $\mathbb{R}$ is differentiable at $(\bar{x}(a), \bar{x}(b))$. Here $\bar{x} \in E$ denotes a local minimum point of f on E. In this case, necessary conditions are (12.8) together with the *natural boundary conditions*

$$p(a) = \gamma_\xi(\bar{x}(a), \bar{x}(b)), \quad p(b) = -\gamma_\eta(\bar{x}(a), \bar{x}(b)).$$

However, within the framework of nonsmooth analysis the functions φ and γ are allowed to attain the value $+\infty$. For instance, let

$$\gamma(\xi, \eta) := \begin{cases} 0 & \text{if } \xi = \alpha \text{ and } \eta = \beta, \\ +\infty & \text{otherwise.} \end{cases}$$

Then (12.13) passes into the fixed end point problem (12.3). Now we consider a more interesting class of problems. Let the function $\gamma : \mathbb{R}^2 \to \mathbb{R} \cup \{+\infty\}$ and the multifunction $F : \mathbb{R}^2 \rightrightarrows \mathbb{R}$ be given. Consider the problem

$$\gamma(x(a), x(b)) \longrightarrow \min, \ x \in E, \ \dot{x}(t) \in F(t, x(t)), \ t \in [a, b] \ (\text{a.e.}). \qquad (12.14)$$

Setting

$$\varphi(t, x, v) := \begin{cases} 0 & \text{if } v \in F(t, x), \\ +\infty & \text{otherwise,} \end{cases}$$

we have

$$\int_a^b \varphi(t, x(t), \dot{x}(t))\, dt = \begin{cases} 0 & \text{if } \dot{x}(t) \in F(t, x(t)),\ t \in [a, b] \text{ (a.e.)}, \\ +\infty & \text{otherwise.} \end{cases}$$

Hence problem (12.14) is of the form (12.13). Of course, the assumptions (Â) are not satisfied for the function φ in this case. Nevertheless, it is possible to reduce problem (12.14) in such a way that Theorem 12.2.3 applies. However, this requires facts on differentiable inclusions that lie beyond the scope of this book (we refer to Clarke [36], Clarke et al. [39], and Loewen [123]).

12.3 Multiplier Rules Involving Upper Convex Approximations

Our aim in this section is to establish necessary optimality conditions for the problem

$$f(x) \to \min, \quad x \in M \cap A,$$

under the assumption that M is described by scalar inequalities and/or an operator equation. First we consider the problem:

(P1) Minimize $f(x)$
　　　subject to $g_i(x) \le 0\ (i = 1, \dots, m), \quad x \in A$.

　　We set

$$I := \{1, \dots, m\}, \quad I(\bar{x}) := \{i \in I \mid g_i(\bar{x}) = 0\} \tag{12.15}$$

and make the following assumptions:

(A1)　E is a normed vector space,
　　　　$A \subset E$ is nonempty and convex, $D \subset E$ is nonempty and open,
　　　　$f, g_i : D \to \mathbb{R}$ for $i = 1, \dots, m$,
　　　　there exist $\varphi \in \mathrm{UC_r}(f, \bar{x})$ and $\psi_i \in \mathrm{UC_r}(g_i, \bar{x})$ for $i \in I(\bar{x})$,
　　　　g_i is upper semicontinuous at $\bar{x}$ for $i \in I \setminus I(\bar{x})$.

Theorem 12.3.1 *Let (A1) be satisfied and assume that $\bar{x}$ is a local solution of (P1).*

(a) *One has:*

$$\exists \lambda \in \mathbb{R}_+\ \exists \mu_i \in \mathbb{R}_+\ (i \in I(\bar{x})): \ \textit{not all equal to zero,}$$

$$\lambda \varphi(y) + \sum_{i \in I(\bar{x})} \mu_i \psi_i(y) \ge 0 \quad \forall y \in A - \bar{x}. \tag{12.16}$$

(b) *Let the following condition be satisfied:*

$$\exists y_0 \in A - \bar{x} \quad \forall i \in I(\bar{x}): \ \psi_i(y_0) < 0. \tag{12.17}$$

Then (a) *holds with* $\lambda = 1$.

Proof.

(a) Let $y \in A - \bar{x}$ and $\psi_i(y) < 0$ for each $i \in I(\bar{x})$. Then it follows from Proposition 11.4.5(a) that $y \in T_r(M_1 \cap A, \bar{x})$, where

$$M_1 := \{x \in D \mid g_i(x) \leq 0 \ (i = 1, \dots, m)\}.$$

In this connection, notice that $A - \bar{x} \subseteq T_r(A, \bar{x})$ because A is convex. Since $\bar{x}$ is a local solution of f on $M_1 \cap A$, Proposition 12.1.1(a) implies that $\varphi(y) \geq 0$. Consequently, the system

$$y \in A - \bar{x}, \ \varphi(y) < 0, \ \psi_i(y) < 0 \ \forall i \in I(\bar{x})$$

has no solution. The assertion now follows by Proposition 10.2.1 with $H := \{o\}$. (Notice that to verify (10.3), we need only set $\bar{y}_i := \psi_i(x_0) + 1$ for each $i \in I(\bar{x})$, with some $x_0 \in A - \bar{x}$.)

(b) This is an immediate consequence of (a).   □

Remark 12.3.2

(a) In terms of radial upper convex approximations, the conditions in (a) and (b) above are *generalized John conditions* and *generalized Karush–Kuhn–Tucker conditions*, respectively (cf. Sect. 5.2). Likewise, (12.17) is a *generalized Slater condition*.

(b) Concerning the existence of (radial) upper convex approximations φ and ψ_i, we refer to Theorem 7.3.2 (cf. Remark 11.4.6(a)).

Now we consider the problem:

(P2) Minimize $f(x)$
subject to $g_i(x) \leq 0 \ (i = 1, \dots, m)$, $h(x) = o$, $x \in A$.

Recall the notation (12.15). We agree on the following assumptions:

(A2) E and F are Banach spaces,
$A \subseteq E$ is nonempty, convex, and closed, $D \subseteq E$ is nonempty and open,
$f, g_i : D \to \mathbb{R}$ for $i = 1, \dots, m$,
there exist $\varphi \in UC(f, \bar{x})$ and $\psi_i \in UC(g_i, \bar{x})$ for $i \in I(\bar{x})$,
g_i is upper semicontinuous at $\bar{x}$ for $i \in I \setminus I(\bar{x})$,
$h : D \to F$ is F-differentiable on a neighborhood of $\bar{x}$, h' is continuous at $\bar{x}$.

Theorem 12.3.3 *Let* (A2) *be satisfied and assume that* $\bar{x}$ *is a local solution of* (P2):

(a) *If* $h'(\bar{x})[\mathbb{R}_+(A - \bar{x})]$ *is closed or* F *is finite dimensional, then*

$$\exists \lambda \in \mathbb{R}_+ \quad \mu_i \in \mathbb{R}_+ \ (i \in I(\bar{x})) \quad \exists v \in F^* : \quad \lambda, \mu_i, v \text{ not all zero},$$

$$\lambda \varphi(y) + \sum_{i \in I(\bar{x})} \mu_i \psi_i(y) + \langle v \circ h'(\bar{x}), y \rangle \geq 0 \quad \forall y \in A - \bar{x}. \quad (12.18)$$

(b) *Let φ and ψ_i, $i \in I(\bar{x})$, be continuous on E and let the following condition be satisfied:*

$$h'(\bar{x})\big[\mathbb{R}_+(A - \bar{x})\big] = F. \tag{12.19}$$

Then (12.18) holds with λ, μ_i, $i \in I(\bar{x})$, not all zero.

(c) *Let (12.19) and the following condition be satisfied:*

$$\exists\, y_0 \in A - \bar{x}: \ \psi_i(y_0) < 0 \ \forall i \in I(\bar{x}), \ h'(\bar{x})y_0 = o. \tag{12.20}$$

Then (12.18) holds with $\lambda = 1$.

Proof.

(a) In view of (b), we may assume that $h'(\bar{x})\big[\mathbb{R}_+(A-\bar{x})\big] \neq F$. Then by a separation theorem (Theorem 1.5.9 in the general case, or Proposition 1.5.7 if F is finite dimensional), there exists $v \in F^*$ satisfying $v \neq o$ and $\langle v, z \rangle \geq 0$ for each $z \in h'(\bar{x})\big[\mathbb{R}_+(A - \bar{x})\big]$. Hence the assertion holds with $\lambda = 0$ and $\mu_i = 0$ for each $i \in I(\bar{x})$.

(b) Let $y \in A - \bar{x}$, $\psi_i(y) < 0$ for each $i \in I(\bar{x})$, and $h'(\bar{x})y = o$. Then it follows from Theorem 11.4.9 that $y \in T(M_2 \cap A, \bar{x})$, where M_2 is defined by the functional constraints. Since $\bar{x}$ is a local solution of f on $M_2 \cap A$, Proposition 12.1.1(b) says that $\varphi(y) \geq 0$. Consequently, the system

$$y \in A - \bar{x}, \ \varphi(y) < 0, \ \psi_i(y) < 0 \ \forall i \in I(\bar{x}), \ h'(\bar{x})y = o$$

has no solution. The assertion now follows by Proposition 10.2.4.

(c) This is an immediate consequence of (b). $\qquad\qquad\qquad\qquad\qquad \Box$

Remark 12.3.4

(a) The optimality condition (12.18) is again a *generalized John condition* or, if $\lambda = 1$, a *generalized Karush–Kuhn–Tucker condition*, with the *Lagrange multipliers* μ_i and v.

(b) If, in particular, f is H-differentiable at $\bar{x}$, we set $\varphi := f'(\bar{x})$, analogously for g_i, $i \in I(\bar{x})$. If, in addition, $A = E$, then (12.18) reduces to

$$\lambda\, f'(\bar{x}) + \sum_{i\in I(\bar{x})} \mu_i\, g_i'(\bar{x}) + v \circ h'(\bar{x}) = o.$$

Moreover, if $F := \mathbb{R}^r$ and so $h = (h_1, \ldots, h_r)$, where $h_k : D \to \mathbb{R}$ for $k = 1, \ldots, r$, then (12.18), (12.19), and (12.20) pass, respectively, into

$$\exists\, \lambda \in \mathbb{R}_+, \quad \exists\, \mu_i \in \mathbb{R}_+ \,(i \in I(\bar{x})), \quad \exists\, \nu_j \in \mathbb{R}\,(j = 1, \ldots, r): \quad \text{not all zero,}$$

$$\lambda\, f'(\bar{x}) + \sum_{i\in I(\bar{x})} \mu_i\, g_i'(\bar{x}) + \sum_{j=1}^{r} \nu_j\, h_j'(\bar{x}) = o, \tag{12.21}$$

$$h_1'(\bar{x}), \ldots, h_r'(\bar{x}) \text{ are linearly independent elements of } E^*, \tag{12.22}$$

$$\exists\, y_0 \in E: \ \langle g_i'(\bar{x}), y_0 \rangle < 0 \ \forall i \in I(\bar{x}), \ \langle h_j'(\bar{x}), y_0 \rangle = 0 \ \forall j = 1, \ldots, r. \tag{12.23}$$

The conditions (12.22) and (12.23) together are the *Mangasarian–Fromowitz regularity condition*.

(c) If, in (P2), the inequalities $g_i(x) \leq 0$, $i \in I$, are absent, then the corresponding data in Theorem 12.3.3 have to be omitted. This is verified by applying Theorem 11.4.1 instead of Theorem 11.4.9.

For finite-dimensional F, we now relax the differentiability hypothesis on the mapping h. We assume that h is F-differentiable *at* the point $\bar{x}$ only. Also we now choose concrete upper convex approximations for the functionals g_i. More precisely, we consider the following nonsmooth optimization problem:

(P3) Minimize $f(x)$
$\qquad$ subject to $g_i(x) \leq 0, \quad i = 1, \ldots, p,$
$\qquad\qquad\quad \tilde{g}_j(x) \leq 0, \quad j = 1, \ldots, q,$
$\qquad\qquad\quad h_k(x) = 0, \quad k = 1, \ldots, r,$
$\qquad\qquad\quad x \in A.$

We want to establish a necessary condition for a point $\bar{x} \in E$ to be a local solution of (P3). We set

$$I := \{1, \ldots, p\}, \quad I(\bar{x}) := \{i \in I \mid g_i(\bar{x}) = 0\},$$
$$J := \{1, \ldots, q\}, \quad J(\bar{x}) := \{j \in J \mid \tilde{g}_j(\bar{x}) = 0\}.$$

The assumptions are:

(A3) E is a Banach space, $A \subseteq E$ is closed and epi-Lipschitz at $\bar{x}$,
$\qquad f : E \to \mathbb{R}$ is H-differentiable at $\bar{x}$ or locally L-continuous around $\bar{x}$,
$\qquad g_i : E \to \mathbb{R}$ is H-differentiable at $\bar{x}$ for $i \in I(\bar{x})$,
$\qquad g_i : E \to \mathbb{R}$ is continuous at $\bar{x}$ for $i \in I \setminus I(\bar{x})$,
$\qquad \tilde{g}_j : E \to \mathbb{R}$ is locally L-continuous around $\bar{x}$ for $j \in J(\bar{x})$,
$\qquad \tilde{g}_j : E \to \mathbb{R}$ is continuous at $\bar{x}$ for $j \in J \setminus J(\bar{x})$,
$\qquad h_k : E \to \mathbb{R}$ is F-differentiable at $\bar{x}$ and continuous in a neighborhood
$\qquad$ of $\bar{x}$ for $k = 1, \ldots, r$.

Further we need certain constraint qualifications:

$h_1'(\bar{x}), \ldots, h_r'(\bar{x})$ are linearly independent elements of E^*. $\hfill$ (12.24)

$\exists y_0 \in \mathrm{T_C}(A, \bar{x}): \langle g_i'(\bar{x}), y_0 \rangle < 0 \, \forall i \in I(\bar{x}), \, \tilde{g}_j^\Diamond(\bar{x}, y_0) < 0 \, \forall j \in J(\bar{x}),$

$\langle h_k'(\bar{x}), y_0 \rangle = 0 \, \forall k = 1, \ldots, r.$ $\hfill$ (12.25)

Theorem 12.3.5 *Let the assumptions* (A3) *be satisfied and let $\bar{x}$ be a local solution of* (P3)*:*

(a) *There exist scalars $\lambda \geq 0$, $\mu_i \geq 0$ for $i \in I(\bar{x})$, $\tilde{\mu}_j \geq 0$ for $j \in J(\bar{x})$, and ν_k for $k = 1, \ldots, r$ that are not all zero, such that for any $y \in \mathrm{T_C}(A, \bar{x})$,*

$$\lambda f^\Diamond(\bar{x}, y) + \sum_{i \in I(\bar{x})} \mu_i \langle g_i'(\bar{x}), y \rangle + \sum_{j \in J(\bar{x})} \tilde{\mu}_j \, \tilde{g}_j^\Diamond(\bar{x}, y) + \sum_{k=1}^r \nu_k \langle h_k'(\bar{x}), y \rangle \geq 0.$$

$\hfill$ (12.26)

(b) *If (12.24) is satisfied, then (a) holds, where* λ, μ_i $(i \in I(\bar{x}))$, *and* $\tilde{\mu}_j$ $(j \in J(\bar{x}))$ *are not all zero.*

(c) *If (12.24) and (12.25) are satisfied, then (a) holds with* $\lambda = 1$.

Proof.

(I) Define $h : E \to \mathbb{R}^r$ by $h := (h_1, \ldots, h_r)$. If (12.24) is not satisfied, then $h'(\bar{x})(E)$ is a proper linear subspace of $\mathbb{R}^r$. Hence there exists $(\nu_1, \ldots, \nu_r) \in \mathbb{R}^r \setminus \{o\}$ such that $\sum_{k=1}^{r} \nu_k \langle h'_k(\bar{x}), y \rangle = 0$ for any $y \in E$. Setting $\lambda = \mu_i = \tilde{\mu}_j = 0$, we see that (12.26) holds.

(II) Assume now that (12.24) is satisfied. We show that (b) holds, which also verifies (a) in this case. Since the F-derivative $h'(\bar{x})$ is surjective, Halkin's correction theorem (Theorem 3.7.5) ensures that there exist a neighborhood U of $\bar{x}$ and a mapping $\rho : U \to E$ that is F-differentiable at $\bar{x}$ and satisfies

$$\rho(\bar{x}) = o, \quad \rho'(\bar{x}) = o, \tag{12.27}$$

$$h_k(x + \rho(x)) = \langle h'_k(\bar{x}), x - \bar{x} \rangle \quad \forall x \in U \quad \forall k = 1, \ldots, r. \tag{12.28}$$

(III) We claim that the following system has no solution:

$$y \in \text{int } T_C(A, \bar{x}), \tag{12.29}$$

$$f^{\Diamond}(\bar{x}, y) < 0, \tag{12.30}$$

$$\langle g'_i(\bar{x}), y \rangle < 0, \quad i \in I(\bar{x}), \tag{12.31}$$

$$\tilde{g}_j^{\Diamond}(\bar{x}, y) < 0, \quad j \in J(\bar{x}), \tag{12.32}$$

$$\langle h'_k(\bar{x}), y \rangle = 0, \quad k = 1, \ldots, r. \tag{12.33}$$

Assume, to the contrary, that the system does have a solution y. Set

$$\theta(\tau) := \bar{x} + \tau y + \rho(\bar{x} + \tau y), \quad \tau \in [0, 1].$$

The relations (12.28) and (12.33) show that $h_k(\theta(\tau)) = 0$ for $k = 1, \ldots, r$ whenever $\tau \in [0, 1]$ is so small that $\bar{x} + \tau y \in U$. The properties of ρ entail $\theta(0) = \bar{x}$ and $\theta'(0) = y$. Hence the chain rule (Proposition 3.4.4) gives

$$\lim_{\tau \downarrow 0} \tau^{-1} [g_i(\theta(\tau)) - g_i(\theta(0))] = \langle g'_i(\bar{x}), y \rangle \underset{(12.31)}{<} 0 \quad \forall i \in I(\bar{x}).$$

Since, for $i \in I(\bar{x})$, $g_i(\theta(0)) = g_i(\bar{x}) = o$, we deduce that

$$g_i(\theta(\tau)) < 0 \quad \text{for } \tau \in (0, 1] \text{ sufficiently small and all } i \in I(\bar{x}). \tag{12.34}$$

Since $\tilde{g}_j$ is locally L-continuous around $\bar{x}$, the Michel–Penot directional derivative can be written as

$$\tilde{g}_j^{\Diamond}(\bar{x}, y) = \sup_{z \in E} \limsup_{\substack{\tau \downarrow 0 \\ y' \to y}} \tau^{-1} [\tilde{g}_j(\bar{x} + \tau(y' + z)) - \tilde{g}_j(\bar{x} + \tau z)] \quad \forall j \in J(\bar{x}).$$

$$\tag{12.35}$$

By (12.27) we have

$$y' := y + \tau^{-1}\rho(\bar{x} + \tau y) \to y \quad \text{as } \tau \downarrow 0. \qquad (12.36)$$

From this, (12.32), and (12.35) we conclude that

$$\tau^{-1}\big[\tilde{g}_j\big(\bar{x} + \tau y + \rho(\bar{x} + \tau y)\big) - \tilde{g}_j(\bar{x})\big] < 0 \quad \forall j \in J(\bar{x})$$

and so

$$\tilde{g}_j\big(\theta(\tau)\big) < 0 \quad \text{for } \tau \in (0,1] \text{ sufficiently small and all } j \in J(\bar{x}). \quad (12.37)$$

Analogously, by (12.30) we obtain

$$f\big(\theta(\tau)\big) < f(\bar{x}) \quad \text{for } \tau \in (0,1] \text{ sufficiently small.} \qquad (12.38)$$

By assumption (A3), the hypertangent cone $H(A, \bar{x})$ is nonempty and so coincides with $\operatorname{int} T_C(A, x)$ (Proposition 11.1.9). Therefore $y \in H(A, \bar{x})$. In view of (12.36) and Lemma 11.1.8, we obtain

$$\theta(\tau) = \bar{x} + \tau\big(y + \tau^{-1}\rho(\bar{x} + \tau y)\big) \in A. \qquad (12.39)$$

Since g_i, $i \notin I(\bar{x})$, and $\tilde{g}_j$, $j \notin J(\bar{x})$, are continuous at $\bar{x}$, we further have

$$g_i\big(\theta(\tau)\big) < 0 \quad \text{for } \tau \in (0,1] \text{ sufficiently small and all } i \in I \setminus I(\bar{x}), \qquad (12.40)$$

$$\tilde{g}_j\big(\theta(\tau)\big) < 0 \quad \text{for } \tau \in (0,1] \text{ sufficiently small and all } j \in J \setminus J(\bar{x}). \qquad (12.41)$$

The relations (12.34)–(12.41) contradict the fact that $\bar{x}$ is a local solution of (P3). Thus we have shown that the system (12.29)–(12.33) has no solution.

(IV) We want to apply Proposition 10.2.4. Define

$$s := \operatorname{card} I(\bar{x}), \quad t := \operatorname{card} J(\bar{x}),$$

$$G := E \times \mathbb{R} \times \mathbb{R}^s \times \mathbb{R}^t, \quad P := -T_C(A, \bar{x}) \times \mathbb{R}_+ \times \mathbb{R}_+^s \times \mathbb{R}_+^t,$$

$$S(y) := \big(y,\, f^{\Diamond}(\bar{x}, y),\, (\langle g_i'(\bar{x}), y\rangle)_{i \in I(\bar{x})},\, (\tilde{g}_j^{\Diamond}(\bar{x}, y))_{j \in J(\bar{x})}\big), \quad y \in E,$$

$$T(y) := \big(\langle h_1'(\bar{x}), y\rangle, \ldots, \langle h_r'(\bar{x}), y\rangle\big), \quad y \in E.$$

Notice that P is a convex cone with nonempty interior in G and that the mapping $S : E \to G$ is P-convex. By step (III), the system

$$y \in E, \; S(y) \in -\operatorname{int} P, \; T(y) = o$$

has no solution. Hence by Proposition 10.2.4 there exist $u \in \mathrm{T_C}(A, \bar{x})^\circ$, $\lambda \geq 0$, $\mu_i \geq 0$, $i \in I(\bar{x})$, and $\tilde{\mu}_j \geq 0$, $j \in J(\bar{x})$, not all zero, as well as $\nu_1, \ldots, \nu_r \in \mathbb{R}$ satisfying

$$
\begin{aligned}
- \langle u, y \rangle - \lambda f^\Diamond(\bar{x}, y) - \sum_{i \in I(\bar{x})} \mu_i \langle g_i'(\bar{x}), y \rangle \\
- \sum_{j \in J(\bar{x})} \tilde{\mu}_j \tilde{g}_j^\Diamond(\bar{x}, y) - \sum_{k=1}^r \nu_k \langle h_k'(\bar{x}), y \rangle \leq 0 \quad \forall y \in E.
\end{aligned}
\tag{12.42}
$$

The multipliers λ, μ_i, $i \in I(\bar{x})$, and $\tilde{\mu}_j$, $j \in J(\bar{x})$, cannot be simultaneously equal to zero since otherwise the inequality (12.42) would imply that u is also zero which is contradictory. The assertion of the theorem now follows from (12.42) since $\langle u, y \rangle \leq 0$ for any $y \in \mathrm{T_C}(A, \bar{x})$.

(V) The assertion (c) is obvious. □

12.4 Clarke's Multiplier Rule

A drawback of the multiplier rules established so far is the differentiability assumption on the operator $h : E \to F$ defining the equality constraint. In this section, we present a multiplier rule under a crucially weakened assumption on h, provided F is finite dimensional. We consider the following problem:

(P4) Minimize $f(x)$
   subject to $g_i(x) \leq 0$ $(i = 1, \ldots, m)$, $h_j(x) = 0$ $(j = 1, \ldots, n)$, $x \in A$.

The assumptions are:

(A4) E is a Banach space, $A \subseteq E$ is nonempty and closed,
   $f, g_i, h_j : E \to \mathbb{R}$ $(i = 1, \ldots, m, j = i, \ldots, n)$ are all locally L-continuous around any $x \in A$.

Theorem 12.4.1 (Clarke's Multiplier Rule) *Let* (A4) *be satisfied and assume that* $\bar{x}$ *is a local solution of* (P4). *Then:*

$$
\exists \lambda \in \mathbb{R}_+ \ \exists \mu_i \in \mathbb{R}_+ \ (i = 1, \ldots, m) \ \exists \nu_j \in \mathbb{R} \ (j = 1, \ldots, n) : \ \textit{not all zero,}
$$
$$
\mu_i g_i(\bar{x}) = 0 \ (i = 1, \ldots, m),
\tag{12.43}
$$
$$
o \in \lambda \partial_\circ f(\bar{x}) + \sum_{i=1}^m \mu_i \partial_\circ g_i(\bar{x}) + \sum_{j=1}^n \nu_j \partial h_j(\bar{x}) + \mathrm{N_C}(A, \bar{x}).
\tag{12.44}
$$

Proof.

(I) We may and do assume that $\bar{x}$ is a *global* solution of (P4) because the following argument can be applied with A replaced by $A \cap \mathrm{B}(\bar{x}, \eta)$, where $\eta > 0$ is sufficiently small.

(II) Let $\epsilon > 0$ be given. Define $\psi : E \to \overline{\mathbb{R}}$ by

$$\psi(x) := \max\{f(x) - f(\bar{x}) + \tfrac{\epsilon}{2}, g_1(x), \ldots, g_m(x), |h_1(x)|, \ldots, |h_n(x)|\}.$$

Then ψ is locally L-continuous and so l.s.c. Since $\bar{x}$ is a solution of (P4), we have

$$\psi(x) > 0 \quad \forall x \in A, \tag{12.45}$$

which, together with $\psi(\bar{x}) = \epsilon/2$, implies

$$\psi(\bar{x}) < \inf_{x \in A} \psi(x) + \epsilon. \tag{12.46}$$

(III) By step (II), Ekeland's variational principle in the form of Corollary 8.2.6 applies to ψ. Choosing $\lambda := \sqrt{\epsilon}$ we conclude that there exists $x_\epsilon \in A$ such that

$$\|x_\epsilon - \bar{x}\| \leq \sqrt{\epsilon}, \tag{12.47}$$

$$\psi(x_\epsilon) \leq \psi(x) + \sqrt{\epsilon}\,\|x_\epsilon - x\| \quad \forall x \in A. \tag{12.48}$$

(IV) Let κ_0 be a common local Lipschitz constant of the functionals f, $g_1, \ldots, g_m$, and $h_1, \ldots, h_n$ around $\bar{x}$. Then, as is easy to see, each $\kappa > \kappa_0$ is a local Lipschitz constant of the functional $x \mapsto \psi(x) + \sqrt{\epsilon}\,\|x_\epsilon - x\|$ around x_ϵ for ϵ sufficiently small. Since by (12.48), this functional has a minimum on A at x_ϵ, Proposition 12.1.2 shows that x_ϵ is a local minimizer of

$$\widetilde{\psi}(x) := \psi(x) + \sqrt{\epsilon}\,\omega_{x_\epsilon}(x) + \kappa d_A(x), \quad \text{where } \omega_{x_\epsilon}(x) := \|x_\epsilon - x\|.$$

Hence the sum rule (Proposition 7.4.3) gives

$$o \in \partial_\circ \widetilde{\psi}(x_\epsilon) \subseteq \partial_\circ \psi(x_\epsilon) + \sqrt{\epsilon}\,\partial_\circ \omega_{x_\epsilon}(x_\epsilon) + \kappa \partial_\circ d_A(x_\epsilon).$$

(V) Now we apply the maximum rule (Proposition 7.4.7) to ψ, Propositions 4.6.2 and 7.3.9 to ω_{x_ϵ}, and Proposition 11.2.4 to d_A. Hence there exist nonnegative numbers $\hat{\lambda}$, $\hat{\mu}_1, \ldots, \hat{\mu}_m$ and $\nu'_1, \ldots, \nu'_n$ such that

$$\hat{\lambda} + \sum_{i=1}^{m} \hat{\mu}_i + \sum_{j=1}^{n} \nu'_j = 1, \tag{12.49}$$

$$o \in \hat{\lambda}\partial_\circ f(x_\epsilon) + \sum_{i=1}^{m} \hat{\mu}_i \partial_\circ f_i(x_\epsilon) + \sum_{j=1}^{n} \nu'_j \partial_\circ |h_j|(x_\epsilon) + \sqrt{\epsilon}\,B_{E^*} + N_C(A, x_\epsilon). \tag{12.50}$$

Since in the maximum rule, the active indices only count and $\psi(x_\epsilon) > 0$ (see (12.45)), we also have

$$\begin{aligned} g_i(x_\epsilon) \leq 0 &\implies \hat{\mu}_i = 0 \ (i = 1, \ldots, m), \\ h_j(x_\epsilon) = 0 &\implies \nu'_j = 0 \ (j = 1, \ldots, n). \end{aligned} \tag{12.51}$$

For $j = 1, \ldots, n$ define

$$\hat{\nu}_j := \begin{cases} (\text{sign}(h_j(x_\epsilon))\nu'_j & \text{if } h_j(x_\epsilon) \neq 0, \\ 0 & \text{if } h_j(x_\epsilon) = 0. \end{cases}$$

By the chain rule (Corollary 7.4.6 with $g := |\cdot|$), we then obtain

$$\nu'_j \, \partial_\circ |h_j|(x_\epsilon) \subseteq \hat{\nu}_j \, \partial_\circ h_j(x_\epsilon) \quad (j = 1, \ldots, n).$$

This and (12.50) imply the existence of some $x^*_\epsilon \in E^*$ satisfying

$$x^*_\epsilon \in \hat{\lambda} \partial_\circ f(x_\epsilon) + \sum_{i=1}^{n} \hat{\mu}_i \, \partial_\circ g_i(x_\epsilon) + \sum_{j=1}^{n} \hat{\nu}_j \, \partial_\circ h_j(x_\epsilon), \tag{12.52}$$

$$x^*_\epsilon \in \sqrt{\epsilon} B_{E^*} - N_C(A, x_\epsilon). \tag{12.53}$$

(VI) Now we replace ϵ by a sequence $\epsilon_k \downarrow 0$. From (12.47) we obtain $x_{\epsilon_k} \to \bar{x}$ as $k \to \infty$. Moreover, in view of (12.52) the sequence $(x^*_{\epsilon_k})$ is contained in a $\sigma(E^*, E)$-compact set (cf. Proposition 7.3.7) and so has a $\sigma(E^*, E)$-cluster point x^*. Observe that the numbers $\hat{\mu}_i$ also depend on ϵ_k, and by (12.49) have cluster points μ_i. An analogous observation yields scalars λ and ν_j, $j = 1, \ldots, n$. By Proposition 7.3.7(c) we have

$$x^* \in \lambda \partial_\circ f(\bar{x}) + \sum_{i=1}^{n} \mu_i \partial_\circ g_i(\bar{x}) + \sum_{j=1}^{n} \nu_j \partial_\circ h_j(\bar{x}) \quad \text{and} \quad x^* \in -N_C(A, \bar{x}),$$

which is equivalent to (12.44).

(VII) The numbers $\hat{\lambda}$ and $\hat{\mu}_i$, $i = 1, \ldots, m$, are nonnegative and so are λ and μ_i. By (12.49) the numbers λ, μ_i, and ν_j are not all zero. It remains to verify (12.43). Let $i \in \{1, \ldots, m\}$ be such that $g_i(\bar{x}) < 0$. Then the continuity of g_i implies that, for all k sufficiently large, $g_i(x_{\epsilon_k}) < 0$ and so (by (12.51)) $\hat{\mu}_i = 0$. Hence $\mu_i = 0$. $\qquad \square$

12.5 Approximate Multiplier Rules

We again consider problem (P4). However, for technical reasons we now adopt a somewhat different notation for the functionals involved. More precisely, we consider the problem:

(P5) Minimize $f(x)$
    subject to $f_i(x) \leq 0 \ (i = 1, \ldots, m)$,
            $f_i(x) = 0 \ (i = m+1, \ldots, n)$,
            $x \in A$.

Our aim now is to derive multiplier rules under the following weak assumptions:

(A5) E is a Fréchet smooth Banach space, $A \subseteq E$ is nonempty and closed,
   $f_i : E \to \mathbb{R}$ is l.s.c. for $i = 0, 1, \dots, m$,
   $f_i : E \to \mathbb{R}$ is continuous for $i = m + 1, \dots, n$.

The principal ingredients of the proof of the following multiplier rule are the weak local approximate sum rule (Theorem 9.2.7) and the representation results of Theorems 11.6.5 and 11.6.6.

Theorem 12.5.1 (Approximate Multiplier Rule) *Let* (A5) *be satisfied and assume that $\bar{x}$ is a local solution of* (P5).

(a) *For any $\epsilon > 0$ and any $\sigma(E^*, E)$-neighborhood V of zero, the following holds:*

$$\exists\, (x_i, f_i(x_i)) \in (\bar{x}, f_i(\bar{x})) + \epsilon B_{E \times \mathbb{R}} \quad (i = 0, 1, \dots, n) \quad \exists\, x_{n+1} \in \bar{x} + \epsilon B_E$$
$$\exists\, \mu_i \geq 0 \ (i = 0, \dots, n): \quad \mu_i \text{ not all zero,}$$
$$0 \in \sum_{i=0}^{m} \mu_i \partial_F f_i(x_i) + \sum_{i=m+1}^{n} \mu_i \big(\partial_F f_i(x_i) \cup \partial_F(-f_i)(x_i)\big) + N_F(A, x_{n+1}) + V.$$

(b) *Assume that, in addition, the following conditions are satisfied:*

$$\liminf_{x \to_f \bar{x}} d(\partial_F f_i(x), o) > 0 \quad \text{for } i = 1, \dots, m,$$
$$\liminf_{x \to \bar{x}} d(\partial_F f_i(x) \cup \partial_F(-f_i)(x), o) > 0 \quad \text{for } i = m+1, \dots, n. \tag{12.54}$$

Then conclusion (a) *holds with $\mu_0 = 1$ and $\mu_i > 0$ for $i = 1, \dots, n$.*

Proof. Let $\epsilon > 0$ and a $\sigma(E^*, E)$-neighborhood V of zero be given:

(I) First we verify (b). By assumption, $\bar{x}$ is a local minimizer of $\psi := f_0 + \delta_{(\cap_{i=1}^{n} S_i) \cap A}$, where

$$S_i := \begin{cases} \{x \in E \mid f_i(x) \leq 0\} & \text{if } i = 1, \dots, m, \\ \{x \in E \mid f_i(x) = 0\} & \text{if } i = m+1, \dots, n. \end{cases}$$

Hence we have

$$0 \in \partial_F \psi(\bar{x}) = \partial_F \Big(f_0 + \sum_{i=1}^{n} \delta_{S_i} + \delta_A\Big)(\bar{x}). \tag{12.55}$$

Let $\eta \in (0, \epsilon/2)$ and let U be a convex $\sigma(E^*, E)$-neighborhood of zero satisfying $U + \eta B_{E^*} \subseteq V$. By Theorem 9.2.7 there exist $(x_0, f_0(x_0)) \in$

$(\bar{x}, f_0(\bar{x})) + \eta B_{E \times \mathbb{R}}$, $(y_i, f_i(y_i)) \in (\bar{x}, f_i(\bar{x})) + \eta B_{E \times \mathbb{R}}$ $(i = 1, \ldots, n)$, and $x_{n+1} \in \eta B_E$ such that

$$o \in \partial_F f(x_0) + \sum_{i=1}^{n} N_F(S_i, y_i) + N_F(A, x_{n+1}) + U;$$

in this connection notice that by definition $\partial_F \delta_{S_i}(y_i) = N_F(S_i, y_i)$. Now let $x_0^* \in \partial_F f(x_0)$, $y_i^* \in N_F(S_i, y_i)$ $(i = 1, \ldots, n)$, and $y_{n+1}^* \in N_F(A, x_{n+1})$ be such that $o \in x_0^* + \sum_{i=1}^{n} y_i^* + y_{n+1}^* + U$. By Theorems 11.6.5 and 11.6.6 there exist $(x_i, f_i(x_i)) \in (y_i, f_i(y_i)) + \eta B_{E \times \mathbb{R}}$, $\lambda_i > 0$, and

$$x_i^* \in \partial_F f_i(x_i) \quad \text{for } i = 1, \ldots, m,$$
$$x_i^* \in \partial_F f_i(x_i) \cup \partial_F(-f_i)(x_i) \quad \text{for } i = m + 1, \ldots, n$$

such that $\|\lambda_i x_i^* - y_i^*\| < \frac{\eta}{n}$ for $i = 1, \ldots, n$. It follows that

$$o \in x_0^* + \sum_{i=1}^{n} \lambda_i x_i^* + y_{n+1}^* + \eta B_{E^*} + U.$$

Setting

$$\mu_0 := \frac{1}{1 + \sum_{i=1}^{n} \lambda_i}, \quad \mu_i := \frac{\lambda_i}{1 + \sum_{i=1}^{n} \lambda_i} \quad \text{for } i = 1, \ldots, n$$

and recalling that U is a convex set containing o, we obtain

$$o \in \sum_{i=0}^{n} \mu_i x_i^* + \left(1 + \sum_{i=1}^{n} \lambda_i\right)^{-1} y_{n+1}^* + \eta B_{E^*} + U.$$

Since $N_F(A, x_{n+1})$ is a cone and $\eta B_{E^*} + U \subseteq V$, conclusion (b) follows.

(II) Now we verify (a). In view of step (I) we may assume that condition (12.54) fails. Suppose that for some $i \in \{1, \ldots, m\}$ we have $\liminf_{x \to_f \bar{x}} d(\partial_F f_i(x), o) = 0$; the argument is analogous if $i \in \{m + 1, \ldots, n\}$. We may assume that $\epsilon \in (0, 1)$ is so small that $B_{E^*}(o, \epsilon) \subseteq V$. Then there exist $x_i \in B_E(\bar{x}, \epsilon)$ and $x_i^* \in \partial_F f_i(x_i)$ such that $|f_i(x_i) - f_i(\bar{x})| < \epsilon$ and $\|x_i^*\| < \epsilon$. Then $-x_i^* \in V$. Setting $\mu_i := 1$ and $\mu_j := 0$ for any $j \neq i$, conclusion (a) follows. $\qquad\square$

Remark 12.5.2

(a) If $\partial_F(-f_i)(x_i) = -\partial_F f_i(x_i)$ (as is the case if, for instance, f_i is a C^2 function), then $\mu_i\big(\partial_F f_i(x_i) \cup \partial_F(-f_i)(x_i)\big)$, where $\mu_i \in \mathbb{R}_+$, corresponds to $\mu_i \partial_F f_i(x_i)$, where $\mu_i \in \mathbb{R}$, and so to the fact that Lagrange multipliers associated with equality constraints have arbitrary sign (cf. Remark 12.3.4(b)).

(b) Conclusion (a) of Theorem 12.5.1 is of John type while conclusion (b) is of Karush–Kuhn–Tucker type. The conditions (12.54) constitute a constraint qualification.

12.6 Bibliographical Notes and Exercises

Theorem 12.2.3 is due to Clarke (see [36]), our presentation follows Loewen [123]. Concerning detailed presentations of the classical calculus of variations we recommend Cesari [30] and Giaquinta and Hildebrandt [71,72].

Theorem 12.4.1 was established by Clarke [35]. In a similar way but with the aid of a more sophisticated maximum rule, Clarke [36] obtained a sharpened multiplier rule. Theorem 12.4.1 provides an optimality condition of the John type (cf. Sects. 5.2 and 12.3). As a regularity condition, Clarke introduced the concept of *calmness*, see [35, 36]. A modification of this concept is due to Rockafellar [187].

Theorem 12.3.5 generalizes a multiplier rule of Halkin [82], where all functionals involved are assumed to be F-differentiable at $\bar{x}$. For results related to Theorem 12.3.5 but in terms of subdifferentials, we refer to Ye [217]. The multiplier rule of Theorem 12.5.1 was established by Borwein et al. [21] (see also Borwein and Zhu [23]).

In addition to the above references, we give a (necessarily subjective) selection of further papers on optimality conditions for nonconvex problems: Bazaraa et al. [10], Degiovanni and Schuricht [46], Hiriart-Urruty [84, 85, 87], Ngai and Théra [152, 153], Ioffe [99], Jourani [107], Loewen [123], Mordukhovich and Wang [146], Neustadt [150], Penot [159, 162], Pshenichnyi [169], Scheffler [190], Scheffler and Schirotzek [192], and Schirotzek [194, 195]. For a detailed discussion and a lot of references on the subject we recommend Mordukhovich [137].

Substantial applications of generalized derivatives are elaborated by Clarke et al. [39], Panagiotopoulos [157], and Papageorgiou and Gasinski [158].

Exercise 12.6.1 Prove Proposition 12.1.4.

Exercise 12.6.2 Apply Theorem 12.3.1 to the following problem (cf. Exercise 11.7.10):

$$\text{minimize } f(x_1, x_2) := x_1 + x_2 \quad \text{subject to} \quad 0 \le x_2 \le x_1^3, \ (x_1, x_2) \in \mathbb{R}^2.$$

Exercise 12.6.3 Define $f, h : \mathbb{R}^2 \to \mathbb{R}$ by $f(x, y) := x$ and

$$h(x, y) := \begin{cases} y & \text{if } x \ge 0, \\ y - x^2 & \text{if } x < 0, \ y \le 0, \\ y + x^2 & \text{if } x < 0, \ y > 0. \end{cases}$$

The problem $f(x, y) \longrightarrow \min$ subject to $h(x, y) = 0$ obviously has the solution $(\bar{x}, \bar{y}) = (0, 0)$. Show that no nonzero Lagrange multipliers exist at $(\bar{x}, \bar{y})$. Which assumption of Theorem 12.3.5 is violated (cf. Fernandez [67])?

Exercise 12.6.4 The aim of this exercise is to generalize Theorem 12.3.3 (cf. Schirotzek [196]):

(a) Applying Theorem 11.7.12 (see Exercise 11.7.11(a)), prove the following:

Theorem 12.6.5 *In addition to the assumptions of Theorem 11.7.12, suppose that $f : E \to \overline{\mathbb{R}}$ is regularly locally convex at $\bar{x}$ and that $\bar{x}$ is a local minimizer of f on $A \cap M$. Then there exist $\lambda \in \mathbb{R}_+$, $y^* \in P^\circ$, and $z^* \in Q^\circ$ such that $(\lambda, y^*) \neq o$ and*

$$\lambda f_H(\bar{x}, y) + \langle y^*, g_H(\bar{x}, y) \rangle + \langle z^*, h'(\bar{x}) y \rangle \geq 0 \quad \forall y \in A - \bar{x},$$
$$\langle y^*, g(\bar{x}) \rangle = 0, \quad \langle z^*, h(\bar{x}) \rangle = 0.$$

(b) Formulate and verify corresponding variants of the other assertions of Theorem 12.3.3.

Exercise 12.6.6 Let $p, q : [a, b] \to \mathbb{R}$ and $k : [a, b] \times [a, b] \to \mathbb{R}$ be continuous functions. Denote

$$A := \{ x \in L^2[a, b] \mid |x(t)| \leq 1 \text{ for almost all } t \in [a, b] \}.$$

Consider the problem (see Tröltzsch [209], cf. Schirotzek [196])

$$\text{minimize } f(x) := \int_a^b p(t) x(t) \, dt,$$

$$\text{subject to } \int_a^b k(s, t) x^2(s) \, ds \leq q(t) \quad \forall t \in [a, b], \quad x \in A.$$

Show that if $\bar{x}$ is a local solution of the problem, then there exist a number $\lambda \in \mathbb{R}_+$ and a nondecreasing function $v : [a, b] \to \mathbb{R}$ such that $(\lambda, v) \neq o$, and

$$\int_a^b y(t) \left[\lambda p(t) + 2 \bar{x}(t) \int_a^b k(t, s) \, dv(s) \right] dt \geq 0 \quad \forall y \in A - \bar{x},$$

$$\int_a^b \left[\int_a^b k(s, t) \bar{x}^2(s) \, ds - q(t) \right] dv(t) = 0.$$

Find a condition ensuring that $\lambda > 0$.
Hint: Choose, among others, $G := C[a, b]$ with the maximum norm and apply Theorem 12.6.5.

13

Extremal Principles and More Normals and Subdifferentials

Fréchet subdifferentials are a flexible tool for nonsmooth analysis, in particular in Fréchet smooth Banach spaces where they coincide with viscosity subdifferentials. But in general they admit approximate calculus rules only. Applying set convergence operations, Mordukhovich developed concepts of subdifferentials and normals that admit a rich *exact* calculus. To a great extent the properties of these objects are based on *extremal principles*, which work especially well in Fréchet smooth Banach spaces (more generally, in Asplund spaces), where the new subdifferentials and normal cones are sequential Painlevé–Kuratowski upper limits of Fréchet subdifferentials and Fréchet normal cones, respectively.

13.1 Mordukhovich Normals and Subdifferentials

We start with a definition.

Definition 13.1.1 Let A be a nonempty subset of E.

(a) If $x \in A$ and $\epsilon \geq 0$, then the set

$$\widehat{N}_\epsilon(A, x) := \left\{ x^* \in E^* \; \limsup_{y \to_A x} \frac{\langle x^*, y - x \rangle}{\|y - x\|} \leq \epsilon \right\}$$

is called *set of ϵ-normals* to A at x.

(b) If $\bar{x} \in A$, then the set

$$N_M(A, \bar{x}) := \operatorname*{sLim\,sup}_{\substack{x \to \bar{x} \\ \epsilon \downarrow 0}} \widehat{N}_\epsilon(A, x)$$

will be called *Mordukhovich normal cone (M-normal cone)* to A at $\bar{x}$.

Remark 13.1.2

(a) In general, the set $\widehat{N}_\epsilon(A, x)$ is not a cone unless $\epsilon = 0$. The definitions immediately yield

$$N_F(A, x) = \partial_F \delta_A(x) = \widehat{N}_0(A, x) \quad \text{and} \quad N_F(A, x) + \epsilon B_{E^*} \subseteq \widehat{N}_\epsilon(A, x) \quad \forall \epsilon > 0.$$

(b) We have $x^* \in \widehat{N}_\epsilon(A, x)$ if and only if for any $\gamma > 0$ the function

$$\varphi(x) := \langle x^*, x - \bar{x} \rangle - (\epsilon + \gamma)\|x - \bar{x}\|, \quad x \in A,$$

attains a local maximum at $\bar{x}$.

(c) The set $N_M(A, \bar{x})$ is easily seen to be a cone. Recalling the definition of the sequential Painlevé–Kuratowski upper limit, we see that $x^* \in N_M(A, \bar{x})$ if and only if there exist sequences $\epsilon_k \downarrow 0$, $x_k \to_A \bar{x}$, and $x_k^* \xrightarrow{w^*} x^*$ such that $x_k^* \in \widehat{N}_{\epsilon_k}(A, x_k)$ for any $k \in \mathbb{N}$.

(d) We always have $N_F(A, x) \subseteq N_M(A, x)$. Example 13.1.3 shows that this inclusion may be strict; it also shows that $N_M(A, \bar{x})$, in contrast to $\widehat{N}_\epsilon(A, x)$, is in general nonconvex and so cannot be the polar of any tangent cone to A at $\bar{x}$.

Example 13.1.3 Let $A := \{(x, y) \in \mathbb{R}^2 \mid y \geq -|x|\}$. Then $N_F(A, (0,0)) = \{(0,0)\}$ but

$$N_M(A, (0,0)) = \{(x^*, x^*) \in \mathbb{R}^2 \mid x^* \leq 0\} \cup \{(x^*, -x^*) \in \mathbb{R}^2 \mid x^* \geq 0\}.$$

In finite-dimensional spaces there is a close relationship between the Mordukhovich and the Fréchet normal cone.

Theorem 13.1.4 *If A is a closed subset of the Euclidean space E, then*

$$N_M(A, \bar{x}) = \text{sLim} \sup_{x \to \bar{x}} N_F(A, x). \tag{13.1}$$

Proof. We identify E^* with E. It is obvious that the right-hand side of (13.1) is contained in the left-hand side. We show the opposite inclusion. Thus let $x^* \in N_M(A, \bar{x})$ be given. Then (cf. Remark 13.1.2(c)) there exist sequences $\epsilon_k \downarrow 0$, $x_k \to_A \bar{x}$, and $x_k^* \xrightarrow{w^*} x^*$ such that $x_k^* \in \widehat{N}_{\epsilon_k}(A, x_k)$ for any $k \in \mathbb{N}$. Let $\eta > 0$ be arbitrary. For any $k \in \mathbb{N}$ choose $y_k \in \text{proj}_A(x_k + \eta x_k^*)$ which exists by Corollary 5.4.5. Since y_k depends on η, we also write $y_k = y_k(\eta)$. We have

$$\|x_k + \eta x_k^* - y_k\|^2 \leq \eta^2 \|x_k^*\|^2. \tag{13.2}$$

Calculating the left-hand side via the inner product gives

$$\|x_k + \eta x_k^* - y_k\|^2 = \|x_k - y_k\|^2 + 2\eta(x_k^* \mid x_k - y_k) + \eta^2 \|x_k^*\|^2.$$

Hence (13.2) passes into

$$\|x_k - y_k\|^2 \le 2\eta(x_k^* \,|\, y_k - x_k). \tag{13.3}$$

By (13.2) it follows that $y_k(\eta) \to x_k$ as $\eta \downarrow 0$. From this and the choice of x_k^*, we deduce that for any k there exists $\eta_k > 0$ such that, with $z_k := y_k(\eta_k)$, we have $(x_k^* \,|\, z_k - x_k) \le 2\epsilon_k\|z_k - x_k\|$ for any $k \in \mathbb{N}$. This inequality together with (13.3) shows that $\|x_k - z_k\| \le 4\eta_k\epsilon_k$ and so $z_k \to \bar{x}$ as $k \to \infty$. Defining $z_k^* := x_k^* + (1/\eta_k)(x_k - z_k)$, we obtain $\|z_k^* - x_k^*\| \le 4\epsilon_k$ and so $z_k^* \to x^*$ as $k \to \infty$. Therefore, to see that $x^* \in \text{sLim}\sup_{x \to \bar{x}} \mathrm{N}_F(A, x)$, it remains to show that $z_k^* \in \mathrm{N}_F(A, z_k)$ for any k. Given $x \in A$, we calculate

$$
\begin{aligned}
0 \le \|x_k + \eta_k x_k^* - x\|^2 &- \|x_k + \eta_k x_k^* - z_k\|^2 \\
= (\eta_k x_k^* + x_k - x \,|\, \eta_k x_k^* &+ x_k - z_k) + (\eta_k x_k^* + x_k - x \,|\, z_k - x) \\
- (\eta_k x_k^* + x_k - z_k \,|\, x - z_k) &- (\eta_k x_k^* + x_k - z_k \,|\, \eta_k x_k^* + x_k - x) \\
= -2\eta_k(z_k^* \,|\, x - z_k) &+ \|x - z_k\|^2.
\end{aligned}
$$

We thus obtain $(z_k^* \,|\, x - z_k) \le (1/2\eta_k)\|x - z_k\|^2$ for all $x \in A$, which implies that in fact $z_k^* \in \mathrm{N}_F(A, z_k)$. $\qquad\square$

If the set A is convex, we have a simple representation of $\widehat{\mathrm{N}}_\epsilon(A, \bar{x})$.

Proposition 13.1.5 *If $A \subseteq E$ is convex and $\bar{x} \in A$, then*

$$\widehat{\mathrm{N}}_\epsilon(A, \bar{x}) = \{x^* \in E^* \,|\, \langle x^*, x - \bar{x}\rangle \le \epsilon\|x - \bar{x}\| \quad \forall\, x \in A\} \quad \forall\, \epsilon \ge 0. \tag{13.4}$$

In particular, $\widehat{\mathrm{N}}_0(A, \bar{x}) = \mathrm{N}_F(A, \bar{x}) = \mathrm{N}(A, \bar{x})$.

Proof. It is clear that the right-hand side of (13.4) is contained in $\widehat{\mathrm{N}}_\epsilon(A, \bar{x})$ (even if A is not convex). Now let $x^* \in \widehat{\mathrm{N}}_\epsilon(A, \bar{x})$ be given and fix $x \in A$. For any $\tau \in (0, 1]$ we then have $x_\tau := \bar{x} + \tau(x - \bar{x}) \in A$ (because A is convex) and $x_\tau \to \bar{x}$ as $\tau \downarrow 0$. The definition of the ϵ-normals now implies that for any $\gamma > 0$ we obtain

$$\langle x^*, x_\tau - \bar{x}\rangle \le (\epsilon + \gamma)\|x_\tau - \bar{x}\| \quad \text{for all sufficiently small } \tau > 0.$$

It follows that x^* is an element of the right-hand side of (13.4). $\qquad\square$

The concept to be introduced now will be important for calculus rules. In this connection, recall Remark 13.1.2(d).

Definition 13.1.6 The nonempty subset A of E is said to be *normally regular* at $\bar{x} \in A$ if $\mathrm{N}_M(A, \bar{x}) = \mathrm{N}_F(A, \bar{x})$.

The next result yields examples of normally regular sets.

Proposition 13.1.7 *Let $A \subseteq E$ and $\bar{x} \in A$. Assume that for some neighborhood U of $\bar{x}$ the set $A \cap U$ is convex. Then A is normally regular at $\bar{x}$ and*

$$N_M(A, \bar{x}) = \{x^* \in E^* \mid \langle x^*, x - \bar{x} \rangle \leq 0 \quad \forall x \in A \cap U\}. \tag{13.5}$$

Proof. By Proposition 13.1.5 and Remark 13.1.2(d) we know that the right-hand side of (13.5) is contained in $N_M(A, \bar{x})$. Now take any $x^* \in N_M(A, \bar{x})$. Then there exist sequences (ϵ_k), (x_k), and (x_k^*) as in Remark 13.1.2(c). For all $k \in \mathbb{N}$ sufficiently large we thus have $x_k \in A \cap U$ and so by Proposition 13.1.5,

$$\langle x_k^*, x - x_k \rangle \leq \epsilon_k \|x - x_k\| \quad \forall x \in A \cap U.$$

Letting $k \to \infty$ and recalling Exercise 10.10.1, we see that x^* is an element of the right-hand side of (13.5). $\qquad\square$

We will use the M-normal cone to define subdifferentials of (arbitrary) proper functions $f : E \to \overline{\mathbb{R}}$. In view of this we first establish a result on the structure of the M-normal cone to an epigraph.

Lemma 13.1.8 *Assume that $f : E \to \overline{\mathbb{R}}$ is proper and $(\bar{x}, \bar{\alpha}) \in \mathrm{epi}\, f$. Then $(x^*, -\lambda) \in N_M(\mathrm{epi}\, f, (\bar{x}, \bar{\alpha}))$ implies $\lambda \geq 0$.*

Proof. Let $(x^*, -\lambda) \in N_M(\mathrm{epi}\, f, (\bar{x}, \bar{\alpha}))$ be given. By the definition of the M-normal cone there exist sequences $\epsilon_k \downarrow 0$, $(x_k, \alpha_k) \to_{\mathrm{epi}\, f} (\bar{x}, \bar{\alpha})$, $x_k^* \xrightarrow{\mathrm{w}^*} x^*$, and $\lambda_k \to \lambda$ such that

$$\limsup_{(x, \alpha) \to_{\mathrm{epi}\, f} (x_k, \alpha_k)} \frac{\langle x_k^*, x - x_k \rangle - \lambda_k(\alpha - \alpha_k)}{\|(x, \alpha) - (x_k, \alpha_k)\|} \leq \epsilon_k \quad \forall k \in \mathbb{N}.$$

Choosing $x := x_k$ we obtain

$$\limsup_{\alpha \to \alpha_k} \frac{-\lambda_k \operatorname{sgn}(\alpha - \alpha_k)}{\|(x_k, 1)\|} \leq \epsilon_k \quad \forall k \in \mathbb{N}.$$

Letting $k \to \infty$, we see that $\lambda \geq 0$. $\qquad\square$

Definition 13.1.9 *Let $f : E \to \overline{\mathbb{R}}$ be proper and $\bar{x} \in \mathrm{dom}\, f$. We call*

$$\partial_M f(\bar{x}) := \{x^* \in E^* \mid (x^*, -1) \in N_M(\mathrm{epi}\, f, (\bar{x}, f(\bar{x})))\}$$

Mordukhovich subdifferential (M-subdifferential) or basic subdifferential of f at $\bar{x}$ and we call

$$\partial_M^\infty f(\bar{x}) := \{x^* \in E^* \mid (x^*, 0) \in N_M(\mathrm{epi}\, f, (\bar{x}, f(\bar{x})))\}$$

singular Mordukhovich subdifferential (singular M-subdifferential) of f at $\bar{x}$.

Remark 13.1.10 It follows immediately from the definitions that the M-normal cone can be regained from the M-subdifferentials. In fact, for any subset A of E and any point $\bar{x} \in A$ one has

$$N_M(A, \bar{x}) = \partial_M \delta_A(\bar{x}) = \partial_M^\infty \delta_A(\bar{x}).$$

As a consequence of Lemma 13.1.8 we have

Lemma 13.1.11 Let $f : E \to \overline{\mathbb{R}}$ be proper and $\bar{x} \in \text{dom} f$. If $N_M(\text{epi} f, (\bar{x}, f(\bar{x}))) \neq \{(o, 0)\}$ and $\partial_M^\infty f(\bar{x}) = \{o\}$, then $\partial_M f(\bar{x})$ is nonempty.

Proof. See Exercise 13.13.1.   □

Below we shall derive alternative descriptions of the Mordukhovich subdifferential in terms of subdifferentials to be defined now that modify the Fréchet subdifferential.

Definition 13.1.12 Let $f : E \to \overline{\mathbb{R}}$ be proper, $\bar{x} \in \text{dom} f$, and $\epsilon \geq 0$. The set

$$\widehat{\partial}_\epsilon f(\bar{x}) := \left\{ x^* \in E^* \mid \liminf_{x \to \bar{x}} \frac{f(x) - f(\bar{x}) - \langle x^*, x - \bar{x}\rangle}{\|x - \bar{x}\|} \geq -\epsilon \right\}$$

is called *(Fréchet) ϵ-subdifferential* of f at $\bar{x}$.

Remark 13.1.13

(a) *For $\epsilon = 0$ we regain the Fréchet subdifferential: $\widehat{\partial}_0 f(\bar{x}) = \partial_F f(\bar{x})$.*
(b) *For every $\epsilon \geq 0$ we have $x^* \in \widehat{\partial}_\epsilon f(\bar{x})$ if and only if for any $\gamma > 0$ the function*

$$\psi(x) := f(x) - f(\bar{x}) - \langle x^*, x - \bar{x}\rangle + (\epsilon + \gamma)\|x - \bar{x}\|, \quad x \in E,$$

attains a local minimum at $\bar{x}$ (cf. Remark 13.1.2(b)).

We define still another sort of ϵ-subdifferentials.

Definition 13.1.14 Let $f : E \to \overline{\mathbb{R}}$ be proper, $\bar{x} \in \text{dom} f$, and $\epsilon \geq 0$. The set

$$\widehat{\partial}_{g\epsilon} f(\bar{x}) := \{x^* \mid (x^*, -1) \in \widehat{N}_\epsilon(\text{epi} f, (\bar{x}, f(\bar{x})))\}$$

is said to be the *geometric ϵ-subdifferential* of f at $\bar{x}$.

The next result describes the relationship between the two ϵ-subdifferentials.

Theorem 13.1.15 Let $f : E \to \overline{\mathbb{R}}$ be proper and $\bar{x} \in \text{dom} f$.

(a) *For any $\epsilon \geq 0$ one has $\widehat{\partial}_\epsilon f(\bar{x}) \subseteq \widehat{\partial}_{g\epsilon} f(\bar{x})$.*
(b) *If $\epsilon \in [0, 1)$ and $x^* \in \widehat{\partial}_{g\epsilon} f(\bar{x})$, then $x^* \in \widehat{\partial}_{\tilde{\epsilon}} f(\bar{x})$, where $\tilde{\epsilon} := \epsilon(1 + \|x^*\|)/(1 - \epsilon)$.*

Proof.

(a) Let $\epsilon \geq 0$, $x^* \in \widehat{\partial}_\epsilon f(\bar{x})$, and $\gamma > 0$ be given. By Remark 13.1.2(b) there exists a neighborhood U of $\bar{x}$ such that

$$f(x) - f(\bar{x}) - \langle x^*, x - \bar{x} \rangle \geq -(\epsilon + \gamma)\|x - \bar{x}\| \quad \forall x \in U.$$

It follows that

$$\langle x^*, x - \bar{x} \rangle + f(\bar{x}) - \alpha \leq (\epsilon + \gamma)\|(x, \alpha) - (\bar{x}, f(\bar{x}))\|.$$

Hence the function

$$\tilde{\varphi}(x, \alpha) := \langle x^*, x - \bar{x} \rangle - (\alpha - f(\bar{x}))$$
$$- (\epsilon + \gamma)\|(x, \alpha) - (\bar{x}, f(\bar{x}))\|, \quad (x, \alpha) \in \text{epi } f,$$

attains a local maximum at $(\bar{x}, f(\bar{x}))$. By Remark 13.1.2(b) we conclude that $(x^*, -1) \in \widehat{N}_\epsilon(\text{epi } f, (\bar{x}, f(\bar{x}))$ and so $x^* \in \widehat{\partial}_{g\epsilon} f(\bar{x})$.

(b) Now let $\epsilon \in [0, 1)$ and $\tilde{\epsilon}$ as above be given. Assume that $x^* \in E$ is not in $\widehat{\partial}_{\tilde{\epsilon}} f(\bar{x})$. Then there are $\gamma > 0$ and a sequence (x_k) converging to $\bar{x}$ as $k \to \infty$ such that

$$f(x_k) - f(\bar{x}) - \langle x^*, x - \bar{x} \rangle + (\tilde{\epsilon} + \gamma)\|x_k - \bar{x}\| < 0 \quad \forall k \in \mathbb{N}.$$

Put $\alpha_k := f(\bar{x}) + \langle x^*, x_k - \bar{x} \rangle - (\tilde{\epsilon} + \gamma)\|x_k - \bar{x}\|$. Then $\alpha_k \to f(\bar{x})$ as $k \to \infty$ and $(x_k, \alpha_k) \in \text{epi } f$ for any $k \in \mathbb{N}$. An elementary consideration shows that for any $k \in \mathbb{N}$,

$$\frac{\langle x^*, x_k - \bar{x} \rangle - (\alpha_k - f(\bar{x}))}{\|(x_k, \alpha_k) - (\bar{x}, f(\bar{x}))\|} = \frac{(\tilde{\epsilon} + \gamma)\|x_k - \bar{x}\|}{\|(x_k - \bar{x}, \langle x^*, x_k - \bar{x} \rangle - (\tilde{\epsilon} + \gamma)\|x_k - \bar{x}\| \|}$$

$$\geq \frac{\tilde{\epsilon} + \gamma}{1 + \|x^*\| + (\tilde{\epsilon} + \gamma)} > \frac{\tilde{\epsilon}}{1 + \|x^*\| + \tilde{\epsilon}} = \epsilon.$$

Hence $(x^*, -1) \notin \widehat{N}_\epsilon(\text{epi } f, (\bar{x}, f(\bar{x})))$ and so $x^* \notin \widehat{\partial}_{g\epsilon} f(\bar{x})$. $\quad\square$

Remark 13.1.16 For $\epsilon = 0$, Theorem 13.1.15 and Remark 13.1.13(a) yield

$$\widehat{\partial}_{g0} f(\bar{x}) = \widehat{\partial}_0 f(\bar{x}) = \partial_F f(\bar{x}).$$

This together with the definition of the geometric ϵ-subdifferential and Remark 13.1.2(a) show that in any Banach space E we have (cf. Remark 11.3.6)

$$\partial_F f(\bar{x}) = \{x^* \mid (x^*, -1) \in N_F(\text{epi } f, (\bar{x}, f(\bar{x})))\}.$$

Now we establish the announced alternative description of the M-subdifferential.

Theorem 13.1.17 *Let $f : E \to \overline{\mathbb{R}}$ be proper and $\bar{x} \in \mathrm{dom}\, f$.*

(a) *One always has*

$$\partial_M f(\bar{x}) = \mathrm{sLimsup}\, \widehat{\partial}_{g\epsilon} f(x) = \mathrm{sLim\, sup}\, \widehat{\partial}_{\epsilon} f(x). \qquad (13.6)$$
$$\phantom{\partial_M f(\bar{x}) = }{}_{x \to_f \bar{x},\, \epsilon \downarrow 0} \qquad {}_{x \to_f \bar{x},\, \epsilon \downarrow 0}$$

(b) *If, in addition, f is l.s.c. and E is finite dimensional, then*

$$\partial_M f(\bar{x}) = \mathrm{sLim\, sup}\, \partial_F f(x). \qquad (13.7)$$
$$\phantom{\partial_M f(\bar{x}) = }{}_{x \to_f \bar{x}}$$

Proof.

(a) The first equation in (13.6) follows immediately from the definitions of the geometric ϵ-subdifferential and the ϵ-normal set. This and Theorem 13.1.15(a) further show that $\mathrm{sLim\, sup}_{x \to_f \bar{x},\, \epsilon \downarrow 0} \widehat{\partial}_\epsilon f(x) \subseteq \partial_M f(\bar{x})$. Now we verify the opposite inclusion. Thus let $x^* \in \partial_M f(\bar{x})$ be given. Then there exist sequences $\epsilon_k \downarrow 0$, $x_k \to_f \bar{x}$, and $x_k^* \xrightarrow{w^*} x^*$ such that $x_k^* \in \widehat{\partial}_{g\epsilon_k} f(x_k)$ for any $k \in \mathbb{N}$. For all sufficiently large k we have $\epsilon_k \in (0,1)$. By Theorem 13.1.15(b) it follows for these k that $x_k^* \in \widehat{\partial}_{\tilde{\epsilon}_k} f(x_k)$, where $\tilde{\epsilon}_k := \epsilon_k(1 + \|x_k^*\|)/(1 - \epsilon_k)$. Since the sequence (x_k^*) is bounded, we have $\tilde{\epsilon}_k \downarrow 0$ as $k \to \infty$. This shows that x^* is an element of the right-hand side of (13.6).

(b) This is a consequence of Theorem 13.1.4 and Remark 13.1.16. $\square$

Since $\partial_F f(\bar{x}) = \widehat{\partial}_0 f(\bar{x}) \subseteq \widehat{\partial}_\epsilon f(\bar{x})$ for any $\epsilon > 0$, it follows from Theorem 13.1.17 that we always have $\partial_F f(\bar{x}) \subseteq \partial_M f(\bar{x})$. The example

$$f(x) := \begin{cases} x^2 \sin(1/x) & \text{if } x \neq 0, \\ 0 & \text{if } x = 0 \end{cases}$$

shows that the inclusion may be strict. In fact, here we have $\partial_F f(0) = \{f'(0)\} = \{0\}$ but $\partial_M f(0) = [-1, 1]$. Below it will turn out that equality in the above inclusion will ensure equality in calculus rules for M-subdifferentials. This motivates the following notion.

Definition 13.1.18 *The proper functional $f : E \to \overline{\mathbb{R}}$ is said to be* lower regular *at $\bar{x} \in \mathrm{dom}\, f$ if $\partial_M f(\bar{x}) = \partial_F f(\bar{x})$.*

Proposition 13.1.19 *Let $A \subseteq E$ and $\bar{x} \in A$. The indicator functional δ_A is lower regular at $\bar{x}$ if and only if the set A is normally regular at $\bar{x}$.*

Proof. See Exercise 13.13.2. $\square$

In Fréchet smooth Banach spaces (and more generally, in Asplund spaces) the M-subdifferential at $\bar{x}$ can be represented with the aid of F-subdifferentials

at nearby points. We prepare this result by an approximate sum rule for
ϵ-subdifferentials.

Lemma 13.1.20 *Assume that E is a Fréchet smooth Banach space and that
$f_1, f_2 : E \to \bar{\mathbb{R}}$ are proper and l.s.c. with f_1 locally L-continuous around
$\bar{x} \in \mathrm{dom}\, f_1 \cap \mathrm{dom}\, f_2$. Then for any $\epsilon \geq 0$ and any $\gamma > 0$ one has*

$$\hat{\partial}_\epsilon(f_1 + f_2)(\bar{x}) \subseteq \bigcup \Big\{ \partial_F f_1(x_1) + \partial_F f_2(x_2) \,\Big|$$

$$x_i \in \mathrm{B}_E(\bar{x}, \gamma),\ |f_i(x_i) - f_i(\bar{x})| \leq \gamma,\ i = 1, 2 \Big\} + (\epsilon + \gamma)\mathrm{B}_{E^*}.$$

Proof.

(I) Fix ϵ and γ as above and choose η such that $0 < \eta < \min\{\gamma/4, \tilde{\eta}\}$, where $\tilde{\eta}$
is the positive solution of $\tilde{\eta}^2 + (2 + \epsilon)\tilde{\eta} - \gamma = 0$. Now let $x^* \in \hat{\partial}_\epsilon(f_1 + f_2)(\bar{x})$
be given and define

$$f_0(x) := f_1(x) - \langle x^*, x - \bar{x} \rangle + (\epsilon + \eta)\|x - \bar{x}\| \quad \forall x \in E.$$

By Remark 13.1.13 the function $f_0 + f_2$ attains a local minimum at $\bar{x}$.
Then Lemma 9.2.5 shows that condition (9.19) is satisfied for $f_0 + f_2$.
Applying the approximate sum rule of Theorem 9.2.6 (with η instead of
ϵ) to the sum $f_0 + f_2$, we find $x_i \in \mathrm{B}(\bar{x}, \eta)$, $i = 1, 2$, $x_0^* \in \partial_F f_0(x_1)$ and
$x_2^* \in \partial_F f_2(x_2)$, such that

$$|f_1(x_1) + (\epsilon + \eta)\|x_1 - \bar{x}\| - f_1(\bar{x})| \leq \eta, \quad |f_2(x_2) - f_2(\bar{x})| \leq \eta,$$
$$o \in x_0^* + x_2^* + \eta \mathrm{B}_{E^*}.$$

By Proposition 9.2.2 we obtain $x_0^* = x_1^* - x^*$ with

$$x_1^* \in \partial_F\big(f_1 + (\epsilon + \eta)\omega_{\bar{x}}\big)(x_1), \quad \text{where } \omega_{\bar{x}}(x) := \|x - \bar{x}\|.$$

It follows that

$$|f_1(x_1) - f_1(\bar{x})| \leq \eta(\epsilon + \eta + 1) \quad \text{and} \quad x^* \in x_1^* + x_2^* + \eta \mathrm{B}_{E^*}.$$

(II) Now we evaluate x_1^*. Applying Remark 13.1.13 to x_1^*, we conclude that,
with

$$g(x) := (\epsilon + \eta)\|x - \bar{x}\| - \langle x_1^*, x - x_1 \rangle + \eta\|x - x_1\|, \quad x \in E,$$

the function $f_1 + g$ attains a local minimum at x_1. By Proposition 4.6.2
the convex function g satisfies $\partial g(x) \subseteq -x_1^* + (\epsilon + 2\eta)\mathrm{B}_{E^*}$ for any $x \in E$.
Now we apply Theorem 9.2.6 to the sum $f_1 + g$ and obtain $\tilde{x}_1 \in \mathrm{B}_E(x_1, \eta)$
such that

$$\|f_1(\tilde{x}_1 - f_1(x_1)\| \leq \eta \quad \text{and} \quad x_1^* \in \partial_F f_1(\tilde{x}_1) + (\epsilon + 3\eta)\mathrm{B}_{E^*}.$$

Summarizing we have

$$x^* \in \partial_F f_1(\tilde{x}_1) + \partial_F f_2(x_2) + (\epsilon + 4\eta)B_{E^*},$$
$$\|\tilde{x}_1 - \bar{x}\| \le 2\eta, \quad |f_1(\tilde{x}_1) - f_1(\bar{x})| \le \eta(\epsilon + \eta + 2).$$

The definition of η shows that the proof is complete.   □

With the aid of Lemma 13.1.20 we obtain, in a Fréchet smooth Banach space, a limiting representation of the M-subdifferential that is simpler than the one in an arbitrary Banach space and that generalizes the corresponding result for finite-dimensional Banach spaces (see Theorem 13.1.17).

Theorem 13.1.21 *Assume that E is a Fréchet smooth Banach space, $f : E \to \overline{\mathbb{R}}$ is proper and l.s.c., and $\bar{x} \in \mathrm{dom}\, f$.*

(a) *For any $\epsilon \ge 0$ and any $\gamma > 0$ one has*

$$\widehat{\partial}_\epsilon f(\bar{x}) \subseteq \bigcup \left\{ \partial_F f(x) : x \in B_E(\bar{x}, \gamma),\ |f(x) - f(\bar{x})| \le \gamma \right\} + (\epsilon + \gamma)B_{E^*}. \tag{13.8}$$

(b) *One has*

$$\partial_M f(\bar{x}) = \mathrm{sLim\,sup}_{x \to_f \bar{x}} \partial_F f(x).$$

Proof. Applying Lemma 13.1.20 with $f_1 := o$ and $f_2 := f$, we immediately obtain (a). Now choose $\gamma := \epsilon$ in (13.8) and pass to the limit $\epsilon \downarrow 0$. This yields (b).   □

Due to Theorem 13.1.21(b) we can reformulate Proposition 9.5.1 on the Clarke subdifferential in the following way.

Proposition 13.1.22 *Let E be a Fréchet smooth Banach space and let $f : E \to \mathbb{R}$ be locally L-continuous on E. Then for any $\bar{x} \in E$ one has*

$$\partial_o f(\bar{x}) = \overline{\mathrm{co}}^* \partial_M f(\bar{x}).$$

In this context, recall the discussion in Remark 9.5.2. The normal cone counterpart of Theorem 13.1.21 is

Theorem 13.1.23 *Assume that E is a Fréchet smooth Banach space, A is a closed subset of E, and $\bar{x} \in A$.*

(a) *For any $\epsilon \ge 0$ and any $\gamma > 0$ one has*

$$\widehat{N}_\epsilon(A, \bar{x}) \subseteq \bigcup \left\{ N_F(A, x) \ \middle|\ x \in A \cap B(\bar{x}, \gamma) \right\} + (\epsilon + \gamma)B_{E^*}. \tag{13.9}$$

(b) *One has*

$$N_M(A, \bar{x}) = \mathrm{sLim\,sup}_{x \to \bar{x}} N_F(A, \bar{x}).$$

Proof. See Exercise 13.13.3.   □

13.2 Coderivatives

Convention. In this section, unless otherwise specified, E and F are Fréchet smooth Banach spaces.

In Sect. 11.5 we considered contingent derivatives as derivative-like objects for multifunctions. Now we study an alternative concept. Since we want to apply the approximate calculus for F-subdifferentials established above, we restrict ourselves to Fréchet smooth Banach spaces.

Definition 13.2.1 Let $\Phi : E \rightrightarrows F$ be a closed multifunction and $(\bar{x}, \bar{y}) \in \text{graph} \, \Phi$.

(a) For any $\epsilon \geq 0$, the multifunction $\widehat{D}_\epsilon^* \Phi(\bar{x}, \bar{y}) : F^* \rightrightarrows E^*$ defined by

$$\widehat{D}_\epsilon^* \Phi(\bar{x}, \bar{y})(y^*) := \left\{ x^* \in E^* \mid (x^*, -y^*) \in \widehat{N}_\epsilon(\text{graph} \, \Phi, (\bar{x}, \bar{y})) \right\} \quad \forall y^* \in F^*$$

is said to be the ϵ-*coderivative* of Φ at $(\bar{x}, \bar{y})$. In particular, $D_F^* \Phi(\bar{x}, \bar{y}) := \widehat{D}_0^* \Phi(\bar{x}, \bar{y})$ is called *Fréchet coderivative* (*F-coderivative*) of Φ at $(\bar{x}, \bar{y})$.

(b) The multifunction $D_M^* \Phi(\bar{x}, \bar{y}) : F^* \rightrightarrows E^*$ defined by

$$D_M^* \Phi(\bar{x}, \bar{y})(y^*) := \left\{ x^* \in E^* \mid (x^*, -y^*) \in N_M(\text{graph} \, \Phi, (\bar{x}, \bar{y})) \right\} \quad \forall y^* \in F^*$$

is said to be the *Mordukhovich coderivative* (*M-coderivative*) of Φ at $(\bar{x}, \bar{y})$.

Remark 13.2.2

(a) By definition of the respective normal cone we have the following characterizations of these coderivatives:

$$D_M^* \Phi(\bar{x}, \bar{y})(\bar{y}^*) = \operatorname*{sLim\,sup}_{\substack{(x,y) \to (\bar{x}, \bar{y}) \\ y^* \xrightarrow[\epsilon \downarrow 0]{w^*} \bar{y}^*}} \widehat{D}_\epsilon^* \Phi(x, y)(y^*),$$

i.e., $D_M^* \Phi(\bar{x}, \bar{y})(\bar{y}^*)$ consists of all $\bar{x}^* \in E^*$ for which there exist sequences $\epsilon_k \downarrow 0$, $(x_k, y_k) \to (\bar{x}, \bar{y})$, and $(x_k^*, y_k^*) \xrightarrow{w^*} (\bar{x}^*, \bar{y}^*)$ satisfying $(x_k, y_k) \in \text{graph} \, \Phi$ and $x_k^* \in \widehat{D}_\epsilon^* \Phi(x_k, y_k)(y_k^*)$.

(b) In particular, we have

$$x^* \in D_F^* \Phi(\bar{x}, \bar{y})(y^*) \quad \Longleftrightarrow \quad (x^*, -y^*) \in \partial_F \delta_{\text{graph} \, \Phi}(\bar{x}, \bar{y}).$$

The viscosity characterization of the F-subdifferential now shows that $x^* \in D_F^* \Phi(\bar{x}, \bar{y})(y^*)$ if and only if there exists a C^1 function $g : E \times F \to \mathbb{R}$ such that $g'(\bar{x}, \bar{y}) = (x^*, -y^*)$ and $\delta_{\text{graph} \, \Phi}(x, y) - g(x, y) \geq 0$ for any $(x, y) \in E \times F$ near $(\bar{x}, \bar{y})$.

If $\varPhi$ is single-valued, we write $D_F^*\varPhi(\bar{x})(y^*)$ instead of $D_F^*\varPhi(\bar{x}, \varPhi(\bar{x}))(y^*)$, analogously for D_M^*. If $f : E \to \overline{\mathbb{R}}$ is a proper functional, we define the *epigraphical multifunction* $E_f : E \rightrightarrows \mathbb{R}$ by

$$E_f(x) := \{\tau \in \mathbb{R} \mid \tau \geq f(x)\} \quad \forall\, x \in E.$$

Proposition 13.2.3 relates coderivatives to familiar concepts.

Proposition 13.2.3

(a) *If $\varPhi : E \to F$ is an F-differentiable function, then*

$$D_F^*\varPhi(\bar{x})(y^*) = \{\varPhi'(\bar{x})^*y^*\} \quad \forall\, y^* \in F^*,$$

i.e., $D_F^*\varPhi(\bar{x})$ can be identified with the adjoint of the continuous linear operator $\varPhi'(\bar{x})$.

(b) *If $\varPhi : E \to F$ is a strictly F-differentiable function, then $D_M^*\varPhi(\bar{x})$ can be identified with the adjoint of the continuous linear operator $\varPhi'(\bar{x})$.*

(c) *If $f : E \to \overline{\mathbb{R}}$ is a proper functional and $\bar{x} \in \mathrm{dom}\, f$, then*

$$\partial_M f(\bar{x}) = D_M^* E_f(\bar{x}, f(\bar{x}))(1) \quad and \quad \partial_M^\infty f(\bar{x}) = D_M^* E_f(\bar{x}, f(\bar{x}))(0).$$

Proof. Leaving (a) and (c) as Exercise 13.13.5, we now verify (b). The proof of (a) is independent of that of (b), thus we may assume that we already have

$$\{\varPhi'(\bar{x})^*y^*\} = D_F^*\varPhi(\bar{x})(y^*) \subseteq D_M^*\varPhi(\bar{x})(y^*) \quad \forall\, y^* \in F^*.$$

It remains to prove the reverse inclusion. Take any $y^* \in F^*$ and $x^* \in D_M^*\varPhi(\bar{x})(y^*)$. By Remark 13.2.2 there exist sequences $\epsilon_k \downarrow 0$, $x_k \to \bar{x}$, and $(x_k^*, y_k^*) \xrightarrow{w^*} (x^*, y^*)$ such that for any x close to $\bar{x}$ and any $k \in \mathbb{N}$ we have

$$\langle x_k^*, x - x_k \rangle - \langle y_k^*, \varPhi(x) - \varPhi(x_k) \rangle \leq \epsilon_k \big(\|x - x_k\| + \|\varPhi(x) - \varPhi(x_k)\| \,.$$

Since $\varPhi$ is strictly F-differentiable, Proposition 3.2.4(iv) shows that for any sequence $\eta_i \downarrow 0$ as $i \to \infty$ there exist $\rho_i > 0$ such that

$$\|\varPhi(x) - \varPhi(z) - \varPhi'(\bar{x})(x - z)\| \leq \eta_i \|x - z\| \quad \forall\, x, z \in B(\bar{x}, \rho_i)\, \forall\, i \in \mathbb{N}.$$

Hence we find a sequence (k_i) in $\mathbb{N}$ such that

$$\langle x_{k_i}^* - \varPhi'(\bar{x})^*y_{k_i}^*, x - x_{k_i} \rangle \leq \tilde{\epsilon}_i \|x - x_{k_i}\| \quad \forall\, x \in B(x_{k_i}, \rho_{k_i})\, \forall\, i \in \mathbb{N};$$

here, $\tilde{\epsilon}_i := (\lambda + 1)(\epsilon_{k_i} + \eta_i \|y_{k_i}^*\|)$ and $\lambda > 0$ denotes a Lipschitz constant of $\varPhi$ around $\bar{x}$. It follows that

$$\|x_{k_i}^* - \varPhi'(\bar{x})^*y_{k_i}^*\| \leq \tilde{\epsilon}_i \quad \text{for all sufficiently large } i \in \mathbb{N}. \tag{13.10}$$

Since

$$\tilde{\epsilon}_i \downarrow 0 \quad \text{and} \quad x_{k_i}^* - \varPhi'(\bar{x})^*y_{k_i}^* \xrightarrow{w^*} x^* - \varPhi'(\bar{x})^*y^* \quad \text{as } i \to \infty,$$

and the norm functional on E^* is weak* l.s.c. (see Exercise 1.8.10), we finally obtain from (13.10) that $x^* = \Phi'(\bar{x})^* y^*$.   □

Next we consider coderivatives of locally L-continuous mappings.

Proposition 13.2.4 *Let E and F be Banach spaces and let $f : E \to F$ be locally L-continuous around $\bar{x} \in E$ with Lipschitz constant $\lambda > 0$.*

(a) *For any $\epsilon \geq 0$ there exists $\eta > 0$ such that*

$$\sup\{\|x^*\| \mid x^* \in \widehat{D}^*_\epsilon f(x)(y^*)\} \leq \lambda \|y^*\| + \epsilon(1 + \lambda)$$

whenever $x \in B(\bar{x}, \eta)$, $\|f(x) - f(\bar{x})\| \leq \eta$, and $y^* \in F^*$.

(b) *If, in particular, F is finite dimensional, then*

$$\sup\{\|x^*\| \mid x^* \in D^*_M f(\bar{x})(y^*)\} \leq \lambda \|y^*\| \quad \forall y^* \in F^*.$$

Proof.

(a) Without loss of generality we may assume that $\lambda \geq 1$. Let $\eta > 0$ be such that

$$\|f(x) - f(u)\| \leq \lambda \|x - u\| \quad \forall x, u \in B(\bar{x}, 2\eta). \tag{13.11}$$

Take $x \in B(\bar{x}, \eta)$ satisfying $\|f(x) - f(\bar{x})\| \leq \eta$, $y^* \in F^*$, and $x^* \in \widehat{D}^*_\epsilon f(x)(y^*)$. Furthermore, take $\gamma > 0$. By the definition of ϵ-coderivatives and ϵ-normals, there exists $\alpha \in (0, \eta)$ such that

$$\langle x^*, u - x \rangle - \langle y^*, f(u) - f(x) \rangle \leq (\epsilon + \gamma)(\|u - x\| + \|f(u) - f(x)\|) \tag{13.12}$$

whenever $u \in B(x, \alpha)$ and $\|f(u) - f(x)\| \leq \alpha$. Take any $u \in B(x, \alpha/\lambda) \subseteq B(x, \alpha)$. Then $u \in B(\bar{x}, 2\eta)$, and (13.11) yields $\|f(x) - f(u)\| \leq \lambda(\alpha/\lambda) = \alpha$. Invoking this estimate into (13.12), we obtain

$$\langle x^*, u - x \rangle \leq \|y^*\| \alpha + (\epsilon + \gamma)(\alpha/\lambda + \alpha).$$

Since this is true for any $u \in B(x, \alpha/\lambda)$, it follows that $\|x^*\| \leq \lambda \|y^*\| + (\epsilon + \gamma)(1 + \lambda)$. Since $\gamma > 0$ was arbitrary, the conclusion follows.

(b) Take any $x^* \in D^*_M f(\bar{x})(y^*)$. Then there exist sequences $\epsilon_k \downarrow 0$, $x_k \to \bar{x}$, $x^*_k \xrightarrow{w^*} x^*$, and $y^*_k \to y^*$ (here we exploit that F is finite dimensional) such that $x^*_k \in D^*_{\epsilon_k} f(x_k)(y^*_k)$ for any $k \in \mathbb{N}$. By (a) we have $\|x^*_k\| \leq \lambda \|y^*_k\| + \epsilon_k(1 + \lambda)$ for all k large enough. Since the norm functional on E^* is weak* l.s.c. (see Exercise 1.8.10), we obtain the assertion by letting $k \to \infty$.   □

Corollary 13.2.5 *If $f : E \to \mathbb{R}$ is locally L-continuous around $\bar{x} \in E$, then*

$$D^*_M f(\bar{x})(0) = \partial^\infty_M f(\bar{x}) = \{o\}. \tag{13.13}$$

Proof. The Lipschitz property of f implies that the set epi f is closed and that for some closed neighborhood U of $(\bar{x}, f(\bar{x}))$ we have $U \cap \text{graph} f = U \cap \text{bd}(\text{epi} f)$. By Exercise 13.13.4 we therefore obtain $\partial_M^\infty f(\bar{x}) \subseteq D_M^* f(\bar{x})(0)$. In view of Proposition 13.2.4(b) the assertion is verified. $\square$

We supplement Corollary 13.2.5 by a result that we state without proof (see Mordukhovich [137]).

Proposition 13.2.6 *If $f : E \to \mathbb{R}$ is locally L-continuous around $\bar{x} \in E$, then*

$$D_M^* f(\bar{x})(\alpha) = \partial_M(\alpha f)(\bar{x}) \quad \forall \alpha \in \mathbb{R}.$$

We briefly turn to calculus rules for F-coderivatives. As in the case of F-subdifferentials, approximate calculus rules for F-coderivatives can be established in weak form and in strong form.

Theorem 13.2.7 (Weak Coderivative Sum Rule) *Let $\Phi_1, \ldots, \Phi_n : E \rightrightarrows F$ and $\Phi := \sum_{i=1}^n \Phi_i$ be closed multifunctions. Further let $\bar{x} \in E$, $\bar{y} \in \Phi(\bar{x})$, $\bar{y}_i \in \Phi_i(\bar{x})$ for $i = 1, \ldots, n$ and assume that $\bar{y} = \sum_{i=1}^n \bar{y}_i$. Suppose that $y^* \in F^*$ and $x^* \in D_F^* \Phi(\bar{x}, \bar{y})(y^*)$.*
Then for any $\epsilon > 0$ and any weak* neighborhoods U of zero in E^* and V of zero in F^*, there exist $(x_i, y_i) \in (\text{graph} \Phi_i) \cap B((\bar{x}, \bar{y}_i), \epsilon)$, $y_i^* \in y^* + V$, and $x_i^* \in D_F^* \Phi_i(x_i, y_i)(y_i^*)$, $i = 1, \ldots, n$, such that

$$\max_{i=1,\ldots,n} \{\|x_i^*\|, \|y_i^*\|\} \cdot \text{diam}\{(x_1, y_1), \ldots, (x_n, y_n)\} < \epsilon, \quad (13.14)$$

$$x^* \in \sum_{i=1}^n x_i^* + U. \quad (13.15)$$

Proof. Since $x^* \in D_F^* \Phi(\bar{x}, \bar{y})(y^*)$, there exists a C^1 function g as in Remark 13.2.2. Observe that

$$\sum_{i=1}^n \delta_{\text{graph} \Phi_i}(x, y_i) \geq \delta_{\text{graph} \Phi}\left(x, \sum_{i=1}^n y_i\right) \quad \text{and}$$

$$\sum_{i=1}^n \delta_{\text{graph} \Phi_i}(\bar{x}, \bar{y}_i) = \delta_{\text{graph} \Phi}(\bar{x}, \bar{y}) = 0.$$

Thus, defining $\tilde{f}, \tilde{g} : E \times F \times \cdots \times F \to \mathbb{R}$ by

$$\tilde{f}(x, y_1, \ldots, y_n) := \sum_{i=1}^n \delta_{\text{graph} \Phi_i}(x, y_i), \quad \tilde{g}(x, y_1, \ldots, y_n) := g\left(x, \sum_{i=1}^n y_i\right),$$

we see that $\tilde{f} - \tilde{g}$ attains a local minimum at $(\bar{x}, \bar{y}_1, \ldots, \bar{y}_n)$. Since $g'(\bar{x}, \bar{y}) = (x^*, -y^*)$ (cf. Remark 13.2.2), it follows that $\tilde{g}'(\bar{x}, \bar{y}_1, \ldots, \bar{y}_n) = (x^*, -y^*, \ldots, -y^*)$. Hence

$$(x^*, -y^*, \ldots, -y^*) \in \partial_F \tilde{f}(\bar{x}, \bar{y}_1, \ldots, \bar{y}_n).$$

By the approximate sum rule of Theorem 9.2.7 there exist $x_i \in \mathrm{B}(\bar{x}, \epsilon)$, $y_i \in \mathrm{B}(\bar{y}_i, \epsilon)$, $y_i^* \in F^*$, and $x_i^* \in \mathrm{D}_F^* \Phi_i(x_i, y_i)(y_i^*)$ satisfying (13.14) and

$$(x^*, -y^*, \ldots, -y^*) \in (x_1^*, -y_1^*, 0, \ldots, 0) + (x_2^*, 0, -y_2^*, 0, \ldots, 0)$$

$$+ \cdots + (x_n^*, 0, \ldots, 0, -y_n^*) + (U \times V \times \cdots \times V).$$

It follows that $y_i^* \in y^* + V$ for $i = 1, \ldots, n$ and that (13.15) holds.   □

Similarly a weak chain rule can be derived for F-coderivatives. We employ the following notation. Given multifunctions $\Psi : E \rightrightarrows F$ and $\Phi : F \rightrightarrows G$, we define the *composition* $\Phi \circ \Psi : E \rightrightarrows G$ by

$$\Phi \circ \Psi(x) := \bigcup \{\Phi(y) \mid y \in \Psi(x)\}.$$

Theorem 13.2.8 (Weak Coderivative Chain Rule) *In addition to E and F let G be a Fréchet smooth Banach space, let $\Psi : E \rightrightarrows F$ and $\Phi : F \rightrightarrows G$ be closed multifunctions, let $\bar{x} \in E$, $\bar{y} \in \Psi(\bar{x})$, and $\bar{z} \in \Phi(\bar{y})$. Suppose that $z^* \in G^*$ and $x^* \in \mathrm{D}_F^*(\Phi \circ \Psi)(\bar{x}, \bar{z})(z^*)$. Then for any $\epsilon > 0$ and any weak* neighborhoods U of zero in E^*, V of zero in F^*, and W of zero in G^*, there exist $x_2 \in \mathrm{B}_E(\bar{x}, \epsilon)$, $y_i \in \mathrm{B}_F(\bar{y}, \epsilon)$, $i = 1, 2$, and $z_1 \in \mathrm{B}_G(\bar{z}, \epsilon)$ as well as $x_2^* \in E^*$, $y_i^* \in F^*$, $i = 1, 2$, and $z_1^* \in G^*$ satisfying*

$$y_1^* - y_2^* \in V, \quad z_1^* \in z^* + W,$$

$$y_1^* \in \mathrm{D}_F^* \Phi(y_1, z_1)(z_1^*), \quad x_2^* \in \mathrm{D}_F^* \Psi(x_2, y_2)(y_2^*),$$

$$\max\{\|x_2^*\|, \|y_1^*\|, \|y_2^*\|, \|z_1^*\|\} \cdot \|y_1 - y_2\| < \epsilon,$$

$$x^* \in x_2^* + U.$$

Proof. Since $x^* \in \mathrm{D}_F^*(\Phi \circ \Psi)(\bar{x}, \bar{z})(z^*)$, there exists a C^1 function $g : E \times G \to \mathbb{R}$ such that $g'(\bar{x}, \bar{z}) = (x^*, -z^*)$ and $\delta_{\mathrm{graph}(\Phi \circ \Psi)}(x, z) - g(x, z) \geq 0$ for any (x, z) near $(\bar{x}, \bar{z})$. We have

$$\delta_{\mathrm{graph}\,\Phi}(y, z) + \delta_{\mathrm{graph}\,\Psi}(x, y) \geq \delta_{\mathrm{graph}(\Phi \circ \Psi)}(x, z) \quad \text{and}$$

$$\delta_{\mathrm{graph}\,\Phi}(\bar{y}, \bar{z}) + \delta_{\mathrm{graph}\,\Psi}(\bar{x}, \bar{y}) = \delta_{\mathrm{graph}(\Phi \circ \Psi)}(\bar{x}, \bar{z}) = 0.$$

Defining $\tilde{f}, \tilde{g} : E \times F \times G \to \bar{\mathbb{R}}$ by

$$\tilde{f}(x, y, z) := \delta_{\mathrm{graph}\,\Phi}(y, z) + \delta_{\mathrm{graph}\,\Psi}(x, y), \quad \tilde{g}(x, y, z) := g(x, z),$$

we thus conclude that $(\bar{x}, \bar{y}, \bar{z})$ is a local minimizer of the functional $\tilde{f} - \tilde{g}$. Moreover, we have $\tilde{g}'(\bar{x}, \bar{y}, \bar{z}) = (x^*, 0, -z^*)$. Therefore

$$(x^*, 0, -z^*) \in \partial_F \tilde{f}(\bar{x}, \bar{y}, \bar{z}).$$

The conclusion of the theorem now follows by applying Theorem 9.2.7 to $\tilde{f}$. The details are left as Exercise 13.13.6. □

Strong calculus rules for F-coderivatives can be obtained similarly by applying the strong local approximate sum rule of Theorem 9.2.6 instead of Theorem 9.2.7. As a consequence of the above results, we can easily derive exact sum and chain rules for M-coderivatives in finite-dimensional spaces. First we introduce another concept.

Definition 13.2.9 The multifunction $\Phi : E \rightrightarrows F$ is said to be *inner semi-compact* at $\bar{x} \in \mathrm{Dom}\,\Phi$ if for any sequence $x_k \to \bar{x}$ there exists a sequence $y_k \in \Phi(x_k)$ that contains a convergent subsequence.

Observe that if $\mathrm{Dom}\,\Phi = E$ and Φ is locally compact at $\bar{x}$ (in particular, if F is finite dimensional and Φ is locally bounded at $\bar{x}$), then Φ is inner semicompact at $\bar{x}$.

Theorem 13.2.10 (Exact Coderivative Sum Rule) *Let E and F be finite-dimensional Banach spaces, let Φ_1 and Φ_2 be closed multifunctions and let $\bar{y} \in \Phi_1(\bar{x}) + \Phi_2(\bar{x})$. Assume that the multifunction $S : E \times F \rightrightarrows F \times F$ defined by*

$$S(x, y) := \{(y_1, y_2) \mid y_1 \in \Phi_1(x)\, y_2 \in \Phi_2(x),\ y_1 + y_2 = y\} \qquad (13.16)$$

is inner semicompact at $(\bar{x}, \bar{y})$. Assume also that

$$\mathrm{D}_M^*\Phi_1(\bar{x}, y_1)(o) \cap \left(-\mathrm{D}_M^*\Phi_2(\bar{x}, y_2)(o)\right) = \{o\} \quad \forall\,(y_1, y_2) \in S(\bar{x}, \bar{y}). \qquad (13.17)$$

Then for any $y^* \in F^*$ one has

$$\mathrm{D}_M^*(\Phi_1 + \Phi_2)(\bar{x}, \bar{y})(y^*) \subseteq \bigcup_{(y_1, y_2) \in S(\bar{x}, \bar{y})} \left(\mathrm{D}_M^*\Phi_1(\bar{x}, y_1)(y^*) + \mathrm{D}_M^*\Phi_2(\bar{x}, y_2)(y^*)\right).$$

Proof. Take any $x^* \in \mathrm{D}_M^*(\Phi_1 + \Phi_2)(\bar{x}, \bar{y})(y^*)$. Then

$$(x^*, -y^*) \in \mathrm{N}_M(\mathrm{graph}(\Phi_1 + \Phi_2), (\bar{x}, \bar{y})).$$

By Theorem 13.1.4 there exist sequences $(x_k, y_k) \to (\bar{x}, \bar{y})$ and $(x_k^*, y_k^*) \to (x^*, y^*)$ as $k \to \infty$ such that $(x_k, y_k) \in \mathrm{graph}(\Phi_1 + \Phi_2)$ and $x_k^* \in \mathrm{D}_F^*(\Phi_1 + \Phi_2)(y_k^*)$ for any $k \in \mathbb{N}$. Since S is inner semicompact, we may assume (omitting relabeling) that there exists a sequence $(y_{1k}, y_{2k}) \in \Psi(x_k, y_k)$ converging to $(y_1, y_2) \in \Psi(\bar{x}, \bar{y})$. Applying Theorem 13.2.7 for any k, we find $y_{ik}^* \in \mathrm{B}(y_k^*, 1/k)$ such that

$$x_{ik}^* \in \mathrm{D}_F^*\Phi_i(x_k, y_{ik})(y_{ik}^*) \quad \text{and} \quad \|x_k^* - x_{1k}^* - x_{2k}^*\| < 1/k, \quad i = 1, 2. \qquad (13.18)$$

(I) If the sequence $(\|x_{1k}^*\|)$ is unbounded, we may assume (passing to a subsequence if necessary) that $\|x_{1k}^*\| \to \infty$ and that $(x_{1k}^*/\|x_{1k}^*\|)$ converges to a unit vector z^*. Dividing the second relation in (13.18) by $\|x_{1k}^*\|$ and letting $k \to \infty$, we conclude that $(x_{2k}^*/\|x_{1k}^*\|)$ converges to $-z^*$. The first relation in (13.18) now shows that

$$z^* \in D_M^* \Phi_1(\bar{x}, y_1)(o) \cap \left(-D_M^* \Phi_2(\bar{x}, y_2)(o)\right),$$

which contradicts the assumption (13.17).

(II) Since by step (I) we know that the sequence $(\|x_{1k}^*\|)$ is bounded, the second relation in (13.18) shows that the sequence $(\|x_{2k}^*\|)$ is also bounded. Thus we may assume that both sequences are convergent, and passing to the limit in (13.18) as $k \to \infty$ we conclude that $x^* \in D_M^* \Phi_1(\bar{x}, y_1)(y^*) + D_M^* \Phi_2(\bar{x}, y_2)(y^*)$, which proves the theorem. $\qquad \square$

In a similar way, an exact coderivative chain rule can be established in finite-dimensional spaces.

Theorem 13.2.11 (Exact Coderivative Chain Rule) *Let E, F, G be finite-dimensional Banach spaces, let $\Psi : E \rightrightarrows F$ and $\Phi : F \rightrightarrows G$ be closed multifunctions, let $\bar{x} \in E$ and $\bar{z} \in (\Phi \circ \Psi)(\bar{x})$. Assume that the multifunction $T : E \times G \rightrightarrows F$ defined by $T(x, z) := \Psi(x) \cap \Phi^{-1}(z)$ is inner semicompact at $(\bar{x}, \bar{z})$. Assume further that for any $\bar{y} \in T(\bar{x}, \bar{z})$ the condition*

$$D_M^* \Phi(\bar{y}, \bar{z})(o) \cap \left(-D_M^* \Psi^{-1}(\bar{y}, \bar{x})(o)\right) = \{o\}$$

is satisfied. Then for any $z^* \in G^*$ one has

$$D_M(\Phi \circ \Psi)(\bar{x}, \bar{z})(z^*) \subseteq \bigcup_{\bar{y} \in T(\bar{x}, \bar{z})} \left(D_M^* \Psi(\bar{x}, \bar{y}) \circ D_M^* \Phi(\bar{y}, \bar{z})(z^*)\right).$$

Proof. See Exercise 13.13.7. $\qquad \square$

The following example shows that in an infinite-dimensional space, exact calculus rules as that of Theorem 13.2.10 may fail.

Example 13.2.12 Let E be a separable infinite-dimensional Banach space, let H_1 and H_2 be closed subspaces of E such that $H_1^\perp \cap H_2^\perp = \{o\}$ and $H_1^\perp + H_2^\perp$ is weak* dense (which implies $H_1 \cap H_2 = \{o\}$) but not closed in E^*; see Exercise 13.13.8 for a concrete example. Now define $\Phi_1, \Phi_2 : E \rightrightarrows \mathbb{R}$ by graph $\Phi_i := H_i \times \mathbb{R}_+$ for $i = 1, 2$. Then we have graph$(\Phi_1 + \Phi_2) = \{o\} \times \mathbb{R}_+$, $S(o, o) = \{(0, 0)\}$, and $S(x, y) = \emptyset$ whenever $(x, y) \neq (o, o)$; here S is as in (13.16). The multifunction S is evidently inner semicompact at (o, o). Moreover, it is easy to see that $D_M^* \Phi_i(o, o)(0) = H_i^\perp$ for $i = 1, 2$ and $D_M^* (\Phi_1 + \Phi_2)(o, o)(0) = E^*$. Hence the regularity condition (13.17) is satisfied but the sum rule

$$D_M^* (\Phi_1 + \Phi_2)(o, o)(0) \subseteq D_M^* \Phi_1(o, o)(0) + D_M^* \Phi_2(o, o)(0)$$

fails.

13.3 Extremal Principles Involving Translations

We start with a definition.

Definition 13.3.1 Let $A_1, \ldots, A_n$ be nonempty subsets of the normed vector space E. A point $\bar{x} \in \bigcap_{i=1}^{n} A_i$ is said to be a *local extremal point* of the system $(A_1, \ldots, A_n)$ if there exist sequences $(z_{1k}), \ldots, (z_{nk})$ in E and a neighborhood U of $\bar{x}$ such that $z_{ik} \to o$ as $k \to \infty$ for $i = 1, \ldots, n$ and

$$\bigcap_{i=1}^{n} (A_i - z_{ik}) \cap U = \emptyset \quad \text{for all sufficiently large } k \in \mathbb{N}.$$

In this case, $(A_1, \ldots, A_n, \bar{x})$ is said to be an *extremal system* in E.

Remark 13.3.2 Geometrically this means that the sets $A_1, \ldots, A_n$ can be locally pushed apart by small translations. In particular, a point $\bar{x} \in A_1 \cap A_2$ is a local extremal point of (A_1, A_2) if and only if there exists a neighborhood U of $\bar{x}$ such that for any $\epsilon > 0$ there is an element $z \in B(o, \epsilon)$ with $(A_1 + z) \cap A_2 \cap U = \emptyset$.

Example 13.3.3 shows the close relationship of this extremality concept to constrained optimization.

Example 13.3.3 Let $f : E \to \overline{\mathbb{R}}$ and let $M \subseteq E$. If $\bar{x}$ is a local solution to the problem

$$\text{minimize } f(x) \text{ subject to } x \in M,$$

then $(\bar{x}, f(\bar{x}))$ is a local extremal point of the system (A_1, A_2), where $A_1 := \text{epi } f$ and $A_2 := M \times \{f(\bar{x})\}$ (see Exercise 13.13.9).

The example $A_1 := \{(\xi, \xi) \mid \xi \in \mathbb{R}\}$, $A_2 := \{(\xi, -\xi) \mid \xi \in \mathbb{R}\}$, $\bar{x} := (0, 0)$ shows that the condition $A_1 \cap A_2 = \{\bar{x}\}$ does not imply that $\bar{x}$ is a local extremal point of (A_1, A_2). On the other hand, we have the following necessary condition.

Lemma 13.3.4 *If* $(A_1, \ldots, A_n, \bar{x})$ *is an extremal system in* E, *then there exists a neighborhood* U *of* $\bar{x}$ *such that*

$$(\text{int } A_1) \cap \cdots \cap (\text{int } A_{n-1}) \cap A_n \cap U = \emptyset.$$

Proof. See Exercise 13.13.10. □

The next result reveals the relationship between extremality and the separation property.

Proposition 13.3.5 *If* A_1 *and* A_2 *are subsets of* E *such that* $A_1 \cap A_2 \neq \emptyset$, *then the following holds:*

(a) *If* A_1 *and* A_2 *are separated, then* $(A_1, A_2, \bar{x})$ *is an extremal system for any* $\bar{x} \in A_1 \cap A_2$.

(b) *Assume, in addition, that A_1 and A_2 are convex and int $A_1 \neq \emptyset$. If $(A_1, A_2, \bar{y})$ is an extremal system for some $\bar{y} \in A_1 \cap A_2$, then A_1 and A_2 are separated (and $(A_1, A_2, \bar{x})$ is an extremal system for any $\bar{x} \in A_1 \cap A_2$).*

Proof.

(a) If A_1 and A_2 are separated, then there exist $x^* \in E^* \setminus \{o\}$ and $\alpha \in \mathbb{R}$ such that

$$\langle x^*, x \rangle \leq \alpha \, \forall x \in A_1 \quad \text{and} \quad \langle x^*, y \rangle \geq \alpha \, \forall y \in A_2. \tag{13.19}$$

Choose $x_0 \in E$ such that $\langle x^*, x_0 \rangle > 0$ and set $z_k := (1/k)x_0$ for all $k \in \mathbb{N}$. We show that $(A_1 - z_k) \cap A_2 = \emptyset$ for any $k \in \mathbb{N}$, which evidently implies the conclusion of (a). Assume, to the contrary, there exists $y \in (A_1 - z_k) \cap A_2$ for any k. Then it follows that

$$\alpha \geq \langle x^*, y + z_k \rangle = \langle x^*, y \rangle + \frac{1}{k} \langle x^*, x_0 \rangle \quad \forall k,$$

which contradicts the second inequality in (13.19).

(b) If $(A_1, A_2, \bar{y})$ is an extremal system, then Lemma 13.3.4 implies that $(\text{int } A_1) \cap (A_2 \cap U)$ is empty for some neighborhood U of $\bar{y}$ which may be assumed to be convex. By the weak separation theorem (Theorem 1.5.3) there exist $x^* \in E^*$ and $\alpha \in \mathbb{R}$ such that

$$\langle x^*, x \rangle \leq \alpha \, \forall x \in A_1 \quad \text{and} \quad \langle x^*, y \rangle \geq \alpha \, \forall y \in A_2 \cap U.$$

It remains to show that $\langle x^*, y \rangle \geq \alpha$ for all $y \in A_2$. Suppose we had $\langle x^*, y_0 \rangle < \alpha$ for some $y_0 \in A_2$. Set $y := \lambda y_0 + (1 - \lambda)\bar{y}$. If $\lambda \in (0,1)$ is sufficiently small, we have $y \in A_2 \cap U$ and so $\alpha \leq \lambda \langle x^*, y_0 \rangle + (1-\lambda)\langle x^*, \bar{y} \rangle$, which is a contradiction because $\langle x^*, y_0 \rangle < \alpha$ and $\langle x^*, \bar{y} \rangle \leq \alpha$. $\qquad \square$

By Proposition 13.3.5 and the weak separation theorem we now obtain:

Corollary 13.3.6 *If A_1 and A_2 are convex subsets of E such that $A_1 \cap A_2 \neq \emptyset$ and int $A_1 \neq \emptyset$, then the following conditions are mutually equivalent:*

(i) $(A_1, A_2, \bar{x})$ *is an extremal system for any $\bar{x} \in A_1 \cap A_2$.*
(ii) A_1 *and A_2 are separated.*
(iii) $(\text{int } A_1) \cap A_2 = \emptyset$.

Definition 13.3.7

(a) An extremal system $(A_1, \ldots, A_n, \bar{x})$ in E is said to satisfy the *exact extremal principle* if there exist $x_i^* \in N_M(A_i, \bar{x})$, $i = 1, \ldots, n$, such that

$$x_1^* + \cdots + x_n^* = o \quad \text{and} \quad \|x_1^*\| + \cdots + \|x_n^*\| = 1. \tag{13.20}$$

(b) An extremal system $(A_1, \ldots, A_n, \bar{x})$ in E is said to satisfy the *approximate extremal principle* if for every $\epsilon > 0$ there exist

$$x_i \in A_i \cap B_E(\bar{x}, \epsilon) \quad \text{and} \quad x_i^* \in N_F(A_i, x_i) + \epsilon B_{E^*}, \quad i = 1, \ldots, n,$$

such that (13.20) holds.

Remark 13.3.8 We consider the case of two sets:

(a) An extremal system $(A_1, A_2, \bar{x})$ satisfies the exact extremal principle if and only if there exists $x^* \in E^* \setminus \{o\}$ such that

$$x^* \in N_M(A_1, \bar{x}) \cap (-N_M(A_2, \bar{x})) . \tag{13.21}$$

If A_1 and A_2 are convex, then by Proposition 13.1.5 the relation (13.21) is equivalent to

$$\langle x^*, y_1 \rangle \leq \langle x^*, y_2 \rangle \quad \forall y_1 \in A_1 \, \forall y_2 \in A_2,$$

which means that A_1 and A_2 are separated. Hence the exact extremal principle can be considered as a statement on local separation of nonconvex sets.

(b) Similarly, an extremal system $(A_1, A_2, \bar{x})$ satisfies the approximate extremal principle if and only if for any $\epsilon > 0$ there exist $x_i \in A_i \cap B(\bar{x}, \epsilon)$, where $i = 1, 2$, and $x^* \in E^*$ such that $\|x^*\| = 1$ and

$$x^* \in (N_F(A_1, x_1) + \epsilon B_{E^*}) \cap (-N_F(A_2, x_2) + \epsilon B_{E^*}). \tag{13.22}$$

This can be analogously interpreted as an approximate separation of A_1 and A_2 near $\bar{x}$.

We consider a first application of the approximate extremal principle. Recall that in terms of the normal cone of convex analysis, $N(A, \bar{x})$, the set of support points of A is

$$\{ x \in \text{bd}\, A \mid N(A, x) \neq \{o\} \}.$$

If A is closed and convex, the Bishop–Phelps theorem (Theorem 1.5.6) ensures that this set is dense in $\text{bd}\, A$. Now we show that the approximate extremal principle yields an approximate analogue of the Bishop–Phelps theorem without the convexity hypothesis if $N(A, \bar{x})$ is replaced by the Fréchet normal cone $N_F(A, \bar{x})$.

Proposition 13.3.9 *Let A be a proper closed subset of E and $\bar{x} \in \text{bd}\, A$. If the approximate extremal principle holds for the system $(A, \{\bar{x}\}, \bar{x})$, then the set*

$$\{ x \in \text{bd}\, A \mid N_F(A, x) \neq \{o\} \}$$

is dense in $\text{bd}\, A$.

Proof. By the approximate extremal principle for $(A, \{\bar{x}\}, \bar{x})$ with $\epsilon \in (0, 1/2)$, we find $x \in A \cap B_E(\bar{x}, \epsilon)$ and $x^* \in N_F(A, x) + \epsilon B_{E^*}$ such that $\|x^*\| = 1/2$. It follows that $x \in \operatorname{bd} A$ because otherwise we had $N_F(A, x) = \{o\}$ and so $\|x^*\| \leq \epsilon < 1/2$: a contradiction. In particular we see that $N_F(A, x) \neq \{o\}$. □

The next result will be crucial for substantial applications of the approximate extremal principle.

Theorem 13.3.10 *If E is a Fréchet smooth Banach space, then the approximate extremal principle holds for any extremal system $(A_1, \ldots, A_n, \bar{x})$ in E, where $n \in \mathbb{N}$ and the sets $A_1, \ldots, A_n$ are closed.*

Proof.

(I) First we consider the case $n = 2$.

(Ia) Let $\bar{x} \in A_1 \cap A_2$ be a local extremal point of the system (A_1, A_2). Then there exists a closed neighborhood U of $\bar{x}$ such that for any $\epsilon > 0$ there is $z \in E$ satisfying $\|z\| \leq \epsilon^3/2$ and

$$(A_1 + z) \cap A_2 \cap U = \emptyset. \tag{13.23}$$

Obviously we may assume that $U = E$ (otherwise replace A_2 by $A_2 \cap U$) and $\epsilon < 1/2$. Equip $E \times E$ with the Euclidean product norm, which is also F-differentiable away from the origin. Define $\varphi : E \times E \to \mathbb{R}$ by

$$\varphi(\mathbf{x}) := \omega(x_1 - x_2 + z) = \|x_1 - x_2 + z\|, \quad \text{where} \quad \mathbf{x} := (x_1, x_2) \in E \times E.$$

By (13.23) we have $\varphi(\mathbf{x}) > 0$ for any $\mathbf{x} \in A_1 \times A_2$. Hence φ is a C^1 function in a neighborhood of each point of $A_1 \times A_2$. Set $\bar{\mathbf{x}} := (\bar{x}, \bar{x})$ and define the set

$$W(\bar{\mathbf{x}}) := \left\{ \mathbf{x} \in A_1 \times A_2 \mid \varphi(\mathbf{x}) + (\epsilon/2)\|\mathbf{x} - \bar{\mathbf{x}}\|^2 \leq \varphi(\bar{\mathbf{x}}) \right\},$$

which is nonempty and closed. For any $\mathbf{x} \in W(\bar{\mathbf{x}})$ we obtain $(\epsilon/2)\|\mathbf{x} - \bar{\mathbf{x}}\|^2 \leq \varphi(\bar{\mathbf{x}})$ and so

$$\|x_1 - \bar{x}\|^2 + \|x_2 - \bar{x}\|^2 \leq (2/\epsilon)\varphi(\bar{\mathbf{x}}) = (2/\epsilon)\|z\| \leq \epsilon^2.$$

We conclude that $W(\bar{\mathbf{x}}) \subseteq B_{E \times E}(\bar{\mathbf{x}}, \epsilon)$.

(Ib) For $k = 0, 1, \ldots$ we inductively define points $\mathbf{x}_k \in A_1 \times A_2$ and sets $W(\mathbf{x}_k)$ as follows. Given $\mathbf{x}_k$ and $W(\mathbf{x}_k)$, choose $\mathbf{x}_{k+1} \in W(\mathbf{x}_k)$ such that

$$\varphi(\mathbf{x}_{k+1}) + \epsilon \sum_{i=0}^{k} \frac{\|\mathbf{x}_{k+1} - \mathbf{x}_i\|^2}{2^{i+1}}$$

$$< \inf_{\mathbf{x} \in W(\mathbf{x}_k)} \left(\varphi(\mathbf{x}) + \epsilon \sum_{i=0}^{k} \frac{\|\mathbf{x} - \mathbf{x}_i\|^2}{2^{i+1}} \right) + \frac{\epsilon^3}{2^{3k+2}}$$

and define

$$W(\mathbf{x}_{k+1}) := \left\{ \mathbf{x} \in A_1 \times A_2 \;\middle|\; \varphi(\mathbf{x}) + \epsilon \sum_{i=0}^{k+1} \frac{\|\mathbf{x} - \mathbf{x}_i\|^2}{2^{i+1}} \right.$$

$$\left. \leq \varphi(\mathbf{x}_{k+1}) + \epsilon \sum_{i=0}^{k} \frac{\|\mathbf{x}_{k+1} - \mathbf{x}_i\|^2}{2^{i+1}} \right\}.$$

For any k, the set $W(\mathbf{x}_k)$ is a nonempty closed subset of $A_1 \times A_2$ and one has $W(\mathbf{x}_{k+1}) \subseteq W(\mathbf{x}_k)$. We show that $\operatorname{diam} W(\mathbf{x}_k) \to 0$ as $k \to \infty$. For all $k \in \mathbb{N}$ and $\mathbf{x} \in W(\mathbf{x}_{k+1})$ we obtain

$$\frac{\epsilon \|\mathbf{x} - \mathbf{x}_{k+1}\|^2}{2^{k+2}} \leq \varphi(\mathbf{x}_{k+1})$$

$$+ \epsilon \sum_{i=0}^{k} \frac{\|\mathbf{x}_{k+1} - \mathbf{x}_i\|^2}{2^{i+1}} - \left(\varphi(\mathbf{x}) + \epsilon \sum_{i=0}^{k} \frac{\|\mathbf{x} - \mathbf{x}_i\|^2}{2^{i+1}} \right)$$

$$\leq \varphi(\mathbf{x}_{k+1}) + \epsilon \sum_{i=0}^{k} \frac{\|\mathbf{x}_{k+1} - \mathbf{x}_i\|^2}{2^{i+1}}$$

$$- \inf_{\mathbf{x} \in W(\mathbf{x}_k)} \left\{ \varphi(\mathbf{x}) + \epsilon \sum_{i=0}^{k} \frac{\|\mathbf{x} - \mathbf{x}_i\|^2}{2^{i+1}} \right\} < \frac{\epsilon^3}{2^{3k+2}}.$$

Thus $\operatorname{diam} W(\mathbf{x}_k) \leq \epsilon/2^{k-1} \to 0$ as $k \to \infty$. By the Cantor intersection theorem the set $\cap_{k=0}^{\infty} W(\mathbf{x}_k)$ consists of exactly one point $\hat{\mathbf{x}} = (\hat{x}_1, \hat{x}_2) \in W(\mathbf{x}_0) \subseteq B_{E \times E}(\overline{\mathbf{x}}, \epsilon)$ and one has $\mathbf{x}_k \to \hat{\mathbf{x}}$ as $k \to \infty$.

(Ic) We show that $\hat{\mathbf{x}}$ is a minimizer of the function

$$\tilde{\varphi}(\mathbf{x}) := \varphi(\mathbf{x}) + \epsilon \sum_{i=0}^{\infty} \frac{\|\mathbf{x} - \mathbf{x}_i\|^2}{2^{i+1}}, \quad \mathbf{x} \in E \times E,$$

over $A_1 \times A_2$. Thus let $\mathbf{x} \in A_1 \times A_2$, $\mathbf{x} \neq \hat{\mathbf{x}}$. The construction of $W(\mathbf{x}_k)$ shows that for some $k \in \mathbb{N}$ we have

$$\varphi(\mathbf{x}) + \epsilon \sum_{i=0}^{k} \frac{\|\mathbf{x} - \mathbf{x}_i\|^2}{2^{i+1}} > \varphi(\mathbf{x}_k) + \epsilon \sum_{i=0}^{k-1} \frac{\|\mathbf{x}_k - \mathbf{x}_i\|^2}{2^{i+1}}.$$

Since the sequence on the right-hand side is nonincreasing as $k \to \infty$, it follows that in fact $\hat{\mathbf{x}}$ is a minimizer of $\tilde{\varphi}$ on $A_1 \times A_2$ and so a minimizer of $\psi := \tilde{\varphi} + \delta_{A_1 \times A_2}$ on $E \times E$. We conclude that $o \in \partial_F \psi(\hat{\mathbf{x}})$ (Proposition 9.1.5). Since φ is a C^1 function in a neighborhood of $\hat{\mathbf{x}}$, so is the function $\tilde{\varphi}$. By Proposition 9.2.2 and the definition of the Fréchet normal cone we obtain

$$-\tilde{\varphi}'(\hat{\mathbf{x}}) \in N_F(A_1 \times A_2, \hat{\mathbf{x}}) = N_F(A_1, \hat{\mathbf{x}}) \times N_F(A_2, \hat{\mathbf{x}}). \tag{13.24}$$

(Id) We calculate $\widetilde{\varphi}'(\hat{\mathbf{x}})$. It follows from the definition of $\widetilde{\varphi}$ that $\widetilde{\varphi}'(\hat{\mathbf{x}}) = (u_1^*, u_2^*) \in E^* \times E^*$, where

$$u_1^* := x^* + \epsilon \sum_{i=0}^{\infty} v_{1i}^* \frac{\|\hat{x}_1 - x_{1i}\|}{2^i}, \quad u_2^* := -x^* + \epsilon \sum_{i=0}^{\infty} v_{2i}^* \frac{\|\hat{x}_2 - x_{2i}\|}{2^i},$$

$$x^* := \omega'(\hat{x}_1 - \hat{x}_2 + z), \quad (x_{1i}, x_{2i}) := \mathbf{x}_i,$$

$$v_{ji}^* := \begin{cases} \omega'(\hat{x}_j - x_{ji}) & \text{if } \hat{x}_j - x_{ji} \neq 0, \\ 0 & \text{otherwise}; \end{cases}$$

here $i = 0, 1, \ldots$ and $j = 1, 2$. We have $\|x^*\| = 1$ and for $j = 1, 2$,

$$\sum_{i=0}^{\infty} \|v_{ji}^*\| \frac{\|\hat{x}_j - x_{ji}\|}{2^i} \leq 1.$$

Thus, putting $x_j := \hat{x}_j$ and $x_j^* := (-1)^j x^*/2$ for $j = 1, 2$, we obtain the conclusion of the theorem in the case $n = 2$.

(II) Now we treat the general case by induction. If $\bar{x}$ is a local extremal point of the system $(A_1, \ldots, A_n)$, where $n > 2$, then $\bar{y} := (\bar{x}, \ldots, \bar{x}) \in E^{n-1}$ is a local extremal point of the system (B_1, B_2), where

$$B_1 := A_1 \times \cdots \times A_{n-1} \quad \text{and} \quad B_2 := \{(x, \ldots, x) \in E^{n-1} \mid x \in A_n\}.$$

By step (I) the approximate extremal principle holds for the system $(B_1, B_2, \bar{y})$, from which the assertion follows. The details are left as Exercise 13.13.12. $\qquad\qquad\square$

Now we are going to discuss the relationship between the approximate extremal principle and other properties of a Banach space. In this connection we adopt the following terminology.

Definition 13.3.11 We say that the *subdifferential variational principle* holds in the Banach space E if for every proper l.s.c. functional $f : E \to \overline{\mathbb{R}}$ bounded below, every $\epsilon > 0$, every $\lambda > 0$, and every $\bar{x} \in E$ such that $f(\bar{x}) < \inf_E f + \epsilon$, there exist $z \in E$ and $z^* \in \partial_F f(z)$ satisfying

$$\|z - \bar{x}\| < \lambda, \quad f(z) < \inf_E f + \epsilon \quad \text{and} \quad \|z^*\| < \epsilon/\lambda.$$

Theorem 13.3.12 (Characterization of Asplund Spaces) *For any Banach space E the following assertions are mutually equivalent:*

(a) *E is an Asplund space.*
(b) *The approximate extremal principle holds for any extremal system $(A_1, \ldots, A_n, \bar{x})$, where $n \in \mathbb{N}$ and $A_1, \ldots, A_n$ are closed sets.*
(c) *The subdifferential variational principle holds in E.*

Proof. Remarks on (a) $\implies$ (b): By Theorem 13.3.10 we know that the approximate extremal principle holds if E is Fréchet smooth. If now E is an arbitrary Asplund space, then every separable subspace is Fréchet smooth (see Deville et al. [50]). By a method called *separable reduction* certain problems concerning F-subdifferentials and F-normal cones can be reduced from an arbitrary Banach space to separable subspaces. In this way, the approximate extremal principle can be transmitted from a Fréchet smooth Banach space to an arbitrary Asplund space. However, the method of separable reduction is quite involved and will not be treated here; we refer to Borwein and Zhu [24] and Mordukhovich [136].

(b) $\implies$ (c): Let f, ϵ, λ, and $\bar{x}$ be as in the definition of the subdifferential variational principle. Choose $\tilde{\epsilon} \in (0, \epsilon)$ such that $f(\bar{x}) < \inf_E f + (\epsilon - \tilde{\epsilon})$ and put $\tilde{\lambda} := (2\epsilon - \tilde{\epsilon})\lambda/(2\epsilon)$. Notice that $\tilde{\lambda} < \lambda$. By Ekeland's variational principle (Theorem 8.2.4) there exists $z \in E$ satisfying $\|z - \bar{x}\| < \tilde{\lambda}$, $f(z) < \inf_E f + (\epsilon - \tilde{\epsilon})$, and

$$f(z) < f(x) + \|x - z\|(\epsilon - \tilde{\epsilon})/\tilde{\lambda} \quad \forall x \in E \setminus \{z\}. \tag{13.25}$$

We equip $E \times \mathbb{R}$ with the norm $(x, \tau) := \|x\| + |\tau|$ and $E^* \times \mathbb{R}$ with the corresponding dual norm $(x^*, \tau^*)\| = \max\{\|x^*\|, |\tau^*|\}$. Put $g(x) := \|x - z\|(\epsilon - \tilde{\epsilon})/\tilde{\lambda}$ and define

$$A_1 := \operatorname{epi} f \quad \text{and} \quad A_2 := \{(x, \tau) \in E \times \mathbb{R} \mid \tau \leq f(z) - g(x)\}.$$

Then A_1 and A_2 are closed in $E \times \mathbb{R}$, and it follows from (13.25) that $(z, f(z))$ is a local extremal point of the system (A_1, A_2). Hence (b) ensures that the approximate extremal principle holds in E and so in $E \times \mathbb{R}$. Consequently, for any $\eta > 0$ we find $(x_i, \tau_i) \in A_i$ and $(x_i^*, \tau_i^*) \in N_F(A_i, (x_i, \tau_i))$, where $i = 1, 2$, such that

$$\|x_i - z\| + |\tau_i - f(z)| < \eta, \tag{13.26}$$

$$1/2 - \eta < \max\{\|x_i^*\|, |\tau_i^*|\} < 1/2 + \eta, \tag{13.27}$$

$$\max\{\|x_1^* + x_2^*\|, |\tau_1^* + \tau_2^*|\} < \eta. \tag{13.28}$$

The definition of the Fréchet normal cone entails that for $i = 1, 2$ we have

$$\limsup_{(x,\tau) \to A_i (x_i, \tau_i)} \frac{\langle x_i^*, x - x_i \rangle + \tau_i^*(\tau - \tau_i)}{\|x - x_i\| + |\tau - \tau_i|} \leq 0. \tag{13.29}$$

It follows from the definition of A_2 and (13.26) that $\tau_2 = f(z) - g(x)$ whenever η is sufficiently small. Moreover, by (13.27) we see that $(x_2^*, \tau_2^*) \neq o$. Thus, (13.29) shows that $\tau_2^* > 0$. Hence we obtain

$$x_2^*/\tau_2^* \in \partial_F g(x_2) \quad \text{and} \quad \|x_2^*\|/\tau_2^* \leq (\epsilon - \tilde{\epsilon})/\tilde{\lambda}.$$

The latter inequality together with (13.27) yields

$$\tau_2^* \geq \min\left\{ \frac{(1 - 2\eta)\tilde{\lambda}}{2(\epsilon - \tilde{\epsilon})}, \frac{1}{2} - \eta \right\}. \tag{13.30}$$

From this estimate and (13.28) we must conclude that $\tau_1^* < 0$ whenever η is sufficiently small. Now it follows that $\tau_1 = f(x_1)$. In fact, by (13.29) with $i = 1$ the inequality $\tau_1 > f(x_1)$ would imply that $\tau_1^* = 0$. Thus we have shown that $-x_1^*/\tau_1^* \in \partial_F f(x_1)$. The estimate (13.30) also yields $\eta/\tau_2^* \to 0$ as $\eta \downarrow 0$. Therefore, for η sufficiently small we have

$$\frac{\|x_1^*\|}{|\tau_1^*|} < \frac{\|x_2^*\| + \eta}{\tau_2^* - \eta} = \left(\frac{\|x_2^*\|}{\tau_2^*} + \frac{\eta}{\tau_2^*}\right) \bigg/ \left(1 - \frac{\eta}{\tau_2^*}\right) < \frac{\epsilon}{\lambda}.$$

Furthermore, the definition of $\tilde{\lambda}$ and (13.26) give

$$\|x_1 - \bar{x}\| < \tilde{\lambda} + \eta \quad \text{and} \quad f(x_1) = \tau_1 < \inf_E f + \epsilon - \tilde{\epsilon} + \eta.$$

Defining $z := x_1$ and $z^* := -x_1^*/\tau_1^*$, the conclusion of (c) follows.

(c) $\Longrightarrow$ (a): Let $\varphi : E \to \mathbb{R}$ be convex and continuous. Then for any $x \in E$ the set $\partial_F \varphi(x)$ coincides with $\partial \varphi(x)$ and is nonempty (Proposition 4.1.6). We show that there exists a dense subset D of E such that $\partial_F(-\varphi)(x)$ is nonempty for any $x \in D$. It is evident that this implies F-differentiability of φ on D. Choose $\bar{x} \in E$ and $\epsilon > 0$. Since $\psi := -\varphi$ is continuous, there exists $\eta \in (0, \epsilon)$ such that $\psi(x) > \psi(\bar{x}) - \epsilon$ for every $x \in B(\bar{x}, \eta)$. The functional $f := \psi + \delta_{B(\bar{x}, \eta)}$ is l.s.c. on E. By applying (c) to f we obtain $z \in E$ satisfying $\|z - \bar{x}\| < \eta$ and $\partial_F f(z) \neq \emptyset$. It follows that $\partial_F(-\varphi)(z) \neq \emptyset$. Therefore, φ is F-differentiable on a dense subset of E.   □

By Theorem 13.3.12 the approximate extremal principle holds in an Asplund space for arbitrary closed sets. Next we shall show that the exact extremal principle holds in an Asplund space for closed sets with an additional property to be defined now.

Definition 13.3.13 The set $A \subseteq E$ is said to be *sequentially normally compact (SNC)* at $\bar{x} \in A$ if for any sequence $((\epsilon_k, x_k, x_k^*))$ in $(0, +\infty) \times A \times E^*$ one has

$$\left[\epsilon_k \downarrow 0, \quad x_k \to \bar{x}, \quad x_k^* \in \widehat{N}_{\epsilon_k}(A, x_k), \quad x_k^* \xrightarrow{w^*} o \quad \text{as} \quad k \to \infty\right]$$
$$\Longrightarrow \quad x_k^* \to o \quad \text{as} \quad k \to \infty.$$

If E is an Asplund space, then due to Theorem 13.1.23 we may choose $\epsilon_k = 0$ for any $k \in \mathbb{N}$ in Definition 13.3.13. Clearly in a finite-dimensional Banach space any nonempty subset is SNC at each of its points. In Sect. 13.4 we shall describe classes of SNC sets in infinite-dimensional Banach spaces.

Theorem 13.3.14 *Let E be an Asplund space and $(A_1, \ldots, A_n, \bar{x})$ be an extremal system in E. Assume that the sets $A_1, \ldots, A_n$ are closed and all but one of them are SNC at $\bar{x}$. Then the exact extremal principle holds for $(A_1, \ldots, A_n, \bar{x})$.*

Proof. Let (ϵ_k) be a sequence of positive numbers such that $\epsilon_k \downarrow 0$ as $k \to \infty$. Since the approximate extremal principle holds for $(A_1, \ldots, A_n, \bar{x})$, for any $k \in \mathbb{N}$ and $i = 1, \ldots, n$ there exist $x_{ik} \in A_i \cap \mathrm{B}(\bar{x}, \epsilon_k)$ and $x_{ik}^* \in \mathrm{N}_F(A_i, x_{ik}) + \epsilon_k \mathrm{B}_{E^*}$ satisfying

$$x_{1k}^* + \cdots + x_{nk}^* = o \quad \text{and} \quad \|x_{1k}^*\| + \cdots + \|x_{nk}^*\| = 1. \tag{13.31}$$

We have $x_{ik} \to \bar{x}$ as $k \to \infty$ for $i = 1, \ldots, n$. Since for $i = 1, \ldots, n$ the sequence $(x_{ik}^*)_{k \in \mathbb{N}}$ is bounded in the dual of the Asplund space E, Theorem 4.3.21 implies that a subsequence, again denoted $(x_{ik}^*)_{k \in \mathbb{N}}$, is weak* convergent to some $x_i^* \in E^*$. Since $x_{ik}^* \in \widehat{\mathrm{N}}_{\epsilon_k}(A_i, x_{ik})$ for any k (cf. Remark 13.1.2(a)), the definition of the M-normal cone shows that $x_i^* \in \mathrm{N}_M(A_i, \bar{x})$. It is evident that $x_1^* + \cdots + x_n^* = o$. It remains to show that the x_i^* are not simultaneously zero. By hypothesis we may assume that $A_1, \ldots, A_{n-1}$ are SNC. We now suppose we had $x_i^* = o$ for $i = 1, \ldots, n-1$. Since

$$\|x_{nk}^*\| \leq \|x_{1k}^*\| + \cdots + \|x_{n-1k}^*\| \quad \forall k \in \mathbb{N},$$

we must conclude, letting $k \to \infty$, that $x_n^* = o$. But this contradicts the second equation in (13.31). $\square$

Applying Theorem 13.3.14 to the system $(A, \{\bar{x}\}, \bar{x})$, we obtain the following result.

Corollary 13.3.15 *Let E be an Asplund space and A be a proper closed subset of E. Then $\mathrm{N}_M(A, \bar{x}) \neq \{o\}$ at every boundary point $\bar{x}$ of A where A is SNC.*

13.4 Sequentially Normally Compact Sets

We will now describe classes of SNC sets. For this we introduce a concept that generalizes that of an epi-Lipschitzian set (cf. Exercise 13.13.13).

Definition 13.4.1 The set $A \subseteq E$ is said to be *compactly epi-Lipschitzian* at $\bar{x} \in A$ if there exist a compact set $C \subseteq E$ and numbers $\eta > 0$ and $\hat{\tau} > 0$ such that

$$A \cap \mathrm{B}(\bar{x}, \eta) + \tau \mathrm{B}(o, \eta) \subseteq A + \tau C \quad \forall \tau \in (0, \hat{\tau}). \tag{13.32}$$

If E is finite dimensional, then choosing $\eta = \hat{\tau} := 1$ and $C := \mathrm{B}(o, 1)$ we see that any subset A of E is compactly epi-Lipschitzian at each point $\bar{x} \in A$ but A need not be epi-Lipschitzian (see Remark 11.1.11).

Proposition 13.4.2 *If $A \subseteq E$ is compactly epi-Lipschitzian at $\bar{x} \in A$, then A is SNC at $\bar{x}$.*

Proof. Let C, η, and $\hat{\tau}$ be as in Definition 13.4.1. We will show that there exists $\alpha > 0$ such that for any $\epsilon > 0$ we have

$$\widehat{N}_\epsilon(A, x) \subseteq \left\{ x^* \in E^* \mid \eta\|x^*\| \leq \epsilon(\alpha + \eta) + \max_{z \in C}\langle x^*, z\rangle \right\} \quad \forall x \in A \cap B(\bar{x}, \eta).$$
(13.33)

Let $x \in A \cap B(\bar{x}, \eta)$ and a sequence (τ_k) in $(0, \hat{\tau})$ with $\tau_k \downarrow 0$ be given. In view of (13.32), for any $k \in \mathbb{N}$ and any $y \in B(o, 1)$ there exists $z_k \in C$ such that $x + \tau_k(\eta y - z_k) \in A$. Since C is compact, a subsequence of (z_k) is convergent to some point $\hat{z} \in C$ as $k \to \infty$. By the definition of $\widehat{N}_\epsilon(A, x)$ it follows that

$$\langle x^*, \eta y - \hat{z}\rangle - \epsilon\|\eta y - \hat{z}\| \leq 0 \quad \forall x^* \in \widehat{N}_\epsilon(A, x).$$

Since this is true for any $y \in B(o, 1)$, it follows that (13.33) holds with $\alpha := \max_{z \in C}\|z\|$.

Now let sequences $\epsilon_k \downarrow 0$, $x_k \to_A \bar{x}$, and $x_k^* \xrightarrow{w^*} o$ be given such that $x_k^* \in \widehat{N}_{\epsilon_k}(A, x_k)$ for all $k \in \mathbb{N}$. Since C is compact, we have $\langle x_k^*, z\rangle \to 0$ as $k \to \infty$ uniformly in $z \in C$. Therefore, (13.33) implies that $\|x_k^*\| \to 0$ as $k \to \infty$. Hence A is SNC. $\square$

Our aim now is to give a sufficient condition for the intersection of SNC sets to be SNC. This result will later be applied to derive optimality conditions. To prepare the result, we first establish an approximate intersection rule for F-normal cones that is also remarkable on its own.

Proposition 13.4.3 *Assume that E is a Fréchet smooth Banach space, A_1 and A_2 are closed subsets of E, and $\bar{x} \in A_1 \cap A_2$. Further let $x^* \in N_F(A_1 \cap A_2, \bar{x})$. Then for any $\epsilon > 0$ there exist $\lambda \geq 0$, $x_i \in A_i \cap B(\bar{x}, \epsilon)$, and $x_i^* \in N_F(A_i, x_i) + \epsilon B_{E^*}$, $i = 1, 2$, such that*

$$\lambda x^* = x_1^* + x_2^* \quad and \quad \max\{\lambda, \|x_1^*\|\} = 1.$$
(13.34)

Proof. Fix $\epsilon > 0$. The definition of the F-normal cone implies that there is a neighborhood U of $\bar{x}$ such that

$$\langle x^*, x - \bar{x}\rangle - \epsilon\|x - \bar{x}\| \leq 0 \quad \forall x \in A_1 \cap A_2 \cap U.$$
(13.35)

Equip $E \times \mathbb{R}$ with the norm $\|(x, \alpha)\| := \|x\| + |\alpha|$. Define closed subsets B_1, B_2 of $E \times \mathbb{R}$ by

$$B_1 := \{(x, \alpha) \mid x \in A_1, \alpha \geq 0\},$$
$$B_2 := \{(x, \alpha) \mid x \in A_2, \alpha \leq \langle x^*, x - \bar{x}\rangle - \epsilon\|x - \bar{x}\|\}.$$

From (13.35) we obtain

$$B_1 \cap (B_2 - (o, \rho)) \cap (U \times \mathbb{R}) = \emptyset \quad \forall \rho > 0.$$

Hence $(\bar{x}, 0) \in B_1 \cap B_2$ is a local extremal point of the system (B_1, B_2). By the approximate extremal principle of Theorem 13.3.10 we find $(x_i, \alpha_i) \in B_i$ and $(\tilde{x}_i^*, \lambda_i) \in N_F(B_i, (x_i, \alpha_i))$, $i = 1, 2$, such that

$$\max\{\|\tilde{x}_1^* + \tilde{x}_2^*\|, |\lambda_1 + \lambda_2|\} < \epsilon, \tag{13.36}$$

$$\frac{1}{2} - \epsilon < \max\{\|\tilde{x}_i^*\|, |\lambda_i|\} < \frac{1}{2} + \epsilon, \quad i = 1, 2, \tag{13.37}$$

$$\|x_i - \bar{x}\| + |\alpha_i| < \epsilon \quad i = 1, 2. \tag{13.38}$$

The definition of the F-normal cone implies that $\tilde{x}_1^* \in N_F(A_1, x_1)$, $\lambda_1 \leq 0$, and

$$\limsup_{(x,\alpha) \to_{B_2}(x_2,\alpha_2)} \frac{\langle \tilde{x}_2^*, x - x_2 \rangle + \lambda_2(\alpha - \alpha_2)}{\|x - x_2\| + |\alpha - \alpha_2|} \leq 0. \tag{13.39}$$

The definition of B_2 shows that $\lambda_2 \geq 0$ and

$$\alpha_2 \leq \langle x^*, x_2 - \bar{x} \rangle - \epsilon \|x_2 - \bar{x}\|. \tag{13.40}$$

Now we distinguish two cases:

Case 1. We assume that the inequality in (13.40) is strict. Then (13.39) implies that $\lambda_2 = 0$ and $\tilde{x}_2^* \in N_F(A_2, x_2)$. Define

$$x_1^* := \tilde{x}_1^* \quad \text{and} \quad x_2^* := -\tilde{x}_1^* = \tilde{x}_2^* - (\tilde{x}_1^* + \tilde{x}_2^*).$$

It then follows from (13.36) to (13.38) that the assertion (13.34) holds with $\lambda = 0$.

Case 2. Now we assume that (13.40) holds as an equation. Take any $(x, \alpha) \in B_2$ such that

$$\alpha = \langle x^*, x - \bar{x} \rangle - \epsilon \|x - \bar{x}\|, \quad x \in A_2 \setminus \{x_2\}. \tag{13.41}$$

Substituting this into (13.39), we conclude that for some neighborhood V of x_2 we have

$$\langle \tilde{x}_2^*, x - x_2 \rangle + \lambda_2(\alpha - \alpha_2) \leq \epsilon(\|x - x_2\| + |\alpha - \alpha_2|) \quad \forall x \in A_2 \cap V; \tag{13.42}$$

here, α satisfies (13.41) and so

$$\alpha - \alpha_2 = \langle x^*, x - x_2 \rangle + \epsilon(\|x_2 - \bar{x}\| - \|x - \bar{x}\|). \tag{13.43}$$

It follows that

$$|\alpha - \alpha_2| \leq (\|x^*\| + \epsilon)\|x - x_2\|. \tag{13.44}$$

Now we substitute the right-hand side of (13.43) for $\alpha - \alpha_2$ into the left-hand side of (13.42), and we apply the estimate (13.44) to the right-hand side of (13.42). In this way, (13.42) passes into

$$\langle \tilde{x}_2^* + \lambda_2 x^*, x - x_2 \rangle \leq \epsilon c \|x - x_2\| \quad \forall x \in A_2 \cap V,$$

where $c := 1 + \|x^*\| + \lambda_2 + \epsilon$. The definition of ϵ-normals shows that

$$\tilde{x}_2^* + \lambda_2 x^* \in \widehat{N}_{\epsilon c}(A_2, x_2).$$

Recall that λ_2 is nonnegative, and by (13.37) we have $\lambda_2 + \epsilon < 1$ if ϵ is sufficiently small. It follows that $1 + \|x^*\| < c < 2 + \|x^*\|$ and so the positive constant c may be chosen depending only on x^*. By virtue of Theorem 13.1.23(a) there exists $z_2 \in A_2 \cap B(x_2, \epsilon)$ such that

$$\tilde{x}_2^* + \lambda_2 x^* \in N_F(A_2, z_2) + 2\epsilon c\, B_{E^*}.$$

Put $\eta := \max\{\lambda_2, \|\tilde{x}_2^*\|\}$. For $\epsilon < \frac{1}{4}$ we obtain from (13.37) that $\frac{1}{4} < \eta < \frac{3}{4}$. Now define

$$\lambda := \lambda_2/\eta, \quad x_1^* := -\tilde{x}_2^*/\eta, \quad x_2^* := (\tilde{x}_2^* + \lambda_2 x^*)/\eta.$$

Then $\lambda \geq 0$, $\max\{\lambda, \|x_1^*\|\} = 1$, and $\lambda x^* = x_1^* + x_2^*$. By (13.36) we further have $\|\tilde{x}_1^* + \tilde{x}_2^*\|/\eta \leq 4\epsilon$. Hence we finally obtain

$$x_1^* = \tilde{x}_1^*/\eta - (\tilde{x}_1^* + \tilde{x}_2^*)/\eta \in N_F(A_1, x_1) + 4\epsilon\, B_{E^*},$$
$$x_2^* \in N_F(A_2, z_2) + 8\epsilon c\, B_{E^*}.$$

Recalling that $c > 0$ depends on x^* only completes the proof. $\square$

In Sect. 13.6 we will describe the M-normal cone of the inverse image of a multifunction. In this context we shall need SNC properties of sets in a product space which we now define.

Definition 13.4.4 Assume that E and F are Fréchet smooth Banach spaces, A is a nonempty closed subset of $E \times F$, and $(\bar{x}, \bar{y}) \in A$.

(a) The set A is said to be *partially sequentially normally compact (PSNC)* at $(\bar{x}, \bar{y})$ with respect to E if for any sequences $(x_k, y_k) \to_A (\bar{x}, \bar{y})$ and $(x_k^*, y_k^*) \in N_F(A, (x_k, y_k))$, one has

$$\left[x_k^* \xrightarrow{w^*} o \text{ and } \|y_k^*\| \to 0 \text{ as } k \to \infty \right] \implies \|x_k^*\| \to 0 \text{ as } k \to \infty.$$

(b) The set A is said to be *strongly partially sequentially normally compact (strongly PSNC)* at $(\bar{x}, \bar{y})$ with respect to E if for any sequences $(x_k, y_k) \to_A (\bar{x}, \bar{y})$ and $(x_k^*, y_k^*) \in N_F(A, (x_k, y_k))$, one has

$$\left[x_k^* \xrightarrow{w^*} o \text{ and } y_k^* \xrightarrow{w^*} o \text{ as } k \to \infty \right] \implies \|x_k^*\| \to 0 \text{ as } k \to \infty.$$

The (strong) PSNC property with respect to F is defined analogously.

It is clear that if the set A is SNC, then it is strongly PSNC and so PSNC. The strong PSNC property will not be needed until Sect. 13.6; for comparison with the PSNC property we already introduced it here. We also need the following concept.

Definition 13.4.5 Assume that E and F are Fréchet smooth Banach spaces, A_1 and A_2 are closed subsets of $E \times F$, and $(\bar{x}, \bar{y}) \in A_1 \cap A_2$. The system (A_1, A_2) is said to satisfy the *mixed qualification condition* at $(\bar{x}, \bar{y})$ with respect to F if for any sequences $(x_{ik}, y_{ik}) \to_{A_i} (\bar{x}, \bar{y})$ and $(x_{ik}^*, y_{ik}^*) \xrightarrow{w^*} (x_i^*, y_i^*)$ as $k \to \infty$ with $(x_{ik}^*, y_{ik}^*) \in N_F(A_i, (x_{ik}, y_{ik}))$, $i = 1, 2$, one has

$$\left[x_{1k}^* + x_{2k}^* \xrightarrow{w^*} o, \; \|y_{1k}^* + y_{2k}^*\| \to 0 \text{ as } k \to \infty \right] \implies (x_1^*, y_1^*) = (x_2^*, y_2^*) = o.$$

Remark 13.4.6 It is left as Exercise 13.13.14 to show that the condition

$$N_M(A_1, (\bar{x}, \bar{y})) \cap \left(-N_M(A_2, (\bar{x}, \bar{y})) \right) = \{(o, o)\}$$

is sufficient for the system (A_1, A_2) to satisfy the mixed qualification condition at $(\bar{x}, \bar{y})$ with respect to F.

Theorem 13.4.7 *Assume that E and F are Fréchet smooth Banach spaces, A_1 and A_2 are closed subsets of $E \times F$, and $(\bar{x}, \bar{y}) \in A_1 \cap A_2$, one of A_1, A_2 is SNC and the other is PSNC at $(\bar{x}, \bar{y})$ with respect to E. Assume further that (A_1, A_2) satisfies the mixed qualification condition at $(\bar{x}, \bar{y})$ with respect to F. Then $A_1 \cap A_2$ is PSNC at $(\bar{x}, \bar{y})$ with respect to E.*

Proof.

(I) Assume that A_1 is SNC at $(\bar{x}, \bar{y})$ and A_2 is PSNC at $(\bar{x}, \bar{y})$ with respect to E. Take any sequences $(x_k, y_k) \in A_1 \cap A_2$ and $(x_k^*, y_k^*) \in N_F(A_1 \cap A_2, (x_k, y_k))$ satisfying $(x_k, y_k) \to (\bar{x}, \bar{y})$, $x_k^* \xrightarrow{w^*} o$, and $\|y_k^*\| \to 0$ as $k \to \infty$. We have to show that $\|x_k^*\| \to 0$ as $k \to \infty$. Observe that it suffices to show that some subsequence of (x_k^*) is norm convergent to zero. In fact, if this is done and the entire sequence (x_k^*) were not norm convergent to zero, we could find a subsequence $(x_{k_\nu}^*)$ and some $\rho > 0$ such that $\|x_{k_\nu}^*\| \geq \rho$ for any $\nu \in \mathbb{N}$. Applying the above to the sequences (x_{k_ν}) and $(x_{k_\nu}^*)$, we would find a subsequence of the latter that is norm convergent to zero, which contradicts the construction of $(x_{k_\nu}^*)$.

(II) Take an arbitrary sequence $\epsilon_k \downarrow 0$ as $k \to \infty$. Applying Proposition 13.4.3 to x_k^* for every k, we obtain sequences $(x_{ik}, y_{ik}) \in A_i$, $(x_{ik}^*, y_{ik}^*) \in N_F(A_i, (x_{ik}, y_{ik}))$, where $i = 1, 2$, and $\lambda_k \geq 0$ such that

$$\|(x_{ik}, y_{ik}) - (x_k, y_k)\| \leq \epsilon_k \quad \forall k \in \mathbb{N}, \quad i = 1, 2, \tag{13.45}$$

$$\|(x_{1k}^*, y_{1k}^*) + (x_{2k}^*, y_{2k}^*) - \lambda_k(x_k^*, y_k^*)\| \leq 2\epsilon_k \quad \forall k \in \mathbb{N}, \tag{13.46}$$

$$1 - \epsilon_k \leq \max\{\lambda_k, \|x_{1k}^*\|, \|y_{1k}^*\|\} \leq 1 + \epsilon_k \quad \forall k \in \mathbb{N}. \tag{13.47}$$

Since the sequence (x_k^*, y_k^*) is weak* convergent, it is bounded. Hence by (13.46) and (13.47) the sequences (x_{ik}^*), (y_{ik}^*), $i = 1, 2$, and (λ_k) are also bounded. Since E and F are Asplund spaces, we conclude by Theorem 4.3.21 that a subsequence of (x_{ik}^*, y_{ik}^*) (which we do not relabel) is weak* convergent as $k \to \infty$ to some $(\tilde{x}_i^*, \tilde{y}_i^*)$, $i = 1, 2$, and that the corresponding subsequence of λ_k is convergent to some $\lambda \geq 0$. From $x_k^* \xrightarrow{w^*} o$, $\|y_k^*\| \to 0$, and (13.46) we obtain

$$x_{1k}^* + x_{2k}^* \xrightarrow{w^*} o \quad \text{and} \quad \|y_{1k}^* + y_{2k}^*\| \to 0 \quad \text{as } k \to \infty. \tag{13.48}$$

The mixed qualification condition implies that $\tilde{x}_i^* = \tilde{y}_i^* = o$ for $i = 1, 2$. Since A_1 is SNC, we conclude that $\|x_{1k}^*\| \to 0$ and $\|y_{1k}^*\| \to 0$. The latter together with (13.48) yields $\|y_{2k}^*\| \to 0$. Moreover, since A_2 is PSNC at $(\bar{x}, \bar{y})$ with respect to E, we obtain $\|x_{2k}^*\| \to 0$. From (13.47) we now see that $\lambda > 0$. Hence (13.46) implies that (x_k^*) is norm convergent to zero as $k \to \infty$. $\qquad \square$

Corollary 13.4.8 *Assume that $A_1, \ldots, A_n$ are closed subsets of the Fréchet smooth Banach space E, that $\bar{x} \in \cap_{i=1}^n A_i$, and that*

$$\left[x_i^* \in \mathrm{N}_M(A_i, \bar{x}), i = 1, \ldots, n, \ x_1^* + \cdots + x_n^* = o \right] \implies x_1^* = \cdots = x_n^* = o. \tag{13.49}$$

If each A_i is SNC at $\bar{x}$, then so is $A_1 \cap \cdots \cap A_n$.

Proof. For $n = 2$ this follows immediately from Theorem 13.4.7 with $F := \{o\}$. In this connection observe that by Remark 13.4.6 the condition (13.49) ensures that the mixed qualification condition is satisfied. For $n > 2$ the assertion follows by induction. $\qquad \square$

Now we turn to the SNC property of sets of the form $\Phi^{-1}(S)$. First we formulate some hypotheses:

(H1) E and F are Fréchet smooth Banach spaces, S is a closed subset of F.
(H2) $\Phi : E \rightrightarrows F$ is a multifunction, $\bar{x} \in \Phi^{-1}(S)$.
(H3) The multifunction $x \mapsto \Phi(x) \cap S$ is inner semicompact at $\bar{x}$.
(H4) graph Φ is closed and is SNC at $(\bar{x}, \bar{y})$ for every $\bar{y} \in \Phi(\bar{x}) \cap S$.
(H5) $\mathrm{N}_M(S, \bar{y}) \cap \ker \mathrm{D}_M^* \Phi(\bar{x}, \bar{y}) = \{o\}$ for every $\bar{y} \in \Phi(\bar{x}) \cap S$.

Theorem 13.4.9 *If the hypotheses (H1)–(H5) are satisfied, then $\Phi^{-1}(S)$ is SNC at $\bar{x}$.*

Proof.

(I) Consider sequences $x_k \to \bar{x}$ and $x_k^* \xrightarrow{w^*} o$, where $x_k^* \in \mathrm{N}_F(\Phi^{-1}(S), x_k)$ for any $k \in \mathbb{N}$. We have to show that $\|x_k^*\| \to 0$ as $k \to \infty$. By (H3) and (H4) we find a sequence $y_k \in \Phi(x_k) \cap S$ that contains a subsequence, again denoted (y_k), converging to some $\bar{y} \in \Phi(x_k) \cap S$. Defining

$$B_1 := \text{graph } \Phi \quad \text{and} \quad B_2 := E \times S,$$

we obtain that $(x_k^*, o) \in \mathrm{N}_F(B_1 \cap B_2, (x_k, y_k))$.

(II) We want to apply Theorem 13.4.7. Therefore, we check the SNC properties of the sets B_1 and B_2. First notice that

$$N_F(B_2, (x_k, y_k)) = N_F(E, x_k) \times N_F(S, y_k) = \{o\} \times N_F(S, y_k). \quad (13.50)$$

Hence B_2 is always PSNC at $(\bar{x}, \bar{y})$ with respect to E. Moreover, by (H4) the set B_1 is SNC at $(\bar{x}, \bar{y})$ or B_1 is PSNC at $(\bar{x}, \bar{y})$ with respect to F. Thus, the SNC properties required in Theorem 13.6.4 are satisfied.

(III) It is left as Exercise 13.13.15 to show that the hypothesis (H5) implies the mixed qualification condition with respect to F.

(IV) By Theorem 13.4.7 the set $B_1 \cap B_2$ is PSNC at $(\bar{x}, \bar{y})$ with respect to E. Hence it follows that (x_k^*) is norm convergent to zero.   □

13.5 Calculus for Mordukhovich Subdifferentials

We start with a special case of a sum rule for M-subdifferentials.

Proposition 13.5.1 *Assume that $f : E \to \overline{\mathbb{R}}$ is proper and l.s.c., $\bar{x} \in \operatorname{dom} f$, and $g : E \to \mathbb{R}$.*

(a) *If g is strictly F-differentiable at $\bar{x}$, then*

$$\partial_M (f + g)(\bar{x}) = \partial_M f(\bar{x}) + \{g'(\bar{x})\}. \quad (13.51)$$

(b) *If g is locally L-continuous around $\bar{x}$, then $\partial_M^\infty g(\bar{x}) = \{o\}$ and*

$$\partial_M^\infty (f + g)(\bar{x}) = \partial_M^\infty f(\bar{x}). \quad (13.52)$$

Proof.

(a) (I) First we show that

$$\partial_M (f + g)(\bar{x}) \subseteq \partial_M f(\bar{x}) + \{g'(\bar{x})\}. \quad (13.53)$$

Since g is strictly F-differentiable, for any sequence $\eta_i \downarrow 0$ as $i \to \infty$ there exists a sequence $\delta_i \downarrow 0$ such that

$$|g(z) - g(x) - \langle g'(\bar{x}), z - x \rangle| \leq \eta_i \|z - x\| \quad \forall x, z \in B(\bar{x}, \delta_i) \quad \forall i \in \mathbb{N}. \quad (13.54)$$

Now let $x^* \in \partial_M (f + g)(\bar{x})$ be given. By Theorem 13.1.17 there are sequences $x_k \to_{f+g} \bar{x}$, $x_k^* \xrightarrow{w^*} x^*$, and $\epsilon_k \downarrow 0$ as $k \to \infty$ satisfying

$$x_k^* \in \widehat{\partial}_{\epsilon_k} (f + g)(x_k) \quad \forall k \in \mathbb{N}. \quad (13.55)$$

Let (k_i) be a strictly increasing sequence of positive integers such that $\|x_{k_i} - \bar{x}\| \leq \delta_i / 2$ for any $i \in \mathbb{N}$. In view of (13.55) we find $\hat{\delta}_i \in (0, \delta_i / 2)$ such that for all $x \in B(x_{k_i}, \hat{\delta}_i)$ and any $i \in \mathbb{N}$ we have

$$f(x) - f(x_{k_i}) + g(x) - g(x_{k_i}) - \langle x_{k_i}^*, x - x_{k_i} \rangle \geq -2\epsilon_{k_i} \|x - x_{k_i}\|. \quad (13.56)$$

Since $x \in B(x_{k_i}, \hat{\delta}_i)$ implies $x \in B(\bar{x}, \delta_i)$, it follows from (13.54) and (13.56) that for all $x \in B(x_{k_i}, \hat{\delta}_i)$ and any $i \in \mathbb{N}$ we obtain

$$g(x) - g(x_{k_i}) - \langle x_{k_i}^* - g'(\bar{x}), x - x_{k_i} \rangle \geq -(2\epsilon_{k_i} + \eta_i) \| x - x_{k_i} \|.$$

We conclude that

$$x_{k_i}^* - g'(\bar{x}) \in \hat{\partial}_{\tilde{\epsilon}_i} f(x_{k_i}), \quad \text{where } \tilde{\epsilon}_i := 2\epsilon_{k_i} + \eta_i \quad \forall i \in \mathbb{N}. \quad (13.57)$$

Since $f(x_{k_i}) + g(x_{k_i}) \to f(\bar{x}) + g(\bar{x})$ and g is continuous at $\bar{x}$, we see that $f(x_{k_i}) \to f(\bar{x})$ as $i \to \infty$. Therefore, by Theorem 13.1.17 it follows from (13.57) that $x^* - g'(\bar{x}) \in \partial_M f(\bar{x})$. This verifies (13.53).

(II) The opposite inclusion to (13.53) is obtained from the latter inclusion in the following way:

$$\partial_M f(\bar{x}) = \partial_M[(f + g) + (-g)](\bar{x}) \subseteq \partial_M(f + g)(\bar{x}) - g'(\bar{x}).$$

(b) Now we prove

$$\partial_M^\infty (f + g)(\bar{x}) \subseteq \partial_M^\infty f(\bar{x}). \quad (13.58)$$

Once this is done, the opposite inclusion follows as in step (II) above. Finally, the relation $\partial_M^\infty g(\bar{x}) = \{o\}$ is the special case $f = o$ of (13.52) (and is already contained in Corollary 13.2.5).

Let $x^* \in \partial_M^\infty(f + g)(\bar{x})$ be given. By definition of the singular M-subdifferential there exist sequences $x_k \to \bar{x}$, $\alpha_k \to (f + g)(\bar{x})$, $\epsilon_k \downarrow 0$, $\eta_k \downarrow 0$, and $\gamma_k \to 0$ such that $\alpha_k \geq f(x_k) + g(x_k)$ and

$$\langle x_k^*, x - x_k \rangle + \gamma_k(\alpha - \alpha_k) \leq 2\epsilon_k(\| x - x_k \| + |\alpha - \alpha_k|) \quad (13.59)$$

for all $(x, \alpha) \in \text{epi}\,(f + g)$ satisfying $x \in B(x_k, \eta_k)$ and $|\alpha - \alpha_k| \leq \eta_k$ for any $k \in \mathbb{N}$. Let $\lambda > 0$ denote a Lipschitz constant of g around $\bar{x}$ and put

$$\tilde{\eta}_k := \frac{\eta_k}{2(\lambda + 1)}, \quad \tilde{\alpha}_k := \alpha_k - g(x_k).$$

It follows that $\tilde{\alpha}_k \geq f(x_k)$ for any k and $\tilde{\alpha}_k \to f(\bar{x})$ as $k \to \infty$. Notice that for any k and any $(x, \tilde{\alpha})$ satisfying

$$(x, \tilde{\alpha}) \in \text{epi}\,f, \quad x \in B(x_k, \tilde{\eta}_k) \quad \text{and} \quad |\tilde{\alpha} - \tilde{\alpha}_k| \leq \tilde{\eta}_k, \quad (13.60)$$

we have

$$(x, \tilde{\alpha} + g(x)) \in \text{epi}\,(f + g) \quad \text{and} \quad |(\tilde{\alpha} + g(x)) - \alpha_k| \leq \eta_k.$$

This and (13.59) give

$$\langle x_k^*, x - x_k \rangle + \alpha_k(\tilde{\alpha} - \tilde{\alpha}_k)$$
$$\leq \tilde{\epsilon}_k(\| x - x_k \| + |\tilde{\alpha} - \tilde{\alpha}_k|), \quad \text{where } \tilde{\epsilon}_k := 2\epsilon_k(\lambda + 1) + |\gamma_k|\lambda$$

for any $(x, \tilde{\alpha})$ satisfying (13.60). It follows that

$$(x_k^*, \gamma_k) \in \hat{N}_{\tilde{\epsilon}_k}(\text{epi}\,f, (x_k, \tilde{\alpha}_k)) \quad \forall k \in \mathbb{N}.$$

This implies $x^* \in \partial_M^\infty f(\bar{x})$ and completes the proof. $\qquad \square$

Next we derive a useful property of M-subdifferentials of locally L-continuous functionals.

Proposition 13.5.2 *Let E be an Asplund space and $f : E \to \overline{\mathbb{R}}$ be a proper l.s.c. functional. Then $\partial_M f(\bar{x}) \neq \emptyset$ for every $\bar{x} \in \mathrm{dom}\, f$ where f is locally L-continuous.*

Proof. The assumptions on f imply that the set $A := \mathrm{epi}\, f$ is closed and epi-Lipschitzian around $(\bar{x}, f(\bar{x}))$. By Proposition 13.4.2, $\mathrm{epi}\, f$ is also SNC. Hence Corollary 13.3.15 implies that $N_M\big(\mathrm{epi}\, f, (\bar{x}, f(\bar{x}))\big) \neq \{(o, 0)\}$. By Proposition 13.5.1 we know that $\partial_M^\infty f(\bar{x}) = \{o\}$. Hence Lemma 13.1.11 completes the proof. $\square$

Applying the approximate extremal principle, we now establish an exact sum rule for M-subdifferentials in a Fréchet smooth Banach space. In this context, the following property will compensate for the absence of finite dimensionality.

Definition 13.5.3 *The proper functional $f : E \to \overline{\mathbb{R}}$ is said to be sequentially normally epi-compact (SNEC) at $\bar{x} \in \mathrm{dom}\, f$ if $\mathrm{epi}\, f$ is SNC at $(\bar{x}, f(\bar{x}))$.*

Remark 13.5.4 The functional f is SNEC at $\bar{x}$ if $\mathrm{epi}\, f$ is compactly epi-Lipschitzian at $(\bar{x}, f(\bar{x}))$ (Proposition 13.4.2) and so, in particular, if $\mathrm{epi}\, f$ is epi-Lipschitzian at $(\bar{x}, f(\bar{x}))$ (Exercise 13.13.13) or E is finite dimensional.

We will also make use of the following qualification condition:

$$\big[x_i^* \in \partial_M^\infty f_i(\bar{x}), i = 1, \ldots, n \text{ and } x_1^* + \cdots + x_n^* = o\big] \implies x_1^* = \cdots = x_n^* = o. \tag{13.61}$$

Theorem 13.5.5 (Sum Rule for M-Subdifferentials) *Let E be a Fréchet smooth Banach space, let $f_1, \ldots, f_n : E \to \overline{\mathbb{R}}$ be l.s.c., and let $\bar{x} \in \cap_{i=1}^n \mathrm{dom}\, f_i$. Assume that all but one of f_i are SNEC at $\bar{x} \in \cap_{i=1}^n \mathrm{dom}\, f_i$ and that condition (13.61) is satisfied. Then*

$$\partial_M(f_1 + \cdots + f_n)(\bar{x}) \subseteq \partial_M f_1(\bar{x}) + \cdots + \partial_M f_n(\bar{x}). \tag{13.62}$$

If, in addition, $f_1, \ldots, f_n$ are all lower regular at $\bar{x}$, then so is $f_1 + \cdots + f_n$ and (13.62) holds with equality.

Proof. We verify the assertions for the case that $n = 2$, leaving the induction proof in the case $n > 2$ as Exercise 13.13.18. Thus let f_1 and f_2 be given and assume that f_1 is SNEC at $\bar{x}$. Take any $x^* \in \partial_M(f_1 + f_2)(\bar{x})$. By the definition of the M-subdifferential there exist sequences $x_k \to \bar{x}$ and $x_k^* \xrightarrow{w^*} x^*$ such that $f_i(x_k) \to f_i(\bar{x})$ as $k \to \infty$, where $i = 1, 2$, and $x_k^* \in \partial_F(f_1 + f_2)(x_k)$ for all $k \in \mathbb{N}$. Let (ϵ_k) be a sequence of positive real numbers such that $\epsilon_k \downarrow 0$ as

$k \to \infty$. The definition of the F-subdifferential implies that for any $k \in \mathbb{N}$ there exists a neighborhood U_k of x_k such that

$$(f_1 + f_2)(x) - (f_1 + f_2)(x_k) - \langle x_k^*, x - x_k \rangle + \epsilon_k \| x - x_k \| \geq 0 \quad \forall x \in U_k. \quad (13.63)$$

Since f_1 and f_2 are l.s.c., the sets

$$A_{1k} := \{(x, \mu) \in E \times \mathbb{R} \mid f_1(x) - f_1(x_k) \leq \mu\} \quad \text{and}$$

$$A_{2k} := \{(x, \mu) \in E \times \mathbb{R} \mid f_2(x) - f_2(x_k) - \langle x_k^*, x - x_k \rangle + \epsilon_k \| x - x_k \| \leq -\mu\}$$

are closed. Applying (13.63) we obtain for any k,

$$(x_k, 0) \in A_{1k} \cap A_{2k} \quad \text{and} \quad A_{1k} \cap (A_{2k} - (0, \alpha)) \cap (U_k \times \mathbb{R}) = \emptyset \quad \forall \alpha > 0.$$

Hence $(A_{1k}, A_{2k}, (x_k, 0))$ is an extremal system and so by Theorem 13.3.10 the extremal principle holds for this system. It follows that for $i = 1, 2$ and any $k \in \mathbb{N}$ there exist $(x_{i,k}, \mu_{i,k}) \in (\text{epi } f_i) \cap \mathrm{B}(x_k, f_i(x_k), \epsilon_k)$, $(\tilde{x}_k^*, \alpha_k) \in E^* \times \mathbb{R}$, and $(\tilde{y}_k^*, \beta_k) \in E^* \times \mathbb{R}$ satisfying

$$1/2 - \epsilon_k \leq \| \tilde{x}_k^* \| + |\alpha_k| \leq 1/2 + \epsilon_k, \quad (13.64)$$

$$1/2 - \epsilon_k \leq \| \tilde{y}_k^* \| + |\beta_k| \leq 1/2 + \epsilon_k, \quad (13.65)$$

$$\| (\tilde{x}_k^*, \alpha_k) + (\tilde{y}_k^*, \beta_k) \| \leq \epsilon_k, \quad (13.66)$$

$$(\tilde{x}_k^*, \alpha_k) \in \mathrm{N}_F \big(A_{1k}, (x_{1,k}, \mu_{1,k} - f_1(x_k)) \big), \quad (13.67)$$

$$(\tilde{y}_k^*, \beta_k) \in \mathrm{N}_F \big(A_{2k}, (x_{2,k}, \gamma_k) \big), \quad (13.68)$$

where $\gamma_k := -\mu_{2,k} + f_2(x_k) + \langle x_k^*, x_{2,k} - x_k \rangle - \epsilon_k \| x_{2,k} - x_k \|$. For $i = 1, 2$ the sequence $(x_{i,k}, \mu_{i,k})$ converges in epi f_i to $(\bar{x}, f_i(\bar{x}))$ as $k \to \infty$. Since the sequences $(\tilde{x}_k^*, \alpha_k)$ and $(\tilde{y}_k^*, \beta_k)$ are bounded and E is an Asplund space (Proposition 4.7.15), the sequences contain subsequences, again denoted $(\tilde{x}_k^*, \alpha_k)$ and $(\tilde{y}_k^*, \beta_k)$, that are weak* convergent to some $(\tilde{x}^*, \alpha)$ and $(\tilde{y}^*, \beta)$, respectively (Theorem 4.3.21). In view of (13.67) and the definition of the Mordukhovich normal cone we conclude that

$$(\tilde{x}^*, \alpha) \in \mathrm{N}_M \big(\text{epi } f_1, (\bar{x}, f_1(\bar{x})) \big). \quad (13.69)$$

Moreover, by (13.68) and the definition of the Fréchet normal cone we obtain

$$\limsup_{\substack{(x, \mu) \to (x_{2,k}, \mu_{2,k}) \\ \text{epi} f_2}} \frac{\langle \tilde{y}_k^*, x - x_{2,k} \rangle - \beta_k \big(\mu - \mu_{2,k} - \langle x_k^*, x - x_{2,k} \rangle + \epsilon_k \| x - x_{2,k} \| \big)}{\| x - x_{2,k} \| + | \mu - \mu_{2,k} | + |\langle x_k^*, x - x_{2,k} \rangle| + \epsilon_k \| x - x_{2,k} \|} \leq 0,$$

which implies

$$(\beta_k x_k^* + \tilde{y}_k^*, -\beta_k) \in \widehat{N}_{\eta_k}\big(\mathrm{epi}\, f_2,\, (x_{2,k}, \mu_{2,k})\big),$$

where $\eta_k := \epsilon_k(1 + \|x_k^*\| + \epsilon_k + |\beta_k|)$ for each k. Letting $k \to \infty$, we deduce that $(\beta x^* + \tilde{y}^*, -\beta) \in N_M\big(\mathrm{epi}\, f_2,\, (\bar{x}, f_2(\bar{x}))\big)$, where by (13.66) we have $\tilde{y}^* = -\tilde{x}^*$ and $\beta = -\alpha$. Thus,

$$(-\alpha x^* - \tilde{x}^*, \alpha) \in N_M\big(\mathrm{epi}\, f_2,\, (\bar{x}, f_2(\bar{x}))\big). \tag{13.70}$$

Now we show that $\alpha \neq 0$. Suppose we have $\alpha = 0$, then (13.69) and (13.70) give

$$(\tilde{x}^*, 0) \in N_M\big(\mathrm{epi}\, f_1,\, (\bar{x}, f_1(\bar{x}))\big) \ \text{and}\ (-\tilde{x}^*, 0) \in N_M\big(\mathrm{epi}\, f_2,\, (\bar{x}, f_2(\bar{x}))\big).$$

Hence the definition of the singular subdifferential and the condition (13.61) imply that $x^* = o$, which entails that $(\tilde{x}_k^*, \alpha_k) \xrightarrow{w^*} (o, 0)$ as $k \to \infty$. Notice that by (13.67) we have $(\tilde{x}_k^*, \alpha_k) \in N_F(\mathrm{epi}\, f_1,\, (x_{1,k}, \mu_{1,k}))$ and that f_1 is SNEC at $\bar{x}$. Therefore the sequence $((\tilde{x}_k^*, \alpha_k))$ is norm convergent to $(o, 0)$ as $k \to \infty$. But this contradicts the inequalities (13.64) and (13.65). Hence we must conclude that $\alpha \neq 0$ and so $\alpha < 0$ because $\alpha_k \leq 0$ for any k. Now (13.69) and (13.70) pass into

$$(\tilde{x}^*/|\alpha|, -1) \in N_M\big(\mathrm{epi}\, f_1,\, (\bar{x}, f_1(\bar{x}))\big) \ \text{and}\ (x^* - \tilde{x}^*/|\alpha|, -1) \in N_M\big(\mathrm{epi}\, f_2,\, (\bar{x}, f_2(\bar{x}))\big).$$

Defining $x_1^* := \tilde{x}^*/|\alpha|$, $x_2^* := x^* - x_1^*$, and recalling that $\partial_M f_i(\bar{x}) = N_M\big(\mathrm{epi}\, f_i,\, (\bar{x}, f_i(\bar{x}))\big)$, we obtain $x^* \in \partial_M f_1(\bar{x}) + \partial_M f_2(\bar{x})$, and the proof of (13.62) is complete.

To verify the statement on equality, notice that we always have

$$\partial_F f_1(\bar{x}) + \cdots + \partial_F f_n(\bar{x}) \subseteq \partial_F(f_1 + \cdots + f_n)(\bar{x}).$$

If each f_i is lower regular at $\bar{x}$, then it follows from the latter inclusion and (13.62) that $f_1 + \cdots + f_n$ is also lower regular at $\bar{x}$ and (13.62) holds with equality. $\square$

If E is finite dimensional, any function $f : E \to \overline{\mathbb{R}}$ is SNEC at any $\bar{x} \in \mathrm{dom}\, f$. Therefore the next result is an immediate consequence of Theorem 13.5.5.

Corollary 13.5.6 *Let E be a finite-dimensional Banach space, let $f_1, \ldots, f_n : E \to \overline{\mathbb{R}}$ be l.s.c., and let $\bar{x} \in \cap_{i=1}^n \mathrm{dom}\, f_i$. Assume that condition (13.61) is satisfied. Then (13.62) holds. If, in addition, $f_1, \ldots, f_n$ are all lower regular at $\bar{x}$, then so is $f_1 + \cdots + f_n$ and (13.62) holds with equality.*

By examples similar to Example 13.2.12 one can show that in an infinite-dimensional Banach space the SNEC property is not dispensable.

13.6 Calculus for Mordukhovich Normals

The sum rule for M-subdifferentials implies an intersection rule for M-normal cones that will be crucial for deriving multiplier rules. We need the following qualification condition:

$$\begin{bmatrix} x_i^* \in \mathrm{N}_M(A_i, \bar{x}), & i = 1, \ldots, n & \text{and} & x_1^* + \cdots + x_n^* = o \end{bmatrix}$$
$$\implies \quad x_1^* = \cdots = x_n^* = o. \tag{13.71}$$

Theorem 13.6.1 (Intersection Rule 1 for M-normal Cones) *Let E be a Fréchet smooth Banach space, let $A_1, \ldots, A_n$ be nonempty closed subsets of E, and let $\bar{x} \in \cap_{i=1}^n A_i$. Assume that all but one of A_i are SNC at $\bar{x}$ and that condition (13.71) is satisfied. Then*

$$\mathrm{N}_M(A_1 \cap \cdots \cap A_n, \bar{x}) \subseteq \mathrm{N}_M(A_1, \bar{x}) + \cdots + \mathrm{N}_M(A_n, \bar{x}). \tag{13.72}$$

If, in addition, $A_1, \ldots, A_n$ are all normally regular at $\bar{x}$, then so is $A_1 \cap \cdots \cap A_n$ and (13.72) holds with equality.

Proof. Apply Theorem 13.5.5 to $f_i := \delta_{A_i}$ and use the statement of Exercise 13.13.17. □

Next we will establish an intersection rule for M-normal cones in a product space $Z := E \times F$ that we equip with the norm $\|(x, y)\| := \|x\| + \|y\|$, $(x, y) \in E \times F$. Recall that if E and F are Fréchet smooth, so is $E \times F$. We will make use of the following concept.

Definition 13.6.2 Let A_1 and A_2 be closed subsets of the Fréchet smooth Banach space Z and let $\bar{z} \in A_1 \cap A_2$. The system (A_1, A_2) is said to satisfy the *limiting qualification condition* at $\bar{z}$ if for any sequences $z_{ik} \to_{A_i} \bar{z}$, $z_{ik}^* \in \mathrm{N}_F(A_i, z_{ik})$, and $z_{ik}^* \xrightarrow{\text{w}^*} z_i^*$ as $k \to \infty$, $i = 1, 2$, one has

$$\begin{bmatrix} \|z_{1k}^* + z_{2k}^*\| \to 0 \text{ as } k \to \infty \end{bmatrix} \implies z_1^* = z_2^* = o.$$

Remark 13.6.3 Observe that the limiting qualification condition is a special case of the mixed qualification condition with respect to E when $F := \{o\}$. Hence by Remark 13.4.6 the condition

$$\mathrm{N}_M(A_1, \bar{z}) \cap \left(-\mathrm{N}_M(A_2, \bar{z})\right) = \{o\} \tag{13.73}$$

is sufficient for the system (A_1, A_2) to satisfy the limiting qualification condition.

Theorem 13.6.4 (Intersection Rule 2 for M-normal Cones) *Let E and F be Fréchet smooth Banach spaces, A_1 and A_2 be closed subsets of $E \times F$, and $(\bar{x}, \bar{y}) \in A_1 \cap A_2$. Assume that (A_1, A_2) satisfies the limiting qualification condition, A_1 is PSNC at $(\bar{x}, \bar{y})$ with respect to E, and A_2 is strongly PSNC at $(\bar{x}, \bar{y})$ with respect to F. Then one has*

$$\mathrm{N}_M(A_1 \cap A_2, (\bar{x}, \bar{y})) \subseteq \mathrm{N}_M(A_1, (\bar{x}, \bar{y})) + \mathrm{N}_M(A_2, (\bar{x}, \bar{y})). \tag{13.74}$$

Proof.

(I) Take any $(x^*, y^*) \in N_M(A_1 \cap A_2, (\bar{x}, \bar{y}))$. By Theorem 13.1.23 there exist sequences $(x_k, y_k) \in A_1 \cap A_2$ and $(x_k^*, y_k^*) \in N_F(A_1 \cap A_2, (x_k, y_k))$ satisfying

$$(x_k, y_k) \to (\bar{x}, \bar{y}) \quad \text{and} \quad (x_k^*, y_k^*) \xrightarrow{\text{w}^*} (x^*, y^*).$$

Take any sequence $\epsilon_k \downarrow 0$ as $k \to \infty$. By applying the approximate intersection rule of Proposition 13.4.3 for every k, we find $(u_{ik}, v_{ik}) \in A_i$, $(u_{ik}^*, v_{ik}^*) \in N_F(A_i, (u_{ik}, v_{ik}))$, where $i = 1, 2$, and $\lambda_k \geq 0$ such that

$$\|(u_{ik}, v_{ik}) - (x_k, y_k)\| \leq \epsilon_k, \quad i = 1, 2, \tag{13.75}$$

$$\|(u_{1k}^*, v_{1k}^*) + (u_{2k}^*, v_{2k}^*) - \lambda_k(x_k^*, y_k^*)\| \leq 2\epsilon_k, \tag{13.76}$$

$$1 - \epsilon_k \leq \max\{\lambda_k, \|(u_{1k}^*, v_{1k}^*)\|\} \leq 1 + \epsilon_k. \tag{13.77}$$

The sequence (x_k, y_k) is weak* convergent and therefore bounded. By (13.76) and (13.77), for $i = 1, 2$ the sequence (u_{ik}^*, v_{ik}^*) is also bounded and so by Theorem 4.3.21 some subsequence (which we do not relabel) is weak* convergent to (u_i^*, v_i^*). We further have $\lambda_k \to \lambda \geq 0$. From (13.76) and (13.77) we obtain for $k \to \infty$ that $(u_i^*, v_i^*) \in N_M(A_i, (\bar{x}, \bar{y}))$, $i = 1, 2$, and $\lambda(x^*, y^*) = (u_1^*, v_1^*) + (u_2^*, v_2^*)$. It remains to show that $\lambda > 0$.

(II) Suppose that $\lambda = 0$. Then (13.76) shows that $\|(u_{1k}^*, v_{1k}^*) + (u_{1k}^*, v_{1k}^*)\| \to 0$ as $k \to \infty$. Since (A_1, A_2) satisfies the limiting qualification condition, it follows that $(u_i^*, v_i^*) = o$ for $i = 1, 2$. Hence

$$(u_{ik}^*, v_{ik}^*) \xrightarrow{\text{w}^*} (o, o) \quad \text{as } k \to \infty, \quad i = 1, 2. \tag{13.78}$$

The strong PSNC property of A_2 implies that $\|v_{2k}^*\| \to 0$ as $k \to \infty$ and (13.76) shows that $\|v_{1k}^*\| \to 0$, too. This, (13.78), and the PSNC property of A_1 now yield $\|u_{1k}^*\| \to 0$ and so $\|(u_{1k}^*, v_{1k}^*)\| \to 0$ as $k \to \infty$, which contradicts the estimate (13.77).

$\square$

Applying Theorem 13.6.4 with $F := \{o\}$ immediately gives the following result.

Corollary 13.6.5 *Let E be a Fréchet smooth Banach space, A_1 and A_2 be closed subsets of E, and $\bar{x} \in A_1 \cap A_2$. Assume that (A_1, A_2) satisfies the limiting qualification condition at $\bar{x}$ and that A_1 or A_2 is SNC at $\bar{x}$. Then (13.74) holds.*

Our aim now is to describe the M-normal cone to a set of the form $\Phi^{-1}(S)$, where $\Phi : E \rightrightarrows F$ is a multifunction and $S \subseteq F$. We make the following assumptions:

(A1) The multifunction $x \mapsto \Phi(x) \cap S$ is inner semicompact at $\bar{x}$.

(A2) For every $\bar{y} \in \Phi(\bar{x}) \cap S$, the set S is SNC at $\bar{y}$ or the set graph Φ is PSNC at $(\bar{x}, \bar{y})$ with respect to F.

(A3) For every $\bar{y} \in \Phi(\bar{x}) \cap S$ one has $N_M(S, \bar{y}) \cap \ker D_M^* \Phi(\bar{x}, \bar{y}) = \{o\}$.

Theorem 13.6.6 *Let E and F be Fréchet smooth Banach spaces, $\Phi : E \rightrightarrows F$ be a multifunction with closed graph, and S be a closed subset of F. Let the assumptions (A1)–(A3) be satisfied. Then one has*

$$N_M(\Phi^{-1}(S), \bar{x}) \subseteq \bigcup \left(D_M^* \Phi(\bar{x}, \bar{y})(y^*) \mid y^* \in N_M(S, \bar{y}), \bar{y} \in \Phi(\bar{x}) \cap S \right).$$
(13.79)

Proof.

(I) Take an arbitrary $x^* \in N_M(\Phi^{-1}(S), \bar{x})$. Then there are sequences $x_k \to \bar{x}$ and $x_k^* \in N_F(\Phi^{-1}(S), x_k)$ such that $x_k \to \bar{x}$ and $x_k^* \xrightarrow{w^*} x^*$. By (A1) we find a subsequence of $y_k \in \Phi(x_k) \cap S$ that converges to some $\bar{y} \in F$. The closedness assumptions ensure that $\bar{y} \in \Phi(\bar{x}) \cap S$. Define $B_1 := \operatorname{graph}\Phi$ and $B_2 := E \times S$, which are closed subsets of the Fréchet smooth Banach space $E \times F$. It is clear that $(x_k, y_k) \in B_1 \cap B_2$ for any k and it is easy to see that $(x_k^*, o) \in N_F(B_1 \cap B_2, (x_k, y_k))$ for any k. Hence $(x^*, o) \in N_M(B_1 \cap B_2, (\bar{x}, \bar{y}))$.

(II) As in the proof of Theorem 13.4.9 it follows that B_1 and B_2 have the SNC properties required in Theorem 13.6.4. We show that condition (13.73) is satisfied for the system (B_1, B_2), which by Remark 13.6.3 implies the limiting qualification condition for this system. Take any

$$(u^*, v^*) \in N_M(B_1, (\bar{x}, \bar{y})) \cap (-N_M(B_2, (\bar{x}, \bar{y}))).$$

By definition of the coderivative we immediately obtain $u^* \in D_M^* \Phi(\bar{x}, \bar{y})$ $(-v^*)$. In view of (13.50) we further have $(-u^*, -v^*) \in \{o\} \times N_M(S, \bar{y})$. It follows that $u^* = o$ and so

$$-v^* \in N_M(S, \bar{y}) \cap \ker D_M^* \Phi(\bar{x}, \bar{y}).$$

Assumption (A3) now implies that $v^* = o$. Hence (13.73) is satisfied. Therefore we may apply Theorem 13.6.4, which yields the existence of $(x_1^*, y_1^*) \in N_M(\operatorname{graph}\Phi, (\bar{x}, \bar{y}))$ and $y_2^* \in N_M(S, \bar{y})$ satisfying $(x^*, o) = (x_1^*, y_1^*) + (o, y_2^*)$ and so

$$(x^*, -y_2^*) = (x_1^*, y_1^*) \in N_M(\operatorname{graph}\Phi, (\bar{x}, \bar{y})).$$

Thus $x^* \in D_M^* \Phi(\bar{x}, \bar{y})(y_2^*)$, and it follows that x^* is an element of the right-hand side of (13.79). □

13.7 Optimality Conditions

Convention. Throughout this section, we assume that E and F are Fréchet smooth Banach spaces and $f : E \to \overline{\mathbb{R}}$ is a proper l.s.c. functional.

We consider the problem

$$\text{minimize } f(x) \text{ subject to } x \in A,$$

where A is a closed subset of E. From the discussion in Sect. 12.1 we know that if $\bar{x}$ is a local minimizer of f on A, then it follows that $o \in \partial_F(f + \delta_A)(\bar{x})$. This implies

$$o \in \partial_M(f + \delta_A)(\bar{x}). \tag{13.80}$$

We now formulate hypotheses (H1) and a qualification condition (Q1) ensuring that we may apply the exact sum rule of Theorem 13.5.5:

(H1) A is a closed subset of E, $\bar{x} \in A$, A is SNC at $\bar{x}$ or f is SNEC at $\bar{x}$.
(Q1) $\partial_M^\infty f(\bar{x}) \cap \left(-\mathrm{N}_M(A, \bar{x})\right) = \{o\}$.

Proposition 13.7.1 *Assume that* (H1) *and* (Q1) *are satisfied. If $\bar{x}$ is a local minimizer of f on A, then*

$$o \in \partial_M f(\bar{x}) + \mathrm{N}_M(A, \bar{x}). \tag{13.81}$$

Proof. Since epi $\delta_A = A \times [0, +\infty)$, the functional δ_A is SNEC if (and only if) the set A is SNC. Moreover, we have $\partial_M^\infty \delta_A(\bar{x}) = \mathrm{N}_M(A, \bar{x})$ (Remark 13.1.10). Therefore, (Q1) implies that the condition (13.61) is satisfied for f and δ_A. Applying Theorem 13.5.5 to (13.80) yields the assertion.   □

From Proposition 13.7.1 we can deduce further optimality conditions. First we formulate the assumptions:

(H2) For $i = 1, \ldots, r$ the set $A_i \subseteq E$ is closed and SNC at $\bar{x} \in \cap_{i=1}^r A_i$.
(Q2) $\left[x^* \in \partial_M^\infty f(\bar{x}), \quad x_i^* \in \mathrm{N}_M(A_i, \bar{x}), \quad x^* + x_1^* + \cdots + x_r^* = o\right]$
$$\implies x^* = x_1^* = \cdots = x_r^* = o.$$

Proposition 13.7.2 *Let the hypotheses* (H2) *and the qualification condition* (Q2) *be satisfied. If $\bar{x}$ is a local minimizer of f on $\cap_{i=1}^r A_i$, then one has*

$$o \in \partial_M f(\bar{x}) + \mathrm{N}_M(A_1, \bar{x}) + \cdots + \mathrm{N}_M(A_r, \bar{x}). \tag{13.82}$$

Proof. We verify the assertion for $r = 2$; it then follows for $r \geq 2$ by induction. We want to apply Proposition 13.7.1 to $A := A_1 \cap A_2$. From (Q2) with $x^* := o$ we obtain

$$\mathrm{N}_M(A_1, \bar{x}) \cap \left(-\mathrm{N}(A_2, \bar{x})\right) = \{o\}. \tag{13.83}$$

This and (H2) imply by Corollary 13.4.8 that $A_1 \cap A_2$ is SNC at $\bar{x}$. By Theorem 13.6.4 and Remark 13.6.3, the condition (13.83) also implies that

$$N_M(A_1 \cap A_2, \bar{x}) \subseteq N_M(A_1, \bar{x}) + N_M(A_2, \bar{x}). \tag{13.84}$$

Now we convince ourselves that condition (Q1) holds. Take any $x^* \in \partial_M^\infty f(\bar{x})$ such that $-x^* \in N_M(A_1 \cap A_2, \bar{x})$. By (13.84) there exist $x_i^* \in N_M(A_i, \bar{x})$, $i = 1, 2$, such that $-x^* = x_1^* + x_2^*$. Hence $x^* = o$ by (Q2). Refering to Proposition 13.7.1 and (13.84) completes the proof.   □

If f is locally L-continuous around $\bar{x}$, then by Proposition 13.5.1 the condition (Q2) reduces to

$$(Q2^*)\ \left[x_i^* \in N_M(A_i, \bar{x}),\ \ x_1^* + \cdots + x_r^* = o\right] \implies x_1^* = \cdots = x_r^* = o,$$

which is a pure *constraint* qualification. We show that (Q2*) is implied by a classical constraint qualification. Let the constraint sets be given as

$$A_i := \{x \in E \mid f_i(x) \le 0\}, \quad i = 1, \ldots, r,$$

where for $i = 1, \ldots, r$ the functional $f_i : E \to \mathbb{R}$ is strictly F-differentiable at $\bar{x} \in \cap_{i=1}^r A_i$. Assume the following variant of the Mangasarian–Fromowitz constraint qualification:

$$f_1'(\bar{x}), \ldots, f_r'(\bar{x}) \quad \text{are positively linearly independent.} \tag{13.85}$$

By Theorem 11.6.1 we have

$$N_M(A_i, \bar{x}) = N_C A_i, \bar{x}) = \{\lambda f_i'(x) \mid \lambda \ge 0\}.$$

Suppose now that $x_i^* \in N_M(A_i, \bar{x})$ for $i = 1, \ldots, r$ and $x_1^* + \cdots + x_r^* = o$. Then there exist $\lambda_i \ge 0$ such that $x_i^* = \lambda_i f_i'(\bar{x})$ for $i = 1, \ldots, r$. Condition (13.85) thus implies that for any i we have $\lambda_i = 0$ and so $x_i^* = o$.

Finally we consider the problem

$$\text{minimize } f(x) \text{ subject to } x \in \Phi^{-1}(S) \cap A.$$

In this connection we make the following assumptions:

(H3) $\Phi : E \rightrightarrows F$ is a multifunction with closed graph, $A \subseteq E$ and $S \subseteq F$ are closed.
    $x \mapsto \Phi(x) \cap S$ is inner semicompact at $\bar{x} \in \Phi^{-1}(S)$.
    A is SNC at $\bar{x}$ and graph Φ is SNC at $(\bar{x}, \bar{y})$ for every $\bar{y} \in \Phi(\bar{x}) \cap S$.
(Q3a) $N_M(S, \bar{y}) \cap \ker D_M^* \Phi(\bar{x}, \bar{y}) = \{o\}$ for every $\bar{y} \in \Phi(\bar{x}) \cap S$.
(Q3b) $\Big[x^* \in \partial_M^\infty f(\bar{x}),\ \ x_1^* \in \bigcup\big(D_M^*\Phi(\bar{x}, \bar{y})(y^*) \mid \bar{y} \in \Phi(\bar{x}) \cap S,\ y^* \in N_M(S, \bar{y})\big), x_2^* \in N_M(A, \bar{x}),\ \ x^* + x_1^* + x_2^* = o\Big]$
    $\implies x^* = x_1^* = x_2^* = o.$

Theorem 13.7.3 *Let the hypotheses* (H3) *and the qualification conditions* (Q3a) *and* (Q3b) *be satisfied. If $\bar{x}$ is a local minimizer of f on $\Phi^{-1}(S) \cap A$, then*

$$o \in \partial_M f(\bar{x}) + \bigcup\left(D_M^* \Phi(\bar{x}, \bar{y})(y^*) \mid \bar{y} \in \Phi(\bar{x}) \cap S,\ y^* \in N_M(S, \bar{y})\right) + N_M(A, \bar{x}).$$

Proof. Define $A_1 := \Phi^{-1}(S)$ and $A_2 := A$. By Theorem 13.6.6 we have

$$N_M(A_1, \bar{x}) \subseteq \bigcup\left(D_M^* \Phi(\bar{x}, \bar{y})(y^*) \mid \bar{y} \in \Phi(\bar{x}) \cap S,\ y^* \in N_M(S, \bar{y})\right). \quad (13.86)$$

Moreover, A is SNC at $\bar{x}$ by hypothesis and $\Phi^{-1}(S)$ is SNC at $(\bar{x}, \bar{y})$ due to Theorem 13.4.9. Finally, the qualification condition (Q3b) together with (13.86) ensures by Corollary 13.4.8 that $\Phi^{-1}(S) \cap A$ is SNC. Hence the assertion follows from Proposition 13.7.2. □

Notice that (Q3a) depends on the constraints only. If f is locally L-continuous around $\bar{x}$ and $\Phi : E \to F$ is strictly F-differentiable at $\bar{x}$, then by Propositions 13.2.3 and 13.5.1 the qualification condition (Q3b) reduces to

$$\left(-N_M(A, \bar{x})\right) \cap \Phi'(\bar{x})^*\left(N_M(S, \Phi(\bar{x}))\right) = \{o\}, \quad (13.87)$$

which is now also a constraint qualification; compare (Q2*). The optimality condition of Theorem 13.7.3 in this case reads

$$o \in \partial_M f(\bar{x}) + \Phi'(\bar{x})^*\left(N_M(S, \Phi(\bar{x}))\right) + N_M(A, \bar{x}).$$

In connection with the hypotheses (H3) recall that in finite-dimensional spaces any nonempty subset is an SNC set.

Theorem 13.7.3 is a very general result from which various specific optimality conditions can be derived. Assume, for instance, that the constraints are functional inequalities and equations of the following form (compare, for instance, problem (P5) in Sect. 12.5):

$$
\begin{aligned}
f_i(x) &\leq 0, & i &= 1, \ldots, r, \\
f_i(x) &= 0, & i &= r+1, \ldots, r+s, \\
x &\in A,
\end{aligned}
\quad (13.88)
$$

where $f_i : E \to \mathbb{R}$ for $i = 1, \ldots, r+s$. We define $\Phi : E \to \mathbb{R}^{r+s}$ by $\Phi := (f_1, \ldots, f_{r+s})$ and put

$$S := \left\{(\alpha_1, \ldots, \alpha_{r+s}) \in \mathbb{R}^{r+s} \mid \alpha_i \leq 0 \text{ for } i = 1, \ldots, r\right.$$
$$\left. \text{and } \alpha_i = 0 \text{ for } i = r+1, \ldots, r+s\right\}.$$

Then (13.88) is of the form $x \in \Phi^{-1}(S) \cap A$ and so Theorem 13.7.3 can be applied. For exploiting $D_M^* \Phi(\bar{x})$ we may write

$$\Phi(x) = (f_1(x), 0, \ldots, 0) + \cdots + (0, \ldots, 0, f_{r+s}(x)), \quad x \in E,$$

and apply one or the other coderivative sum rule.

So far we considered optimality conditions of the Karush–Kuhn–Tucker type. It is not difficult using, for instance, Theorem 13.7.3 to derive optimality conditions of the John type. We shall not pursue this way. Rather we will demonstrate how to obtain optimality conditions of the John type by a direct application of the exact extremal principle. We consider the problem

$$\text{minimize } f(x) \text{ subject to the constraints (13.88).}$$

In addition to the convention at the beginning of this section, we make the following hypotheses:

(H4) The functions $f_1, \ldots, f_{r+s}$ are continuous.
   All but one of the sets epi f, epi f_i $(i = 1, \ldots, r)$, graph f_i $(i = r + 1, \ldots, r + s)$, and A are SNC at $(\bar{x}, f(\bar{x}))$, $(\bar{x}, 0)$, and $\bar{x}$, respectively.

Theorem 13.7.4 *Let the assumptions* (H4) *be satisfied. Assume that $\bar{x}$ is a local minimizer of f subject to the constraints* (13.88).

(a) *There exist*

$$(x^*, -\lambda) \in N_M\big(\text{epi } f, ((\bar{x}, f(\bar{x})))\big), \quad y^* \in N_M(A, \bar{x}),$$
$$(x_i^*, -\lambda_i) \in N_M\big(\text{epi } f_i, ((\bar{x}, 0))\big), \quad i = 1, \ldots, r,$$
$$(x_i^*, -\lambda_i) \in N_M\big(\text{graph } f_i, ((\bar{x}, 0))\big), \quad i = r+1, \ldots, r+s$$

satisfying

$$x^* + x_1^* + \cdots + x_{r+s}^* + y^* = o,$$
$$\|(x^*, \lambda)\| + \|(x_1^*, \lambda_1)\| + \cdots + \|(x_{r+s}^*, \lambda_{r+s})\| + \|y^*\| = 1, \qquad (13.89)$$
$$\lambda_i f_i(\bar{x}) = 0, \quad i = 1, \ldots, r.$$

(b) *Assume, in addition, that the functions $f, f_1, \ldots, f_{r+s}$ are locally L-continuous around $\bar{x}$. Then there exist nonnegative real numbers λ, $\lambda_1, \ldots, \lambda_{r+s}$ such that $\lambda_i f_i(\bar{x}) = 0$ for $i = 1, \ldots, r$ and*

$$o \in \lambda \partial_M f(\bar{x}) + \sum_{i=r+1}^{r+s} \lambda_i \Big(\partial_M f_i(\bar{x}) \cup \partial_M(-f_i)(\bar{x})\Big) + N_M(A, \bar{x}).$$

Proof.

(a) Define the following subsets of $E \times \mathbb{R}^{r+s+1}$:

$$B := \{(x, \alpha, \alpha_1, \ldots, \alpha_{r+s}) \mid \alpha \geq f(x)\},$$
$$B_i := \{(x, \alpha, \alpha_1, \ldots, \alpha_{r+s}) \mid \alpha_i \geq f_i(x)\}, \quad i = 1, \ldots, r,$$
$$B_i := \{(x, \alpha, \alpha_1, \ldots, \alpha_{r+s}) \mid \alpha_i = f_i(x)\}, \quad i = r+1, \ldots, r+s,$$
$$B_{r+s+1} := A \times \{o\}.$$

Without loss of generality we may assume that $f(\bar{x}) = 0$. Then $(\bar{x}, o) \in E \times \mathbb{R}^{r+s+1}$ is a local extremal point of the system of closed sets B, $B_1, \ldots, B_{r+s+1}$. Therefore the hypotheses (H4) ensure that by Theorem 13.3.14 the exact extremal principle holds for the above system, which immediately yields the assertion except for the equations $\lambda_i f_i(\bar{x}) = 0$, $i = 1, \ldots, r$. Assume we have $f_i(\bar{x}) < 0$ for some $i \in \{1, \ldots, r\}$. Then by continuity, it follows that $f_i(x) < 0$ for all x in a neighborhood of $\bar{x}$. Hence $(\bar{x}, o)$ is an interior point of epi f_i. It follows that $\mathrm{N}_M(\mathrm{epi}\, f_i, (\bar{x}, o))) = \{o\}$ and so $\lambda_i = 0$.

(b) By the definition of the M-subdifferential and by Lemma 13.1.8 we have

$$(x^*, -\lambda) \in \mathrm{N}_M(\mathrm{epi}\, f, (\bar{x}, f(\bar{x}))) \iff x^* \in \lambda \partial f(\bar{x}) \text{ and } \lambda \geq 0.$$

Moreover, the definition of the M-coderivative and Proposition 13.2.6 imply that

$$(x^*, -\lambda) \in \mathrm{N}_M(\mathrm{graph}\, f, (\bar{x}, f(\bar{x}))) \iff x^* \in D_M^* f(\bar{x})(\lambda) = \partial_M(\lambda f)(\bar{x}).$$

Notice that $\partial_M(\lambda f)(\bar{x}) \subseteq |\lambda|(\partial_M f(\bar{x}) \cup \partial(-f)(\bar{x}))$ for any $\lambda \in \mathbb{R}$. The assertion now follows from (a). $\qquad\square$

13.8 The Mordukhovich Subdifferential of Marginal Functions

Convention. In this section, E and F denote Banach spaces.

In Sect. 9.7 we derived representations of the F-subdifferential of a marginal function f of the form $f(x) := \inf_{y \in F} \varphi(x, y)$, $x \in E$ (see Propositions 9.7.1 and 9.7.2). In this section we establish a representation of the (singular) M-subdifferential of the more general marginal function $f : E \to \overline{\mathbb{R}}$ defined by

$$f(x) := \inf\{\varphi(x, y) \mid y \in \Phi(x)\}, \tag{13.90}$$

where the function $\varphi : E \times F \to \overline{\mathbb{R}}$ and the multifunction $\Phi : E \rightrightarrows F$ are given. Marginal functions like f appear as value functions in parametric optimization problems of the form

$$\text{minimize } \varphi(x, y) \quad \text{subject to } y \in \Phi(x),$$

where x denotes a parameter. We will make use of the multifunction $\Theta : E \rightrightarrows F$ defined by

$$\Theta(x) := \{y \in \Phi(x) \mid \varphi(x, y) = f(x)\}. \tag{13.91}$$

In terms of parametric optimization, $\Theta(x)$ consists of all $y \in \Phi(x)$ at which the infimum of $\varphi(x, \cdot)$ is attained. Recall the notion of an inner semicompact multifunction. We also need the following auxiliary function $\vartheta : E \times F \to \overline{\mathbb{R}}$:

$$\vartheta(x, y) := \varphi(x, y) + \delta_{\mathrm{graph}\, \Phi}(x, y).$$

Theorem 13.8.1 *Assume that $\varphi : E \times F \to \overline{\mathbb{R}}$ is l.s.c., Φ is closed, and Θ is inner semicompact at $\bar{x} \in \operatorname{dom} f \cap \operatorname{Dom} \Theta$.*

(a) *One has*

$$\partial_M f(\bar{x}) \subseteq \left\{ x^* \in E^* \mid (x^*, o) \in \bigcup_{\bar{y} \in \Theta(\bar{x})} \partial_M \vartheta(\bar{x}, \bar{y}) \right\}, \tag{13.92}$$

$$\partial_M^\infty f(\bar{x}) \subseteq \left\{ x^* \in E^* \mid (x^*, o) \in \bigcup_{\bar{y} \in \Theta(\bar{x})} \partial_M^\infty \vartheta(\bar{x}, \bar{y}) \right\}. \tag{13.93}$$

(b) *If, in addition, φ is strictly F-differentiable at any $(\bar{x}, \bar{y})$, where $\bar{y} \in \Theta(\bar{x})$, then*

$$\partial_M f(\bar{x}) \subseteq \bigcup_{\bar{y} \in \Theta(\bar{x})} \left(\varphi_{11}(\bar{x}, \bar{y}) + D_M^* \Phi(\bar{x}, \bar{y})(\varphi_{12}(\bar{x}, \bar{y})) \right), \tag{13.94}$$

$$\partial_M^\infty f(\bar{x}) \subseteq \bigcup_{\bar{y} \in \Theta(\bar{x})} D_M^* \Phi(\bar{x}, \bar{y})(o). \tag{13.95}$$

Proof.

(a) (I) We verify (13.92). Take any $x^* \in \partial_M f(\bar{x})$. By Theorem 13.1.17 there exist sequences $x_k \to_f \bar{x}$, $x_k^* \xrightarrow{w^*} x^*$, and $\epsilon_k \downarrow 0$ as $k \to \infty$ satisfying $x_k^* \in \widehat{\partial}_{\epsilon_k} f(x_k)$ for any $k \in \mathbb{N}$. Hence there exists a sequence $\eta_k \downarrow 0$ such that

$$\langle x_k^*, x - x_k \rangle \leq f(x) - f(x_k) + 2\epsilon_k \|x - x_k\| \quad \forall x \in B(x_k, \eta_k).$$

Recalling the definition of f, Θ, and ϑ, we obtain for all $y_k \in \Theta(x_k)$, all $(x, y) \in B((x_k, y_k), \eta_k)$, and all $k \in \mathbb{N}$ the estimate

$$\langle (x_k^*, o), (x, y) - (x_k, y_k) \rangle \leq \vartheta(x, y) - \vartheta(x_k, y_k) + 2\epsilon_k (\|x - x_k\| + \|y - y_k\|).$$

From this we conclude that

$$(x_k^*, o) \in \widehat{\partial}_{2\epsilon_k} \vartheta(x_k, y_k) \quad \forall k \in \mathbb{N}. \tag{13.96}$$

Since Θ is inner semicompact at $\bar{x}$, we find a sequence $y_k \in \Theta(x_k)$ that contains a subsequence, again denoted (y_k), that converges to some $\bar{y} \in F$. Since $x_k \to \bar{x}$, $y_k \in \Phi(x_k)$ for any k, and Φ has a closed graph, it follows that $\bar{y} \in \Phi(\bar{x})$. The lower semicontinuity of φ implies

$$\varphi(\bar{x}, \bar{y}) \leq \liminf_{k \to \infty} \varphi(x_k, y_k) = \liminf_{k \to \infty} f(x_k) = f(\bar{x}),$$

which together with the definition of f gives $\varphi(\bar{x}, \bar{y}) = f(\bar{x})$ and so $\bar{y} \in \Theta(\bar{x})$. This, (13.96), and Theorem 13.1.17 show that inclusion (13.92) holds.

(II) The verification of (13.93) is left as Exercise 13.13.19.

(b) The representations (13.94) and (13.95) follow from (a) by applying the sum rule of Proposition 13.5.1 to the function ϑ and recalling Remark 13.1.10 and the definition of the M-coderivative.   □

Now let Φ be a single-valued mapping which we denote $T : E \to F$. Then (13.90) passes into

$$f(x) = \varphi(x, T(x)) =: (\varphi \circ T)(x), \quad x \in E. \tag{13.97}$$

If, in particular, φ does not explicitly depend on x, then $\varphi \circ T$ is the usual composition of φ and T. Recall that if φ is (strictly) F-differentiable at $(\bar{x}, \bar{y})$, then the partial derivative $\varphi_{|2}(\bar{x}, \bar{y})$ is an element of F^* and $\langle \varphi_{|2}(\bar{x}, \bar{y}), T \rangle : E \to \mathbb{R}$ denotes the scalarization of T.

Applying Theorem 13.8.1 we now establish another chain rule (cf. Theorem 9.2.9).

Theorem 13.8.2 (Chain Rule) *Let* $T : E \to F$ *be locally L-continuous around* $\bar{x} \in E$ *and* $\varphi : E \times F \to \mathbb{R}$ *be strictly F-differentiable at* $(\bar{x}, T(\bar{x}))$. *Then*

$$\partial_M (\varphi \circ T)(\bar{x}) = \varphi_{|1}(\bar{x}, \bar{y}) + \partial_M \langle \varphi_{|2}(\bar{x}, \bar{y}), T \rangle (\bar{x}), \quad \text{where } \bar{y} := T(\bar{x}). \tag{13.98}$$

Proof. Since φ is strictly F-differentiable at $(\bar{x}, \bar{y})$, for any sequence $\eta_i \downarrow 0$ there exists a sequence $\delta_i \downarrow 0$ such that

$$|\varphi(z, T(z)) - \varphi(x, T(x)) - \langle \varphi_{|1}(\bar{x}, \bar{y}), z - x \rangle - \langle \varphi_{|2}(\bar{x}, \bar{y}), T(z) - T(x) \rangle|$$
$$\le \eta_i (\|z - x\| + \|T(z) - T(x)\|) \quad \forall x, z \in B(\bar{x}, \delta_i) \quad \forall i \in \mathbb{N}. \tag{13.99}$$

Now let $x^* \in \partial_M (\varphi \circ T)(\bar{x})$ be given. By Theorem 13.1.17 there exist sequences $x_k \to \bar{x}$, $x_k^* \xrightarrow{w^*} x^*$, and $\epsilon_k \downarrow 0$ as $k \to \infty$ satisfying $(\varphi \circ T)(x_k) \to (\varphi \circ T)(\bar{x})$ and

$$x_k^* \in \widehat{\partial}_{\epsilon_k} (\varphi \circ T)(x_k) \quad \forall k \in \mathbb{N}. \tag{13.100}$$

Choose a strictly increasing sequence (k_i) of positive integers such that $\|x_{k_i} - \bar{x}\| \le \delta_i / 2$ for any $i \in \mathbb{N}$. In view of (13.100), for any i we can further choose $\tilde{\delta}_i \in (0, \delta_i / 2)$ such that for all $x \in B(x_{k_i}, \tilde{\delta}_i)$ and any $i \in \mathbb{N}$ we obtain

$$\varphi(x, T(x)) - \varphi(x_{k_i}, T(x_{k_i})) - \langle x_{k_i}^*, x - x_{k_i} \rangle \ge -2\epsilon_{k_i} \|x - x_{k_i}\|. \tag{13.101}$$

Let $\lambda > 0$ be a Lipschitz constant of φ in a neighborhood of $\bar{x}$ containing x_k for all sufficiently large $k \in \mathbb{N}$. Since $x \in B(x_{k_i}, \tilde{\delta}_i)$ implies $x \in B(\bar{x}, \delta_i)$, the estimates (13.99) and (13.101) give

$$\langle \varphi_{|2}(\bar{x}, \bar{y}), T(x) \rangle - \langle \varphi_{|2}(\bar{x}, \bar{y}), T(x_{k_i}) \rangle - \langle x_{k_i}^* - \varphi_{|1}(\bar{x}, \bar{y}), x - x_{k_i} \rangle$$
$$\ge -(2\epsilon_{k_i} + \eta_i(\lambda + 1)) \|x - x_{k_i}\| \quad \forall x \in B(x_{k_i}, \tilde{\delta}_i) \quad \forall i \in \mathbb{N}.$$

Hence the definition of the ϵ-subdifferential shows that, with $\tilde{\epsilon}_i := 2\epsilon_{k_i} + \eta_i(\lambda + 1)$, we have

$$x_{k_i}^* - \varphi_{|1}(\bar{x}, \bar{y}) \in \partial_{\tilde{\epsilon}_i}\langle \varphi_{|2}(\bar{x}, \bar{y}), T\rangle(x_{k_i}) \quad \forall i \in \mathbb{N}.$$

Letting $i \to \infty$ and recalling Theorem 13.1.17, we obtain

$$x^* - \varphi_{|1}(\bar{x}, \bar{y}) \in \partial_M\langle \varphi_{|2}(\bar{x}, \bar{y}), T\rangle(\bar{x}).$$

Thus we have verified the inclusion $\subseteq$ of (13.98). The verification of the opposite inclusion is left as Exercise 13.13.20. □

13.9 A Nonsmooth Implicit Function Theorem

Consider a multifunction $\Phi : E \times F \rightrightarrows G$ between Banach spaces E, F, and G. Let $(\bar{x}, \bar{y}) \in E \times F$ be such that $o \in \Phi(\bar{x}, \bar{y})$. As in the classical implicit function theorem (see Theorem 3.7.2), we want to *locally solve the generalized equation* $o \in \Phi(x, y)$ *for* y, i.e., we seek conditions ensuring that, for y near $\bar{y}$, the set $\{x \in E \mid o \in \Phi(x, y)\}$ is nonempty. Let

$$f(x, y) := \mathrm{d}(\Phi(x, y), o).$$

Our approach is based on the observation that $o \in \Phi(x, y)$ if and only if $f(x, y) = 0$ or equivalently, $f(x, y) \leq 0$. Consequently, we can treat the set-valued problem by a single-valued one. Therefore we start with an implicit function theorem associated with the inequality $f(x, y) \leq 0$, where for the time being the functional f is arbitrary. We set

$$\Psi(y) := \{x \in E \mid f(x, y) \leq 0\} \quad \text{and} \quad f_+(x, y) := \max\{0, f(x, y)\}.$$

Furthermore, $\partial_{F,1}f(\bar{x}, \bar{y})$ denotes the F-subdifferential of $x \mapsto f(x, \bar{y})$ at $(\bar{x}, \bar{y})$. We consider the following assumptions:

(A1) E and F are Fréchet smooth Banach spaces, U is a nonempty open subset of $E \times F$, and $(\bar{x}, \bar{y}) \in U$.

(A2) $f : E \times F \to \overline{\mathbb{R}}$ is proper and such that $f(\bar{x}, \bar{y}) \leq 0$.

(A3) The functional $y \mapsto f(\bar{x}, y)$ is u.s.c. at $\bar{y}$.

(A4) For any y near $\bar{y}$ the functional $x \mapsto f(x, y)$ is l.s.c.

(A5) There exists $\sigma > 0$ such that for any $(x, y) \in U$ with $f(x, y) > 0$, $x^* \in \partial_{F,1}f(x, y)$ implies $\|x^*\| \geq \sigma$.

Assumption (A5) is a nonsmooth substitute for the bijectivity requirement on $f_{11}(x, y)$ in the classical implicit function theorem (see Theorem 3.7.2 with x and y interchanged). This classical theorem yields, among others, a representation of the F-derivative of the implicit function in terms of the partial F-derivatives of the given function. In the following remarkable result, the coderivative of the multifunction Ψ is represented in terms of F-subderivatives of f_+.

Theorem 13.9.1 (Nonsmooth Implicit Function Theorem) *Suppose that the assumptions (A1)–(A5) are satisfied. Then there exist open sets $V \subseteq E$ and $W \subseteq F$ such that $\bar{x} \in V$ and $\bar{y} \in W$ and that the following holds:*

(a) *For any $y \in W$ the set $V \cap \Psi(y)$ is nonempty.*
(b) *For any $(x, y) \in V \times W$ one has*

$$d(x, \Psi(y)) \le \frac{f_+(x, y)}{\sigma}.$$

(c) *For any $(x, y) \in V \times W$ such that $x \in \Psi(y)$ and any $x^* \in E^*$ one has*

$$D_F^* \Psi(y, x)(x^*) = \left\{ y^* \in F^* \,\middle|\, (-x^*, y^*) \in \bigcup_{\lambda > 0} \lambda \partial_F f_+(x, y) \right\}.$$

Proof. Let $\rho' > 0$ be such that $B(\bar{x}, \rho') \times B(\bar{y}, \rho') \subseteq U$ and set $\rho := \rho'/3$. Since $f(\bar{x}, \bar{y}) \le 0$, it follows from (A3) that there exists an open neighborhood W of $\bar{y}$ with $W \subseteq B(\bar{y}, \rho)$ such that $f(\bar{x}, y) < \rho\sigma$ for any $y \in W$. We show that W and $V := \mathring{B}(\bar{x}, \rho)$ have the required properties.

Ad (a). Let $y \in W$ be given. Suppose we had $V \cap \Psi(y) = \emptyset$. Then for any $\tau \in (0, \rho)$ and any $x \in B(\bar{x}, \tau)$ it would follow that $f(x, y) > 0$. Choosing τ close enough to ρ, we may assume that $f(\bar{x}, y) < \tau\sigma$. The decrease principle (Theorem 9.6.3) then implies

$$\inf\{f(x, y) \mid x \in B(\bar{x}, \rho)\} \le f(\bar{x}, y) - \tau\sigma < 0,$$

which is a contradiction to $\inf\{f(x, y) \mid x \in B(\bar{x}, \rho)\} \ge 0$. Therefore $V \cap \Psi(y) \ne \emptyset$.

Ad (b). Let $(x, y) \in V \times W$ be given. First we assume that $B(x, f_+(x, y)/\sigma)$ is not a subset of $\mathring{B}(\bar{x}, \rho')$. Then $\|x - \bar{x}\| + f_+(x, y)/\sigma \ge \rho'$ and so

$$f_+(x, y)/\sigma \ge \rho' - \rho = 2\rho > d(x, \Psi(y)),$$

where the latter inequality follows from (a). Now assume that $B(x, f_+(x, y)/\sigma)$ is a subset of $\mathring{B}(\bar{x}, \rho')$. Let $\tau > f_+(x, y)/\sigma$ be such that $B(x, \tau) \subseteq \mathring{B}(\bar{x}, \rho')$. Since $f(x, y) < \tau\sigma$, we can conclude arguing similarly as in the proof of (a) that there exists $\hat{x} \in B(x, \tau)$ such that $f(\hat{x}, y) \le 0$. Hence $d(x, \Psi(y)) < \tau$. Letting $\tau \downarrow f_+(x, y)/\sigma$ we again obtain (b).

Ad (c). Take any $(x, y) \in V \times W$ such that $x \in \Psi(y)$ and any $x^* \in E^*$. We will show that

$$D_F^* \Psi(y, x)(x^*) \subseteq \left\{ y^* \in Y^* \,\middle|\, (-x^*, y^*) \in \bigcup_{\lambda > 0} \lambda \partial_F f_+(x, y) \right\}. \qquad (13.102)$$

Thus let $y^* \in D_F^* \Psi(y, x)(x^*)$ be given. Then

$$(y^*, -x^*) \in N_F(\text{graph}\,\Psi, (y, x)) = \bigcup_{\lambda > 0} \lambda \partial_F d\,(\text{graph}\,\Psi, (y, x));$$

in this connection we write $\partial_F \, \mathrm{d}(\mathrm{graph}\, \Psi, (y, x))$ for the F-subdifferential of the functional $(v, u) \mapsto \mathrm{d}((v, u), \mathrm{graph}\, \Psi)$ at (y, x). It follows that there exist $\lambda > 0$ and a C^1 functional $g : F \times E \to \mathbb{R}$ with $g'(y, x) = (y^*, -x^*)$ such that for any $(v, u) \in F \times E$ we have (noticing that $\mathrm{d}((y, x), \mathrm{graph}\, \Psi) = 0$),

$$g(v, u) \leq \lambda \, \mathrm{d}((v, u), \mathrm{graph}\, \Psi) + g(y, x)$$
$$\leq \lambda \, \mathrm{d}(u, \Psi(v)) + g(y, x) \leq (\lambda/\sigma) f_+(u, v) + g(y, x).$$

Since $f_+(x, y) = 0$, it thus follows that the functional $(v, u) \mapsto (\lambda/\sigma) f_+(u, v) - g(v, u)$ attains a minimum at (y, x). Hence $(-x^*, y^*) \in (\lambda/\sigma) \partial_F f_+(x, y)$ (observe the order of variables). This verifies (13.102). Since the reverse inclusion follows immediately from $\delta_{\mathrm{graph}\, \Psi} \geq \lambda f_+$ for any $\lambda > 0$, the proof is complete. $\qquad \square$

We apply Theorem 13.9.1 to the special case

$$f(x, y) := \|T(x, y)\|, \tag{13.103}$$

where $T : E \times F \to G$ is a continuously differentiable mapping. We consider the following conditions:

(C1) E, F, and G are Fréchet smooth Banach spaces, U is a nonempty open subset of $E \times F$, and $(\bar{x}, \bar{y}) \in U$.
(C2) $T : E \times F \to G$ is a C^1 mapping such that $T(\bar{x}, \bar{y}) = o$.
(C3) There exists $\sigma > 0$ such that $\sigma \, \mathrm{B}_G \subseteq T_{|1}(x, y)(\mathrm{B}_E)$ for any $(x, y) \in U$.

Proposition 13.9.2 *Let the conditions* (C1)–(C3) *be satisfied and let*

$$\Psi(y) := \{x \in E \mid T(x, y) = o\}.$$

Then there exist open sets $V \subseteq E$ *and* $W \subseteq F$ *such that* $\bar{x} \in V$ *and* $\bar{y} \in W$ *and that the following holds:*

(a) *For any* $y \in W$ *the set* $V \cap \Psi(y)$ *is nonempty.*
(b) *For any* $(x, y) \in V \times W$ *one has*

$$\mathrm{d}(x, \Psi(y)) \leq \frac{\|T(x, y)\|}{\sigma}.$$

(c) *For any* $x^* \in E^*$ *one has*

$$\mathrm{D}_F^* \Psi(\bar{y}, \bar{x})(x^*) = \left\{ -\big(T_{|2}(\bar{x}, \bar{y})\big)^* z^* \mid z^* \in G^*, \, \big(T_{|1}(\bar{x}, \bar{y})\big)^* z^* = x^* \right\}. \tag{13.104}$$

If, in particular, $T_{|1}(\bar{x}, \bar{y})$ *is invertible, then* $\mathrm{D}_F^* \Psi(\bar{y}, \bar{x})(x^*)$ *is a singleton consisting of* $-\big((T_{|1}(\bar{x}, \bar{y}))^{-1} T_{|2}(\bar{x}, \bar{y})\big)^* x^*$.

Proof. Ad (a), (b). Clearly the functional $f : E \times F \to \mathbb{R}$ defined by (13.103) satisfies the assumptions (A1)–(A4). We show that it also satisfies (A5). Let $(x, y) \in U$ be such that $T(x, y) \neq o$. Applying the chain rule we conclude that $x^* \in \partial_{F,1} f(x, y)$ if and only if

$$\langle x^*, u \rangle = \frac{(T(x,y) \mid T_{|1}(x,y)u)}{\|T(x,y)\|} \quad \forall u \in E,$$

in other words,

$$x^* = T_{|1}(x,y)^* v, \quad \text{where } v := \frac{T(x,y)}{\|T(x,y)\|}. \tag{13.105}$$

By (C3) there exists $u \in B_E$ such that $T_{|1}(x,y)u = \sigma v$. Consequently,

$$\|x^*\| = \|T_{|1}(x,y)^* v\| \geq \langle T_{|1}(x,y)^* v, u \rangle = \langle v, T_{|1}(x,y)u \rangle = \langle v, \sigma v \rangle = \sigma.$$

Therefore (A5) is also fulfilled. Conclusions (a) and (b) now follow immediately from Theorem 13.9.1.

Ad (c). We calculate the coderivative of the multifunction Ψ. Recall that $T(\bar{x}, \bar{y}) = o$. Let $\lambda > 0$ and $(-x^*, y^*) \in \lambda \partial_F f(x, y)$ be given. By Proposition 9.1.9(a) we have

$$\langle -x^*, u \rangle + \langle y^*, v \rangle \leq \lambda f_G\big((\bar{x}, \bar{y}), (u, v)\big) \quad \forall (u, v) \in E \times F. \tag{13.106}$$

Observe that

$$\begin{aligned} f_G\big((\bar{x}, \bar{y}), (u, v)\big) &= \|T_{|1}(\bar{x}, \bar{y})u + T_{|2}(\bar{x}, \bar{y})v\| \\ &= \max_{z^* \in B_G^*} \langle z^*, T_{|1}(\bar{x}, \bar{y})u + T_{|2}(\bar{x}, \bar{y})v \rangle; \end{aligned} \tag{13.107}$$

concerning the latter equation see Example 2.2.6. From (13.106) and (13.107) we conclude that

$$\min_{\substack{u \in B_E \\ v \in B_F}} \max_{z^* \in \lambda B_{G^*}} \left(\langle \big(T_{|1}(\bar{x}, \bar{y})\big)^* z^* - x^*, u \rangle + \langle \big(T_{|2}(\bar{x}, \bar{y})\big)^* z^* + y^*, v \rangle \right) \geq 0.$$

Hence there exists $z^* \in \lambda B_{G^*}$ satisfying

$$x^* = \big(T_{|1}(\bar{x}, \bar{y})\big)^* z^* \quad \text{and} \quad y^* = -\big(T_{|2}(\bar{x}, \bar{y})\big)^* z^*.$$

Now Theorem 13.9.1(c) implies (13.104), from which the special case where $T_{|1}(\bar{x}, \bar{y})$ is invertible is immediate. $\qquad \square$

As a consequence of Proposition 13.9.2 we obtain a classical result of Lyusternik [128] and Graves [79].

Corollary 13.9.3 *Assume that (A1) holds and that $T : E \times F \to F$ is a C^1 mapping such that $T(\bar{x}, \bar{y}) = o$ and the partial derivative $T_{|1}(\bar{x}, \bar{y})$ is surjective. Then the conclusions of Proposition 13.9.2 hold.*

Proof. In view of Proposition 13.9.2 we only have to show that condition (C3) is satisfied. Since $T_{|1}(\bar{x}, \bar{y})$ is surjective, the classical open mapping theorem implies that there exists $\sigma > 0$ such that $2\sigma B_F \subseteq T_{|1}(\bar{x}, \bar{y})(B_E)$. Since $T_{|1}$ is continuous, there further exist open neighborhoods V of $\bar{x}$ and W of $\bar{y}$ such that

$$\|T_{|1}(\bar{x}, \bar{y}) - T_{|1}(x, y)\| \le \sigma \quad \forall (x, y) \in V \times W.$$

It follows that

$$2\sigma B_F \subseteq \left[T_{|1}(\bar{x}, \bar{y}) - T_{|1}(x, y)\right](B_E) + T_{|1}(x, y)(B_E) \subseteq \sigma B_F + T_{|1}(x, y)(B_E)$$

and so $\sigma B_F \subseteq T_{|1}(x, y)(B_E)$ for any $(x, y) \in V \times W$. Hence (C3) is satisfied. $\qquad \square$

13.10 An Implicit Multifunction Theorem

We start with an auxiliary result. Let E, G be Fréchet smooth Banach spaces and U an open subset of E. With a given multifunction $\Gamma : U \rightrightarrows G$ we associate the function $g : U \to \overline{\mathbb{R}}$ defined by

$$g(x) := \mathrm{d}(\Gamma(x), o) = \inf\{\|z\| \mid z \in \Gamma(x)\}. \tag{13.108}$$

It will be crucial for the following that g can be written as infimum over a *fixed* set:

$$g(x) = \inf_{z \in G} \gamma(x, z), \quad \text{where } \gamma(x, z) := \|z\| + \delta_{\mathrm{graph}\,\Gamma}(x, z). \tag{13.109}$$

The lemma below establishes the relationship between the F–subdifferential of g and the coderivative of Γ. In this connection, if $S \subseteq G$, $\bar{z} \in G$ and $\eta > 0$, we call

$$\mathrm{proj}_\eta(\bar{z}, S) := \{z \in G \mid \|z - \bar{z}\| \le \mathrm{d}(\bar{z}, S) + \eta\}$$

the η–*approximate projection* of $\bar{z}$ to S.

Lemma 13.10.1 *Let E and G be Fréchet smooth Banach spaces and U an open subset of E. Further let the multifunction $\Gamma : U \rightrightarrows G$ be closed-valued and u.s.c. and let $g : U \to \overline{\mathbb{R}}$ be defined by (13.108). Assume that there exists $\sigma > 0$ such that for any $x \in U$ with $o \notin \Gamma(x)$, one has*

$$\sigma \le \liminf_{\eta \to 0} \Big\{ \|x^*\| \mid x^* \in \mathrm{D}_F^* \Gamma(\hat{x}, \hat{z})(z^*), \ z^* \in G^*, \ \|z^*\| = 1,$$

$$\hat{x} \in \mathrm{B}(x, \eta), \ \hat{z} \in \mathrm{proj}_\eta(o, \Gamma(\hat{x})) \Big\}.$$

Then for any $x \in U$ with $g(x) > 0$, $x^* \in \partial_F g(x)$ implies $\|x^*\| \ge \sigma$.

Proof. Since Γ is closed–valued and u.s.c., graph Γ is a closed set. Hence γ is l.s.c. Moreover, since Γ is u.s.c., the function g is l.s.c. (Exercise 13.13.11) and so coincides with its l.s.c. closure $\underline{g}$. Let $x \in U$ with $g(x) > 0$ and $x^* \in \partial_F g(x)$ be given. Choose $\eta > 0$ such that $g(\hat{x}) \geq g(x)/2$ for any $\hat{x} \in \mathrm{B}(x, \eta)$. By Proposition 9.7.1 there exist (u_η, w_η) and $(u_\eta^*, w_\eta^*) \in \partial_F \gamma(u_\eta, w_\eta)$ satisfying

$$0 < g(u_\eta) < \gamma(u_\eta, w_\eta) < g(u_\eta) + \eta, \tag{13.110}$$

$$\|u_\eta - x\| < \eta, \quad \|u_\eta^* - x^*\| < \eta, \quad \|w_\eta^*\| < \eta. \tag{13.111}$$

From (13.110) we see that $w_\eta \in \mathrm{proj}_\eta(o, \Gamma(u_\eta))$. By the sum rule of Theorem 9.2.6 there exist (x_η, z_η), $(\hat{x}_\eta, \hat{z}_\eta)$ close to (u_η, w_η) and $z_\eta^* \in \partial\omega(\hat{z}_\eta)$, where $\omega(z) := \|z\|$, such that

$$z_\eta \in \mathrm{proj}_\eta\big(o, \Gamma(x_\eta)\big), \quad \|\hat{z}_\eta\| > 0,$$

$$(u_\eta^*, w_\eta^*) \in (o, z_\eta^*) + \mathrm{N}_F\big((x_\eta, z_\eta), \mathrm{graph}\,\Gamma\big) + \eta(\mathrm{B}_{E^*} \times \mathrm{B}_{G^*}).$$

Hence there exist $(\hat{u}_\eta^*, \hat{w}_\eta^*) \in \eta\,(\mathrm{B}_{E^*} \times \mathrm{B}_{G^*})$ satisfying

$$u_\eta^* - \hat{u}_\eta^* \in \mathrm{D}^*\Gamma\big((x_\eta, z_\eta)\big)(z_\eta^* - w_\eta^* + \hat{w}_\eta^*)$$

and so

$$\frac{u_\eta^* - \hat{u}_\eta^*}{\|z_\eta^* - w_\eta^* + \hat{w}_\eta^*\|} \in \mathrm{D}^*\Gamma(x_\eta, z_\eta)\left(\frac{z_\eta^* - w_\eta^* + \hat{w}_\eta^*}{\|z_\eta^* - w_\eta^* + \hat{w}_\eta^*\|}\right).$$

Since $\|z_\eta^* - w_\eta^* + \hat{w}_\eta^*\| \geq 1 - 2\eta$ (recall that $\|z_\eta^*\| = 1$), we obtain

$$\liminf_{\eta \downarrow 0} \|u_\eta^*\| = \liminf_{\eta \downarrow 0} \frac{\|u_\eta^* - \hat{u}_\eta^*\|}{\|z_\eta^* - w_\eta^* + \hat{w}_\eta^*\|} \geq \sigma;$$

here the last inequality follows from the assumption concerning σ. Hence (13.111) shows that $\|x^*\| \geq \sigma$. □

Now we turn to the announced implicit multifunction theorem. For this, we need the following hypotheses.

(H 1) E, F and G are Fréchet smooth Banach spaces, U is a nonempty open subset of $E \times F$ and $(\bar{x}, \bar{y}) \in U$.

(H 2) $\Phi : U \rightrightarrows G$ is a closed–valued multifunction such that $o \in \Phi(\bar{x}, \bar{y})$.

(H 3) The multifunction $\Phi(\bar{x}, \cdot)$ is l.s.c. at $\bar{y}$.

(H 4) For any y near $\bar{y}$ the multifunction $\Phi(\cdot, y)$ is u.s.c..

(H 5) There exists $\sigma > 0$ such that for any $(x, y) \in U$ with $o \notin \Phi(x, y)$, one has

$$\sigma \leq \liminf_{\eta \to 0}\Big\{\|x^*\| \;\Big|\; x^* \in \mathrm{D}_F^*\Phi\big((\hat{x}, y), \hat{z}\big)(z^*), \; z^* \in G^*, \; \|z^*\| = 1,$$

$$\hat{x} \in \mathrm{B}(x, \eta), \; \hat{z} \in \mathrm{proj}_\eta\big(o, \Phi(\hat{x}, y)\big)\Big\}.$$

Further let

$$\Psi(y) := \{x \in E \mid o \in \Phi(x,y)\}, \tag{13.112}$$

that is, $\Psi : F \rightrightarrows E$ is the implicit multifunction defined by the generalized equation $o \in \Phi(x,y)$. Applying Theorem 13.9.1 and making use of Lemma 13.10.1, we immediately obtain the following result.

Theorem 13.10.2 (Implicit Multifunction Theorem) *Assume that the hypotheses* (H1)–(H5) *are satisfied. Then there exist open sets* $V \subseteq E$ *and* $W \subseteq F$ *with* $\bar{x} \in V$ *and* $\bar{y} \in W$ *such that the following holds:*

(a) *For any* $y \in W$ *the set* $V \cap \Psi(y)$ *is nonempty.*
(b) *For any* $(x,y) \in V \times W$ *one has*

$$d(\Psi(y), x) \le \frac{d(\Phi(x,y), o)}{\sigma}.$$

(c) *For any* $(x,y) \in V \times W$ *with* $x \in \Psi(y)$ *and any* $x^* \in E^*$, *one has*

$$D_F^*\Psi(y,x)(x^*) = \left\{ y^* \in F^* \,\middle|\, (-x^*, y^*) \in \bigcup_{\lambda>0} \lambda \,\partial_F d(\Phi(x,y), o) \right\}. \tag{13.113}$$

This theorem characterizes the coderivative of the implicit multifunction Ψ in terms of the Fréchet subdifferential of the scalar function $(x,y) \mapsto d(\Phi(x,y), 0)$. It is natural to ask how the coderivative of Ψ can be characterized directly in terms of Φ. The following result gives a partial answer.

Proposition 13.10.3 *Assume that the hypotheses* (H1)–(H5) *are satisfied. Then for any* $(x,y) \in V \times W$ *with* $x \in \Psi(y)$ *(notation as in Theorem 13.10.2), the following holds:*

(a) *For any* $x^* \in E^*$ *one has*

$$\bigcup_{z^* \in G^*} \left\{ y^* \in F^* \,\middle|\, (-x^*, y^*) \in D_F^*\Phi(x,y,o)(z^*) \right\} \subseteq D_F^*\Psi(y,x)(x^*). \tag{13.114}$$

(b) *For any* $x^* \in E^*$, $y^* \in D_F^*\Psi(y,x)(x^*)$, *and* $\epsilon > 0$ *there exist* $(x_\epsilon, y_\epsilon, z_\epsilon) \in$ *graph* Φ *and* $(x_\epsilon^*, y_\epsilon^*, z_\epsilon^*) \in E^* \times F^* \times G^*$ *such that*

$$\|x - x_\epsilon\| < \epsilon, \quad \|y - y_\epsilon\| < \epsilon, \quad \|z_\epsilon\| < \epsilon,$$
$$\|x^* - x_\epsilon^*\| < \epsilon, \quad \|y^* - y_\epsilon^*\| < \epsilon,$$
$$(-x_\epsilon^*, y_\epsilon^*) \in D_F^*\Phi(x_\epsilon, y_\epsilon, z_\epsilon)(z_\epsilon^*). \tag{13.115}$$

(c) *If, in addition, graph* Φ *is normally regular at* (x,y,o), *then* (13.114) *holds with equality.*

Proof.

(a) Let y^* be an element of the left-hand side of (13.114). Then there exists $z^* \in G^*$ such that

$$(-x^*, y^*, z^*) \in N_F(\operatorname{graph} \Phi, (x, y, o)).$$

Hence there further exists a C^1 function g such that $g'(x, y, o) = (-x^*, y^*, z^*)$ and $\delta_{\operatorname{graph}\Phi} - g$ has a local minimum at (x, y, o). Since $\delta_{\operatorname{graph}\Phi}(\hat{x}, \hat{y}, o) = \delta_{\operatorname{graph}\Psi}(\hat{y}, \hat{x})$ for any $\hat{x} \in E$ and any $\hat{y} \in F$, it follows that the function

$$(\hat{y}, \hat{x}) \mapsto \delta_{\operatorname{graph}\Psi}(\hat{y}, \hat{x}) - g(\hat{x}, \hat{y}, o)$$

attains a local minimum at (y, x). Therefore, $y^* \in D_F^*(y, x)(x^*)$.

(b) Let $y^* \in D_F^* \Psi(y, x)(x^*)$ be given. Then

$$(y^*, -x^*) \in \bigcup_{\lambda > 0} \lambda \partial_F d(\operatorname{graph} \Psi, (y, x)).$$

Thus there exist a C^1 function h satisfying $h'(y, x) = (y^*, -x^*)$ and a constant $\lambda > 0$ such that

$$\begin{aligned} h(y, x) &\le h(\hat{y}, \hat{x}) + \lambda d(\operatorname{graph} \Psi, (\hat{y}, \hat{x})) \\ &\le h(\hat{y}, \hat{x}) + \lambda d(\Psi(\hat{y}, \hat{x}) \le h(\hat{y}, \hat{x}) + (\lambda/\sigma) d(\Phi(\hat{x}, \hat{y}), o); \end{aligned} \tag{13.116}$$

here the last inequality follows by Theorem 13.10.2(b). Since $d(\Phi(\hat{x}, \hat{y}), o) = \inf_{\hat{z} \in G}\{\|\hat{z}\| + \delta_{\operatorname{graph}\Phi}(\hat{x}, \hat{y}, \hat{z})\}$, the estimate (13.116) shows that

$$(y^*, -x^*) \in \partial_F \left(\inf_{\hat{z} \in G} \{ (\lambda/\sigma)\|\hat{z}\| + \delta_{\operatorname{graph}\Phi}(\hat{x}, \hat{y}, \hat{z}) \} \right).$$

Applying Proposition 9.7.1, we obtain the conclusion of (b).

(c) This follows from (13.115) by letting $\epsilon \downarrow 0$. $\qquad\square$

We now derive a sufficient condition for metric regularity. In view of Theorems 10.3.3 and 10.5.2 this is at the same time a sufficient condition for linear openness and pseudo-Lipschitz continuity (Theorem 13.10.2)

Given the multifunction $\widehat{\Phi} : E \rightrightarrows F$, define $\Phi : E \times F \rightrightarrows F$ by

$$\Phi(x, y) := \widehat{\Phi}(x) - y, \quad x \in E, \quad y \in F. \tag{13.117}$$

Proposition 13.10.4 *Assume that* $\widehat{\Phi} : E \rightrightarrows F$ *is a multifunction such that, with* $G := F$ *and* Φ *according to (13.117), the hypotheses (H1)–(H5) are satisfied. Then* $\widehat{\Phi}$ *is metrically regular around* $(\bar{x}, \bar{y})$ *with constant* $1/\sigma$.

Proof. Notice that in this case we have $\Psi = \widehat{\Phi}^{-1}$ and $d(\Phi(x, y), o) = d(\widehat{\Phi}(x), y)$. Hence the assertion follows from Theorem 13.10.2(b). $\qquad\square$

13.11 An Extremal Principle Involving Deformations

The concept of an extremal system introduced in Sect. 13.3 refers to the *translation* of sets (cf. Remark 13.3.2). Now we introduce the concept of an extended extremal system which refers to the *deformation* of sets and so applies to multifunctions.

Definition 13.11.1 Assume that S_i, $i = 1, \ldots, n$, are metric spaces with metrics ρ_i, E is a Banach space, and $\Phi_i : S_i \rightrightarrows E$ are closed-valued multifunctions. A point $\bar{x} \in E$ is said to be a *local extremal point* of the system $(\Phi_1, \ldots, \Phi_n)$ at $(\bar{s}_1, \ldots, \bar{s}_n) \in S_1 \times \cdots \times S_n$ if $\bar{x} \in \Phi_1(\bar{s}_1) \cap \cdots \cap \Phi_n(\bar{s}_n)$ and there is a neighborhood U of $\bar{x}$ such that for any $\epsilon > 0$ there exist $(s_1, \ldots, s_n) \neq (\bar{s}_1, \ldots, \bar{s}_n)$ with

$$\rho_i(s_i, \bar{s}_i) \leq \epsilon, \quad \mathrm{d}(\Phi_i(s_i), \bar{x}) < \epsilon, \quad i = 1, \ldots, n, \quad \Phi_1(s_1) \cap \cdots \cap \Phi_n(s_n) \cap U = \emptyset.$$

If the system $(\Phi_1, \ldots, \Phi_n)$ admits a local extremal point, it is said to be an *extended extremal system*.

Remark 13.11.2 Let $(A_1, A_2, \bar{x})$ be an extremal system in the sense of Definition 13.3.1. Defining

$$S_1 := E, \quad \Phi_1(s_1) := s_1 + A_1 \, \forall \, s_1 \in E, \quad S_2 := \{o\}, \quad \Phi_2(o) := A_2,$$

we see that $\bar{x}$ is a local extremal point of the system (Φ_1, Φ_2) at (o, o) and so (Φ_1, Φ_2) is an extended extremal system.

We now establish an extremal principle for extended extremal systems that corresponds to the approximate extremal principle of Theorem 13.3.10. The result will be applied in Sect. 13.12 to multiobjective optimization.

Theorem 13.11.3 *Let E be a Fréchet smooth Banach space. Assume that, for $i = 1, \ldots, n$, S_i is a metric space with metric ρ_i and $\Phi_i : S_i \rightrightarrows E$ is a closed-valued multifunction. Further let $\bar{x}$ be a local extremal point of the system $(\Phi_1, \ldots, \Phi_n)$ at $(\bar{s}_1, \ldots, \bar{s}_n)$. Then for any $\epsilon > 0$ there exist $s_i \in S_i$, $x_i \in \mathrm{B}_E(\bar{x}, \epsilon)$, and $x_i^* \in E^*$ such that*

$$\rho_i(s_i, \bar{s}_i) \leq \epsilon, \quad x_i \in \Phi_i(s_i), \quad x_i^* \in \mathrm{N}_F(\Phi_i(s_i), x_i) + \epsilon \mathrm{B}_{E^*},$$
$$\max\{\|x_i^*\|, \ldots, \|x_n^*\|\} \geq 1, \quad x_1^* + \cdots + x_n^* = o. \tag{13.118}$$

Proof.

(I) We equip E^n with the Euclidean product norm. Choose $\rho > 0$ such that $U := \mathrm{B}(\bar{x}, \rho)$ is a neighborhood of $\bar{x}$ as in Definition 13.11.1. Now let $\epsilon > 0$ be given and choose η such that

$$0 < \eta < \min\left\{ \frac{\epsilon^2}{5\epsilon + \epsilon^2 + 12n^2}, \frac{\rho^2}{4} \right\}.$$

Further let $s_1, \ldots, s_n$ be as in Definition 13.11.1 with ϵ replaced by η so that, in particular, $\mathrm{d}(\Phi_i(s_i), \bar{x}) < \eta$. Put $A := \Phi_1(s_1) \times \cdots \times \Phi_n(s_n)$ and define $\varphi : U^n \to \overline{\mathbb{R}}$ by

$$\varphi(y_1, \ldots, y_n) := \sum_{i,k=1}^{n} \|y_i - y_k\| + \delta_A(y_1, \ldots, y_n).$$

Since every set $\Phi_i(s_i)$ is closed, the function φ is l.s.c. Moreover, since $\bar{x}$ is a local extremal point of the system $(\Phi_1, \ldots, \Phi_n)$, we conclude that φ is positive on U^n. Choose $\tilde{y}_i \in \Phi_i(s_i)$, $i = 1, \ldots, n$, satisfying

$$\|\tilde{y}_i - \tilde{y}_k\| \leq \mathrm{d}(\Phi_i(s_i), \bar{x}) + \mathrm{d}(\Phi_k(s_k), \bar{x}) + \eta \leq 3\eta.$$

It follows that $\varphi(y_1, \ldots, y_n) \leq 3n^2\eta < \epsilon^2/4$.

(II) Applying Ekeland's variational principle in the form of Corollary 8.2.6, we find $\tilde{x}_i \in \mathrm{B}(\tilde{y}_i, \epsilon/2) \subseteq \mathrm{B}(\bar{x}, \epsilon)$ such that the function

$$f(y_1, \ldots, y_n) := \sum_{i,k=1}^{n} \|y_i - y_k\| + \frac{\epsilon}{2} \sum_{i=1}^{n} \|y_i - \tilde{x}_i\| + \delta_A(y_1, \ldots, y_n)$$

attains its minimum over U^n at $(\tilde{x}_1, \ldots, \tilde{x}_n)$. Obviously we may assume that $U^n = E^n$. Define $\psi : E^n \to \bar{\mathbb{R}}$ by

$$\psi(y_1, \ldots, y_n) := \sum_{i,k=1}^{n} \|y_i - y_k\|.$$

We have $\psi(\tilde{x}_1, \ldots, \tilde{x}_n) = \varphi(\tilde{x}_1, \ldots, \tilde{x}_n) > 0$.

(III) Applying the approximate sum rule of Theorem 9.2.6 (in connection with Lemma 9.2.5) to the function f, we find

$$x_i \in \Phi(s_i) \cap \mathrm{B}(\tilde{x}_i, \eta) \subseteq \mathrm{B}(\bar{x}, \epsilon), \quad z_i \in \mathrm{B}(\tilde{x}_i, \eta),$$
$$(z_1^*, \ldots, z_n^*) \in \partial_F \psi(z_1, \ldots, z_n)$$

satisfying

$$0 \in (z_1^*, \ldots, z_n^*) + N_F(\Phi_1(s_1), x_1) \times \cdots \times N_F(\Phi_n(s_n), x_n)$$
$$+ \mathrm{B}_{(E^n)^*}(0, \eta(n+1)); \tag{13.119}$$

here we made use of the representation

$$\partial_F \delta_A(y_1, \ldots, y_n) = N_F(\Phi_1(y_1), y_1) \times \cdots \times N_F(\Phi_n(y_n), y_n)$$
$$\forall\, y_i \in \Phi_i(s_i), \tag{13.120}$$

the verification of which is left as Exercise 13.13.21. Putting $x_i^* := -z_i^*$ for $i = 1, \ldots, n$ we derive from (13.119) that $x_i^* \in N_F(\Phi_i(s_i), x_i) + \epsilon \mathrm{B}_{E^*}$. From $(z_1^*, \ldots, z_n^*) \in \partial_F \psi(z_1, \ldots, z_n)$ we conclude that

$$(z_1^*, \ldots, z_n^*)$$
$$\leq \liminf_{\tau \to 0} \frac{\psi(z_1 + \tau h, \ldots, z_n + \tau h) - \psi(z_1, \ldots, z_n)}{\tau} = 0 \quad \forall\, h \in E,$$

where the latter equation follows from the definition of ψ by symmetry. Hence we obtain

$$x_1^* + \cdots + x_n^* = -(z_1^* + \cdots + z_n^*) = 0. \tag{13.121}$$

It remains to verify that $\max\{\|x_1^*\|, \ldots, \|x_n^*\|\} \geq 1$. We first obtain

$$\sum_{i=1}^{n} \langle z_i^*, -z_i \rangle$$

$$\leq \liminf_{\tau \to 0} \frac{\psi(z_1 - \tau z_1, \ldots, z_n - \tau z_n) - \psi(z_1, \ldots, z_n)}{\tau} = -\psi(z_1, \ldots, z_n).$$

In view of (13.121) we have $z_1^* = -\sum_{i=2}^{n} z_i^*$ and it follows that

$$\psi(z_1, \ldots, z_n) \leq \sum_{i=1}^{n} \langle z_i^*, z_i \rangle = \sum_{i=2}^{n} \langle z_i^*, z_i - z_1 \rangle \tag{13.122}$$

$$\leq \max\{\|z_1^*\|, \ldots, \|z_n^*\|\} \, \psi(z_1, \ldots, z_n).$$

Since $\psi(\tilde{x}_1, \ldots, \tilde{x}_n) > 0$ and $\|z_i - \tilde{x}_i\| \leq \eta$, we may assume (shrinking η further if necessary) that $\psi(z_1, \ldots, z_n) > 0$. Hence (13.122) implies that $\max\{\|z_1^*\|, \ldots, \|z_n^*\|\} \geq 1$, which also holds with z_i^* replaced by x_i^*. $\quad\square$

13.12 Application to Multiobjective Optimization

We want to apply the extremal principle of Theorem 13.11.3 to multiobjective optimization problems. Let P be a nonempty subset of a Banach space F. We define a *preference relation* $\prec$ on F by writing $z_1 \prec z_2$ (read z_1 *is preferred to* z_2) if and only if $(z_1, z_2) \in P$. Given $z \in F$, we denote the level set of $\prec$ at z by $\mathcal{L}(z)$, i.e.,

$$\mathcal{L}(z) := \{y \in F \mid y \prec z\}.$$

Definition 13.12.1 The preference relation $\prec$ is said to be:

- *nonreflexive* if $z \prec z$ does not hold for any $z \in F$.
- *locally satiated* at $\bar{z} \in F$ if $z \in \operatorname{cl} \mathcal{L}(z)$ for any z in a neighborhood of $\bar{z}$.
- *almost transitive* on F if $y \in \operatorname{cl} \mathcal{L}(z)$ and $z \prec z'$ imply $y \prec z'$.

Example 13.12.2 Let Q be a closed cone in F. The *generalized Pareto preference* $\prec$ is defined by $z_1 \prec z_2$ if and only if $z_1 - z_2 \in Q$ and $z_1 \neq z_2$. We have $\mathcal{L}(z) = z + (Q \setminus \{o\})$. Hence $\prec$ is a nonreflexive preference relation that is locally satiated at any $\bar{z} \in F$. It is left as Exercise 13.13.22 to show that $\prec$ is almost transitive on F if and only if the cone Q is pointed (i.e., $Q \cap (-Q) = \{o\}$) and convex.

Now we make the following assumptions:

(A) E and F are Banach spaces, $f : E \to F$, $M \subseteq E$,
  $\prec$ is a nonreflexive, satiated, almost transitive preference relation on F.

We consider the *multiobjective optimization problem*:

(MOP) Minimize $f(x)$ with respect to $\prec$
    subject to $x \in M$.

A point $\bar{x} \in E$ is said to be a *local solution* of (MOP) if there is no $x \in M$ near $\bar{x}$ such that $f(x) \prec f(\bar{x})$.

Lemma 13.12.3 establishes the relationship between the problem (MOP) and extended extremal systems.

Lemma 13.12.3 *Let the assumptions* (A) *be satisfied. Define*

$$S_1 := \mathcal{L}(f(\bar{x})) \cup \{f(\bar{x})\}, \quad \Phi_1(s_1) := M \times \mathrm{cl}\,\mathcal{L}(s_1) \quad \forall\, s_1 \in S_1,$$
$$S_2 := \{o\}, \quad \Phi_2(o) := \{(x, f(x)) \mid x \in E\}. \tag{13.123}$$

If $\bar{x}$ is a local solution of (MOP), *then $(\bar{x}, f(\bar{x}))$ is a local extremal point of the system (Φ_1, Φ_2).*

Proof. See Exercise 13.13.23. □

Applying the extremal principle of Theorem 13.11.3, we now derive a necessary optimality condition for (MOP) in terms of F-subdifferentials. Recall that for $y^* \in F^*$, the scalarization $\langle y^*, f \rangle$ of f is defined by $\langle y^*, f \rangle(x) := \langle y^*, f(x) \rangle$ for any $x \in E$.

Theorem 13.12.4 *In addition to the assumptions* (A) *let E and F be Fréchet smooth Banach spaces and let f be locally L-continuous around $\bar{x} \in M$. If $\bar{x}$ is a local solution of* (MOP), *then for any $\epsilon > 0$ there exist $x_0, x_1 \in B_E(\bar{x}, \epsilon)$, $y_0, y_1 \in B_F(f(\bar{x}), \epsilon)$, $x^* \in N_F(M, x_1)$, and $y^* \in N_F(\mathrm{cl}\,\mathcal{L}(y_0), y_1)$ such that $\|y^*\| = 1$ and*

$$o \in x^* + \partial_F \langle y^*, f \rangle(x_0) + \epsilon B_{E^*}. \tag{13.124}$$

Proof.

(I) We equip $E \times F$ with the norm $\|(x, y)\| := \|x\| + \|y\|$ and $E^* \times F^*$ with the corresponding dual norm $\|(x^*, y^*)\| = \max\{\|x^*\|, \|y^*\|\}$. Let $\rho \in (0, 1)$ be such that f is Lipschitz continuous on $B(\bar{x}, \rho)$ with Lipschitz constant $\lambda > 0$. Let any $\epsilon > 0$ be given and choose

$$\eta := \min\Big\{ \frac{2\epsilon\lambda}{1 + \lambda}, \frac{1}{8(2 + \lambda)}, \frac{\epsilon}{2}, \frac{\rho}{4} \Big\}.$$

By Lemma 13.12.3, $(\bar{x}, f(\bar{x}))$ is a local extremal point of the system (Φ_1, Φ_2) defined in that lemma. Therefore, Theorem 13.11.3 (with ϵ replaced by η) implies that there exist $y_0 \in B_F(f(\bar{x}), \eta)$, $(x_i, y_i) \in B_{E \times \mathbb{R}}((\bar{x}, f(\bar{x})), \eta)$, $i = 1, 2$, $(x_1^*, y_1^*) \in N_F(\Phi_1(y_0), (x_1, y_1))$, and $(x_2^*, y_2^*) \in N_F(\Phi_2(o), (x_2, y_2))$ such that

$$\max\{\|(x_1^*, y_1^*)\|, \|(x_2^*, y_2^*)\|\} > 1 - \eta \quad \text{and} \quad \|(x_1^*, y_1^*) + (x_2^*, y_2^*)\| < \eta. \tag{13.125}$$

It follows that

$$\|(x_i^*, y_i^*)\| > 1 - 2\eta \geq 1/2, \quad i = 1, 2. \tag{13.126}$$

The definition of the F-normal cone implies that for any $(x, y) \in \Phi_2(o)$ sufficiently close to (x_2, y_2) we obtain

$$\langle x_2^*, x - x_2 \rangle + \langle y_2^*, y - y_2 \rangle - \eta \|(x - x_2, y - y_2)\| \leq 0.$$

In this connection we have $y = f(x)$ and $y_2 = f(x_2)$. Hence the function $g : E \to \mathbb{R}$ defined by

$$g(x) := -\left(\langle x_2^*, x - x_2 \rangle + \langle y_2^*, f(x) - f(x_2) \rangle - \eta \|(x - x_2, f(x) - f(x_2))\| \right)$$

attains the local minimum 0 at $x = x_2$. It follows that $o \in \partial_F g(x_2)$.

(II) Applying the approximate sum rule of Theorem 9.2.6 and the chain rule of Theorem 9.2.9 to g, we conclude that there exists $x_0 \in B_E(x_2, \eta) \subseteq B_E(\bar{x}, 2\eta)$ such that

$$o \in -x_2^* - \partial_F \langle y_2^*, f \rangle(x_0) + (1 + \lambda)\eta B_{E^*}. \tag{13.127}$$

From this and the second relation in (13.125) we obtain

$$o \in x_1^* + \partial_F \langle y_1^*, f \rangle(x_0) + 2(2 + \lambda)\eta B_{E^*}. \tag{13.128}$$

We claim that $\|y_1^*\| \geq 1/(4(1 + \lambda))$. To see this, take $z^* \in \partial_F \langle y_1^*, f \rangle(x_0)$ and $u^* \in \eta B_{E^*}$ such that $o = x_1^* + z^* + 2(2 + \lambda)u^*$. It follows that

$$1/2 < \|(x_1^*, y_1^*)\| \leq \|x_1^*\| + \|y_1^*\| = \|z^* + 2(2 + \lambda)u^*\| + \|y_1^*\| \\ \leq \|z^*\| + 2(2 + \lambda)\eta + \|y_1^*\|. \tag{13.129}$$

From the choice of z^* and the L-continuity of f on $B(\bar{x}, \rho)$ we obtain

$$\langle z^*, h \rangle \leq \langle y_1^*, f(x_0 + h) - f(x_0) \rangle \leq \|y_1^*\|\lambda\|h\| \quad \forall h \in B(o, \rho/2)$$

and so $\|z^*\| \leq \lambda \|y_1^*\|$. This, (13.129), and the choice of η imply

$$\|y_1^*\| \geq \frac{\frac{1}{2} - 2(2 + \lambda)\eta}{1 + \lambda} \geq \frac{1}{4(1 + \lambda)}$$

as claimed. Now dividing (13.128) by $\|y_1^*\|$ and defining $x^* := x_1^*/\|y_1^*\|$ and $y^* := y_1^*/\|y_1^*\|$, the conclusion of the theorem follows. □

Remark 13.12.5 Theorem 13.12.4 is the starting point for deriving necessary conditions for multiobjective optimization problems with specified constraints. For instance, let the set M in (MOP) be of the form $M = A_1 \cap A_2 \cap A$, where A is any subset of E and

$$A_1 := \{x \in E \mid g(x) \leq 0\}, \ A_2 := \{x \in E \mid h(x) = 0\}$$

with functionals $g, h : E \to \overline{\mathbb{R}}$. Let $\bar{x}$ be a local solution of (MOP) in this case. Then Theorem 13.12.4 implies the condition (13.124). Here, $x^* \in \mathrm{N}_F(M, \bar{x}_1)$ and the definition of the F-normal cone shows that

$$\varphi(y) := -\langle x^*, y - x_1 \rangle + (\epsilon/3)\|y - x_1\| \geq 0$$

for all $y \in M$ close to x_1. Hence x_1 is a local solution to the *scalar* optimization problem

$$\begin{aligned} \text{minimize} \quad & \varphi(y) \\ \text{subject to} \quad & g(y) \leq 0, \quad h(y) = 0, \quad y \in A. \end{aligned}$$

A necessary condition for this problem can be derived, for instance, with the aid of the approximate multiplier rule of Theorem 12.5.1. It is left as Exercise 13.13.24 to elaborate the details.

13.13 Bibliographical Notes and Exercises

The presentation of this chapter was strongly influenced by the seminal two-volume monograph [136, 137] of Mordukhovich, in particular by the first volume. The second volume contains profound investigations of optimal control governed by ordinary differential equations and by functional–differential relations. Concerning variational analysis in finite-dimensional spaces we recommend the comprehensive monograph [189] of Rockafellar and Wets.

The limiting objects, which we call here *Mordukhovich normal cone* and *Mordukhovich subdifferential*, are called by Mordukhovich *basic normal cone* and *basic subdifferential*, respectively. These objects as well as the idea of extremal principles first appear in [132]. We refer to the monograph cited above for a detailed discussion of the development of this theory as well as a lot of references. The exact sum rule of Theorem 13.5.5 was obtained by Mordukhovich and Shao [142] in an arbitrary Asplund space. Further calculus rules for F- and M-subdifferentials as well as necessary optimality conditions were obtained, among others, by Ngai et al. [151] and Ngai and Théra [154].

Proposition 13.4.3 is due to Mordukhovich and Wang [145]. Mordukhovich coderivatives (called *normal coderivatives* by Mordukhovich) were introduced in [133]. They were systematically studied in finite-dimensional spaces by Mordukhovich [134] and Ioffe [95]. The approximate calculus rules for F-coderivatives in Fréchet smooth Banach spaces (Theorems 13.2.7 and 13.2.8) and their exact counterparts for M-coderivatives in finite-dimensional Banach spaces were taken from Borwein and Zhu [24]. Mordukhovich and Shao [141, 143] derived exact calculus rules for M-coderivatives in Asplund spaces. Example 13.2.12 (together with the concrete data of Exercise 13.13.8

and further similar examples) was constructed by Borwein and Zhu [23]. Weak* sequential limits of proximal subgradients are studied by Clarke et al. [39].

The results of Sects. 13.9 and 13.10, in particular the remarkable implicit multifunction theorem (Theorem 13.10.2), are due to Ledyaev and Zhu [120].

Investigating multiobjective optimal control problems, Zhu [227] observed that the approximate extremal principle holds for more general deformations than translations of the set system. This observation finally led to the general concept of an extended extremal system that was developed and applied by Mordukhovich et al. [144]. Readers interested in multiobjective optimization (also known as *vector optimization*) are referred to Göpfert et al. [77], Jahn [102], Pallaschke and Rolewicz [156], and the literature cited in these books.

Exercise 13.13.1 Verify Lemma 13.1.11.

Exercise 13.13.2 Prove Proposition 13.1.19.

Exercise 13.13.3 Verify Theorem 13.1.23.

Exercise 13.13.4 Let A be a closed subset of E and $\bar{x} \in A$. Prove the following assertions:

(a) If $\bar{x} \in A$ and U is a closed neighborhood of $\bar{x}$, then $N_M(A \cap U, \bar{x}) = N_M(A, \bar{x})$.
(b) If $\bar{x} \in \operatorname{bd} A$, then $N_M(A, \bar{x}) \subseteq N_M(\operatorname{bd} A, \bar{x})$.

Hint: Take any nonzero $x^* \in N_M(A, \bar{x})$ and find a sequence (ϵ_k, x_k, x_k^*), such that, in particular, $x_k^* \xrightarrow{w^*} x^*$. Since it follows that $\liminf \|x_k^*\| > 0$, conclude that $x_k \notin \operatorname{int} A$ for k large enough.

Exercise 13.13.5 Verify the assertions (a) and (c) of Proposition 13.2.3.

Exercise 13.13.6 Elaborate the details of the proof of Theorem 13.2.8.

Exercise 13.13.7 Prove Theorem 13.2.11.

Exercise 13.13.8 Let E be the sequence space l^2. For any $k \in \mathbb{N}$, let u_k denote the kth unit vector of l^2, let α_k be a positive number such that $\sqrt{1 - \frac{1}{k^2}} \le \alpha_k < 1$ and define

$$y_k := u_{2k}, \quad z_k := \sqrt{1 - \alpha_k^2}\, u_{2k-1} - \alpha_k u_{2k}.$$

Finally define $H_1 := \operatorname{cl} \operatorname{span}\{y_k \mid k \in \mathbb{N}\}$ and $H_2 := \operatorname{cl} \operatorname{span}\{z_k \mid k \in \mathbb{N}\}$. Show that $H_1^\circ \cap H_2^\circ = \{o\}$ and $H_1^\circ + H_2^\circ$ is weak* dense but not closed in E^* (cf. Example 13.2.12).
Hint: Verify that

$$H_1^\circ = \operatorname{cl} \operatorname{span}\{y_k^* \mid k \in \mathbb{N}\}, \quad H_2^\circ = \operatorname{cl} \operatorname{span}\{z_k^* \mid k \in \mathbb{N}\},$$

where $y_k^* := u_{2k-1}$ and $z_k^* := \alpha_k u_{2k-1} + \sqrt{1 - \alpha_k^2}\, u_{2k}$.

Exercise 13.13.9 Verify the statement in Example 13.3.3

Exercise 13.13.10 Prove Lemma 13.3.4.

Exercise 13.13.11 Let $\Gamma : E \rightrightarrows G$ be a multifunction between the Banach spaces E and G. Define $g : E \to \bar{\mathbb{R}}$ by $g(x) := \mathrm{d}(\Gamma(x), o)$. Show that if Γ is u.s.c., then g is l.s.c.

Exercise 13.13.12 Elaborate step (II) of the proof of Theorem 13.3.10.

Exercise 13.13.13 Show that the set $A \subseteq E$ is epi-Lipschitzian at $\bar{x} \in A$ if and only if it is compactly epi-Lipschitzian at $\bar{x}$, where the associated compact set C can be chosen as a singleton.

Exercise 13.13.14 Verify Remark 13.4.6.

Exercise 13.13.15 Carry out step (III) in the proof of Theorem 13.4.9.

Exercise 13.13.16 For $i = 1, 2$ let A_i be a nonempty subset of the normed vector space E_i and $\bar{x}_i \in A_i$. Prove that

$$\mathrm{N}_M(A_1 \times A_2, (\bar{x}_1, \bar{x}_2)) = \mathrm{N}_M(A_1, \bar{x}_1) \times \mathrm{N}_M(A_2, \bar{x}_2).$$

Exercise 13.13.17 Show that for any nonempty subset A of E and $\bar{x} \in A$ one has

$$\partial_M \delta_A(\bar{x}) = \partial_M^\infty \delta_A(\bar{x}) = \mathrm{N}_M(A, \bar{x}).$$

Exercise 13.13.18 Carry out the induction proof for Theorem 13.5.5.

Exercise 13.13.19 Prove the assertion (13.93) of Theorem 13.8.1.

Exercise 13.13.20 Verify the inclusion $\supseteq$ of (13.98) under the assumptions of Theorem 13.8.2.

Exercise 13.13.21 Verify the representation of $\partial_F \delta_A(y_1, \ldots, y_n)$ in (13.120).

Exercise 13.13.22 Verify the assertion in Example 13.12.2.

Exercise 13.13.23 Verify Lemma 13.12.3.

Exercise 13.13.24 Carry out the program indicated in Remark 13.12.5.

A

Appendix: Further Topics

In this final chapter we briefly indicate some concepts and developments in nonsmooth analysis that have not been treated in the preceding text. For a comprehensive discussion we refer to the monograph [189] by Rockafellar and Wets (finite-dimensional theory) and the monograph [136, 137] by Mordukhovich (infinite-dimensional theory).

In the sequel, unless otherwise specified, let E be a normed vector space, $f : E \to \overline{\mathbb{R}}$ a proper functional, and $\bar{x} \in \text{dom } f$.

(I) Clarke [34] defines the subdifferential

$$\partial_\uparrow f(\bar{x}) := \{x^* \in E^* \mid (x^*, -1) \in \text{N}_C\big(\text{epi } f, (\bar{x}, f(\bar{x}))\big)\}.$$

This set is always $\sigma(E^*, E)$-closed but may be empty. However, if $\bar{x}$ is a local minimizer of f, then $o \in \partial_\uparrow f(\bar{x})$. Moreover, Corollary 11.3.2 shows that $\partial_\uparrow f(\bar{x}) = \partial_o f(\bar{x})$ for any $\bar{x} \in E$ whenever f is locally L-continuous on E.

(II) Rockafellar [184] (see also [186, 189]) defines the *(regular) subderivative* $f^\uparrow(\bar{x}, \cdot) : E \to \overline{\mathbb{R}}$ of f at $\bar{x}$ as

$$f^\uparrow(\bar{x}, y) := \lim_{\epsilon \downarrow 0} \ \limsup_{(x,\alpha) \to_f \bar{x}} \ \inf_{z \in \text{B}(y,\epsilon)} \frac{f(x + \tau z) - \alpha}{\tau}.$$

In this context, $(x, \alpha) \to_f \bar{x}$ means $(x, \alpha) \in \text{epi } f$, $x \to \bar{x}$, and $\alpha \to f(\bar{x})$. Rockafellar shows that $f^\uparrow(\bar{x}, \cdot)$ is l.s.c. and that $f^\uparrow(\bar{x}, o) = -\infty$ if and only if $\partial_\uparrow f(\bar{x}) = \emptyset$. If $f^\uparrow(\bar{x}, o) > -\infty$, then $f^\uparrow(\bar{x}, o) = 0$, $f^\uparrow(\bar{x}, \cdot)$ is sublinear, and

$$\text{epi } f^\uparrow(\bar{x}, \cdot) = \text{T}_C\big(\text{epi } f, (\bar{x}, f(\bar{x}))\big).$$

It then follows that $f^\uparrow(\bar{x}, \cdot)$ is the support functional of the Clarke subdifferential, i.e.,

$$\partial_\uparrow f(\bar{x}) = \{x^* \in E^* \mid \langle x^*, y \rangle \leq f^\uparrow(\bar{x}, y) \quad \forall y \in E\}.$$

The functional f is said to be *directionally Lipschitz* at $\bar{x}$ if epi f is epi-Lipschitz at $(\bar{x}, f(\bar{x}))$. Moreover, f is said to be *subdifferentially regular* if

$$f^\uparrow(\bar{x}, y) = \underline{f}_H(\bar{x}, y) \quad \forall y \in E;$$

this is equivalent to epi f being a tangentially regular set. If f is locally L-continuous around $\bar{x}$, then f is subdifferentially regular if and only if f is regular in the sense of Clarke. Rockafellar shows that the sum rule

$$\partial_\uparrow (f + g)(\bar{x}) \subseteq \partial_\uparrow f(\bar{x}) + \partial_\uparrow g(\bar{x}) \tag{$\diamond$}$$

holds if g is directionally Lipschitz at $\bar{x}$ and $\mathrm{dom}\, f^\uparrow(\bar{x}, \cdot) \cap \mathrm{int}\,\mathrm{dom}\,g^\uparrow(\bar{x}, \cdot)$ is nonempty. The sum rule holds with equality if, in addition, f and g are subdifferentially regular. Certain chain rules are also established.

(III) Michel and Penot [129, 130] study several types of directional derivatives and associated subdifferentials. Their aim is to generalize the G-derivative rather than the strict H-derivative as does the Clarke subdifferential (cf. Remark 7.3.10).

Using the epi-limit convergence concept, Michel and Penot define, among others, the *pseudo-strict derivative* of f at $\bar{x}$ as

$$f^\wedge(\bar{x}, y) := \sup_{z \in E} \; \underset{\substack{\tau \downarrow 0, z' \to z \\ (\bar{x}+\tau z', \alpha) \to_f \bar{x}}}{\mathrm{eLim\,sup}} \; \tfrac{1}{\tau}\big(f(\bar{x} + \tau y + \tau z') - \alpha\big).$$

The functional $f^\wedge(\bar{x}, \cdot)$ is shown to be l.s.c. and sublinear. If f is locally L-continuous around $\bar{x}$, then $f^\wedge(\bar{x}, \cdot)$ coincides with $f^\diamond(\bar{x}, \cdot)$. If the directional G-derivative $f_G(\bar{x}, \cdot)$ exists, is finite and sublinear, then $f^\wedge(\bar{x}, \cdot) = f_G(\bar{x}, \cdot)$. The associated subdifferential, defined as

$$\partial^\wedge f(\bar{x}) := \{x^* \in E^* \mid \langle x^*, y \rangle \leq f^\wedge(\bar{x}, y) \; \forall y \in E\},$$

thus satisfies $\partial^\wedge f(\bar{x}) = \{f'(\bar{x})\}$ whenever f is G-differentiable at $\bar{x}$. If $\bar{x}$ is a local minimizer of f, then $o \in \partial^\wedge f(\bar{x})$. Under certain regularity assumptions, the sum rule $\partial^\wedge (f+g)(\bar{x}) \subseteq \partial^\wedge f(\bar{x}) + \partial^\wedge g(\bar{x})$ and a chain rule are established. Finally, multiplier rules are derived in [130].

(IV) Various other generalized derivative concepts were introduced and studied. Some of them are Halkin's *screen* [83], Treiman's *B-derivatives* [206, 207], and Warga's *derivate containers* [213, 214]. Much was done by Hiriart-Urruty [84, 85, 87] in clarifying and refining these derivative concepts.

(V) Concerning sensitivity analysis, we refer in the finite-dimensional case to Klatte and Kummer [111], Luderer et al. [126], and Rockafellar and Wets [189], and in the infinite-dimensional case to Bonnans and Shapiro [16], Clarke et al. [39], and Mordukhovich [136] as well as to the literature cited in these books.

(VI) Following Borwein and Zhu [23], we started the differential analysis of lower semicontinuous functionals with Zhu's nonlocal fuzzy sum rule which, in turn, is based on the Borwein–Preiss smooth variational principle. The basis of Mordukhovich's work constitutes his extremal principle. Another fundamental result in nonsmooth analysis is a multidirectional mean value theorem (such as Theorem 9.6.2) that originally goes back to Clarke and Ledyaev [37, 38]. In [39] this result is applied to establish, among others, nonsmooth implicit and inverse function theorems. Mordukhovich and Shao [140] showed the equivalence of the extremal principle and the local fuzzy sum rule, and Zhu [226] proved the equivalence of the latter to the nonlocal fuzzy sum rule and to the multidirectional mean value theorem.

(VII) Ioffe systematically investigated *approximate subdifferentials*; see, among others, [94–98, 100].

Let $\mathcal{F}$ denote the collection of all finite-dimensional vector subspaces of E. If S is a nonempty subset of E, we write

$$f_{(S)}(x) := \begin{cases} f(x) & \text{if } x \in S, \\ +\infty & \text{if } x \in E \setminus S. \end{cases}$$

Ioffe starts with the lower directional Hadamard derivative, which he calls *lower directional Dini derivative*,

$$\underline{f}_H(\bar{x}, y) := \liminf_{\tau \downarrow 0,\, z \to y} \tfrac{1}{\tau}\big(f(\bar{x} + \tau z) - f(\bar{x})\big) \quad \forall y \in E.$$

He then defines the *Hadamard subdifferential* (or *Dini subdifferential*)

$$\partial^- f(\bar{x}) := \{x^* \in E^* \mid \langle x^*, y\rangle \le \underline{f}_H(\bar{x}, y)\ \forall y \in E\}$$

and the *A-subdifferential*

$$\partial_A f(\bar{x}) := \bigcap_{L \in \mathcal{F}} \operatorname*{Lim\,sup}_{x \to_f \bar{x}} \partial^- f_{(x+L)}(x).$$

Here, the Painlevé–Kuratowski upper limit is taken with respect to the norm topology in E and the weak star topology in E^*. Observe that Ioffe utilizes the topological form of the Painlevé–Kuratowski upper limit whereas Mordukhovich utilizes the sequential form. The interrelation between the concepts of Ioffe and Mordukhovich is discussed by Mordukhovich and Shao [142], see also Mordukhovich [136].

Ioffe further defines the *G-normal cone* to the set $M \subseteq E$ at $\bar{x} \in M$ as

$$N_G(M, \bar{x}) := \operatorname{cl}^* \bigcup_{\lambda > 0} \lambda \partial_A \mathrm{d}_M(\bar{x})$$

and, respectively, the *G-subdifferential* and the *singular G-subdifferential* of f at $\bar{x}$ as

$$\partial_G f(\bar{x}) := \left\{ x^* \in E^* \mid (x^*, -1) \in N_G\big(\text{epi}\, f, (\bar{x}, f(\bar{x}))\big) \right\},$$

$$\partial_G^\infty f(\bar{x}) := \left\{ x^* \in E^* \mid (x^*, 0) \in N_G\big(\text{epi}\, f, (\bar{x}, f(\bar{x}))\big) \right\}.$$

The G-subdifferential and the A-subdifferential coincide for any function on finite-dimensional normed vector spaces and for directionally Lipschitz functions on arbitrary normed vector spaces. For convex functionals on any normed vector space, the G-subdifferential (in contrast to the A-subdifferential) coincides with the subdifferential of convex analysis. In general, the sets $\partial_A f(\bar{x})$ and $\partial_G f(\bar{x})$ are not convex. However, they are minimal in a certain sense among all "reasonable" subdifferential constructions. In particular, for any proper functional f one has $\partial_C f(\bar{x}) = \overline{\text{co}}^*\big(\partial_G f(\bar{x}) + \partial_G^\infty f(\bar{x})\big)$.

Ioffe develops an extensive calculus for these objects. For instance, if M_1 and M_2 are closed sets one of them being epi-Lipschitz at $\bar{x}$, then an appropriate regularity assumption ensures that

$$N_G(M_1 \cap M_2, \bar{x}) \subseteq N_G(M_1, \bar{x}) + N_G(M_2, \bar{x}),$$

with equality if $N_G(M_i, \bar{x}) = T(M_i, \bar{x})^\circ$ for $i = 1, 2$. This result is verified with the aid of Ekeland's variational principle. It is then applied to derive the following sum rule. If the functionals f and g are l.s.c. and one of them is directionally Lipschitz at $\bar{x}$, then under the regularity assumption $\partial_G^\infty f(\bar{x}) \cap (-\partial_G^\infty g(\bar{x})) = \{o\}$, the sum rule ($\diamond$) holds with $\partial_\uparrow$ replaced by ∂_G. An analogous result is obtained for the A-subdifferential. For an extended chain rule in terms of ∂_A see Jourani and Thibault [106]. We also refer to the survey paper by Hiriart-Urruty et al. [90].

Multiplier rules in terms of G-subdifferentials were established by Glover et al. [74]. For an alternative proof of these multiplier rules and an interesting discussion of the absence of constraint qualifications we refer to Pühl [172]. Modified multiplier rules were obtained by Glover and Craven [73].

(VIII) In these lectures, we confined ourselves to *first-order* necessary optimality conditions (which are also sufficient in the convex case). For *second-order* necessary and/or sufficient conditions we refer in the smooth case to Bonnans and Shapiro [16] and the literature cited therein. Second-order optimality conditions in terms of generalized derivatives are also treated in many papers. As a small selection we refer to Ben-Tal and Zowe [14], Casas and Tröltzsch [29], Chaney [31], Cominetti [40], Mordukhovich [135], Mordukhovich [136], Mordukhovich and Outrata [138], and Rockafellar [188].

References

1. R. A. Adams. *Sobolev Spaces*. Academic, Boston, 1978
2. C. D. Aliprantis and K. C. Border. *Infinite Dimensional Analysis*. Springer, Berlin Heidelberg New York, 1994
3. E. Asplund. Fréchet differentiability of convex functions. *Acta Math.*, 121: 31–47, 1968
4. K. Atkinson and W. Han. *Theoretical Numerical Analysis*. Springer, Berlin Heidelberg New York, 2001
5. J. -P. Aubin. Contingent derivatives of set-valued maps and existence of solutions to nonlinear inclusions and differential inclusions. In L. Nachbin, editor, *Advances in Mathematics, Supplementary Studies*, pages 160–232. Academic, New York, 1981
6. J. -P. Aubin. *Optima and Equilibria*. Springer, Berlin Heidelberg New York, 1993
7. J. -P. Aubin and I. Ekeland. *Applied Nonlinear Analysis*. Wiley, New York, 1984
8. J. -P. Aubin and H. Frankowska. *Set-Valued Analysis*. Birkhäuser, Boston, 1990
9. V. Barbu and Th. Precupanu. *Convexity and Optimization in Banach Spaces*. Sijthoff and Noordhoff, Alphen aan de Rijn, 1978
10. M. S. Bazaraa, J. J. Goode, and M. Z. Nashed. On the cones of tangents with applications to mathematical programming. *J. Optim. Theory Appl.*, 13: 389–426, 1974
11. M. S. Bazaraa, H. D. Sherali, and C. M. Shetty. *Nonlinear Programming: Theory and Algorithms*. Wiley, New York, 1993
12. M. S. Bazaraa and C. M. Shetty. *Foundations of Optimization*, volume 122 of *Lecture Notes in Economics and Mathematical Systems*. Springer, Berlin Heidelberg New York, 1976
13. B. Beauzamy. *Introduction to Banach Spaces and Their Geometry*. North-Holland, Amsterdam, 1985
14. A. Ben-Tal and J. Zowe. A unified theory of first and second order conditions for extremum problems in topological vector spaces. *Math. Progr. Stud.*, 19: 39–76, 1982
15. E. Bishop and R. R. Phelps. A proof that every Banach space is subreflexive. *Bull. Am. Math. Soc.*, 67:97–98, 1961

16. J. F. Bonnans and A. Shapiro. *Perturbation Analysis of Optimization Problems.* Springer, Berlin Heidelberg New York, 2000

17. J. M. Borwein and A. D. Ioffe. Proximal analysis in smooth spaces. *Set-Valued Anal.*, 4:1–24, 1996

18. J. M. Borwein and A. S. Lewis. *Convex Analysis and Nonlinear Optimization.* Springer, Berlin Heidelberg New York, 2000

19. J. M. Borwein and D. Preiss. A smooth variational principle with applications to subdifferentiability and to differentiability of convex functions. *Trans. Am. Math. Soc.*, 303:517–527, 1987

20. J. M. Borwein and H. M. Strójwas. Proximal analysis and boundaries of closed sets in Banach space. I: Theory. *Can. J. Math.*, 38:431–452, 1986

21. J. M. Borwein, J. S. Treiman, and Q. J. Zhu. Necessary conditions for constrained optimization problems with semicontinuous and continuous data. *Trans. Am. Math. Soc.*, 350:2409–2429, 1998

22. J. M. Borwein and Q. J. Zhu. Viscosity solutions and viscosity subderivatives in smooth Banach spaces with applications to metric regularity. *SIAM J. Control Optim.*, 34:1568–1591, 1996

23. J. M. Borwein and Q. J. Zhu. A survey of subdifferential calculus with applications. *Nonlinear Anal.*, 38:687–773, 1999

24. J. M. Borwein and Q. J. Zhu. *Techniques of Variational Analysis.* Springer, Berlin Heidelberg New York, 2005

25. J. M. Borwein and D. M. Zhuang. Verifiable necessary and sufficient conditions for openness and regularity of set-valued and single-valued maps. *J. Math. Anal. Appl.*, 134:441–459, 1988

26. G. Bouligand. Sur les surfaces dépourvues de points hyperlimites. *Ann. Soc. Polon. Math.*, 9:32–41, 1930

27. D. Braess. *Nonlinear Approximation Theory.* Springer, Berlin Heidelberg New York, 1986

28. A. Brøndsted. Conjugate convex functions in topological vector spaces. *Fys. Medd. Dans. Vid. Selsk.*, 34:1–26, 1964

29. E. Casas and F. Tröltzsch. Second-order necessary and sufficient optimality conditions for optimization problems and applications to control theory. *SIAM J. Optim.*, 13:406–431, 2002

30. L. Cesari. *Optimization – Theory and Applications.* Springer, Berlin Heidelberg New York, 1983

31. R. W. Chaney. A general sufficiency theorem for nonsmooth nonlinear programming. *Trans. Am. Math. Soc.*, 276:235–245, 1983

32. I. Cioranescu. *Geometry of Banach Spaces, Duality Mappings and Nonlinear Problems.* Kluwer, Dordrecht, 1990

33. F. H. Clarke. *Necessary Conditions for Nonsmooth Problems in Optimal Control and the Calculus of Variations.* Ph.D. Thesis, University of Washington, 1973

34. F. H. Clarke. Generalized gradients and applications. *Trans. Am. Math. Soc.*, 205:247–262, 1975

35. F. H. Clarke. A new approach to Lagrange multipliers. *Math. Oper. Res.*, 1: 165–174, 1976

36. F. H. Clarke. *Optimization and Nonsmooth Analysis.* SIAM, Philadelphia, 1990

37. F. H. Clarke and Yu. S. Ledyaev. Mean value inequalities. *Proc. Am. Math. Soc.*, 122:1075–1083, 1994

38. F. H. Clarke and Yu. S. Ledyaev. Mean value inequalities in Hilbert space. *Trans. Am. Math. Soc.*, 344:307–324, 1994

39. F. H. Clarke, Yu. S. Ledyaev, R. J. Stern, and P. R. Wolenski. *Nonsmooth Analysis and Control Theory*. Springer, Berlin Heidelberg New York, 1998

40. R. Cominetti. Metric regularity, tangent sets and second-order optimality conditions. *Appl. Math. Optim.*, 21:265–287, 1990

41. M. G. Crandall, L. C. Evans, and P. -L. Lions. Some properties of viscosity solutions of Hamilton–Jacobi equations. *Trans. Am. Math. Soc.*, 282:487–502, 1984

42. M. G. Crandall and P. -L. Lions. Viscosity solutions of Hamilton–Jacobi equations. *Trans. Am. Math. Soc.*, 277:1–41, 1983

43. B. D. Craven. *Mathematical Programming and Control Theory*. Chapman and Hall, London, 1978

44. B. D. Craven. *Control and Optimization*. Chapman and Hall, London, 1995

45. B. D. Craven, J. Gwinner, and V. Jeyakumar. Nonconvex theorems of the alternative and minimization. *Optimization*, 18:151–163, 1987

46. M. Degiovanni and F. Schuricht. Buckling of nonlinearly elastic rods in the presence of obstacles treated by nonsmooth critical point theory. Math. Ann. 311, 675–728, 1998

47. V. F. Demyanov and A. M. Rubinov. *Quasidifferentiable Calculus*. Optimization Software, Publications Division, New York, 1986

48. V. F. Demyanov and A. M. Rubinov. *Foundations of Nonsmooth Analysis and Quasidifferentiable Calculus*. Nauka, Moscow, 1990 (in Russian)

49. R. Deville, G. Godefroy, and V. Zizler. A smooth variational principle with applications to Hamilton–Jacobi equations in infinite dimensions. *J. Funct. Anal.*, 111:197–212, 1993

50. R. Deville, G. Godefroy, and V. Zizler. *Smoothness and Renormings in Banach Spaces*. Pitman Monographs and Surveys in Pure and Applied Mathematics, No. 64. Wiley, New York, 1993

51. J. Diestel. *Geometry of Banach Spaces. Lecture Notes in Mathematics*. Springer, Berlin Heidelberg New York, 1974

52. J. Diestel. *Sequences and Series in Banach Spaces*. Springer, Berlin Heidelberg New York, 1983

53. J. Dieudonné. *Foundations of Modern Analysis*. Academic, New York, 1960

54. A. V. Dmitruk, A. A. Milyutin, and N. P. Osmolovskii. Lyusternik's theorem and the theory of extrema. *Russ. Math. Surv.*, 35:11–51, 1980

55. S. Dolecki. A general theory of necessary optimality conditions. *J. Math. Anal. Appl.*, 78:267–308, 1980

56. I. Ekeland. On the variational principle. *J. Math. Anal. Appl.*, 47:324–353, 1974

57. I. Ekeland. Nonconvex minimization problems. *Bull. Am. Math. Soc. (N.S.)*, 1:443–474, 1979

58. I. Ekeland and R. Temam. *Convex Analysis and Variational Problems*. North-Holland, Amsterdam, 1976

59. K. -H. Elster, R. Reinhardt, M. Schäuble, and G. Donath. *Einführung in die nichtlineare Optimierung*. Teubner, Leipzig, 1977

60. J. Elstrodt. *Maß- und Integrationstheorie*. Springer, Berlin Heidelberg New York, 1999

61. E. Ernst, M. Théra, and C. Zalinescu. Slice-continuous sets in reflexive Banach spaces: convex constrained optimization and strict convex separation. *J. Funct. Anal.*, 223:179–203, 2005

354　References

62. H. Eschrig. *The Fundamentals of Density Functional Theory*. Edition am Gutenbergplatz, Leipzig, 2003
63. L. C. Evans and R. F. Gariepy. *Measure Theory and Fine Properties of Functions*. CRC, Boca Raton, 1992
64. K. Fan. Asymptotic cones and duality of linear relations. In O. Shisha, editor, *Inequalities*, volume 2, pages 179–186. Academic, London, 1970
65. W. Fenchel. On conjugate convex functions. *Can. J. Math.*, 1:73–77, 1949
66. W. Fenchel. *Convex Cones, Sets and Functions. Lecture Notes*. Princeton University, Princeton, 1951
67. L. A. Fernandez. On the limits of the Lagrange multiplier rule. *SIAM Rev.*, 39:292–297, 1997
68. O. Ferrero. Theorems of the alternative for set-valued functions in infinite-dimensional spaces. *Optimization*, 20:167–175, 1989
69. D. G. Figueiredo. *Lectures on the Ekeland Variational Principle with Applications and Detours*. Springer (Published for the Tata Institute of Fundamental Research, Bombay), Berlin Heidelberg New York, 1989
70. F. Giannessi. Theorems of the alternative for multifunctions with applications to optimization: general results. *J. Optim. Theory Appl.*, 55:233–256, 1987
71. M. Giaquinta and S. Hildebrandt. *Calculus of Variations I*. Springer, Berlin Heidelberg New York, 1996
72. M. Giaquinta and S. Hildebrandt. *Calculus of Variations II*. Springer, Berlin Heidelberg New York, 1996
73. B. M. Glover and B. D. Craven. A Fritz John optimality condition using the approximate subdifferential. *J. Optim. Theory Appl.*, 82:253–265, 1994
74. B. M. Glover, B. D. Craven, and S. D. Flåm. A generalized Karush–Kuhn–Tucker optimality condition without constraint qualification using the approximate subdifferential. *Numer. Funct. Anal. Optim.*, 14:333–353, 1993
75. B. M. Glover, V. Jeyakumar, and W. Oettli. A Farkas lemma for difference sublinear systems and quasidifferentiable programming. *Math. Oper. Res.*, 63: 109–125, 1994
76. A. Göpfert. *Mathematische Optimierung in allgemeinen Vektorräumen*. Teubner, Leipzig, 1973
77. A. Göpfert, Chr. Tammer, H. Riahi, and C. Zalinescu. *Variational Methods in Partially Ordered Spaces*. Springer, Berlin Heidelberg New York, 2003
78. A. Göpfert, Chr. Tammer, and C. Zalinescu. On the vectorial Ekeland's variational principle and minimal points in product spaces. *Nonlinear Anal.*, 39: 909–922, 2000
79. L. M. Graves. Some mapping theorems. *Duke Math. J.*, 17:111–114, 1950
80. K. Groh. On monotone operators and forms. *J. Convex Anal.*, 12:417–429, 2005
81. Ch. Großmann and H. -G. Roos. *Numerical Treatment of Partial Differential Equations*. Springer, Berlin Heidelberg, New York, to appear
82. H. Halkin. Implicit functions and optimization problems without continuous differentiability of the data. *SIAM J. Control*, 12:229–236, 1974
83. H. Halkin. Mathematical programming without differentiability. In D. L. Russell, editor, *Calculus of Variations and Control Theory*, pages 279–288. Academic, New York, 1976
84. J. -B. Hiriart-Urruty. Refinements of necessary optimality conditions in non-differentiable programming I. *Appl. Math. Optim.*, 5:63–82, 1979
85. J. -B. Hiriart-Urruty. Tangent cones, generalized gradients and mathematical programming in Banach spaces. *Math. Oper. Res.*, 4:79–97, 1979

86. J. -B. Hiriart-Urruty. Mean value theorems in nonsmooth analysis. *Numer. Funct. Anal. Optim.*, 2:1–30, 1980

87. J. -B. Hiriart-Urruty. Refinements of necessary optimality conditions in non-differentiable programming II. *Math. Progr. Stud.*, 19:120–139, 1982

88. J. -B. Hiriart-Urruty and C. Lemaréchal. *Convex Analysis and Minimization Algorithms. I: Fundamentals.* Springer, Berlin Heidelberg New York, 1993

89. J. -B. Hiriart-Urruty and C. Lemaréchal. *Convex Analysis and Minimization Algorithms. II: Advanced Theory and Bundle Methods.* Springer, Berlin Heidelberg New York, 1993

90. J. -B. Hiriart-Urruty, M. Moussaoui, A. Seeger, and M. Volle. Subdifferential calculus without constraint qualification, using approximate subdifferentials: a survey. *Nonlinear Anal.*, 24:1727–1754, 1995

91. R. B. Holmes. *A Course on Optimization and Best Approximation.* Springer, Berlin Heidelberg New York, 1972

92. R. B. Holmes. *Geometric Functional Analysis and Its Applications.* Springer, Berlin Heidelberg New York, 1975

93. A. D. Ioffe. Regular points of Lipschitz functions. *Trans. Am. Math. Soc.*, 251:61–69, 1979

94. A. D. Ioffe. Nonsmooth analysis: differential calculus of nondifferentiable mappings. *Trans. Am. Math. Soc.*, 266:1–56, 1981

95. A. D. Ioffe. Approximate subdifferentials and applications. I. The finite dimensional theory. *Trans. Am. Math. Soc.*, 281(1):389–416, 1984

96. A. D. Ioffe. Approximate subdifferentials and applications. II. *Mathematika*, 33:11–128, 1986

97. A. D. Ioffe. Approximate subdifferentials and applications. III. The metric theory. *Mathematika*, 36:1–38, 1989

98. A. D. Ioffe. Proximal analysis and approximate subdifferentials. *J. Lond. Math. Soc.*, 41:175–192, 1990

99. A. D. Ioffe. A Lagrange multiplier rule with small convex-valued subdifferentials for nonsmooth problems of mathematical programming involving equality and nonfunctional constraints. *Math. Progr. Stud.*, 58:137–145, 1993

100. A. D. Ioffe. Nonsmooth subdifferentials: their calculus and applications. In V. Lakshmikantham, editor, *Proceedings of the First World Congress of Nonlinear Analysts*, pages 2299–2310. De Gruyter, Berlin, 1996

101. A. D. Ioffe and V. Tikhomirov. *Theory of Extremal Problems.* North-Holland, New York, 1978 (German ed.: Deutscher Verlag der Wissenschaften, Berlin, 1979; original Russian ed.: Nauka, Moskow, 1974)

102. J. Jahn. *Vector Optimization: Theory, Applications and Extensions.* Springer, Berlin Heidelberg New York, 2004

103. G. J. O. Jameson. *Topology and Normed Spaces.* Chapman and Hall, London, 1974

104. V. Jeyakumar. Convexlike alternative theorems and mathematical programming. *Optimization*, 16:643 652, 1985

105. F. John. Extremum problems with inequalities as side conditions. In K. O. Friedrichs, O. E. Neugebauer, and J. J. Stoker, editors, *Studies and Essays: Courant Anniversary Volume*, pages 187–204. Wiley-Interscience, New York, 1948

106. A. Jourani and L. Thibault. The approximate subdifferential of composite functions. *Bull. Aust. Math. Soc.*, 47:443–455, 1993

356   References

107. A. Jourani and L. Thibault. Metric regularity for strongly compactly Lipschitzian mappings. *Nonlinear Anal.*, 24:229–240, 1995

108. M. Kadec. Spaces isomorphic to a locally uniformly convex space. *Izv. Vysš. Učebn. Zaved. Mat.*, 6:51–57, 1959 (in Russian)

109. W. Karush. *Minima of Functions of Several Variables with Inequalities as Side Conditions.* Master's Thesis, University of Chicago, Chicago, 1939

110. D. Klatte and B. Kummer. Contingent derivatives of implicit (multi-)functions and stationary points. *Ann. Oper. Res.*, 101:313–331, 2001

111. D. Klatte and B. Kummer. *Nonsmooth Equations in Optimization. Regularity, Calculus, Methods and Applications.* Kluwer, Dordrecht, 2002

112. W. Krabs. *Optimierung und Approximation.* Teubner, Stuttgart, 1975

113. A. Y. Kruger and B. S. Mordukhovich. Extremal points and Euler equations in nonsmooth optimization. *Dokl. Akad. Nauk BSSR*, 24:684–687, 1980 (in Russian)

114. H. W. Kuhn and A. W. Tucker. Nonlinear programming. In J. Neyman, editor, *Proceedings of the Second Berkeley Symposium on Mathematical Statistics and Probability*, pages 481–492. University of California Press, Berkeley, 1951

115. G. Köthe. *Topologische lineare Räume I.* Springer, Berlin Heidelberg New York, 1960

116. M. Landsberg and W. Schirotzek. Mazur–Orlicz type theorems with some applications. *Math. Nachr.*, 79:331–341, 1977

117. J. B. Lasserre. A Farkas lemma without a standard closure condition. *SIAM J. Control Optim.*, 35:265–272, 1997

118. P. -J. Laurent. *Approximation et Optimisation.* Hermann, Paris, 1972

119. G. Lebourg. Valeur moyenne pour gradient généralisé. *C. R. Acad. Sci. Paris*, 281:795–797, 1975

120. Yu. S. Ledyaev and Q. J. Zhu. Implicit multifunction theorems. *Set-Valued Anal.*, 7:209–238, 1999

121. Yu. S. Ledyaev and Q. J. Zhu. Implicit multifunction theorems. Preprint (revised version), http://homepages.wmich.edu/~zhu/papers/implicit. html, 2006

122. U. Ledzewicz and S. Walczak. On the Lyusternik theorem for nonsmooth operators. *Nonlinear Anal.*, 22:121–128, 1994

123. P. D. Loewen. *Optimal Control via Nonsmooth Analysis.* American Mathematical Society, Providence, 1993

124. P. D. Loewen. A mean value theorem for Fréchet subgradients. *Nonlinear Anal.*, 23:1365–1381, 1994

125. P. D. Loewen and X. Wang. A generalized variational principle. *Can. J. Math.*, 53(6):1174–1193, 2001

126. B. Luderer, L. Minchenko, and T. Satsura. *Multivalued Analysis and Nonlinear Programming Problems with Perturbations.* Kluwer, Dordrecht, 2002

127. D. Luenberger. *Optimization by Vector Space Methods.* Wiley, New York, 1969

128. L. A. Lyusternik. On constrained extrema of functionals. *Mat. Sb.*, 41:390–401, 1934 (in Russian)

129. P. Michel and J. -P. Penot. Calcul sous-différentiel pour les fonctions lipschitziennes et non lipschitziennes. *C. R. Acad. Sci. Paris*, 298:269–272, 1984

130. P. Michel and J. -P. Penot. A generalized derivative for calm and stable functions. *Diff. Int. Eqs.*, 5:433–454, 1992

131. H. Minkowski. *Theorie der konvexen Körper, insbesondere Begründung ihres Oberflächenbegriffs.* Teubner, Leipzig, 1911

132. B. S. Mordukhovich. Maximum principle in the problem of time optimal control with nonsmooth constraints. *J. Appl. Math. Mech.*, 40:960–969, 1976

133. B. S. Mordukhovich. Metric approximations and necessary optimality conditions for general classes of nonsmooth extremal problems. *Sov. Math. Dokl.*, 22:526–530, 1980

134. B. S. Mordukhovich. *Approximation Methods in Problems of Optimization and Control.* Nauka, Moscow, 1988 (in Russian)

135. B. S. Mordukhovich. Sensitivity analysis in nonsmooth optimization. In D. A. Field and V. Komkov, editors, *Theoretical Aspects of Industrial Design*, volume 58 of *Proceedings in Applied Mathematics*, pages 32–46. SIAM, Philadelphia, 1992

136. B. S. Mordukhovich. *Variational Analysis and Generalized Differentiation I: Basic Theory.* Springer, Berlin Heidelberg New York, 2006

137. B. S. Mordukhovich. *Variational Analysis and Generalized Differentiation II: Applications.* Springer, Berlin Heidelberg New York, 2006

138. B. S. Mordukhovich and J. V. Outrata. On second-order subdifferentials and their applications. *SIAM J. Optim.*, 12:139–169, 2001

139. B. S. Mordukhovich and Y. Shao. Differential characterizations of covering, metric regularity, and Lipschitzian properties of multifunctions between Banach spaces. *Nonlinear Anal.*, 25:1401–1424, 1995

140. B. S. Mordukhovich and Y. Shao. Extremal characterizations of Asplund spaces. *Proc. Am. Math. Soc.*, 124:197–205, 1996

141. B. S. Mordukhovich and Y. Shao. Nonconvex coderivative calculus for infinite-dimensional multifunctions. *Set-Valued Anal.*, 4:205–236, 1996

142. B. S. Mordukhovich and Y. Shao. Nonsmooth sequential analysis in Asplund spaces. *Trans. Am. Math. Soc.*, 348:1235–1280, 1996

143. B. S. Mordukhovich and Y. Shao. Mixed coderivatives of set-valued mappings in variational analysis. *J. Appl. Anal.*, 4:269–294, 1998

144. B. S. Mordukhovich, J. S. Treiman, and Q. J. Zhu. An extended extremal principle with applications to multiobjective optimization. *SIAM J. Optim.*, 14:359–379, 2003

145. B. S. Mordukhovich and B. Wang. Extensions of generalized differential calculus in Asplund spaces. *J. Math. Anal. Appl.*, 272:164–186, 2002

146. B. S. Mordukhovich and B. Wang. Necessary suboptimality and optimality conditions via variational principles. *SIAM J. Control Optim.*, 41:623–640, 2002

147. J. -J. Moreau. Functions convexes duales et points proximaux dans un espace hilbertian. *C. R. Acad. Sci. Paris*, 255:2897–2899, 1962

148. J. -J. Moreau. Propriété des applications 'prox'. *C. R. Acad. Sci. Paris*, 256:1069–1071, 1963

149. M. M. Mäkelä and P. Neittaanmäki. *Nonsmooth Optimization.* World Scientific, Singapore, 1992

150. L. W. Neustadt. *Optimization: A Theory of Necessary Conditions.* Princeton University Press, Princeton, 1976

151. H. V. Ngai, D. T. Luc, and M. Théra. Extensions of Fréchet ε-subdifferential calculus and applications. *J. Math. Anal. Appl.*, 268:266–290, 2002

152. N. V. Ngai and M. Théra. Metric inequality, subdifferential calculus and applications. *Set-Valued Anal.*, 9:187–216, 2001

153. N. V. Ngai and M. Théra. A fuzzy necessary optimality condition for non-Lipschitz optimization in Asplund spaces. *SIAM J. Optim.*, 12:656–668, 2002

154. N. V. Ngai and M. Théra. Error bounds and implicit multifunction theorem in smooth Banach spaces and applications to optimization. *Set-Valued Anal.*, 12:195–223, 2004

155. D. Pallaschke. Ekeland's variational principle, convex functions and Asplund spaces. In W. Krabs and J. Zowe, editors, *Modern Methods of Optimization*, pages 274–312. Springer, Berlin Heidelberg New York, 1992

156. D. Pallaschke and S. Rolewicz. *Foundations of Mathematical Optimization: Convex Analysis Without Linearity.* Kluwer, Dordrecht, 1997

157. P. D. Panagiotopoulos. *Hemivariational Inequalities.* Springer, Berlin Heidelberg New York, 1993

158. N. S. Papageorgiou and L. Gasinski. *Nonsmooth Critical Point Theory and Nonlinear Boundary Value Problems.* Chapman and Hall, Boca Raton, 2004

159. J. -P. Penot. Calcul sous-differentiel et optimisation. *J. Funct. Anal.*, 27: 248–276, 1978

160. J. -P. Penot. On regularity conditions in mathematical programming. *Math. Progr. Stud.*, 19:167–199, 1982

161. J. -P. Penot. Open mapping theorems and linearization stability. *Numer. Funct. Anal. Optim.*, 8:21–35, 1985

162. J. -P. Penot. The drop theorem, the petal theorem and Ekeland's variational principle. *Nonlinear Anal.*, 10:813–822, 1986

163. J. -P. Penot. On the mean value theorem. *Optimization*, 19:147–156, 1988

164. J. -P. Penot. Metric regularity, openness and Lipschitzian behavior of multi-functions. *Nonlinear Anal.*, 13:629–643, 1989

165. R. R. Phelps. *Convex Functions, Monotone Operators and Differentiability*, volume 1364 of *Lecture Notes in Mathematics.* Springer, Berlin Heidelberg New York, 1993

166. D. Preiss. Fréchet derivatives of Lipschitzian functions. *J. Funct. Anal.*, 91:312–345, 1990

167. D. Preiss and L. Zajíček. Fréchet differentiation of convex functions in a Banach space with a separable dual. *Proc. Am. Math. Soc.*, 91(2):202–204, 1984

168. B. N. Pšeničnyi. *Convex Analysis and Extremal Problems.* Nauka, Moscow, 1969 (in Russian)

169. B. N. Pšeničnyj. *Notwendige Optimalitätsbedingungen.* Teubner, Leipzig, 1972

170. B. N. Pshenichnyi. *Necessary Conditions for an Extremum.* Dekker, New York, 1971 (German ed.: Teubner, Leipzig, 1972; original Russian ed.: Nauka, Moscow, 1969)

171. H. Pühl. Convexity and openness with linear rate. *J. Math. Anal. Appl.*, 227:382–395, 1998

172. H. Pühl. *Nichtdifferenzierbare Extremalprobleme in Banachräumen: Regularitätsbedingungen sowie die Approximation von Niveaumengen.* Doctoral Thesis, Technische Universität Dresden, Dresden, 1999

173. H. Pühl and W. Schirotzek. Linear semi-openness and the Lyusternik theorem. *Eur. J. Oper. Res.*, 157:16–27, 2004

174. A. Roberts and D. Varberg. *Convex Functions.* Academic, New York, 1973

175. S. M. Robinson. Regularity and stability for convex multivalued functions. *Math. Oper. Res.*, 1:130–143, 1976

176. S. M. Robinson. Stability theory for systems of inequalities. Part II. Differentiable nonlinear systems. *SIAM J. Numer. Anal.*, 13:497–513, 1976

177. R. T. Rockafellar. *Convex Functions and Dual Extremum Problems.* Ph.D. Thesis, Harvard University, Cambridge, 1963

178. R. T. Rockafellar. Level sets and continuity of conjugate convex functions. *Trans. Am. Math. Soc.*, 123:46–63, 1966

179. R. T. Rockafellar. Duality and stability in extremum problems involving convex functions. *Pac. J. Math.*, 21:167–187, 1967

180. R. T. Rockafellar. *Convex Analysis*. Princeton University Press, Princeton, 1970

181. R. T. Rockafellar. On the maximal monotonicity of subdifferential mappings. *Pac. J. Math.*, 33:209–216, 1970

182. R. T. Rockafellar. On the maximality of sums of nonlinear monotone operators. *Trans. Am. Math. Soc.*, 149:75–88, 1970

183. R. T. Rockafellar. Directionally Lipschitzian functions and subdifferential calculus. *Proc. Lond. Math. Soc.*, 39:331–355, 1979

184. R. T. Rockafellar. Generalized directional derivatives and subgradients of nonconvex functions. *Can. J. Math.*, 32:157–180, 1980

185. R. T. Rockafellar. Proximal subgradients, marginal values and augmented Lagrangians in nonconvex optimization. *Math. Oper. Res.*, 6:424–436, 1981

186. R. T. Rockafellar. *The Theory of Subgradients and Its Applications to Problems of Optimization: Convex and Nonconvex Functions*. Heldermann, Berlin, 1981

187. R. T. Rockafellar. Extensions of subgradient calculus with applications to optimization. *Nonlinear Anal.*, 9:665–698, 1985

188. R. T. Rockafellar. Second-order optimality conditions in nonlinear programming obtained by way of epi-derivatives. *Math. Oper. Res.*, 14:462–484, 1989

189. R. T. Rockafellar and R. J. -B. Wets. *Variational Analysis*. Springer, Berlin Heidelberg New York, 1998

190. H. -P. Scheffler. *Beiträge zur Analysis von nichtglatten Optimierungsproblemen*. Doctoral Thesis, Technische Universität Dresden, Dresden, 1987

191. H. -P. Scheffler. Mean value properties of nondifferentiable functions and their application in nonsmooth analysis. *Optimization*, 20:743–759, 1989

192. H. -P. Scheffler and W. Schirotzek. Necessary optimality conditions for nonsmooth problems with operator constraints. *Z. Anal. Anwend.*, 7:419–430, 1988

193. W. Schirotzek. On Farkas type theorems. *Commentat. Math. Univ. Carol.*, 22:1–14, 1981

194. W. Schirotzek. On a theorem of Ky Fan and its application to nondifferentiable optimization. *Optimization*, 16:353–366, 1985

195. W. Schirotzek. Nonasymptotic necessary conditions for nonsmooth infinite optimization problems. *J. Math. Anal. Appl.*, 118:535–546, 1986

196. W. Schirotzek. *Differenzierbare Extremalprobleme*. Teubner, Leipzig, 1989

197. L. Schwartz. *Analyse II. Calcul Différentiel et Équations Différentielles*. Hermann, Paris, 1997

198. S. Simons. The least slope of a convex function and the maximal monotonicity of its subdifferential. *J. Optim. Theory Appl.*, 71:127–136, 1991

199. I. Singer. Generalizations of methods of best approximation to convex optimization in locally convex spaces I: extension of continuous linear functionals and characterizations of solutions of continuous convex programs. *Rev. Roum. Math. Pures Appl.*, 19:65–77, 1974

200. C. Stegall. The Radon–Nikodým property in conjugate Banach spaces. *Trans. Am. Math. Soc.*, 264:507–519, 1984

201. J. Stoer and C. Witzgall. *Convexity and Optimization in Finite Dimensions*. Springer, Berlin Heidelberg New York, 1970

360   References

202. M. Studniarski. Mean value theorems and sufficient optimality conditions for nonsmooth functions. *J. Math. Anal. Appl.*, 111:313–326, 1985
203. M. Studniarski. Mean value theorems for functions possessing first order convex approximations. Applications in optimization theory. *Z. Anal. Anwend.*, 4:125–132, 1985
204. A. I. Subbotin. *Generalized Solutions of First-Order PDEs. The Dynamical Optimization Perspective.* Birkhäuser, Basel, 1994
205. K. Sundaresan and S. Swaminathan. *Geometry and Nonlinear Analysis in Banach Spaces. Lecture Notes in Mathematics.* Springer, Berlin Heidelberg New York, 1985
206. J. S. Treiman. Finite dimensional optimality conditions: B-gradients. *J. Optim. Theory Appl.*, 62:139–150, 1989
207. J. S. Treiman. Optimal control with small generalized gradients. *SIAM J. Control Optim.*, 28:720–732, 1990
208. S. Troyanski. On locally uniformly convex and differentiable norms in certain non-separable spaces. *Stud. Math.*, 37:173–180, 1971
209. F. Tröltzsch. *Optimality Conditions for Parabolic Control Problems and Applications.* Teubner, Leipzig, 1984
210. H. Tuy. *Convex Inequalities and the Hahn–Banach Theorem.* Number 97 in Dissertationes Mathematicae (Rozprawy Matematyczne), Warszawa, 1972
211. C. Ursescu. Multifunctions with convex closed graph. *Czech. Math. J.*, 25:438–441, 1975
212. W. Walter. *Analysis 1.* Springer, Berlin Heidelberg New York, 2004
213. J. Warga. Derivate containers, inverse functions, and controllability. In D. L. Russell, editor, *Calculus of Variations and Control Theory*, pages 13–46. Academic, New York, 1976
214. J. Warga. Fat homeomorphisms and unbounded derivate containers. *J. Math. Anal. Appl.*, 81:545–560, 1981
215. D. Werner. *Funktionalanalysis.* Springer, Berlin Heidelberg New York, 1997
216. S. Yamamuro. *Differential Calculus in Topological Linear Spaces*, volume 374 of *Lecture Notes in Mathematics.* Springer, Berlin Heidelberg New York, 1974
217. J. J. Ye. Multiplier rules under mixed assumptions of differentiability and Lipschitz continuity. *SIAM J. Control Optim.*, 39:1441–1460, 1999
218. K. Yosida. *Functional Analysis.* Springer, Berlin Heidelberg New York, 1965
219. D. Yost. Asplund spaces for beginners. *Acta Univ. Carol. Math. Phys.*, 34:159–177, 1993
220. D. Zagrodny. Approximate mean value theorem for upper subderivatives. *Nonlinear Anal.*, 12:1413–1428, 1988
221. E. Zeidler. *Nonlinear Functional Analysis and Its Applications III: Variational Methods and Optimization.* Springer, Berlin Heidelberg New York, 1984
222. E. Zeidler. *Nonlinear Functional Analysis and Its Applications I. Fixed-Point Theory.* Springer, Berlin Heidelberg New York, 1986
223. E. Zeidler. *Nonlinear Functional Analysis and Its Applications II A: Linear Monotone Operators.* Springer, Berlin Heidelberg New York, 1990
224. E. Zeidler. *Nonlinear Functional Analysis and Its Applications II B: Nonlinear Monotone Operators.* Springer, Berlin Heidelberg New York, 1990
225. Q. J. Zhu. Clarke–Ledyaev mean value inequality in smooth Banach spaces. *Nonlinear Anal.*, 32:315–324, 1998
226. Q. J. Zhu. The equivalence of several basic theorems for subdifferentials. *Set-Valued Anal.*, 6:171–185, 1998

227. Q. J. Zhu. Hamiltonian necessary conditions for a multiobjective optimal control problem with endpoint constraints. *SIAM J. Control Optim.*, 39:97–112, 2000

228. W. P. Ziemer. *Weakly Differentiable Functions.* Springer, Berlin Heidelberg New York, 1989

229. J. Zowe and S. Kurcyusz. Regularity and stability for the mathematical programming problem in Banach spaces. *Appl. Math. Optim.*, 5:49–62, 1979

Notation

Index

Universitext

Hurwitz, A.; Kritikos, N.: Lectures on Number Theory

Huybrechts, D.: Complex Geometry: An Introduction

Isaev, A.: Introduction to Mathematical Methods in Bioinformatics

Istas, J.: Mathematical Modeling for the Life Sciences

Iversen, B.: Cohomology of Sheaves

Jacod, J.; Protter, P.: Probability Essentials

Jennings, G. A.: Modern Geometry with Applications

Jones, A.; Morris, S. A.; Pearson, K. R.: Abstract Algebra and Famous Inpossibilities

Jost, J.: Compact Riemann Surfaces

Jost, J.: Dynamical Systems. Examples of Complex Behaviour

Jost, J.: Postmodern Analysis

Jost, J.: Riemannian Geometry and Geometric Analysis

Kac, V.; Cheung, P.: Quantum Calculus

Kannan, R.; Krueger, C. K.: Advanced Analysis on the Real Line

Kelly, P.; Matthews, G.: The Non-Euclidean Hyperbolic Plane

Kempf, G.: Complex Abelian Varieties and Theta Functions

Kitchens, B. P.: Symbolic Dynamics

Kloeden, P.; Ombach, J.; Cyganowski, S.: From Elementary Probability to Stochastic Differential Equations with MAPLE

Kloeden, P. E.; Platen; E.; Schurz, H.: Numerical Solution of SDE Through Computer Experiments

Koralov, L.; Sina, Ya. G.: Theory of Probability and Random Processes

Kostrikin, A. I.: Introduction to Algebra

Krasnoselskii, M. A.; Pokrovskii, A. V.: Systems with Hysteresis

Kuo, H.-H.: Introduction to Stochastic Integration

Kurzweil, H.; Stellmacher, B.: The Theory of Finite Groups. An Introduction

Kyprianou, A. E.: Introductory Lectures on Fluctuations of Lévy Processes with Applications

Lang, S.: Introduction to Differentiable Manifolds

Lefebvre, M.: Applied Stochastic Processes

Lorenz, F.: Algebra I: Fields and Galois Theory

Luecking, D. H., Rubel, L. A.: Complex Analysis. A Functional Analysis Approach

Ma, Zhi-Ming; Roeckner, M.: Introduction to the Theory of (non-symmetric) Dirichlet Forms

Mac Lane, S.; Moerdijk, I.: Sheaves in Geometry and Logic

Marcus, D. A.: Number Fields

Martinez, A.: An Introduction to Semiclassical and Microlocal Analysis

Matoušek, J.: Using the Borsuk-Ulam Theorem

Matsuki, K.: Introduction to the Mori Program

Mazzola, G.; Milmeister G.; Weissman J.: Comprehensive Mathematics for Computer Scientists 1

Mazzola, G.; Milmeister G.; Weissman J.: Comprehensive Mathematics for Computer Scientists 2

Mc Carthy, P. J.: Introduction to Arithmetical Functions

McCrimmon, K.: A Taste of Jordan Algebras

Meyer, R. M.: Essential Mathematics for Applied Field

Meyer-Nieberg, P.: Banach Lattices

Mikosch, T.: Non-Life Insurance Mathematics

Mines, R.; Richman, F.; Ruitenburg, W.: A Course in Constructive Algebra

Moise, E. E.: Introductory Problem Courses in Analysis and Topology

Montesinos-Amilibia, J. M.: Classical Tessellations and Three Manifolds

Morris, P.: Introduction to Game Theory

Nicolaescu, L.: An Invitation to Morse Theory